The Military Balance

2000•2001

OXFORD
UNIVERSITY PRESS

Published by **Oxford University Press** for
The International Institute for Strategic Studies
Arundel House, 13–15 Arundel Street,
London WC2R 3DX, UK

The Military Balance 2000•**2001**

Published by Oxford University Press for
The International Institute for Strategic Studies
Arundel House, 13–15 Arundel Street,
London WC2R 3DX, UK
http://www.iiss.org

Director Dr John Chipman

Assistant Director and Editor Colonel Terence Taylor

Defence Analysts
Ground Forces Phillip Mitchell
Aerospace Wg Cdr Andrew Brookes
Naval Forces Joanna Kidd
Defence Economist Mark Stoker

Editorial Susan Bevan

Project Manager, Design and Production Mark Taylor

Production Assistant Anna Clarke
Research Assistants Charlotte Fielder, Micaela Gustavsson, Isabelle Williams
Cartographer Jillian Luff

This publication has been prepared by the Director of The Institute and his staff, who accept full responsibility for its contents.

First published October 2000

ISBN 0-19-929003-2
ISSN 0459-7222

The Military Balance (ISSN 0459-7222) is published annually in October by Oxford University Press, Great Clarendon Street, Oxford OX2 6DP, UK. The 2000 annual subscription rate is: UK£75 (individual rate), UK£90 (institution rate); overseas US$126(individual rate), US$155 (institution rate).

Payment is required with all orders and subscriptions are accepted and entered by the volume (one issue). Please add sales tax to the prices quoted. Prices include air-speeded delivery to Australia, Canada, India, Japan, New Zealand and the USA. Delivery elsewhere is by surface mail. Air-mail rates are available on request. Payment may be made by cheque or Eurocheque (payable to Oxford University Press), National Girobank (account 500 1056), credit card (Mastercard, Visa, American Express, Diners', JCB), direct debit (please send for details) or UNESCO coupons. Bankers: Barclays Bank plc, PO Box 333, Oxford, UK, code 20-65-18, account 00715654. Claims for non-receipt must be made within four months of dispatch/order (whichever is later).

Please send subscription orders to the Journals Subscription Department, Oxford University Press, Great Clarendon Street, Oxford, OX2 6DP, UK *tel* +44 (0)1865 267907 *fax* +44 (0)1865 267485 *e-mail* jnl.orders@oup.co.uk.

In North America, *The Military Balance* is distributed by Mercury International, 365 Blair Road, Avenel, NJ 07001, USA. Periodical postage paid at Rahway, NJ, and additional entry points.

US POSTMASTER: Send address corrections to *The Military Balance*, c/o Mercury International, 365 Blair Road, Avenel, NJ 07001, USA.

Printed in Great Britain by Bell & Bain Ltd, Glasgow.

Contents

The Military Balance is updated each year to provide an accurate assessment of the military forces and defence expenditures of 169 countries. The data in the current edition is according to IISS assessments as at 1 August 2000.

GENERAL ARRANGEMENT

Part I of *The Military Balance* comprises country entries grouped by region. Regional groupings are preceded by a short introduction describing the military issues facing the region, and significant changes in the defence economics, weapons and other military equipment holdings and acquisitions of the countries concerned. Inclusion of a country or state in *The Military Balance* does not imply legal recognition or indicate support for a particular government.

Part II contains analyses and tables. New elements in this edition include analyses of the developments in information technology in the armed forces and of other issues with a strategic impact. New tables this year include one on mines and another on unmanned aerial vehicles (UAVs).

The loose wall-map is updated from 1999 to show data on recent and current armed conflicts, including fatalities and costs.

USING THE MILITARY BALANCE

The country entries in *The Military Balance* are a quantitative assessment of the personnel strengths and equipment holdings of the world's armed forces. The strengths of forces and the numbers of weapons held are based on the most accurate data available, or, failing that, on the best estimate that can be made with reasonable confidence. The data presented each year reflect judgements based on information available to the IISS at the time the book is compiled. Where information differs from previous editions, this is mainly because of substantive changes in national forces, but it is sometimes because the IISS has reassessed the evidence supporting past entries. An attempt is made to distinguish between these reasons for change in the text that introduces each regional section, but care must be taken in constructing time-series comparisons from information given in successive editions.

In order to interpret the data in the country entries correctly, it is essential to read the explanatory notes beginning on page 5.

The large quantity of data in *The Military Balance* has been compressed into a portable volume by extensive employment of abbreviations. An essential tool is therefore the alphabetical index of abbreviations, which appears on the laminated card at the back of the book. For ease of reference, this may be detached and used as a bookmark.

ATTRIBUTION AND ACKNOWLEDGEMENTS

The International Institute for Strategic Studies owes no allegiance to any government, group of governments, or any political or other organisation. Its assessments are its own, based on the material available to it from a wide variety of sources. The cooperation of governments has been sought and, in many cases, received. However, some data in *The Military Balance* are estimates. Care is taken to ensure that these are as accurate and free from bias as possible. The Institute owes a considerable debt to a number of its own members, consultants and all those who helped compile and check material. The Director and staff of the Institute assume full responsibility for the data and judgements in this book. Comments and suggestions on the data presented are welcomed. Suggestions on the style and method of presentation are also much appreciated.

Readers may use data from *The Military Balance* without applying for permission from the Institute on condition that the IISS and *The Military Balance* are cited as the source in any published work. However, applications to reproduce portions of text, complete country entries or complete tables of *The Military Balance* must be referred to the publishers. Prior to publication, applications should be addressed to: Journals Rights and Permissions, Oxford University Press, Great Clarendon Street, Oxford OX2 6DP, UK, with a copy to the Editor of *The Military Balance.*

Explanatory Notes

ABBREVIATIONS AND DEFINITIONS

Abbreviations are used throughout to save space and avoid repetition. The abbreviations may have both singular or plural meanings; for example, 'elm' = 'element' or 'elements'. The qualification 'some' means *up to*, while 'about' means *the total could be higher than given*. In financial data, '$' refers to US dollars unless otherwise stated; billion (bn) signifies 1,000 million (m). Footnotes particular to a country entry or table are indicated by letters, while those that apply throughout the book are marked by symbols (* for training aircraft counted by the IISS as combat-capable, and † where serviceability of equipment is in doubt). A full list of abbreviations appears on the detachable laminated card at the back of the book.

COUNTRY ENTRIES

Information on each country is shown in a standard format, although the differing availability of information results in some variations. Each entry includes economic, demographic and military data. Military data include manpower, length of conscript service, outline organisation, number of formations and units and an inventory of the major equipment of each service. This is followed, where applicable, by a description of the deployment of each service. Details of national forces stationed abroad and of foreign-stationed forces are also given.

GENERAL MILITARY DATA

Manpower

The 'Active' total comprises all servicemen and women on full-time duty (including conscripts and long-term assignments from the Reserves). Under the heading 'Terms of Service', only the length of conscript service is shown; where service is voluntary there is no entry. 'Reserve' describes formations and units not fully manned or operational in peacetime, but which can be mobilised by recalling reservists in an emergency. Unless otherwise indicated, the 'Reserves' entry includes all reservists committed to rejoining the armed forces in an emergency, except when national reserve service obligations following conscription last almost a lifetime. *The Military Balance* bases its estimates of effective reservist strengths on the numbers available within five years of completing full-time service, unless there is good evidence that obligations are enforced for longer. Some countries have more than one category of 'Reserves', often kept at varying degrees of readiness. Where possible, these differences are denoted using the national descriptive title, but always under the heading of 'Reserves' to distinguish them from full-time active forces.

Other Forces

Many countries maintain paramilitary forces whose training, organisation, equipment and control suggest they may be used to support or replace regular military forces. These are listed, and their roles described, after the military forces of each country. Their manpower is not normally included in the Armed Forces totals at the start of each entry. Home Guard units are counted as paramilitary. Where paramilitary groups are not on full-time active duty, '(R)' is added after the title to indicate that they have reserve status. When internal opposition forces are armed and appear to pose a significant threat to a state's security, their details are listed separately after national paramilitary forces.

Equipment

Quantities are shown by function and type, and represent what are believed to be total holdings, including active and reserve operational and training units and 'in store' stocks. Inventory totals for missile systems – such as surface-to-surface missiles (SSM), surface-to-air missiles (SAM) and anti-tank guided weapons (ATGW) – relate to launchers and not to missiles.

Stocks of equipment held in reserve and not assigned to either active or reserve units are listed as 'in store'. However, aircraft in excess of unit establishment holdings, held to allow for repair and modification or immediate replacement, are not shown 'in store'. This accounts for apparent disparities between unit strengths and aircraft inventory strengths.

Operational Deployments

Where deployments are overseas, *The Military Balance* lists permanent bases and does not normally list short-term operational deployments, particularly where military operations are in progress. An exception is made in the case of peacekeeping operations. Recent developments are also described in the text for each regional section.

GROUND FORCES

The national designation is normally used for army formations. The term 'regiment' can be misleading. It can mean essentially a brigade of all arms; a grouping of battalions of a single arm; or (as in some instances in the UK) a battalion-sized unit. The sense intended is indicated in each case. Where there is no standard organisation, the intermediate levels of command are shown as headquarters (HQs), followed by the total numbers of units that could be allocated to them. Where a unit's title overstates its real capability, the title is given in inverted commas, with an estimate given in parentheses of the comparable unit size typical of countries with substantial armed forces. Guidelines for unit and formation strengths are: **Company** 100–200 • **Battalion** 500–800 • **Brigade (Regiment)** 3,000–5,000 • **Division** 15,000–20,000 • **Corps (Army)** 60,000–80,000.

Equipment

The Military Balance uses the following definitions of equipment:

Main Battle Tank (MBT) An armoured, tracked combat vehicle, weighing at least 16.5 metric tonnes unladen, that may be armed with a 360° traverse gun of at least 75mm calibre. Any new-wheeled combat vehicles that meet the latter two criteria will be considered MBTs.

Armoured Combat Vehicle (ACV) A self-propelled vehicle with armoured protection and cross-country capability. ACVs include:

Heavy Armoured Combat Vehicle (HACV) An armoured combat vehicle weighing more than six metric tonnes unladen, with an integral/organic direct-fire gun of at least 75mm (which does not fall within the definitions of APC, AIFV or MBT). *The Military Balance* does not list

HACVs separately, but under their equipment type (light tank, reconnaissance or assault gun), and where appropriate annotates them as HACV.

Armoured Infantry Fighting Vehicle (AIFV) An armoured combat vehicle designed and equipped to transport an infantry squad, armed with an integral/organic cannon of at least 20mm calibre. Variants of AIFVs are also included and indicated as such.

Armoured Personnel Carrier (APC) A lightly armoured combat vehicle, designed and equipped to transport an infantry squad and armed with integral/organic weapons of less than 20mm calibre. Variants of APCs converted for other uses (such as weapons platforms, command posts and communications vehicles) are included and indicated as such.

Artillery A weapon with a calibre of 100mm and above, capable of engaging ground targets by delivering primarily indirect fire. The definition covers guns, howitzers, gun/howitzers, multiple-rocket launchers and mortars.

Military Formation Strengths

The manpower strength, equipment holdings and organisation of formations such as brigades and divisions differ widely from country to country. Where possible, the normal composition of formations is given in parentheses. It should be noted that where both divisions and brigades are listed, only separate brigades are counted and not those included in divisions.

NAVAL FORCES

Categorisation is based on operational role, weapon fit and displacement. Ship classes are identified by the name of the first ship of that class, except where a class is recognised by another name (such as *Udalay*, *Petya*). Where the class is based on a foreign design or has been acquired from another country, the original class name is added in parentheses. Each class is given an acronym. All such designators are included in the list of abbreviations.

The term 'ship' refers to vessels with over 1,000 tonnes full-load displacement that are more than 60 metres in overall length; vessels of lesser displacement, but of 16m or more overall length, are termed 'craft'. Vessels of less than 16m overall length are not included. The term 'commissioning' of a ship is used to mean the ship has completed fitting out and initial sea trials, and has a naval crew; operational training may not have been completed, but otherwise the ship is available for service. 'Decommissioning' means that a ship has been removed from operational duty and the bulk of its naval crew transferred. Removing equipment and stores and dismantling weapons, however, may not have started. Where known, ships in long-term refit are shown as such.

Definitions

To aid comparison between fleets, the following definitions, which do not necessarily conform to national definitions, are used:

Submarines All vessels equipped for military operations and designed to operate primarily below the surface. Those vessels with submarine-launched ballistic missiles are listed separately under 'Strategic Nuclear Forces'.

Principal Surface Combatant This term includes all surface ships with both 1,000 tonnes full-load displacement and a weapons system for other than self-protection. All such ships are assumed to have an anti-surface ship capability. They comprise: aircraft carriers (defined below); cruisers (over 8,000 tonnes) and destroyers (less than 8,000 tonnes), both of which normally have an anti-air role and may also have an anti-submarine capability; and frigates (less than 8,000 tonnes) which normally have an anti-submarine role. Only ships with a flight deck that extends beyond two-thirds of the vessel's length are classified as aircraft carriers. Ships with shorter flight decks are shown as helicopter carriers.

Patrol and Coastal Combatants These are ships and craft whose primary role is protecting a state's sea approaches and coastline. Included are corvettes (500–1,500 tonnes with an attack capability), missile craft (with permanently fitted missile-launcher ramps and control equipment) and torpedo craft (with anti-surface-ship torpedoes). Ships and craft that fall outside these definitions are classified as 'patrol' and divided into 'offshore' (over 500 tonnes), 'coastal' (75–500 tonnes), 'inshore' (less than 75 tonnes) and 'riverine'. The adjective 'fast' indicates that the ship's speed is greater than 30 knots.

Mine Warfare This term covers surface vessels configured primarily for mine laying or mine countermeasures (such as mine-hunters, minesweepers or dual-capable vessels). They are further classified into 'offshore', 'coastal', 'inshore' and 'riverine' with the same tonnage definitions as for 'patrol' vessels shown above.

Amphibious This term includes ships specifically procured and employed to disembark troops and their equipment onto unprepared beachheads by means such as landing craft or helicopters, or directly supporting amphibious operations. The term 'Landing Ship' (as opposed to 'Landing Craft') refers to vessels capable of an ocean passage that can deliver their troops and equipment in a fit state to fight. Vessels with an amphibious capability but not assigned to amphibious duties are not included. Amphibious craft are listed at the end of each entry.

Support and Miscellaneous This term covers auxiliary military ships. It covers four broad categories: underway support' (e.g. tankers and stores ships), 'maintenance and logistic' (e.g. sealift ships), 'special purposes' (e.g. intelligence collection ships) and 'survey and research' ships.

Merchant Fleet This category is included in a state's inventory when it can make a significant contribution to the state's military sealift capability.

Weapons Systems Weapons are listed in the following order: land-attack missiles, anti-surface-ship missiles, surface-to-air missiles, guns, torpedo tubes, other anti-submarine weapons, helicopters. Missiles with a range of less than 5km, and guns with a calibre of less than 76mm, are not included. Exceptions may be made in the case of some minor combatants with a primary gun armament of a lesser calibre.

Aircraft All armed aircraft, including anti-submarine warfare and maritime-reconnaissance aircraft, are included as combat aircraft in naval inventories.

Organisations Naval groupings such as fleets and squadrons frequently change and are often temporary; organisations are shown only where it is meaningful.

AIR FORCES

The term 'combat aircraft' refers to aircraft normally equipped to deliver air-to-air or air-to-surface ordnance. The 'combat' totals include aircraft in operational conversion units whose main role is weapons training, and training aircraft of the same type as those in front-line squadrons that are assumed to be available for operations at short notice. Training aircraft considered to be combat-capable are marked with an asterisk (*). Armed maritime aircraft are included in combat aircraft totals. Operational groupings of air forces are shown where known. Squadron aircraft strengths vary with aircraft types and from country to country.

Definitions

Different countries often use the same basic aircraft in different roles; the key to determining these roles lies mainly in aircrew training. In *The Military Balance* the following definitions are used as a guide:

Fixed Wing Aircraft

Fighter This term is used to describe aircraft with the weapons, avionics and performance capacity for aerial combat. Multi-role aircraft are shown as fighter ground attack (FGA), fighter, reconnaissance and so on, according to the role in which they are deployed.

Bombers These aircraft are categorised according to their designed range and payload as follows:

Long-range Capable of delivering a weapons payload of more than 10,000kg over an unrefuelled radius of action of over 5,000km;

Medium-range Capable of delivering weapons of more than 10,000kg over an unrefuelled radius of action of between 1,000km and 5,000km;

Short-range Capable of delivering a weapons payload of more than 10,000kg over an unrefuelled radius of action of less than 1,000km.

A few bombers with the radius of action described above, but designed to deliver a payload of less than 10,000kg, and which do not fall into the category of FGA, are described as **light bombers**.

Helicopters

Armed Helicopters This term is used to cover helicopters equipped to deliver ordnance, including for anti-submarine warfare. They may be further defined as:

Attack Helicopters with an integrated fire control and aiming system, designed to deliver anti-armour, air-to-ground or air-to-air weapons;

Combat Support Helicopters equipped with area suppression or self-defence weapons, but without an integrated fire control and aiming system;

Assault Armed helicopters designed to deliver troops to the battlefield.

Transport Helicopters The term describes unarmed helicopters designed to transport personnel or cargo in support of military operations.

ARMS ORDERS AND DELIVERIES

Tables in the regional texts show arms orders and deliveries listed by country buyer for the past and current years, together with country supplier and delivery dates, if known. Every effort has been made to ensure accuracy, but some transactions may not be fulfilled or may differ from those reported.

DEFENCE ECONOMICS

Country entries in **Part I** show defence expenditure, selected economic performance indicators and demographic aggregates. **Part II**, *Analyses and Tables*, contains an international comparison of defence expenditure and military manpower, giving expenditure figures for the past two years against a bench-mark year in constant US dollars. The aim is to provide an accurate measure of military expenditure and of the allocation of economic resources to defence. All country entries are subject to revision each year, as new information, particularly that regarding defence expenditure becomes available. The information is necessarily selective. A wider range of statistics is available to IISS members on request.

In **Part I**, individual country entries typically show economic performance over the past two years, and current-year demographic data. Where these data are unavailable, information from the last available year is provided. Defence expenditure is generally shown for the past two years where official outlays are available, or sufficient data for reliable estimates exist. Current-year defence budgets and, where available, defence budgets for the following year are also listed. Foreign Military Assistance (FMA) data cover outlays for the past year, and budgetary estimates for the current and subsequent years. Unless otherwise indicated, the US is the donor country. All financial data in the country entries are shown both in national currency and US dollars at current-year, not constant, prices. US dollar conversions are generally, but not invariably, calculated from the exchange rates listed in the entry. In a few cases, notably Russia and China, purchasing-power-parity (PPP) rates are used in preference to official or market-exchange rates.

Definitions of terms

To avoid errors in interpretation, an understanding of the definition of defence expenditure is important. Both the UN and NATO have developed standardised definitions, but in many cases countries prefer to use their own definitions (which are not in the public domain). For consistency, the IISS uses the NATO definition (which is also the most comprehensive) throughout.

In *The Military Balance,* military expenditure is defined as the cash outlays of central or federal government to meet the costs of national armed forces. The term 'armed forces' includes strategic, land, naval, air, command, administration and support forces. It also includes paramilitary forces such as the *gendarmerie,* customs service and border guard if these are trained in military tactics, equipped as a military force and operate under military authority in the event of war. Defence expenditures are reported in four categories: Operating Costs, Procurement and Construction, Research and Development (R&D) and Other Expenditure. Operating Costs include: salaries and pensions for military and civilian personnel; the cost of maintaining and training units, service organisations, headquarters and support elements; and the cost of servicing and repairing military equipment and infrastructure. Procurement and Construction expenditure covers national equipment and infrastructure spending, as well as common infrastructure programmes. It also includes financial contributions to multinational military organisations, host-nation support in cash and in kind, and payments made to other countries under bilateral agreements. FMA counts as expenditure by the donor, and not the recipient, government. R&D is defence expenditure up to the point at which new equipment can be put in service, regardless of whether new equipment is actually procured. The fact that the IISS definitions of military expenditure are generally more inclusive than those applied by national governments and the standardised UN format means that our calculated expenditure figures may be higher than national and UN equivalents.

The issue of transparency in reporting military expenditures is a fundamental one. Only a minority of the governments of UN member-states report defence expenditures to their electorates, the UN, the International Monetary Fund (IMF) and other multilateral organisations. In the case of governments with a proven record of transparency, official figures generally conform to a standardised definition of defence expenditure, and consistency problems are not usually a major issue. Where these conditions of transparency and consistency are met, the IISS cites official defence budgets and outlays as reported by national governments, NATO, the UN, the Organisation for Security and Cooperation in Europe (OSCE) and the IMF. On the other hand, some governments do not report defence expenditures until several years have elapsed, while others understate these expenditures in their reports. Where these reporting conditions exist, *The Military Balance* gives IISS estimates of military expenditures for the country concerned. Official defence budgets are also shown, in order to provide a measure of the discrepancy between official figures and what the IISS estimates real defence outlays to be. In these cases *The Military Balance* does not cite official defence expenditures (actual outlays), as these rarely differ significantly from official budgetary data. The IISS defence-expenditure estimates are based on information from several sources, and are marked 'ε'. The most frequent instances of budgetary manipulation or falsification typically involve equipment procurement, R&D, defence industrial investment, covert weapons programmes, pensions for retired military and civilian personnel, paramilitary forces, and non-budgetary sources of revenue for the military arising from ownership of industrial, property and land assets.

The principal sources for economic statistics cited in the country entries are the IMF, the Organisation for Economic Cooperation and Development (OECD), the World Bank and three regional banks (the Inter-American, Asian and African Development Banks). For some countries basic economic data are difficult to obtain. This is the case in a few former command economies in transition and countries currently or recently involved in armed conflict. The Gross Domestic Product (GDP) figures are nominal (current) values at market prices, but GDP per capita figures are nominal values at PPP prices. GDP growth is real not nominal growth, and inflation is the year-on-year change in consumer prices. Two different measures of debt are used to distinguish between OECD and non-OECD countries: for OECD countries, debt is gross public debt (or, more exactly, general government gross financial liabilities) expressed as a proportion of GDP. For all other countries, debt is gross foreign debt denominated in current US dollars. Dollar exchange rates relate to the last two years plus the current year. Values for the past two years are annual averages, while current values are the latest monthly value.

Calculating exchange rates

Typically, but not invariably, the exchange rates shown in the country entries are also used to calculate GDP and defence-expenditure dollar conversions. Where they are not used, it is because the use of exchange rate dollar conversions can misrepresent both GDP and defence expenditure. This may arise when: the official exchange rate is overvalued (as with some Latin American and African countries); relatively large currency fluctuations occur over the short-to-medium term; or when a substantial medium-to-long-term discrepancy between the exchange rate and the dollar PPP exists. Where exchange rate fluctuations are the problem, dollar values are converted using lagged exchange rates (generally by no more than six months). The GDP estimates of the Inter-American Development Bank, usually lower than those derived from official exchange rates, are used for Latin American countries. For former communist countries, PPP rather than market exchange rates are sometimes used for dollar conversions of both GDP and defence expenditures, and this is marked.

The arguments for using PPP are strongest for Russia and China. Both the UN and IMF have issued caveats concerning the reliability of official economic statistics on transitional economies, particularly those of Russia and some Eastern European and Central Asian countries. Non-reporting, lags in the publication of current statistics and frequent revisions of recent data (not always accompanied by timely revision of previously published figures in the same series) pose transparency and consistency problems. Another problem arises with certain transitional economies whose productive capabilities are similar to those of developed economies, but where cost and price structures are often much lower than world levels. PPP dollar values are used in preference to market exchange rates in cases where using such exchange rates may result in excessively low dollar-conversion values for GDP and defence expenditure.

Demographic data

Population aggregates are based on the most recent official census data or, in their absence, demographic statistics taken from *World Population Projections* published annually by the World Bank. Data on ethnic and religious minorities are also provided under country entries where a related security issue exists.

United States

MILITARY CAPABILITY

The signs in 1999 that the long-term decline in US defence spending was coming to an end were confirmed in 2000, with more funds being allocated for, in particular, qualitative improvements in personnel and readiness. The 'people first' policy, announced by the Department of Defense (DoD) in 1999, remained emphatically at the head of the policy agenda. In his budget statement for 2000, Secretary of Defense William Cohen said his department's 'number one priority' was recruiting, retaining and training military personnel. He added that the defence plan for 2000–01 included short- and long-term measures to ensure 'robust military readiness well into the twenty-first century'. The new pressures of overseas deployments have put a priority on high levels of training, equipment maintenance and morale. For example, the armed forces overall lost more than 1,000 trained pilots over the year to mid-2000. Erosion of trained manpower on this scale cannot persist for very long without serious deterioration in operational capabilities. Over the past few years, just when America's permanent overseas presence has declined, the number and frequency of overseas troop deployments have increased. The result is a further increase in use of reserve personnel – even to the extent of National Guard formations being dispatched to the Balkans. This longer-term deployment of reserve personnel has caused serious problems for them and for their employers. The practical measures taken to reverse the trend and its deleterious effect on operational capability include increases in pay and pensions and more spending on recruiting campaigns.

The 2000 defence budget maintains all the major equipment programmes. However, uncertainty over the F-22 fighter programme continues and technological risk-reduction measures for the existing fleet of combat aircraft have been approved at the expense of delaying full development of the Joint Strike Fighter (JSF). More money has been allocated to the controversial National Missile Defense (NMD) programme. The major equipment programmes for land and naval forces continue more or less as planned, with increased emphasis on combat support, mobility and readiness.

Readiness

All four services have been required to develop systems of tracking and reporting the time personnel spend away from their home bases and levels of operational readiness. The Army, Air Force, Navy and Marines have each developed their own means of recording the status of units; all are based on some combination of:

- Time away from home base or port;
- Time between out-of-station deployments; and
- Operating tempo expressed in terms of resources expended over time; for example, flying hours, track hours/mileage (in the case of armoured vehicles) and steaming days (for ships out of home port).

Thresholds have been set by the heads of the Services and can only be exceeded on their personal authority. This rigorous reporting system extends to assets whose capabilities are in heavy demand to support worldwide joint operations, including, in particular, those aircraft involved in electronic- and imagery-surveillance and reconnaissance missions, such as the EA6B *Prowler*, RC-135 *Rivet Joint* and the U-2 high-altitude reconnaissance aircraft. In addition to the individual services' 'tempo management' policies, a Global Military Force Policy has been developed for these so-called 'Low Density/High Demand' units to ensure efficient prioritisation and allocation of assets.

Force Structure

The force structure being developed for the US armed forces, set out in the 2000 budget statement is designed in accordance with *Joint Vision 2010* – a template for joint operations including, stated as new concepts, the following:

- employment of information, mobility and engagement capabilities in order to apply the decisive use of force at the right place and time;
- application of precision engagement with full exploitation of target-acquisition and battle-damage assessment capabilities to deliver lethal and non-lethal weapons accurately;
- defence of US forces and facilities against all forms of attack in peacetime, through all stages of a crisis, at all levels of conflict; and
- enhancement of logistic practices and doctrine to assure force mobility and sustainability.

This template strongly reflects the lessons of the 1990s and strengthens the trend towards lighter, smaller, rapidly deployable and better-protected forces. This trend was heralded at the end of the Cold War in the early 1990s, but was not fully implemented at that time, as demonstrated in the laborious deployments to the Balkans in 1995 and in 1999. The services have all engaged in vigorous trials and experimental programmes to implement the concept set out in *Joint Vision 2010*. Since the February 2000 budget statement, the DoD has published *Joint Vision 2020*, confirming and extending the concepts set out in the earlier document. The new document incorporates lessons from recent operations and trials and expands the vision to include, in particular, coalition and alliance operations. While the concept is some years from being fully realised, the prospects for its implementation are much brighter than before.

The main elements of the structure envisaged for US reserve and regular forces in 2001 are set out in Table 1 below.

Table 1 Summary of conventional force structure for 2001

ARMY		NAVY	
Active corps	4	Aircraft carriers	12
Divisions (active/National Guard)	10/8	Air wings (active/reserve)	10/1[1]
Active armoured cavalry regiments	2	Amphibious ready groups	12
Enhanced separate brigades (National Guard)	15	Attack submarines	55
Separate brigades (National Guard)	3	Surface combatants (active/reserve)	108/8
AIR FORCE		**MARINE CORPS**	
Active fighter wings	12.6[2]	Marine expeditionary forces	3
Reserve fighter wings	7.6[2]	Divisions (active/reserve)	3/1
Reserve air-defense squadrons	4	Air Wings (active/reserve)	3/1
Bombers (total inventory)	190[3]	Force-service-support groups (active/reserve)	3/1

Notes [1] Air wing = 48 aircraft
[2] Fighter wing = 72 aircraft
[3] Reflects a planned reduction of 18 B–52 aircraft

Aviation

The Air Force will complete its transition to an expeditionary deployment concept during 2001, having begun changing its operational deployment planning for the majority of its conventional forces in October 1999. Under this new approach, fighter/attack aircraft and other selected force elements are being grouped into ten Aerospace Expeditionary Force (AEF) groups for deployment-planning purposes. The aim is to improve the quality of life for Air Force personnel by minimising unexpected contingency deployments. Each AEF unit will be prepared to deploy for 90 days every 15 months on a regular cycle. Bombers will be an integral part of the expeditionary air force, with both B-1s and B-52s available for AEF deployments. The B-52 force will decline while the number of B-1s will rise to 70 over the period to 2004, as increasingly capable precision weapons become available for these aircraft. The AEFs do not have dedicated combat-support elements, reflecting the heavy demand for these resources and the stress placed on the crews of air-to-air tankers, command-and-control, and surveillance and target-acquisition aircraft. Some steps are being taken to remedy this situation, such as retaining the E-130 airborne command, control and communications force through to 2005 instead of disbanding it in 2003 as previously planned. Other means of easing the strain, such as increasing the number of crews for air-combat support units, will also be implemented.

Aviation equipment modernisation remains the most costly element in the defence budget absorbing, for example, more than 10% of planned spending in 2001. The principal programmes include the JSF, potentially the most expensive aviation programme, which remains in the concept phase. It is not scheduled to enter the engineering and manufacturing development (EMD) phase until 2001. Procurement is still scheduled to begin in 2005. The JSF will replace the F-16 in the Air Force, the F/A-18C in the Navy and the F/A-18C/D and the AV-8B in the Marine Corps. However, these aircraft, with upgrades, will remain in service for the next decade. For example, about 50% of the Air Force's operational combat aircraft are likely to be F-16s until about 2010. The F-22, which is due to replace the F-15C/D in the air-superiority role, entered the EMD phase in 2000. Nine operational EMD aircraft are planned by the end of 2001. While Congress moved funds from production to research and development, the testing and development schedule is unchanged and the DoD still has a plan to procure 333 aircraft.

Naval Forces

In addition to pressing ahead with its modernisation and acquisition programmes, the Navy has embarked on a process aimed at reducing manning levels in its major surface combatants. This programme will not only save money, but also, it is hoped, reduce stress on personnel from too many overseas deployments. The maritime force will decline from 316 ships at the end of 2001 to stabilise at just over 300 by 2005. This will still provide a sufficient number and variety of vessels to maintain 12 aircraft-carrier battle groups (CVBGs), 12 amphibious ready groups (ARGs), 116 surface combatants, 55 attack submarines, and associated logistics and support forces. The 2001–05 programme supports a force structure of 12 fully deployable carriers.

At the end of 2001, the carrier force will consist of nine nuclear-powered vessels – eight of the CVN-68 *Nimitz*-class plus the *Enterprise* (CVN-65) – and three conventionally-powered units. The newest *Nimitz*-class aircraft carrier, *Ronald Reagan* (CVN-76), will join the fleet in 2003, replacing the *Constellation* (CV-64). Construction of the first of the *Nimitz*-class follow-on ships, designated CVNX, begins in 2003, and it should join the fleet in around 2013, replacing the *Enterprise*, which will then be more than 50 years old. The second CVNX will replace the *J. F. Kennedy* about five years later, when that carrier reaches 50 years old.

Programmes for modernising the fleet of other surface combatants and the submarine force remain on track. The DD-21 *Zumwalt*-class land-attack destroyer is to replace the *Spruance*-class destroyers and the *Perry*-class frigates. The acquisition of 32 ships is still planned, but the start of construction has been postponed from 2005 to 2006, with an in-service date of 2009. In April 2001, a decision is expected on which of the two competing teams of shipbuilders will be awarded the main construction contract. However, the government has already decided that electric drive rather than gas turbines will power the DD-21. Potentially, this change could have consequences for the design and capabilities of surface ships as important as the change from steam power to gas turbines. In particular, there will be more space for weapons, greater endurance and smaller crews will be required. Where the *Iowa*-class battleships had a crew of more than 1,500 and the crew of the current *Arleigh Burke* destroyers is about 300, the DD-21's crew will be less than 100, with twice the firepower of its predecessors. The Navy has started to upgrade its fleet of H-60 anti-submarine helicopters in order to increase its ability to meet the threats of mines and diesel submarines in coastal waters. It is also experimenting with mine-countermeasure (MCM) systems on surface ships and helicopters, so that its naval battle groups will have an integral MCM capability, rather than having to rely on a separate, dedicated MCM fleet.

Plans to acquire 30 *Virginia*-class submarines continue with the first expected to enter service in 2004. However, this will not alleviate the current extreme over-commitment of the nuclear-powered submarine (SSN) fleet. To increase SSN numbers in the short-term, the Navy asked for $1.1billion in its 2001 budget request, in order either to refuel four *Los Angeles*-class SSNs or to start the conversion of several *Ohio*-class nuclear-powered ballistic-missile submarines (SSBNs). It is likely that the former option will be taken because of the difficulties of implementing the Strategic Arms Reduction (START) II Treaty.

There is currently insufficient shipping space to support amphibious operations. The addition of 12 LPD-17s will ease this problem when they are all in service by 2004. Each LPD-17 can carry 700 troops, 25,000 square feet of vehicle-stowage space, 36,000 cubic feet of cargo space, four CH-46 helicopters and two air-cushion vehicles, each capable of carrying 60 tonnes of cargo at over 40 knots.

Land Forces

The Army will maintain four active corps headquarters, ten active divisions (six heavy and four light), and two active armoured-cavalry regiments until 2005. Development of lighter more mobile forces, as envisaged in the last Quadrennial Defense Review (QDR), has been slow. The current restructuring of the heavy Army divisions involves a number of changes. There will be one fewer combat company per combat battalion, a dedicated reconnaissance troop assigned to each brigade, a shift of organic combat-service-support assets from combat battalions to forward-support battalions, and an increased emphasis on command-, control-, and information-support structures. In 2000, the Army has accelerated the process by establishing new, more responsive, light brigades that will initially use readily available commercial equipment and borrowed vehicles. Off-the-shelf medium-armoured vehicles will then be procured to extend this capability until technology allows a new family of combat vehicles to be fielded. The long-term goal is to remove the distinction between traditional heavy and light forces, and create a single 'Objective Force', which is more suitable to the type of missions expected over the next decade.

The Army contributed to integrating the reserve forces into the active element by setting up two integrated division headquarters on 1 October 1999. Each integrated division consists of an active component headquarters and three separate Army National Guard-enhanced brigades. There is one light division, the 7th Infantry Division, at Fort Carson, Colorado, and one heavy division, the 24th Infantry Division (Mechanised), at Fort Riley, Kansas.

Mobility Forces

Mobility forces – airlift, sealift, and land- and sea-based pre-positioning – are being enhanced. These forces include transport aircraft, cargo ships, and ground transportation systems operated by the DoD and commercial carriers. There is now greater use of commercial resources to enhance military mobility and avoid the high cost of maintaining military systems that duplicate capabilities readily attainable from the civil sector. The DoD has established an inter-theatre-airlift objective of about 50 million tonne-miles per day (MTM/D) of cargo capacity. Of this, about 20 MTM/D is provided by commercial aircraft, which contribute to military missions as participants in the Civil Reserve Air Fleet (CRAF). The remaining 30 MTM/D of inter-theatre-airlift capacity is provided by military aircraft, which are designed to perform missions that cannot be flown by commercial aircraft. The DoD plans to have an organic strategic-airlift capacity of 27 MTM/D by the end of 2001.

The DoD will attain a surge sealift capacity of 9.6m square feet by the end of 2001. The eventual goal is 10m square feet. Fast sealift ships, large medium-speed roll-on/roll-off (LMSR) vessels, and the Ready Reserve Force (RRF) provide surge sealift capacity.

The summary of military-mobility forces in Table 2 below demonstrates the US Armed Forces' unrivalled rapid-deployment capabilities.

Table 2 **Military-mobility forces, 2001**

Airlift (operational)		**Aerial Refueling** (operational)[1]		**Sealift**	
C-17	58	KC-135	472	Ready Reserve Force ships	90
C-141	88	KC-10	54	Fast sealift ships	8
C-5	104			Large medium-speed RO/ROs	18
C-130	418				

[1] These ac also perform airlift missions.

Table 3 **Aggregate numbers of strategic-offensive delivery vehicles**

As declared on 1 Jan 2000	US		Russia		Belarus		Kazakstan		Ukraine		Totals	
ICBM	687	*701*	756	*756*	0	*0*	0	*0*	27[2]	*44*	1,470	*1,501*
SLBM	464	*464*	504	*592*	0	*0*	0	*0*	968	*1,056*		
Bombers	300	*315*	78	*74*	0	*0*	0	*0*	32	*43*	410	*432*

Notes [1] The figures in italics are the numbers declared on 1 January 1999.
[2] There are no warheads with these missiles.

Strategic Weapons

Although Russia has now ratified the Strategic Arms Reduction Treaty (START) II, the defence budget provides sufficient funds ($104m) to maintain START I levels of strategic nuclear forces in case implementation does not proceed as planned. If START II is implemented, accountable warheads will be reduced to 3,000–3,500 by 2007. Current plans also provide for those nuclear-delivery vehicles eliminated under START II to be 'deactivated' by the end of 2003. No new intercontinental ballistic missiles (ICBMs) are under development or in production; the only new plan being examined at present is the possibility of replacing the Air Force's *Minuteman* III missiles in around 2020. The US will have difficulty in maintaining the industrial base to maintain and modify strategic missiles, but the budget provision for funding to replace guidance and propulsion systems as necessary for ICBMs and submarine-launched ballistic missiles (SLBMs) will go some way towards keeping a core of expertise.

Following Russia's ratification of START II, the US and Russia have opened discussions on START III, under which the number of accountable warheads would be reduced to 2,000–2,500 on each side. Progress in these negotiations will depend heavily on how US plans for NMD proceed. Future policy will not be clear until the new administration assumes office after the November 2000 presidential election.

The number of delivery vehicles held according to declarations made under the START I Memorandum of Understanding, and associated 1999 Lisbon Protocols, is as in Table 3.

Missile Defence

Although no decision on NMD deployment will be made until after the new US administration is in office, the DoD is preparing to field a system in 2007 with an initial capability in 2005, designed to meet the most immediate perceived threat – currently from North Korea. It is intended to be capable of defending all of the US against a launch of several tens of warheads. The first phase of deployment of NMD architecture includes plans for:

- 100 ground-based interceptors deployed in Alaska;
- An X-band radar deployed in Alaska;
- Upgrades to five existing ballistic-missile-early-warning radars, and
- A combination of satellites from the Defense Support and the Space-Based Infrared Satellite-high (SBIRS-high) programmes.

NMD cost estimates have risen sharply over the past year. The DoD now estimates that the full cost of the planned initial NMD system would be $26bn (at current prices) over the 20 years 1996–2015. The Congressional Budget Office estimated in April 2000 that the same system would cost $30bn up to 2015, and that a phase-two NMD system with 250 interceptors, if it were included in the deployment plan, would cost more than $60bn over the same period. The testing programme is far from complete, with two of the first three of a proposed series of 19 tests having failed. The DoD is due to report to the President on the technical feasibility of the NMD programme in September 2000.

DEFENCE SPENDING

There has been a modest 1–2% real increase in defence spending during the Clinton administration's second term, despite the government's original intention to make cuts of around 5%. There were a number of reasons for this:

- The Republican-led Congress was increasingly successful in augmenting the administration's budget request;
- The cost of contingency operations in the former Yugoslavia and around Iraq was much higher than anticipated;
- The booming US economy and consequent pressure on the labour market necessitated larger-than-planned increases in military salaries;
- The same factor also dictated increases in other personnel costs, including housing and pensions;
- At the same time, the economic phenomenon resulted in much higher-than-expected tax revenues, enabling the government to complete a virtuous circle by eliminating the 1999 budget deficit and raising defence spending simultaneously.

In each budget of President Bill Clinton's second term, Congress, with its Republican majority, has authorised higher levels of defence spending than requested by the administration. The overall increase over the original request for fiscal years 1997–2000 has totalled $55bn, which has provided funds for spending on personnel and readiness, contingency operations in Bosnia and Kosovo, intelligence, Year-2000 related work on information technology, ballistic-missile defence and certain weapon programmes favoured by congressmen. These funds have translated into a slight rise in US defence spending over a period when the country's major allies have been cutting back in real terms. Even so, the US defence burden on the economy continued to fall during Clinton's second term as a proportion of gross domestic product (GDP), to around 3% in 1999–2000 compared with 3.5% in 1996.

Both military personnel and the DoD's civilian manpower continued to decline during the period, as did employment in the defence industry. The latter is now set for a small increase in line with increasing DoD procurement spending over the next few years.

The Defence Budgets for 2000 and 2001

With 2000 a US presidential-election year and a QDR due in 2001, the FY2001 request and forward planning inevitably have an element of marking-time about them. The major change in defence-resource allocation took place in the FY2000 budget, when the administration's request, proposing an additional $112bn to become available to the DoD over the fiscal years 2000–2005, indicated the first sustained long-term increase in defence funding since the end of the Cold War. The overall increase in funds was composed of an $84bn rise from the six-year spending plans proposed in FY1999, with a further $28bn from savings from lower inflation, lower fuel prices and other adjustments.

This increase is modest in real terms, since most of the additional $112bn serves only to cancel out the large defence-spending cuts envisaged in earlier administration plans. In fact, defence

Table 4 US National Defense Budget Function and other selected budgets, 1992, 1995–2005

(US$bn)	National Defense Budget Function[1]		Department of Defense		Atomic Energy Defense Activities	International Security Assistance	Veterans Administration	Total Federal Government Expenditure[2]	Total Federal Budget Surplus[3]
FY	BA	Outlay	BA	Outlay	Outlay	Outlay	Outlay	Outlay	Outlay
1992	295.1	302.3	282.1	286.9	10.6	7.5	33.9	1,129.3	(340.5)
1995	266.3	273.6	255.7	259.4	11.8	5.3	37.8	1,227.2	(226.4)
1996	266.0	266.0	254.4	253.2	11.6	4.6	36.9	1,259.7	(174.1)
1997	270.3	271.7	258.0	258.3	11.3	4.6	39.3	1,290.7	(103.4)
1998	271.3	270.2	258.5	256.1	11.3	5.1	41.8	1,336.0	(30.0)
1999	292.1	275.5	278.4	261.4	12.4	5.5	43.2	1,382.3	0.7
2000	293.3	291.2	279.9	277.5	11.9	5.4	46.7	1,460.6	18.9
2001R	305.4	292.1	291.1	277.5	12.5	6.7	46.4	1,494.8	8.6
2002P	309.2	299.2	294.8	284.3	12.9	6.5	49.1	1,545.2	1.2
2003P	315.6	308.2	300.9	293.0	13.1	6.5	50.9	1,602.9	0.3
2004P	323.4	317.2	308.3	301.9	13.4	6.1	52.6	1,669.1	0.3
2005P	331.7	331.4	316.4	315.8	13.6	6.2	56.2	1,740.5	1.8

Notes

FY = Fiscal Year (1 October–30 September)

R = Request

P = Projection

[1] The National Defense Budget Function subsumes funding for the DoD, the DoE Atomic Energy Defense Activities and some smaller support agencies (including Federal Emergency Management and Selective Service System). It does not include funding for International Security Assistance (under International Affairs), the Veterans Administration, the US Coast Guard (Department of Transport), nor for the National Aeronautics and Space Administration (NASA). Funding for civil projects administered by the DoD is excluded from the figures cited here.

[2] The figures for Federal Government Expenditure and Federal Budget Deficit differ from figures cited in *The Military Balance* because US Government Trust Funds are no longer included. If Trust Funds were included, net US Government expenditure would show a surplus rather than deficit from 1998 [+$69.2bn] through to 2005 [+$215.4bn].

[3] Brackets indicate a deficit, no brackets indicate a surplus.

[4] Early in each calendar year, the US government presents its defence budget to Congress for the next fiscal year which begins on 1 October. It also presents its Future Years' Defense Program (FYDP), which covers the next fiscal year plus the following five. Until approved by Congress, the Budget is called the Budget Request; after approval, it becomes the Budget Authority.

[5] Definitions of US budget terms: **Authorisation** establishes or maintains a government programme or agency by defining its scope. Authorising legislation is normally a prerequisite for appropriations and may set specific limits on the amount that may be appropriated. An authorisation, however, does not make money available. **Budget Authority** is the legal authority for an agency to enter into obligations for the provision of goods or services. It may be available for one or more years. **Appropriation** is one form of Budget Authority provided by Congress for funding an agency, department or programme for a given length of time and for specific purposes. Funds will not necessarily all be spent in the year in which they are initially provided. **Obligation** is an order placed, contract awarded, service agreement undertaken or other commitment made by federal agencies during a given period which will require outlays during the same or some future period. **Outlays** are money spent by a federal agency from funds provided by Congress. Outlays in a given fiscal year are a result of obligations that in turn follow the provision of Budget Authority.

spending is projected to continue declining in 2000, while for 2001–05, the budget would increase by an overall 4.5% in real terms, or less than 1% each year. Even so, this must be measured against the 34% decline in defence spending in real terms between 1985 and 2000.

About $37bn of the additional $112bn funding was allocated to personnel budgets to support recruitment and retention. The six-year budget plan included a 4.8% increase in military pay – the largest rise since 1982 – and improved retirement benefits for long-service personnel. The operations and maintenance budgets were also increased to improve readiness, with increased funding for training (including more flying hours), maintenance and spare parts. Procurement budgets were set to exceed the target levels of the 1997 QDR, which planned for spending on equipment to rise to $60bn by 2001.

Table 5 **National Defense Budget Authority, FY1998–2005**

(US$m)	1998	1999	2000[1] Estimate	2001 Request	2002 Plan	2003 Plan	2004 Plan	2005 Plan
Military personnel	69,822	70,649	73,692	75,801	78,449	80,390	83,085	85,585
Operations & maintenance	97,214	104,990	104,862	109,285	107,485	109,112	112,230	114,780
Procurement	44,772	50,920	54,208	60,272	63,021	66,710	67,652	70,931
RDT&E	37,090	38,290	38,357	37,863	38,371	37,564	37,452	36,351
Military construction	5,463	5,406	4,794	4,549	4,275	3,805	4,576	5,368
Family housing	3,829	3,591	3,597	3,484	3,708	3,863	3,983	4,085
Other incl net receipts	346	4,552	414	-167	-512	-512	-688	-695
Total DoD	**258,536**	**278,398**	**279,924**	**291,087**	**294,797**	**300,932**	**308,290**	**316,405**
Department of Energy (defence-related)	11,704	12,600	12,157	13,084	13,169	13,429	13,771	13,920
Other (defence-related)	1,014	1,149	1,202	1,250	1,262	1,284	1,325	1,357
Total	**271,254**	**292,147**	**293,283**	**305,421**	**309,228**	**315,645**	**323,386**	**331,682**
Total (US$ 2000)	**276,743**	**292,147**	**287,346**	**293,301**	**291,180**	**291,550**	**293,106**	**295,099**
Real growth (%)	-1.6	5.6	-1.6	2.1	-0.7	0.1	0.5	0.7

Note [1] Supplemental budget request for FY2000 is $2.2bn.

The FY2000 budget authorisation is subject to a further increase as a result of a FY2000 supplemental request for $2.2bn submitted to Congress in February 2000.

The administration's 2001 budget request and projections to 2005 are in line with the 2000 plans, with minor changes. The request for FY2001 is some $305bn, compared with last year's projected $301bn, while the projected increase of nearly 3% in defence spending between 2001–05 is lower than in the FY2000 plan. This is because spending in the baseline FY2000 was substantially increased, and may be increased further as a result of the February 2000 supplemental request. There is provision in the request for a further higher-than-inflation pay rise for the military and higher pensions to improve retention levels. There is also more funding for readiness in the light of the expensive contingency operations over the past two years that have been financed through supplemental requests.

The procurement budget for 2001 rises to some $60bn, which meets the target set in the 1997 QDR. All major programmes are confirmed, but considerable uncertainty continues to surround the F-22 programme. The DoD is once again trying to secure production funding for the first 10 F-22s, which Congress denied in FY2000. If the programme proceeds, the planned 333 F-22s are now expected to cost some $63bn, of which $23bn has been committed. The RAH-66 *Comanche* helicopter programme, which has also been subject to prolonged uncertainty, now seems poised to go into production. Another 13 prototypes have been ordered and the US Army still intends to procure over 1,200 of the type, with the first entering service in 2006.

By May 2000, it became clear that Congress was going to add around $4.5bn to the administration's 2001 request, raising the budget authorisation to $310bn. Programmes securing additional funding include NMD, the *Virginia*-class SSN, the DDG-51 destroyer, and the LHD-8 amphibious-assault ship. Congress has moved to prevent the JSF programme from moving into the full development phase, and instead allocated funds for further technological risk-reduction efforts. Congress has authorised advance funding for a sixteenth Joint Surveillance Target Acquisition Radar System (JSTARS) aircraft.

Ballistic Missile Defence

Funding for ballistic missile defence (BMD) rises to $4.7bn in FY2001 from $4.2bn in FY2000. Development and procurement funding for the *Patriot* Advanced Capability 3 (PAC-3) missile is $447m. NMD receives $75m for production funding and $1.9bn for R&D, up from $1.5bn in FY2000.

Table 6 **NMD: Basic characteristics and cost implications**

	Initial capability	Expanded capability 1	Capability 2	Capability 3
Deployment date	2005	2007	2010	2011
Interceptors	20	100	100	250
Launch sites	1	1	1	2
X-band radars	1	1	4	9
Upgraded early warning radars	5	5	5	6
Communications facilities	3	3	4	5
Early warning satellites SBIRS-high	2	4	5	5
Early warning satellites SBIRS-low	0	6	243	24
Cost (US$bn)	**26.1**	**30.1**	**46.9**	**60.5**

Source: Congressional Budget Office

Table 7 **US Funding for contingency operations, FY1999–2001**

(US$m)	1999	2000	2001
Kosovo	3,132	2,025	1,713
Bosnia	1,587	1,603	1,409
South-west Asia	1,261	1,051	1,059
East Timor	2	25	4
Total	**5,982**	**4,704**	**4,185**

Funding for Contingency Operations

The incremental cost of the US contribution to the 1999 NATO-led *Operation Allied Force* against Yugoslavia was $4.9bn, according to DoD and General Accounting Office figures. Some $3.1bn of this sum was accounted for by manpower and operational expenses, while the remaining $1.8bn was spent on munitions replenishment. However, these figures fail to cover the full cost of the operation, as they do not include costs arising from greatly increased equipment utilisation, which particularly curtails aircraft lives. (The US deployed around 720 aircraft by the end of the campaign.) *The Military Balance* estimates the full cost of *Operation Allied Force*, together with associated naval and ground-force support, as $11bn, with three-quarters carried by the US.

Table 8 US Agency for International Development: International Affairs Budget selected programmes, 1998–2001

(US$m)	1998 Actual	1999 Actual	2000 Estimate	2001 Request
Assistance to the Newly Independent States of the Former Soviet Union	771	847	836	830
Support for East European democracy including FY2000 supplement	485	550	1,158	610
Voluntary peacekeeping operations	78	77	152	134
Contributions to UN and other peacekeeping operations	211	219	605	739
Economic support fund	2,420	2,593	2,792	2,313
International military education and training	50	50	50	55
Foreign military financing	3,361	3,400	4,798	3,538
Non-proliferation, anti-terrorism and related programmes	133	218	216	312
Wye Accord: Middle East Peace Process with FY2000 supplement		1,000	2,325	
International narcotics and crime with FY2000 supplement (*Plan Colombia*)	230	517	1,122	568
International disaster assistance	190	388	202	220
UN arrearage payments	100	475	351	
Total	8,028	10,334	14,607	9,319

Foreign Military Assistance

The FY2001 request for International Security Assistance (funded under the US Agency for International Development, now part of the State Department) is an estimated $5.8bn, compared with a revised $7.6bn for FY2000 after enlargement by the Wye Accord and *Plan Colombia* supplements. Israel and Egypt continue to take most of the military-equipment grant allocation. Israel will receive $1.92bn and Egypt $1.3bn in FY2001 as in previous years. The economic-assistance allocation to Israel is cut to $840m, while that to Egypt falls to $695m. Jordan's allocation of military aid rises to $140m. The administration is requesting, in a supplement to the FY2000 budget, some $1.8bn in assistance under the Wye Peace Accord programme for the three parties to the Accord – Israel, Jordan and the Palestinian National Authority (PNA). Israel is set to receive $1.2bn, the PNA $400m, Jordan $200m and Egypt $25m, provided Israel carries out its troop withdrawal commitments.

Table 9 Budget Authority for the Expanded Threat Reduction Initiative in the Former Soviet Union

(US$m)	FY1998	FY1999 Actual	FY2000 Request	FY2001 Request
Department of Defense	382	440	458	458
Department of Energy	212	237	276	364
Department of State	20	41	251	141
Total	614	718	985	963

Expanded Threat Reduction Initiative with the Former Soviet Union

The budget plans for FY2000–2005 called for spending of $4.2bn on threat-reduction assistance to the Former Soviet Union. This included $2.9bn for the DoD's 'Nunn–Lugar' Cooperative Threat Reduction (CTR) programme. The FY2001 request is $458m. The Department of Energy's Former Soviet Union-related Weapons of Mass Destruction (WMD) programmes will receive $364m in FY2001, including $150m (2000: $145m) for International Materials Protection Control and Accounting (MPC&A) and $100m initial funding for the new Long-Term Russian Program.

Table 10 Selected US Department of Energy and DoD programmes for nuclear non-proliferation and demilitarisation in the Former Soviet Union

Programme	Started in	Agency	Funds to 1998	FY1999 Actual	FY2000 Estimate	FY2001 Request
Export control and proliferation prevention	1994	DoE	114	14	13	14
Cooperative Threat Reduction	1992	DoD	1,346	440	458	458
Materials control, protection and accounting	1994	DoE	428	140	145	150
Long-Term Russian Program	2001	DoE	0	0	0	100

Table 11 Major US research and development programmes, FY1998–2001

(US$m)	Designation	1998	1999	2000	2001	Comment
DoD						
BMD	Total R&D spending	3,983	4,208	3,830	4,180	
	of which					
	THAAD	387	432	603	550	Theater High-Altitude Area Defense
	Navy Area	292	242	307	274	
	Navy Theatre	438	366	376	383	
	Joint TMD	684	204	197	0	
	Space-based laser	118	33	73	63	
	Airborne laser	154	253	304	149	
	Family of systems		94	146	231	
	Technical operations		188	214	271	
	MEADS	50	12	49	63	With Ge,It
	***Patriot* PAC-3**	243	237	179	81	
BMD	**NMD**	936	1,092	1,544	1,916	National Missile Defense
sat	***Discoverer* 2**			60		$1bn spent on sats since 1994
JOINT						
FGA	**JSF**	913	925	489	857	Joint Strike Fighter
hel	**V-22 *Osprey***	488	336	182	148	
UAV	**UAV**	517	284	196	252	unmanned aerial vehicles
ASM	**JASSM**	167	123	166	122	Joint Air-to-Surface Stand-off Missile
AAM	**AIM-9X**	106	106	117	43	*Sidewinder*, production in FY2001
AIR FORCE						
FGA	**F-22**	2,011	1,562	1,945	1,412	339 units
FGA	**JSF**	444	454	249	429	Joint Strike Fighter
laser	**YAL-1A ABL**	154	252	304	149	Airborne laser
BMD	**SBIRS-high**	338	509	421	569	Space-Based Infra-Red System (high)
BMD	**SBIRS-low**	214	181	226	241	Space-Based Infra-Red System (low)
ARMY						
lt tk	**TRACER**					With UK in PD phase
SPA	***Crusader***	301	301	266	356	Requirement 480 units from 2003
hel	**RAH-66 *Comanche***	263	352	463	614	Requirement 1,213 from 2007
hel	**CH-47 *Chinook***					Upgrade, $2.1bn for delivery 2003–2012

(US$m)	Designation	1998	1999	2000	2001	Comment
C^3I	**FBCB2**					$20bn planned for digitisation 2000–05
NAVY and MARINE CORPS						
CVN	**CVN-77**		137	246	287	For delivery in 2006
CVN	**CVNX**					Construction to begin in FY2006
SSN	**NSSN**	465	343	372	320	Plans for 20 units
DD	**DD-21**		84		555	$25bn for 32 units from FY2005
ADCC	**ADCC**		6	12		Auxiliary dry cargo carrier
FGA	**JSF**	448	471	240	428	Joint Strike Fighter
arty	**Xm777 155mm**					Requirement for 724 FY2003–06
IT	**IT-21**					$3bn IT programme to 2003

Table 12 Major US equipment orders, FY1998–2001

(US$m)	Designation	FY1998		FY1999		FY2000		FY2001	
		Units	Value	Units	Value	Units	Value	Units	Value
DoD									
BMD	**PAC-3**	48	317	60	187	60	344	60	366
BMD	**TMD**		14.2		23				3.9
BMD	**Navy Area**		14.9		43		18.1		
BMD	**NMD**								75
JOINT									
tpt	**C-130J**	9	545	7	559	2	250	4	436
trg	**JPATS**	22	125	22	108	41	167	48	188
hel	**V-22**	7	701	7	709	11	1,025	20	1,695
PGM	**JDAM**	2,202	101	4,523	137	10,122	271	9,770	244
ASM	**JSOW**	180	167	414	166	528	155	810	263
AAM	**AMRAAM**	293	202	280	141	287	136	279	138
AAM	**AIM-9X**		106		106		117	155	56
UAV	***Predator***	20	136	7	130	7	57	7	45
UAV	***Global Hawk***								22
AIR FORCE									
bbr	**B-2**		650		406		494		145
FGA	**F-16**	3	210	1	157	10	422		149
FGA	**F-15**				101	5	418		61
FGA	**F-22**		73	2	795		281	10	2,546
PGM	**CBU-97/B**			386	126	214	97	300	107
tpt	**C-17**	9	2,367	13	2,983	15	3,355	12	2,891
C^3	**JSTARS**	1	433	2	667	1	507	1	427
sat	**NAVSTAR**	3	259	3	189		233		461
sat	**MILSTAR**		610		514		357		237
sat	**DSP**		103		101		116		116
launcher	***Titan***		516		605		474		496
launcher	***Atlas***			5	176		64		56
launcher	**EELV**				242	1	386	3	621

(US$m)	Designation	FY1998		FY1999		FY2000		FY2001	
		Units	Value	Units	Value	Units	Value	Units	Value
ARMY									
MBT	**M1A2**	120	622	120	699	120	643	80	528
AIFV	**M2A3**	72	302	73	369	72	389	72	391
LAV	**MAV**								646
MRL	**MLRS**	21	176	24	126	39	147	66	205
MRL	**ATACMS**	109	173	120	240	158	319	55	245
MRL	**BAT**			304	95	609	143	741	140
arty	**SADARM**	300	76	30	62		39		68
ATGW	***Javelin***	894	146	3,569	342	2,525	349	3,754	379
veh	**FHTV**	286	112	489	199	450	195	400	166
veh	**FMTV**	1,179	205	1,439	336	2,179	426	2,577	440
veh	**HMMWV**	1,768	121	675	74	841	99	1,002	121
SAM	***Avenger***			15	35	15	34	7	33
hel	**AH-64D**	16	506	20	633	24	790	24	758
ATGM	**Longbow**	1,100	231	2,000	343	2,200	293	2,200	285
hel	**UH-60**	28	283	29	273	19	216	6	87
sat	**DSCS**		102		124		89		93
NAVY and MARINES									
SLBM	***Trident 2***	5	306	5	353	12	526	12	496
SLCM	***Tomahawk***	65	129	250	442	148	141	*100*	91
SSN	**SSN-21**		222		58		66		
SSN	**NSSN**	1	2,510	1	2,293		1,119	1	2,032
CVN	**CVN-68**	1	1,708						
CVN	**CVN-77**				285		998	1	4,377
DDG	**DDG-51**	4	3,622	3	2,899	3	3,006	3	3,386
LPD	**LPD-17**		231	1	638	1	1,517	2	1,534
AK	**ADCC (X)**				6	1	451	1	339
FGA	**F/A-18E/F**	20	2,424	30	3,000	36	2,918	42	3,061
FGA	**AV-8B**	12	334	11	356	11	319	10	235
ELINT	**EA-6B**		117		209		265		203
recce	**EC-2C**	4	376	3	408	3	396	5	334
trg	**T-45TS**	15	296	15	312	15	344	12	281
hel	**MV-22**	7	677	7	662	10	867	16	1,315
hel	**CH-60**	1	59	5	139	17	366	15	258
hel	**SH-60R**		82		226	7	231	4	177
SAM	***Standard***	100	178	114	221	86	211	96	186
SAM	**RAM**			95	44	90	45		23
ATGW	***Javelin***	380	58	741	83	998	93	293	30
UAV	***Predator***				15		11	698	43

United States US

United States dollar US$

		1998	1999	2000	2001
GDP	US$	8.8tr	9.2tr		
per capita	US$	31,500	33,100		
Growth	%	4.3	4.2		
Inflation	%	1.6	2.1		
Publ debt	%	62.4	59.3		
Def bdgt					
BA	US$	271.3bn	292.1bn	293.3bn	
Outlay	US$	270.2bn	275.5bn	291.2bn	
Request					
BA	US$			305.4bn	
Outlay	US$			292.1bn	

Population			275,636,000
Age	13–17	18–22	23–32
Men	9,609,000	9,292,000	19,396,000
Women	9,159,000	8,870,000	18,544,000

Total Armed Forces

ACTIVE 1,365,800

(incl 199,850 women, excl Coast Guard)

RESERVES 1,211,500

(incl Stand-by Reserve)

READY RESERVE 1,181,700

Selected Reserve and Individual Ready Reserve to augment active units and provide reserve formations and units

NATIONAL GUARD 456,600

Army (ARNG) 350,000 **Air Force** (ANG) 106,600

RESERVE 725,100

Army 378,200 **Navy** 178,200 **Marines** 95,000 **Air Force** 73,700

STAND-BY RESERVE 29,800

Trained individuals for mob **Army** 740 **Navy** 11,700 **Marines** 260 **Air Force** 17,100

US Strategic Command (US STRATCOM)

HQ: Offutt AFB, NE (manpower incl in Navy and Air Force totals)

NAVY up to 432 SLBM in 18 SSBN

(Plus 16 *Poseidon* C-3 launchers in one op ex-SSBN redesignated SSN (32 msl), START accountable)

SSBN 18 *Ohio*

10 (SSBN-734) with up to 24 UGM-133A *Trident* D-5 (240 msl)

8 (SSBN-726) with up to 24 UGM-93A *Trident* C-4 (192 msl)

AIR FORCE

ICBM (Air Force Space Command (AFSPC)) 550

11 missile sqn

500 *Minuteman* III (LGM-30G)

50 *Peacekeeper* (MX; LGM-118A) in mod

AC (Air Combat Command (ACC)): 208 active hy bbr (300 START-accountable)

14 bbr sqn (7 B-1, 5 B-52, 2 B-2A)

7 sqn (2 ANG) with 91 B-1B

5 sqn (1 AFR) with 92 B-52H (57 combat ready)

2 sqn with 20 B-2A

FLIGHT TEST CENTRE 5

2 B-52, 2 B-1, 1 B-2

Strategic Recce/Intelligence Collection (Satellites)

IMAGERY Improved *Crystal* (advanced **KH-11**) visible and infra-red imagery (perhaps 3 op, resolution 6in) ***Lacrosse*** (formerly ***Indigo***) radar-imaging satellite (resolution 1–2m)

ELECTRONIC OCEAN RECCE SATELLITE (EORSAT) to detect ships by infra-red and radar

NAVIGATIONAL SATELLITE TIMING AND RANGING (NAVSTAR) 24 satellites, components of Global Positioning System (GPS); block 2R system with accuracy to 1m replacing expired satellites

ELINT/SIGINT 2 ***Orion*** (formerly ***Magnum***), 2 ***Trumpet*** (successor to ***Jumpseat***), 3 name unknown, launched August 1994, May 1995, April 1996

NUCLEAR DETONATION DETECTION SYSTEM detects and evaluates nuclear detonations; sensors to be deployed in NAVSTAR satellites

Strategic Defences

US Space Command (HQ: Peterson AFB, CO)

North American Aerospace Defense Command (NORAD), a combined US–Canadian org (HQ: Peterson AFB, CO)

US Strategic Command (HQ: Offutt AFB, NE)

EARLY WARNING

DEFENSE SUPPORT PROGRAM (DSP) infra-red surveillance and warning system. Detects missile launches, nuclear detonations, aircraft in after burner, spacecraft and terrestrial infra-red events. Approved constellation: 3 op satellites and 1 op on-orbit spare

BALLISTIC-MISSILE EARLY-WARNING SYSTEM (BMEWS) 3 stations: Clear (AK), Thule (Greenland), Fylingdales Moor (UK). Primary mission to track ICBM and SLBM; also used to track satellites

SPACETRACK USAF radars at Incirlik (Turkey), Eglin (FL), Cavalier AFS (ND), Clear, Thule, Fylingdales Moor, Beale AFB (CA), Cape Cod (MA); optical tracking systems in Socorro (NM), Maui (HI), Diego Garcia (Indian Ocean)

USN SPACE SURVEILLANCE SYSTEM (NAVSPASUR) 3 transmitting, 6 receiving-site field stations in south-east US

PERIMETER ACQUISITION RADAR ATTACK CHARACTERISATION SYSTEM (PARCS) 1 north-facing phased-array system at Cavalier AFS (ND); 2,800km range

PAVE PAWS phased-array radars in Massachusetts, GA; 5,500km range

MISCELLANEOUS DETECTION AND TRACKING RADARS US Army Kwajalein Atoll (Pacific) **USAF** Ascension Island (Atlantic), Antigua (Caribbean), Kaena Point (HI), MIT Lincoln Laboratory (MA)

GROUND-BASED ELECTRO-OPTICAL DEEP SPACE SURVEILLANCE SYSTEM (GEODSS) Socorro, Maui, Diego Garcia

AIR DEFENCE

RADARS

OVER-THE-HORIZON-BACKSCATTER RADAR (OTH-B) 1 in Maine (mothballed), 1 in Mountain Home AFB (mothballed); range 500nm (minimum) to 3,000nm

NORTH WARNING SYSTEM to replace DEW line 15 automated long-range (200nm) radar stations 40 short-range (110–150km) stations

DEW LINE system deactivated

AC ANG 60

4 sqn: 3 with 45 F-15A/B, 1 with 15 F-16C/D

ac also on call from Navy, USMC and Air Force

AAM *Sidewinder*, *Sparrow*, AMRAAM

Army 471,700

(incl 71,400 women)

3 Army HQ, 4 Corps HQ (1 AB)

2 armd div (3 bde HQ, 5 tk, 4 mech inf, 3 SP arty bn; 1 MLRS bn, 1 AD bn; 1 avn bde)

2 mech div (3 bde HQ, 5 tk, 4 mech inf, 3 SP arty bn; 1 MLRS bn, 1 ADA bn; 1 avn bde)

1 mech div (3 bde HQ, 4 tk, 5 mech inf, 3 SP arty bn; 1 MLRS bn, 1 ADA bn; 1 avn bde)

1 mech div (3 bde HQ, 4 tk, 3 mech inf, 2 air aslt inf, 3 SP arty bn; 1 AD bn; 1 avn bde)

2 lt inf div (3 bde HQ, 9 inf, 3 arty, 1 AD bn; 1 avn bde)

1 air aslt div (3 bde HQ, 9 air aslt, 3 arty bn; 2 avn bde (7 hel bn: 3 ATK, 2 aslt, 1 comd, 1 med tpt))

1 AB div (3 bde HQ, 9 AB, 3 arty, 1 AD, 1 air cav, 1 avn bde)

5 avn bde (1 army, 3 corps, 1 trg)

3 armd cav regt (1 hy, 1 lt, 1 trg)

6 arty bde (3 with 1 SP arty, 2 MLRS bn; 1 with 3 arty, 1 MLRS bn; 1 with 3 MLRS bn; 1 with 1 MLRS bn)

1 indep inf bn, 1 AB Task Force

9 *Patriot* SAM bn (5 with 6 bty, 2 with 4 bty, 2 with 3 bty)

2 *Avenger* SAM bn

2 Integrated Div HQ (peacetime trg with 6 enhanced ARNG bde)

READY RESERVE

ARMY NATIONAL GUARD (ARNG) 350,000 (incl 37,900 women): capable after mob of manning 8 div (3 armd, 2 mech, 2 med, 1 lt inf) • 18 indep bde, incl 15 enhanced (2 armd, 5 mech, 7 inf, 1 armd cav) • 16 fd arty bde HQ • Indep bn: 3 inf, 36 arty, 19 avn, 9 AD (1 *Patriot*, 8 *Avenger*), 37 engr

ARMY RESERVE (AR) 378,200 (incl 45,100 women): 7 trg div, 5 exercise div, 13 AR/Regional Spt Cmd, 4 hel bn (2 AH-64, 2 CH-47), 4 hel coy (CH-47), 2 ac bn (Of these, 205,000 Standing Reservists receive regular trg and have mob assignment; the remainder receive limited trg, but as former active-duty soldiers could be recalled in an emergency)

EQUIPMENT

MBT some 7,900 M-1 *Abrams* incl M-1A1, M-1A2

RECCE 110 Tpz-1 *Fuchs*

AIFV 6,710 M-2/-3 *Bradley*

APC 15,200 M-113A2/A3 incl variants

TOTAL ARTY 6,074

TOWED 1,591: **105mm**: 458 M-102, 418 M-119; **155mm**: 715 M-198

SP 2,512: **155mm**: 2,512 M-109A1/A2/A6

MRL 227mm: 1,075 MLRS (all ATACMS-capable)

MOR 896: **120mm**: 896 M-120/121; plus **81mm**: 345 M-252

ATGW 8,686 TOW (incl 1,380 HMMWV, 523 M-901, 6,710 M-2/M-3 *Bradley*), 20,000 *Dragon*, 332 *Javelin*

RL 84mm: AT-4

SAM FIM-92A *Stinger*, 785 *Avenger* (veh-mounted *Stinger*), 99 *Linebacker* (4 *Stinger* plus 25mm gun), 485 *Patriot*

SURV Ground 122 AN/TPQ-36 (arty), 70 AN/TPQ-37 (arty), 66 AN/TRQ-32 (COMINT), 29 AN/TSQ-138 (COMINT), 39 AN/TSQ-138A, 67 AN/TLQ-17A (EW) **Airborne** 4 *Guardrail* (RC-12D/H/K, 3 RU-21H ac), 6 EO-5ARL (DHC-7), 35 OV/RV-1D

AMPH 51 ships:

6 *Frank Besson* LST: capacity 32 tk

34 LCU-2000

11 LCU-1600

Plus craft: some 82 LCM-8

UAV 7 *Hunter* (5 in store)

AC some 292: 47 **C-12C/R**, 88 **C-12D/-J**, 3 **C-20**, 48 **C-23A/B**, 11 **C-26**, 2 **C-31**, 1 **C-37**, 2 **C-182**, 2 **O-2**, 1 **PA-31**, 45 **RC-12D/H/K**, 12 **RC-12P/Q**, 2 **T-34**, 22 **UC-35**, 4 **UV-18A**, 2 **UV-20A**

HEL some 5,039 (1,502 armed): 511 **AH-1S**, 743 **AH-64A/D**, 54 **AH-6/MH-6**, 809 **UH-1H/V**, 1,395 **UH-60/MH-60A/L/K**, 4 **UH-60Q**, 66 **EH-60A (ECM)**, 466 **CH/MH-47D/E**, 468 **OH-58A/C** (67 in store), 385 **OH-58D** (incl 194 armed), 136 **TH-67** *Creek*, 2 **RAH-66**

Navy (USN) 370,700

(incl 52,050 women)

2 Fleets: Pacific, Atlantic
Surface combatants further divided into:
5 Fleets: **2nd** Atlantic, **3rd** Pacific, **5th** Indian Ocean, Persian Gulf, Red Sea, **6th** Mediterranean, **7th** W. Pacific; plus Military Sealift Command (MSC), Naval Special Warfare Command, Naval Reserve Force (NRF)

SUBMARINES 74

STRATEGIC SUBMARINES 18 (see p. 25)
TACTICAL SUBMARINES 55 (incl about 8 in refit)
SSGN 33
2 *Seawolf* (SSN-21) with up to 45 *Tomahawk* SLCM plus 8 × 660mm TT; about 50 tube-launched msl and Mk 48 HWT
23 imp *Los Angeles* (SSN-751) with 12 *Tomahawk* SLCM (VLS), 4 × 533mm TT (Mk 48 HWT, *Harpoon*)
8 mod *Los Angeles* (SSN-719) with 12 *Tomahawk* SLCM (VLS), 4 × 533mm TT (Mk 48 HWT, *Harpoon*)
SSN 22
20 *Los Angeles* (SSN-688) with 4 × 533mm TT (Mk 48 HWT, *Harpoon*, *Tomahawk* SLCM)
2 *Sturgeon* (SSN-637) with 4 × 533mm TT (Mk 48 HWT, *Tomahawk* SLCM)
OTHER ROLES 1 ex-SSBN (SSBN 642) (special ops, included in the START-accountable launcher figures)

PRINCIPAL SURFACE COMBATANTS 126

AIRCRAFT CARRIERS 12
CVN 9
8 *Nimitz* (CVN-68) (one in refit)
1 *Enterprise* (CVN-65)
CV 3
2 *Kitty Hawk* (CV-63)
1 *J. F. Kennedy* (CV-67) (in reserve)
AIR WING 11 (10 active, 1 reserve); average Air Wing comprises 9 sqn
3 with 12 F/A-18C, 1 with 14 F-14, 1 with 8 S-3B and 2 ES-3, 1 with 6 SH-60, 1 with 4 EA-6B, 1 with 4 E-2C, 1 spt with C-2
CRUISERS 27
CG 27 *Ticonderoga* (CG-47 *Aegis*)
5 *Baseline* 1 (CG-47–51) with 2 × 2 SM-2 MR SAM/ASROC, 2 × 4 *Harpoon* SSM, 2 × 127mm guns, 2 × 3 ASTT (Mk 46 LWT), 2 SH-2F or SH-60B hel
22 *Baseline* 2/3 (CG-52) with 2 VLS Mk 41 (61 tubes each) for combination of SM-2 ER SAM, and *Tomahawk*; other weapons as *Baseline* 1
DESTROYERS 52
DDG 52
28 *Arleigh Burke* (DDG-51 *Aegis*) with 2 VLS Mk 41 (32 tubes fwd, 64 tubes aft) for combination of *Tomahawk*, SM-2 ER SAM and ASROC 2 × 4 *Harpoon* SSM, 1 × 127mm gun, 2 × 3 ASTT (Mk 46 LWT), 1 SH-60B hel
24 *Spruance* (DD-963) with 2 VLS Mark 41 for combination of *Tomahawk* and ASROC *Harpoon* SSM, *Sea Sparrow* SAM, 2 × 127mm gun, 2 × 3 ASTT, 2 SH-60B hel
FRIGATES 35
FFG 35 *Oliver Hazard Perry* (FFG-7) (10 in reserve) all with *Harpoon* SSM, 1 SM-1 MR SAM, 2 × 3 ASTT (Mk 46), 1 × 76mm gun; plus either 2 × SH-60 or 1 × SH-2F hel

PATROL AND COASTAL COMBATANTS 21

(mainly responsibility of Coast Guard)
PATROL, COASTAL 13 *Cyclone* PFC with SEAL team
PATROL, INSHORE 8<

MINE WARFARE 27

MINELAYERS none dedicated, but mines can be laid from attack submarines, aircraft and surface ships.
MCM 27
1 *Inchon* MCCS in reserve
1 *Osprey* (MHC-51) MHC, 11 *Osprey* in reserve
10 *Avenger* (MCM-1) MCO, 4 *Avenger* in reserve

AMPHIBIOUS 41

LCC 2 *Blue Ridge*, capacity 700 tps
LHD 6 *Wasp*, capacity 1,894 tps, 60 tk; with 5 AV-8B ac, 42 CH-46E, 6 SH-60B hel; plus 3 LCAC, 2 RAM
LHA 5 *Tarawa*, capacity 1,900 tps, 100 tk, 4 LCU or 1 LCAC, 6 AV-8B ac, 12 CH-46E, 9 CH-53E hel, 2 RAM
LPD 11 *Austin*, capacity 900 tps, 4 tk, with 6 CH-46E hel
LSD 15
8 *Whidbey Island* with 4 LCAC, capacity 500 tps, 40 tk
4 *Harpers Ferry* with 2 LCAC, capacity 500 tps, 40tk
3 *Anchorage* with 3 LCAC, capacity 330 tps, 38 tk
LST 2 *Newport*, capacity 347 tps, 10 tk, both in reserve
CRAFT 202
82 LCAC, capacity 1 MBT; about 37 LCU-1610, capacity 1 MBT; 8 LCVP; 75 LCM; plus numerous LCU

NAVAL RESERVE SURFACE FORCES 27

1 CV (*J. F. Kennedy*) fully op with assigned air wg, 8 FFG, 4 MCM, 11 MHC, 2 LST, 1 MCCS (*Inchon*) generally crewed by 70% active and 30% reserve, plus 28 MIUW units

NAVAL INACTIVE FLEET about 23

2 CV, 1 FFG, 2 BB, 4 LST, 4 LKA, 5 AO, 3 AF, 1 AK, 1 AG plus 35 misc service craft

MILITARY SEALIFT COMMAND (MSC)

MSC operates about 110 ships around the world carrying the designation 'USNS' (United States Naval Ships). They are not commissioned ships and are manned by civilians. Some also have small military departments assigned to carry out specialised military functions such as communications and supply operations. MSC ships carry the prefix "T" before their normal hull numbers.

NAVAL AUXILIARY FORCE 34
7 AE • 6 AF • 2 AH • 12 AO • 7 AT/F

Special Mission Ships 26
3 AG • 1 AR/C • 3 AGOS (counter-drug ops) • 11

AGOS • 8 AGHS

Prepositioning Program/Maritime Prepositioning Program 32

1 ro-ro AK • 3 heavy ro-ro AK • 1 flo-flo AK • 4 AK • 13 MPS AK • 4 LASH • 2 AOT • 2 AVB • 1 AG • 1 AO

Sealift Force 24

8 AKR • 11 ro-ro AKR • 5 AOT

Additional Military Sealift

Ready Reserve Force (RRF) 80

27 breakbulk, 31 ro-ro, 4 LASH, 3 'Seebee', 4 tkr, 2 tps, 9 AG (crane at 4–10 days' readiness, maintained by Department of Transport)

National Defence Reserve Fleet (NDRF) 48

38 dry cargo, 8 tkr, 2 tps

Augment Forces

14 cargo handling bn (12 in reserve)

COMMERCIAL SEALIFT about 327

US-flag (152) and effective US-controlled (EUSC, 175) ships potentially available to augment military sealift

NAVAL AVIATION 63,200

(incl 5,700 women)

incl 12 carriers, 11 air wg (10 active, 1 reserve) **Flying hours** F-14: 216; F-18: 252

Average air wg comprises 9 sqn

3 with 12 F/A-18C, 1 with 10 F-14, 1 with 8 S-3B, 1 with 6 SH-60, 1 with 4 EA-6B, 1 with 4 E-2C, 1 spt with C-2

AIRCRAFT

FTR 12 sqn

4 with F-14A, 5 with F-14B, 3 with F-14D

FGA/ATTACK 24 sqn

23 with F/A-18C, 1 with F/A-18A

ELINT 4 sqn

2 with EP-3, 2 with EA-6B

ECM 14 sqn with EA-6B

MR 12 land-based sqn with P-3CIII

ASW 10 sqn with S-3B

AEW 10 sqn with E-2C

COMD 2 sqn with E-6A (TACAMO)

OTHER 2 sqn

2 with C-2A

TRG 16 sqn

2 *'Aggressor'* with F/A-18

14 trg with T-2C, T-34C, T-44, T-45A

HELICOPTERS

ASW 20 sqn

10 with SH-60B (LAMPS Mk III)

10 with SH-60F/HH-60H

MCM 2 sqn with MH-53E

MISC 5 sqn

4 with CH-46, 1 with MH-53E

TRG 2 sqn with TH-57B/C

NAVAL AVIATION RESERVE 23,480

(incl 3,840 women)

FTR ATTACK 3 sqn with F-18

AEW 2 sqn with E-2C

ECM 1 sqn with EA-6B

MPA 8 sqn with P-3C/EP-3J

FLEET LOG SPT 1 wg

10 sqn with C-9B/DC-9, 4 sqn with C-130T

TRG 2 aggressor sqn (1 with F/A-18, 1 with F-5E/F)

HEL 1 wg

ASW 3 sqn: 2 with SH-2G, 1 with SH-60F/HH-60F

MSC 4 sqn: 2 with HH-60H, 1 with UH-3H, 1 with CH-46D

AIRCRAFT

(Naval Inventory includes Marine Corps ac and hel)

1,456 cbt ac; 543 armed hel

193 **F-14** (74 **-A** (ftr, incl 14 NR) plus 26 in store, 73 **-B** (ftr), 46 **-D** (ftr) plus 1 in store) • 771 **F/A-18** (191 **-A** (FGA, incl 34 NR, 80 MC (47 MCR)), 31 **-B** (FGA, incl 2 NR, 4 MC), 405 **-C** (FGA, incl 79 MC), 130 **-D** (FGA, incl 85 MC), 9 **-E** (FGA), 5 **-F** (FGA plus 2 in store) • 36 **F-5E/F** (trg, incl 22 NR and 2 MCR) • 9 **TA-4J** (trg) plus 29 in store • 120 **EA-6B** (incl 4 NR, 20 MC) plus 3 in store • 1 **A6-E** (FGA) plus 88 in store • 119 **AV-8B** (FGA, MC) plus 29 in store • 15 **TAV-8B** (trg, incl 14 MC) plus 3 in store • 69 **E-2** (67 **-C** (AEW, incl 11 NR) plus 7 in store, 2 **TE-2C** (trg) • 260 **P-3** (2 **-B**, plus 27 in store, 226* **-C** (incl 70 NR) plus 18 in store, 11 **EP-3** (ELINT), 12 **NP-3D** (trials), 9 **U/VP-3A** (utl/VIP) • 112 **S-3** plus 15 in store (114 **-B** (ASW) plus 10 in store) • 100 **C-130** (20 **-T** (tpt NR), 78 **-KC-130F/R/T** (incl 77 MC (28 MCR)), 1 **-TC-130G/Q** (tpt/trg), 1 **-DC-130** plus 2 in store) • 2 **CT-39G** (incl 1 MC) • 38 **C-2A** (tpt) • 17 **C-9B** (tpt, 15 NR, 2 MC) • 12 **DC-9** (incl 10 NR) (tpt) • 6 **C-20** (2 **-D** (VIP/NR), 4 **-G** (tpt, 1 MCR)) • 61 **UC-12** (utl) (39 **-B** (incl 12 MC, 3 MCR), 12 **-F** (incl 6 MC), 10 **-M**) • 1 **NU-1B** (utl) • 2 **U-6A** (utl) • 102 **T-2C** (trg) plus 12 in store • 1 **T-39D** (trg) • 17 **T-39N** (trg) • 55 **T-44** (trg) • 104 **T-45** (trg 74 **-A**, 30 **-C**) plus 1 in store • 309 **T-34C** (incl 2 MC) plus 5 in store • 11 **T-38A/B** (trg) • 20 **TC-12B** • 2 **TC-18F** (trg) • 9 **TA-4J** (trg) plus 29 in store

HELICOPTERS

101 **UH-1N** (utl, incl 98 MC (20 MCR)) • 25 **HH-1H** (utl, incl 7 MC) plus 10 in store • 154 **CH-53E** (tpt, incl 150 MC (16 MCR)) plus 11 in store • 44 **CH-53D** (tpt MC) plus 25 in store • 41 **MH-53E** (tpt, incl 12 NR, 5 MC) plus 2 in store • 239 **SH-60** (165 **-B**, 74 **-F**) • 39 **HH-60H** (cbt spt, incl 16 NR) plus 1 in store • 12 **SH-2G** (ASW, 12 NR) plus 2 in store • 8 **VH-60** (ASW/SAR MC) • 2 **SH-3H** (ASW/SAR) plus 16 in store • 50 **UH-3H** (ASW/SAR incl 10 NR) plus 3 in store • 26 **CH-46D** (tpt, trg) • 230 **CH-46E** (tpt, incl 230 MC (25 MCR)) • 53 **UH/HH-46D** (utl incl 9 MC) • 120 **TH-57** (44 **-B** (trg), 76 **-C** (trg) (plus 9 **-C** in store)) • 13 **VH-3A/D** (VIP, incl 11 MC) • 195 **AH-1W** (atk, incl 188 MC (37 MCR)) plus 11 in store • **44 CH-53D** (tpt, MC) plus 25 in store

TILT ROTOR 1 V-22 (MC)

MISSILES

AAM AIM-120 AMRAAM, AIM-7 *Sparrow*, AIM-54A/C *Phoenix*, AIM-9 *Sidewinder*
ASM AGM-45 *Shrike*, AGM-88A HARM; AGM-84 *Harpoon*, AGM-119 *Penguin* Mk-3, AGM-114 *Hellfire*

Marine Corps (USMC) 169,800

(incl 10,100 women)

GROUND

3 div
1 with 3 inf regt (9 bn), 1 tk, 2 lt armd recce (LAV-25), 1 aslt amph, 1 cbt engr bn, 1 arty regt (4 bn), 1 recce coy
1 with 3 inf regt (9 bn), 1 tk, 1 lt armd recce (LAV-25), 1 aslt amph, 1 cbt engr bn, 1 arty regt (4 bn), 1 recce bn
1 with 2 inf regt (6 bn), 1 cbt spt bn (1 AAV, 1 LAR coy), 1 arty regt (2 bn), 1 cbt engr bn, 1 recce coy
3 Force Service Spt Gp
1 bn Marine Corps Security Force (Atlantic and Pacific)
Marine Security Guard bn (1 HQ, 7 region coy)

RESERVES (MCR)

1 div (3 inf (9 bn), 1 arty regt (5 bn); 2 tk, 1 lt armd recce (LAV-25), 1 aslt amph, 1 recce, 1 cbt engr bn)
1 Force Service Spt Gp

EQUIPMENT

MBT 403 M-1A1 *Abrams*
LAV 400 LAV-25 (**25mm** gun) plus 334 variants incl 50 Mor, 95 ATGW (see below)
AAV 1,321 AAV-7A1 (all roles)
TOWED ARTY 105mm: 331 M-101A1; **155mm**: 596 M-198
MOR 81mm: 586 M-252 (incl 50 LAV-M)
ATGW 1,083 TOW, 1,121 *Dragon*, 95 LAV-TOW
RL 83mm: 1,650 SMAW; **84mm**: 1,300 AT-4
SURV 23 AN/TPQ-36 (arty)

AVIATION 36,400

(incl 1,860 women)
Flying hours 230 fixed wing (non-tpt), 327 fixed wing (tpt), 216 (hel)
3 active air wg and 1 MCR air wg
Flying hours cbt aircrew: 264
AIR WING no standard org, but a notional wg comprises
AC 130 fixed-wing: 48 **F/A-18A/C/D**, 60 **AV-8B**, 10 **EA-6B**, 12 **KC-130**
HEL 167: 12 **CH-53D**, 32 **CH-53E**, 36 **AH-1W**, 27 **UH-1N**, 60 **CH-46E**
1 MC C² system, wg support gp

AIRCRAFT

FTR/ATTACK 18 sqn with 240 F/A-18A/C/D (4 MCR sqn)
FGA 7 sqn with 119 AV-8B
ECM 4 sqn with 20 EA-6B
TKR 5 sqn with 77 KC-130F/R/T (2 MCR sqn)
TRG 4 sqn
1 with 12 AV-8B, 14 TAV-8B
1 with 42 F/A-18A/B/C/D, 3 T-34C
1 with 2 F-5E
1 with 8 KC-130F

HELICOPTERS

ARMED 6 lt attack/utl with 188 AH-1W and 98 UH-1N
TPT 15 **med** sqn with 230 CH-46E, 4 sqn with 44 CH-53D; 6 **hy** sqn with 150 CH-53E
TRG 4 sqn
1 with 29 AH-1W, 12 UH-1N, 4 HH-1N
1 with 20 CH-46
1 with 6 CH-53D
1 with 15 CH-53E, 6 MH-53E
SAM 3+ bn
2+ bn (5 bty), 1 MCR bn with *Stinger* and *Avenger*
UAV 2 sqn with *Pioneer*

RESERVES 11,760

(560 women); 1 air wg

AIRCRAFT

FTR/ATTACK 4 sqn with 47 F-18A
1 *Aggressor* sqn with 12 F5-E/F
TKR 2 tkr/tpt sqn with 28 KC-130T

HELICOPTERS

ARMED 2 attack/utl sqn with 47 AH-1W, 20 UH-1N
TPT 4 sqn
2 **med** with 26 CH-46E, 2 **hy** with 19 CH-53E
SAM 1 bn (2 bty) with *Stinger* and *Avenger*

EQUIPMENT (incl MCR): 402 cbt ac; 188 armed hel

Totals included in the Navy inventory

AIRCRAFT

248 **F-18A/-B/-C/-D** (FGA incl 48 MCR) • 119 **AV-8B**, 15* **TAV-8B** (trg) • 20 **EA-6B** (ECM) • 2* **F-5E/F** (trg, MCR) • 77 **KC-130F/R/T** (tkr, incl 28 MCR) • 2 **C-9B** (tpt) • 1 **C-20G** (MCR) (tpt) • 1 **CT-39G** (MCR) • 18 **UC-12B/F** (utl, incl 3 MCR) • 2 **T-34C** (trg)

HELICOPTERS

188 **AH-1W** (GA, incl 37 MCR) • 58 **UH-1N** (utl, incl 20 MCR) • 7 **HH-1H** (utl) • 230 **CH-46E** (tpt incl 25 MCR) • 9 **UH/HH-46D** (utl) • 150 **CH-53-E** (tpt, incl 16 MCR) • 5 **MH-53E**, 49 **CH-53D** (tpt) • 8 **VH-60** (VIP tpt) • 11 **VH-3A/D** (VIP tpt)
TILT ROTOR 12 MV-22B

MISSILES

SAM 1,929 *Stinger*, 235 *Avenger*
AAM *Sparrow* AMRAAM, *Sidewinder*
ASM *Maverick*, *Hellfire*, TOW

Coast Guard (active duty) 36,230 military, 5,910 civilian

(incl 3,540 women)
By law a branch of the Armed Forces; in peacetime

operates under, and is funded by, the Department of Transport

Bdgt Authority

1995	$3.7bn	*1998*	$3.9bn
1996	$3.7bn	*1999*	$3.9bn
1997	$3.8bn	*2000*	expected $4.1bn
		2001	request $4.5bn

PATROL VESSELS 126

OFFSHORE 41

12 *Hamilton* high-endurance with HH-60J LAMPS HU-65A *Dolphin* hel, all with 76mm gun
13 *Bear* med-endurance with HH-65A hel
16 *Reliance* med-endurance with 25mm gun, hel deck
plus 21 sea-going buoy tenders

COASTAL 85

49 *Farallon*, 30 *Point Hope*, 6 *Baracuda*, plus 10 coastal buoy tenders

INLAND, tenders only

15 inland construction tenders, 4 small inland buoy tenders, 18 small river buoy tenders

SPT AND OTHER 24

4 icebreakers, 19 icebreaking tugs, 1 trg

AVIATION (1,050 incl 50 women)

AC 23 HU-25 (plus 21 support or in store), 26 HC-130H (plus 4 support), 2 RU-38A, 35 HH-60J (plus 7 support), 80 HH-65A (plus 13 support), 1 VC-4A, 1 C-20B

RESERVES 8,000 incl 1,080 women

Air Force 353,600

(incl 66,300 women) **Flying hours** ftr 212, bbr 215
Air Combat Comd (ACC) 4 air force, 23 ac wg **Air Mobility Comd** (AMC) 2 air force, 13 ac wg
The US Air Force introduced its Aerospace Expeditionary Force (AEF) concept on 1 October 1999. Almost the entire USAF – active force, reserve force and Air National Guard – is being divided into 10 AEFs. Each AEF will be on call for 90 days every 15 months, and at least 2 of the 10 AEFs will be on call at any one time. Each AEF, with 10,000–15,000 personnel and up to 200 ac, will comprise an air superiority, air-to-ground, precision attack and air mobility capability.

TACTICAL 52 ftr sqn

incl active duty sqn ACC, USAFE and PACAF (sqn may be 12–24 ac)

14 with F-15, 6 with F-15E, 23 with F-16C/D (incl 3 AD), 7 with A-10/OA-10, 2 with F-117

SUPPORT

RECCE 3 sqn with U-2R and RC-135
AEW 1 Airborne Warning and Control wg, 6 sqn (incl 1 trg) with E-3
EW 2 sqn with EC-130
FAC 7 tac air control sqn, mixed A-10A/OA-10A
TRG 36 sqn
1 *Aggressor* with F-16
35 trg with **ac** F-15, F-16, A-10/OA-10, T-37, T-38, AT-38, T-1A, -3A, C-5, -130, -141 **hel** HH-60, U/TH-1
TPT 28 sqn
17 strategic: 5 with C-5 (1 trg), 9 with C-141 (2 trg), 3 with C-17
11 tac airlift with C-130
Units with C-135, VC-137, C-9, C-12, C-20, C-21
TKR 23 sqn
19 with KC-135 (1 trg), 4 with KC-10A
SAR 8 sqn (incl STRATCOM msl spt), HH-60, HC-130N/P
MEDICAL 3 medical evacuation sqn with C-9A
WEATHER RECCE WC-135
TRIALS weapons trg units with **ac** A-10, F-4, F-15, F-16, F-111, T-38, C-141 **hel** UH-1
UAV *Global Hawk* and *Darkstar*, 2 sqn with *Predators*

RESERVES

AIR NATIONAL GUARD (ANG) 106,600

(incl 17,000 women)

BBR 2 sqn with B-1B
FTR 6 AD sqn with F-15, F-16
FGA 45 sqn
6 with A-10/ OA-10
33 with F-16 (incl 1 AD)
6 with F-15A/B (incl 3 AD)
TPT 27 sqn
24 tac (1 trg) with C-130E/H
3 strategic: 1 with C-5, 2 with C-141B
TKR 23 sqn with KC-135E/R (11 with KC-135E, 12 with KC-135R)
SPECIAL OPS 1 sqn (AFSOC) with EC-130E
SAR 3 sqn with **ac** HC-130 **hel** HH-60
TRG 7 sqn

AIR FORCE RESERVE (AFR) 73,700

(incl 14,800 women), 35 wg

BBR 1 sqn with B-52H
FGA 7 sqn
4 with F-16C/D, 3 with A-10/OA-10 (incl 1 trg)
TPT 19 sqn
7 strategic: 2 with C-5A, 5 with C-141B
11 tac: 8 with C-130H, 3 C-130E
1 weather recce with WC-130E/H
TKR 7 sqn with KC-135E/R (5 KC-135R, 2 KC-135E)
SAR 3 sqn (ACC) with **ac** HC-130N/P **hel** HH-60
ASSOCIATE 26 sqn (personnel only)
4 for C-5, 5 for C-141, 1 aero-medical for C-9, 4 C-17A, 4 for KC-10, 1 for KC-135, 1 for Aggressor (F-16), 6 for Trg (T-37, T-38, T-1)

AIRCRAFT

LONG-RANGE STRIKE/ATTACK 208 cbt ac: 94 **B-52H** (57 in service, 35 in store, 2 test) • 93 **B-1B** (91 in service, 2 test) • 21 **B-2A** (20 in service, 1 test)
RECCE 32 **U-2R/S** • 4 **TU-2 R/S** • 8 **E-8C** (JSTARS), 2 **E-9A** • 3 **RC-135** (*Cobra Balls*), 2 **RC-135U** (*Combat*

Sent), 16 **RC-135V/W** (*Rivet Joint*) • 16 **SR-71** • (162 **RF-4C** in store)
COMD 33 **E-3B/C** • 4 **E-4B** • 3 **EC-135** (plus 25 in store)
TAC 2,529 cbt ac (incl ANG, AFR plus 1,004 in store); no armed hel: **F-4** (218 **-D**, **-E**, **-G** models in store) • 717 **F-15** (494 **-A/B/C/D** (ftr incl 100 ANG, 13 test), 210 **-E** (FGA) plus 3 F-15A/B/C/D/E in store) • 1,420 **F-16** (94 **-A** (incl 91 ANG), 47 **-B** (incl 30 ANG), 1,095 **-C** (incl 390 ANG, 61 AFR), 184 **-D** (incl 44 ANG, 11 AFR) (plus 384 F-16A/B in store)) • 6 **F-22A** (2 **YF-22A** in store) • (260 **F-111**/34 **EF-111A** in store) • 52 **F-117** (incl 6* (trg), plus 1 test) • 225 **A-10A** (incl 72 ANG, 39 AFR), plus 13 in store • 109* **OA-10A** (FAC incl 18 ANG, 10 AFR) • 3 **EC-18B/D** (Advanced Range Instrumentation)(4 in store) • 21* **AC-130H/U** (special ops, USAF) • 33 **HC-130N/P** (9 ANG) • 31 **EC-130E/H/J** (special ops incl 9 ANG SOF) • 66 **MC-130E/H/P** (special ops incl 45 SOF (4-Ps ANG)) • 8 **WC-130H/J** (weather recce, (AFR)) (plus 6 in store) • 3 **WC-135W** • 3 **OC-135** ('Open Skies' Treaty) • 1 **EC-137D**
TPT 126 **C-5** (74 **-A** (strategic tpt, incl 14 ANG, 32 AFR), 50 **-B**, 2 **-C**) • 23 **C-9A/C** • 38 **C-12C/-D/-J** (liaison) • 59 **C-17A** • 1 **C-18B** in store • 12 **C-20** (2 **-A**, 5 **-B**, 3 **-C**, 2 **-H**) • 78 **C-21A** (2 ANG) • 3 **C-22B** (ANG) • 3 **C-23A** • 2 **VC-25A** • 11 **C-26A/B** (ANG) • 10 **C-27** in store • 4 **C-32A** • 2 **C-37A** • 2 **C-38A** (ANG) • 521 **C-130B/E/H/J** (incl 224 ANG, 114 AFR), plus 26 in store • 5 **C-135B/C/E** • 4 **C-137B/C** (VIP tpt) • 125 **C-141B** (incl 18 ANG, 45 AFR) plus 49 in store
TKR 546 **KC-135D/E/R/T** (incl 219 ANG, 69 AFR) plus 58 in store • 59 **KC-10A** tkr/tpt
TRG 180 **T-1A** • 111 **T-3A** • 1 **T-6A** • 1 **TE-8A** • 2 **TC-18E** • 3 **UV-18B** • 411 **T-37B** • 408 **T-38** (168 in store) • 81 **AT-38B** (23 in store) • 3 **T-41** • 10 **T-43A** • 1 **CT-43A**• 2 **TC-135S/W**

HELICOPTERS

38 **MH-53-J** *Pave Low* (special ops) • 3 **MH-60G** • 4 **HH-1H** (7 in store) • 108 **HH-60G** (incl 17 ANG, 23 AFR) • 63 **UH-1N**, 6 **TH-53A**

TILT ROTOR 1 **V-22**

UAV

High Level – **RQ-4A** *Global Hawk* prototype
Tactical – 10 **RQ-1A** *Predator*

MISSILES

AAM 9,200+ AIM-9P/L/M *Sidewinder*, 4,300+ AIM-7E/F/M *Sparrow*, 4,500+ AIM-120 A/B AMRAAM
ASM 27,000+ AGM-65A/B/D/G *Maverick*, 8,000+ AGM-88A/B HARM, 70+ AGM-84B *Harpoon*, 1,100 AGM-86B ALCM, 200+ AGM-86C ALCM, 400+ AGM-129A, 100+ AGM-130A, 110+ AGM-142A/B/C/D, AGM-154 *JSOW*

CIVIL RESERVE AIR FLEET (CRAF) 683

commercial ac (numbers fluctuate)

LONG-RANGE 501
passenger 271 (A-310, B-747, B -757, B-767, DC-10, L-1011, MD-11)
cargo 230 (B-747, DC-8, DC-10, L-1011, MD-11)
SHORT-RANGE 95
passenger 81 (A-300, B-727, B-737, MD-80/83)
cargo 14 (L-100, B-727, DC-9)
DOMESTIC AND AERO-MEDICAL 34 (B-767)

Special Operations Forces (SOF)

Units only listed

ARMY (15,300)

5 SF gp (each 3 bn) • 1 Ranger inf regt (3 bn) • 1 special ops avn regt (3 bn) • 1 Psychological Ops gp (5 bn) • 1 Civil Affairs bn (5 coy) • 1 sigs, 1 spt bn

RESERVES (2,800 ARNG, 7,800 AR)

2 ARNG SF gp (3 bn) • 12 AR Civil Affairs HQ (4 comd, 8 bde) • 2 AR Psychological Ops gp • 36 AR Civil Affairs 'bn' (coy)

NAVY (4,000)

1 Naval Special Warfare Comd • 1 Naval Special Warfare Centre • 3 Naval Special Warfare gp • 6 Naval Special Warfare units • 6 SEAL teams • 2 SEAL delivery veh teams • 2 Special Boat sqn • 6 DDS

RESERVES (1,400)

1 Naval Special Warfare Comd det • 6 Naval Special Warfare gp det • 3 Naval Special Warfare unit det • 5 SEAL team det • 2 Special Boat unit • 2 Special Boat sqn • 1 SEAL delivery veh det • 1 CINCSOC det

AIR FORCE (9,320): (AFRC 1,260, ANG 1,040)

1 air force HQ, 1 wg, 14 sqn
7 with AC-130H, 12 AC-130U, 5 MC-130E, 21 MC-130H, 19 MC-130P, 34 MH53J
7 with MH-60G, 5 C-130E. AETC (Air Education and Training Command) 1 wg, 2 sqn: 3 MC-130H, 4 MC-130P, 5 MH-53J, 4 TH-53A

RESERVES

1 wg, 2 sqn:
8 MC-130E, 3 MC-130P, 1 C-130E

ANG

1 wg, 1 sqn:
5 EC-130E

Deployment

Commanders' NATO appointments also shown (e.g., COMEUCOM is also SACEUR)

EUROPEAN COMMAND (EUCOM)

some 100,000. Plus 14,000 Mediterranean 6th Fleet: HQ Stuttgart-Vaihingen (Commander is SACEUR)
ARMY (54,700) HQ US Army Europe (USAREUR), Heidelberg
NAVY HQ US Navy Europe (USNAVEUR), London (Commander is also CINCAFSOUTH)
AIR FORCE (35,500) HQ US Air Force Europe (USAFE),

Ramstein (Commander is COMAIRCENT)
USMC 950

GERMANY

ARMY 42,200
V Corps with 1 armd(-), 1 mech inf div(-), 1 arty, 1 AD (1 *Patriot* (6 bty), 1 *Avenger* bn), 1 engr, 1 avn bde
Army Prepositioned Stocks (APS) for 4 armd/mech bde, approx 57% stored in Ge
EQPT (incl APS in Ge, Be, Lux and Nl)
some 785 MBT, 715 AIFV, 852 APC, 512 arty/MRL/mor, 136 ATK hel
AIR FORCE 14,880, 72 cbt ac
1 air force HQ: USAFE
1 ftr wg: 3 sqn (2 with 54 F-16C/D, 1 with 12 A-10 and 6 OA-10)
1 airlift wg: incl 16 C-130E and 9 C-9A, 9C-21, 2C-20, 1CT-43
NAVY 300
USMC 200

BELGIUM

ARMY 170; approx 22% of POMCUS
NAVY 100
AIR FORCE 520

GREECE

NAVY 240; base facilities at Soudha Bay, Makri (Crete)
AIR FORCE 180; air base gp. Facilities at Iraklion (Crete)

ITALY

ARMY 1,700; HQ: Vicenza. 1 inf bn gp, 1 arty bty
EQPT for Theater Reserve Unit/Army Readiness Package South (TRU/ARPS), incl 116 MBT, 125 AIFV, 59 APC, 15 arty/MLRS/mor
NAVY 4,400; HQ: Gaeta; bases at Naples, La Maddalena, 1 MR sqn with 9 P-3C at Sigonella
AIR FORCE 4,200; 1 AF HQ (16th Air Force), 1 ftr wg, 2 sqn with 36 F-16C/D
Deliberate Force Component 86 F-16C, 4 AC-130, 8 EC-130, 26 F-15, 18 F-15C, 21 EA-6B, 10 KC-135, 12 F-117, 7 UH-60, 22 A-10, 4 U-2, 3 P-3, 9 MH-53, 3 MC-130, 4 MH-60
USMC 200

LUXEMBOURG

ARMY approx 21% of APS

MEDITERRANEAN

NAVY some 14,000 (incl 2,100 Marines). 6th Fleet (HQ: Gaeta, Italy): typically 4 SSN, 1 CVBG (1 CV, 5 surface combatants, 1 fast spt ship), 2 AO, 1 AE, 1 AF, 1 AT/F. MPS-1 (4 ships with equipment for 1 MEF (fwd)). Marine personnel: some 2,000. MEU (SOC) embarked aboard Amphibious Ready Group ships

NETHERLANDS

ARMY 60; approx 7% of APS
AIR FORCE 290
NAVY 10

NORWAY

prepositioning incl 18 M-109, 12 M-198 arty, no aviation assets
AIR FORCE 50
NAVY 10

PORTUGAL

(for Azores, see Atlantic Command)
NAVY 50
AIR FORCE 930

SPAIN

NAVY 1,760; base at Rota
AIR FORCE 250
USMC 120

TURKEY

NAVY 20, spt facilities at Izmir and Ankara
AIR FORCE 1,800; facilities at Incirlik. 1 wg (ac on det only), numbers vary (incl F-15, F-16, EA-6B, KC-135, E-3B/C, C-12, HC-130, HH-60)
Installations for SIGINT, space tracking and seismic monitoring
USMC 220

UNITED KINGDOM

ARMY 450
NAVY 1,220; HQ: London, admin and spt facilities
1 SEAL det
AIR FORCE 9,500
1 air force HQ (3rd Air Force): 1 ftr wg, 53 cbt ac, 2 sqn with 26 F-15E, 1 sqn with 27 F-15C/D
1 special ops gp, 1 air refuelling wg with 15 KC-135, 1 recce sqn, 1 naval air flt
USMC 170

PACIFIC COMMAND (USPACOM)

HQ: Hawaii

ALASKA

ARMY 6,600; 1 lt inf bde
AIR FORCE 9,450; 1 air force HQ (11th Air Force): 1 ftr wg with 2 sqn (1 with 18 F-16, 1 with 6 A-10, 6 OA-10), 1 wg with 2 sqn with 36 F-15C/D, 1 sqn with 18 F-15E, 1 sqn with 10 C-130H, 2 E-3B, 3 C-12, 1 air tkr wg with 8 KC-135R

HAWAII

ARMY 15,500; HQ: US Army Pacific (USARPAC): 1 lt inf div (2 lt inf bde)
AIR FORCE 4,500; HQ: Pacific Air Forces (PACAF): 1 wg with 2 C-135B/C, 1 wg (ANG) with 15 F-15A/B, 4 C-130H and 8 KC-135R
NAVY 7,500; HQ: US Pacific Fleet
Homeport for some 22 SSN, 3 CG, 4 DDG, 2 FFG, 4 spt and misc ships
USMC 7,000; HQ: Marine Forces Pacific

SINGAPORE

NAVY 90; log facilities
AIR FORCE 40 det spt sqn

USMC 20

JAPAN

ARMY 1,800; 1 corps HQ, base and spt units

AIR FORCE 13,550; 1 air force HQ (5th Air Force): 90 cbt ac
1 ftr wg, 2 sqn with 36 F-16, 1 wg, 3 sqn with 54 F-15C/D, 1 sqn with 15 KC-135, 1 SAR sqn with 8 HH-60, 1 sqn with 2 E-3 AWACS, 1 Airlift Wg with 16 C-130 E/H, 4 C-21, 3 C-9, 1 special ops gp with 4 MC-130P and 4 MC-130E

NAVY 5,200; bases: **Yokosuka** (HQ 7th Fleet) homeport for 1 CV, 6 surface combatants **Sasebo** homeport for 4 amph ships, 1 MCM sqn

USMC 19,200; 1 MEF

SOUTH KOREA

ARMY 27,500; 1 Army HQ (UN command), 1 inf div with 2 bde (2 mech inf, 2 air aslt, 2 tk bn), 2 SP arty, 2 MLRS, 1 AD bn, 1 avn, 1 engr bde, 1 air cav bde (2 ATK hel bn), 1 *Patriot* SAM bn (Army tps)

EQPT incl 116 MBT, 126 AIFV, 111 APC, 45 arty/MRL/mor

AIR FORCE 8,700; 1 air force HQ (7th Air Force): 2 ftr wg, 90 cbt ac; 3 sqn with 70 F-16, 1 sqn with 20 A-10, 12 OA-10, 1 special ops sqn, 5 MH-53J, 1 U-2

NAVY 300

USMC 130

GUAM

ARMY 40

AIR FORCE 1,850; 1 air force HQ (13th Air Force)

NAVY 1,850; MPS-3 (4 ships with eqpt for 1 MEB) Naval air station, comms and spt facilities

AUSTRALIA

AIR FORCE 260

NAVY some 40; comms facility at NW Cape, SEWS/SIGINT station at Pine Gap, and SEWS station at Nurrungar

DIEGO GARCIA

NAVY 650; MPS-2 (5 ships with eqpt for 1 MEB) Naval air station, spt facilities

AIR FORCE 20

THAILAND

ARMY 40

NAVY 10

AIR FORCE 30

USMC 40

US WEST COAST

MARINES 1 MEF

AT SEA

PACIFIC FLEET 132,300 USN, 12,850 reserve, 30,450 civilians (HQ: Pearl Harbor) **Main base**: Pearl Harbor **Other bases**: Bangor, Everett, Bremerton (WA), San Diego (CA)

Submarines 8 SSBN, 30 SSN

Surface Combatants 3 CVN, 2 CV, 13 CG/CGN, 13 DDG, 11 DD, 16 FFG

Amph 1 comd, 3 LHA, 3 LPH, 6 LPD, 8 LSD, 1 LST, plus 1 AG, 59 MSC ships

Surface Combatants divided between two fleets

3rd Fleet (HQ: San Diego) covers Eastern and Central Pacific, Aleutian Islands, Bering Sea; typically 3 CVBG, 4 URG, amph gp

7th Fleet (HQ: Yokosuka) covers Western Pacific, Japan, Philippines, ANZUS responsibilities, Indian Ocean; typically 2 CVBG, 1 URG, amph ready gp (1 MEU embarked), 2 MCM

Aircraft 363 tactical, 203 helicopter, 77 P-3, 162 other

CENTRAL COMMAND (USCENTCOM)

commands all deployed forces in its region; HQ: MacDill AFB, FL

ARMY 2,070

AT SEA

5th Fleet HQ: Manama. Average US Naval Forces deployed in Indian Ocean, Persian Gulf, Red Sea: 1 CVBG, 1 URG (forces provided from Atlantic and Pacific), 2 MCM

BAHRAIN

NAVY 680

USMC 220

KUWAIT

ARMY 3,000; 1 bde HQ; prepositioned eqpt for 1 armd bde (2 tk, 1 mech bn, 1 arty bn)

NAVY 10

AIR FORCE 2,100 (force structure varies)

USMC 80

OMAN

AIR FORCE 630

NAVY 60

QATAR

ARMY 30; prepositioned eqpt for 1 armd bde (forming)

SAUDI ARABIA

ARMY 650; 1 *Patriot* SAM, 1 sigs unit incl those on short-term (6 months) duty

AIR FORCE 4,800. Units on rotational detachment, **ac** numbers vary (incl F-15, F-16, F-117, C-130, KC-135, U-2, E-3)

NAVY 20

USMC 250

UAE

AIR FORCE 390

TRAINING ADVISORS

NIGERIA 50 (to be 200)

SOUTHERN COMMAND (USSOUTHCOM)

HQ: Miami, FL

ARMY 2,100; HQ: US Army South, Fort Buchanan, PR: 1 inf, 1 avn bn

USMC 100

AIR FORCE 1,600; 1 wg (1 C-21, 9 C-27, 1 CT-43)

HONDURAS

ARMY 160
USMC 70
AIR FORCE 180

JOINT FORCES COMMAND (USJFCOM)

HQ: Norfolk, VA (CINC has op control of all CONUS-based army and air forces)

US EAST COAST

USMC 19,200; 1 MEF

BERMUDA

NAVY 800

CUBA

NAVY 590 (Guantánamo)
USMC 490 (Guantánamo)

HAITI

USMC 230

ICELAND

NAVY 960; 1 MR sqn with 6 P-3, 1 UP-3
USMC 80
AIR FORCE 600; 6 F-15C/D, 1 KC-135, 1 HC-130, 4 HH-60G

PORTUGAL (AZORES)

NAVY 10; limited facilities at Lajes
AIR FORCE periodic SAR detachments to spt space shuttle ops

UNITED KINGDOM

NAVY 1,220; comms and int facilities at Edzell, Thurso

AT SEA

ATLANTIC FLEET (HQ: Norfolk, VA) **Other main bases** Groton (CT), King's Bay (GA), Mayport (FL)
Submarines 10 SSBN, 16 SSGN, 35 SSN
Surface Combatants 6 CV/CVN, 23 CG/CGN, 21 DDG, 23 FFG. Amph: 1 LCC, 2 LHA, 4 LPH, 6 LPD, 5 LSD, 6 LST, 1 LKA
Surface Forces divided into 2 fleets:
2nd Fleet (HQ: Norfolk) covers Atlantic; typically 4–5 CVBG, amph gp, 4 URG
6th Fleet (HQ: Gaeta, Italy) under op comd of EUCOM, typically 1 CG, 3 DDG, 2 FFG, 3 SSN, amph gp

Continental United States (CONUS)

major units/formations only listed

ARMY (USACOM) 340,300

provides general reserve of cbt-ready ground forces for other comd
Active 1 Army HQ, 3 Corps HQ (1 AB), 1 armd, 2 mech, 1 lt inf, 1 AB, 1 air aslt div; 6 arty bde; 2 armd cav regt, 6 AD bn (1 *Avenger*, 5 *Patriot*)
Reserve (ARNG): 3 armd, 2 mech, 2 med, 1 lt inf div;18 indep bde

NAVY 186,200
AIR FORCE 276,200
USMC 128,100

US STRATEGIC COMMAND (USSTRATCOM)

HQ: Offutt AFB, NE
See entry on p. 25

AIR COMBAT COMMAND (ACC)

HQ: Langley AFB, VA. Provides strategic AD units and cbt-ready Air Force units for rapid deployment

SPACE COMMAND (AFSPACECOM)

HQ: Peterson AFB, CO. Provides ballistic-missile warning, space control, satellite operations around the world, and maintains ICBM force

US SPECIAL OPERATIONS COMMAND (USSOCOM)

HQ: MacDill AFB, FL. Comd all active, reserve and National Guard special ops forces of all services based in CONUS. See p. 31

US TRANSPORTATION COMMAND (USTRANSCOM)

HQ: Scott AFB, IL. Provides all common-user airlift, sealift and land transport to deploy and maintain US forces on a global basis

AIR MOBILITY COMMAND (AMC)

HQ: Scott AFB, IL. Provides strategic, tac and special op airlift, aero-medical evacuation, SAR and weather recce

Forces Abroad

UN AND PEACEKEEPING

BOSNIA (SFOR II): 4,600; 1 div HQ, 1 inf bde plus spt tps **CROATIA** (SFOR): 150 **SFOR AIR ELEMENT** (OP JOINT GUARD) 3,200. Forces are deployed to **Bosnia**, **Croatia**, **Hungary**, **Italy**, **France**, **Germany** and the **United Kingdom**. Aircraft include F/A-16, A-10, AC-130, MC-130, C-130, E-3, U-2, EC-130, RC-135, EA-6B, MH-53J and *Predator* UAV. **EAST TIMOR** (UNTAET): 3 obs **EGYPT** (MFO): 918; 1 inf, 1 spt bn **FYROM** (KFOR): 450 **GEORGIA** (UNOMIG): 2 obs **HUNGARY** (SFOR) 640; 230 Air Force *Predator* UAV **IRAQ/KUWAIT** (UNIKOM): 19 obs **MIDDLE EAST** (UNTSO): 2 obs **WESTERN SAHARA** (MINURSO): 15 obs **SAUDI ARABIA** (*Southern Watch*) **Air Force** units on rotation, numbers vary (incl F-15, F-16, F-117, C-130, KC-135, E-3) **TURKEY** (*Northern Watch*) **Air Force** 1,400; 1 tac, 1 Air Base gp (ac on det only), numbers vary but include F-16, F-15, EA-6B, KC-135, E3B/C, C-12, HC-130 **YUGOSLAVIA** (KFOR): 5,500

Paramilitary

CIVIL AIR PATROL (CAP) 53,000

(incl 1,900 cadets); HQ, 8 geographical regions, 52 wg, 1,700 units, 535 CAP ac, plus 4,700 private ac

MILITARY DEVELOPMENTS

Regional trends

While great strides have been made in the institutional arrangements for the development of an independent European military capability, this progress has not been matched by military spending and weapon-acquisition programmes. In most Western European countries, military spending remains in decline. The December 2000 EU ministerial meeting in Nice will reveal whether the plan for a 60,000-strong independent European force can be achieved by 2003. NATO countries in Europe are feeling the strain of what looks like an increasingly lengthy commitment in Kosovo. Even those with all-professional forces are finding they are over-committed in Europe and beyond. There has been an increase in terrorism in Western Europe carried out by longstanding groups in the UK, Spain and Greece. In Turkey, however, Kurdish terrorism has declined sharply.

European Security and Defence Policy

Since the Cologne European Council in June 1999, the EU has been firmly committed to developing its military and civilian crisis-management capabilities as part of a larger common European policy on security and defence. At the Helsinki European Council in December 1999, the EU made a commitment to develop an autonomous capacity to take decisions and to conduct EU-led military operations, with a 'headline goal' of an independent rapid-reaction force of 50,000–60,000 troops by 2003. The force is to be able to deploy within 60 days and to be sustained for at least a year. The relevant institutional machinery began operating on an interim basis on 1 March 2000, with the establishment of a political and security committee, a military committee and an embryonic European military staff. At the Santa Maria da Feira European Council meeting in June 2000, the EU reaffirmed its determination to meet the 'headline goal' by 2003 and stated that, 'improving military capabilities remains central to the credibility and effectiveness of the Common European Security and Defence Policy'. Four areas were identified in which cooperation with NATO would be sought: security issues, capability goals, the modalities for EU access to NATO assets and the definition of permanent consultation arrangements. Principles were also outlined to allow non-EU European NATO members and other EU accession candidates to contribute to EU military crisis management. Bearing in mind the experience of the Balkans, EU members recognised the deployment of an effective civilian police force as an essential complement to peacekeeping operations. They agreed to make available up to 5,000 police officers by 2003 for international missions across the range of conflict prevention and crisis-management operations. The French presidency of the EU, covering the period July–December 2000, has called for a conference on capability commitments, to be held on 20–21 November. At this meeting, EU members will announce initial national troop and equipment commitments to the 'headline goal'. It should be clear, from the result of this conference, how close the EU will come to meeting its 2003 deadline. This conference will be an important step to prepare the ground for the next EU ministerial meeting in Nice in December 2000, when the key decisions will have to be taken to make the European force a reality. Institutional progress has been impressive, but whether the military capability will have substance and credibility is not yet clear.

Conventional Forces in Europe (CFE) Treaty

On 19 November 1999, the members of the CFE Treaty signed the Agreement on the Adaptation of the Conventional Forces in Europe Treaty (Agreement on Adaptation) in Istanbul. This agreement removed the outdated bloc-to-bloc structure of the original treaty while continuing to regulate both the number and location of weapons held by each member state. The purpose of the treaty remains to prevent a destabilising concentration of military equipment by any one state or group of states in the treaty's area of application. This area stretches from the Atlantic Ocean to the Ural Mountains and covers all or part of the territories of all 30 member states except the US, whose territory is not affected but has forces stationed in the treaty area (see Table 39 on page 303 for the list of member states). The five categories of equipment limited by the treaty are: tanks; armoured combat vehicles (ACV); artillery; combat aircraft; and attack helicopters. The new elements introduced by the Agreement on Adaptation include:

- National Ceilings limiting the total equipment in all five categories that may be held by each state. These are set at or below the ceilings allowed by the original treaty. Equipment limits for NATO countries will fall by an average of 14%.
- Territorial Ceilings that put a cap on the total of national and foreign (stationed) tanks, ACVs and artillery. This arrangement will impose greater constraints than the original treaty on deployment flexibility. In Germany, for example, the original treaty permitted the deployment of 7,500 tanks. Under the Agreement on Adaptation, this is reduced to 4,704. There are arrangements whereby a state can increase its ceiling for a particular category, but there must be a compensating reduction by another state on the same territory. There are special provisions for the 'Flank Zones' (north-west and south-east Europe)[1] under which they can exchange entitlements only with each other, so that the overall total of treaty-limited equipment in the 'Flank Zones' does not increase. Under a special arrangement, Russia can deploy 2,140 armoured combat vehicles (ACV) (up from 1,380) in its 'Flank Zones' in north-west Russia and the North Caucasus.
- An enhanced system of information exchange, verification and transparency, with built-in flexibility to allow temporary deployments for exercises and in the event of crises.

While Russia has broadly complied with its overall obligations under the equipment limits, it has consistently failed to meet its limits within the 'Flank Zone' in the North Caucasus. To help ease the passage of the Adaptation Agreement, Russia announced at the Istanbul meeting measures it would take for its deployments in Georgia and Moldova. It had reached agreement with Georgia on a basic level of temporary deployment and is to close two of its four bases there, Vasiani and Gudauta, by 1 July 2001. Withdrawals began on 5 August 2000. With Moldova, Russia has agreed to remove all its treaty-limited equipment by the end of 2001 and all its forces by the end of 2003. While all 30 member states signed the agreement in Istanbul, the NATO allies agreed not to proceed with ratification until Russia verifiably complied with the adapted treaty. Russia has pledged to do so 'as soon as circumstances permit'. The next treaty review conference is due to take place in May 2001.

Tensions and conflict

Tense situations remain in the Balkans and in the Caucasus region. The NATO-led Kosovo Force (KFOR) is faced with continuing conflict between the ethnic Serb and Albanian populations. The violence is limited to the local level, but there is no easy military solution. The build-up of the UN-led civilian administration has been slow and hampered in particular by the difficulties of creating an effective civilian policing operation. In neighbouring Bosnia, it has been possible to reduce significantly the numbers of military personnel in the Stabilisation Force (SFOR); however, this is still far from the case in Kosovo. The lengthy deployment is putting great stress on NATO forces, including those of countries with full regular forces like the UK and the US. Too many commitments for the available forces have been taken on in the past two years; as a result, professional forces are having difficulty in retaining their personnel because of repeated overseas tours and separations from family. As more European armed forces complete their transition to fully professional forces, in theory there should be more well-trained armed forces available to share the burden. For example, within two years, France's armed forces will be all professional.

Conflicts and tensions in other parts of Europe are a familiar roll-call from the past decade and longer ago, including Nagorno Karabakh, Georgia (Abkhazia) and Cyprus. Although these areas have been relatively stable in 2000 compared to previous years, they still defy efforts at resolution. In Georgia, serious funding problems have resulted in the armed forces being reduced. On 28 June, the Georgian parliament voted to cut the armed forces, including Interior and Border Troops, from 47,500 to about 38,500. Calls for an even bigger reduction were resisted. UN-sponsored talks on Cyprus in August 2000 failed to bring progress and further attempts will be made in September. While levels of Cypriot, Greek and Turkish forces on the island remain unchanged, there is little likelihood of conflict recurring.

A positive development in the Black Sea region is the establishment of a multinational naval force to provide support for peacekeeping and humanitarian operations. Known as BlackSeaFor, it comprises units from the navies of Bulgaria, Georgia, Romania, Russia, Turkey and Ukraine, and was set up under a letter of intent signed in Istanbul on 28 June 2000.

Terrorism

Domestic terrorism in Western Europe has increased since late 1999. This is due to the heightened activity of anti-British splinter groups in Northern Ireland and the Basque separatist group *Euskadi ta Askatasuna* (ETA)'s return to violence after a 14-month cease-fire. In Northern Ireland, the Provisional Irish Republican Army (IRA), the largest and most formidable anti-British republican group, still observes the cease-fire and is allowing limited weapons inspections to further the peace process. The Ulster Freedom Fighters, one of the two main pro-British Loyalist groups, threatened briefly to break its cease-fire in June 2000. Loyalists have increasingly feuded and engaged in violent political protest. Most significantly, the dissident Real IRA and Continuity IRA appear to have consolidated and, within three months of the IRA agreeing to arms inspections on 6 May, have increased their combined strength from 100 to perhaps 140. They are believed to have new weapons from Eastern European sources, including mortars and a grenade launcher, and up to half a tonne of commercial explosives, and to have recruited seasoned bomb experts. Between early May and late July 2000, the dissident groups carried out several bombings or bomb scares, including two in London. None were fatal, but dissident Republican activity will probably rise.

ETA ended its cease-fire on 28 November 1999, and in the subsequent eight months killed at least eight people in a series of bombings and shootings, including two politicians, a Spanish army officer, two police officers and a journalist. In nearly 40 years of violent action, ETA has killed about 800 people. Supported by its own illegal operations and by the Basque diaspora, ETA appears to have sufficient supplies of explosives and small arms to continue its campaign at the level attained in 2000. Prospects for another cease-fire are dim, as Madrid is unwilling to increase the degree of autonomy already granted to the three Basque provinces and enjoys strong public support for this hard line. In Turkey, terrorism has declined substantially since Abdullah Öcalan, the leader of the separatist Kurdistan Workers' Party (PKK), was captured in February 1999 and, six months later, called for an end to the PKK's armed campaign against the Turkish government. In June 2000, Turkey ended 13 years of emergency rule in the easternmost of the five mainly Kurdish provinces. Several thousand armed PKK guerrillas remain based in training camps in northern Iraq and Iran, but reportedly no more than 500 remain on Turkish soil. Fundraising and narcotics sales for financing terrorism have consequently decreased. Given the Turkish government's substantial territorial advances prior to Öcalan's capitulation, combined with the

PKK's low morale, its lack of urban cell networks, and its desire for credibility in Europe, an increase in terrorist violence in Turkey appears unlikely. This will provide the opportunity for the Turkish government to ease its draconian regulations in Eastern Anatolia and to take a more moderate approach to the educational, cultural and linguistic rights sought by the Kurdish minority. On 8 June 2000, the anti-Western 'November 17' group assassinated British military attaché Brigadier Stephen Saunders in Greece, apparently in retaliation for NATO's invasion of Serbia. Although the urban terrorist group has killed only 23 people in 25 years, Saunders' murder highlighted the Greek authorities' general lack of anti-terrorist cooperation with the US and the UK. This has raised fears that the November 17 group and others might perpetrate high-profile attacks during the 2004 Summer Olympics, to be held in Athens. Following Saunders' death, the Greek government took steps to strengthen its joint anti-terrorist enforcement efforts with Washington and London.

DEFENCE SPENDING

NATO

In 1999, defence expenditure by European NATO countries continued the declining trend of recent years, falling by 5% in real terms from $184bn in 1998 to $174bn in 1999 (measured in constant 1999 US dollars). Budgets set for 2000 indicate that the trend will continue, compounded by the weak euro, with a forecast decline of around 6%. Although the three new NATO members increased their budgets in local currency terms – Poland by 8%, Hungary by 19% and the Czech Republic by 7.3% – this was largely erased in dollar terms by domestic currency weakness. Defence spending by the European NATO countries remains at about 40% of total NATO outlay, while NATO European research and development spending is just 25% of that of the United States. On the procurement front, European NATO spending has fallen by 2.2% since 1995, while US procurement has risen by 6.5% (measured in constant 1999 dollars). Budget realities diverge from the EU countries' political commitment to develop a credible independent 60,000 strong force by 2003. The UK is one European country bucking the downward trend: it plans to increase its defence spending from £23bn ($35bn) in 2000 to £25bn ($38bn) in 2003–04. However, this increase poses the question of whether the UK will be able to afford the plans set out in its 1998 Strategic Defence Review.

Equipment programmes

Aircraft Seven European countries have made commitments for 225 Airbus Military Company (AMC) A-400M transport aircraft. Despite the strong political statements, however, these are merely 'firm intentions to procure', not firm orders. At the earliest, contracts will only be signed in 2001. AMC plans to test-fly the first aircraft 51 months after contract signature, with initial deliveries 20 months later. AMC believes it can meet this strict time-scale by using proven Airbus management and construction techniques. This precise contract term is extraordinary, given that, for example, no engine has yet been chosen for the A400M. This choice will become a highly-charged political decision in which the inevitable haggling over industrial shares of the project will take precedence over choosing the best engine design for the aircraft. Germany has indicated it will be the biggest customer by signalling an intention to acquire 73 A400M out of the total commitment of 225, nearly three times the initial order placed by the British, whose forces have a far larger out-of-area remit than the *Bundeswehr*. Given that reductions in the German defence budget are set to continue over the next five years, the German 'intention to procure' 73 A400M could well be a bid to wrest a greater share of the industrial cake from its partners, or even an attempt to take the wing-design leadership from the British, who enjoy that status within Airbus. Despite the appeal of the Raytheon AIM-120 Extended Range Air-to-Air Missile (ERAAM), the UK has selected the European *Meteor* Beyond-Visual-Range Air-to-Air Missile (BVRAAM) for its *Eurofighter* multi-role combat aircraft, although this missile has yet to be developed, much less produced. Like the A400M order, the *Meteor* decision was designed, in the words of UK Defence Minister Geoffrey Hoon, to give 'clear evidence to our friends on both sides of the Atlantic of our strong commitment to enhance Europe's defence capability'. Since then, however, air defence specialists in the German Federal Ministry of Defence (FMOD) have stated that they prefer an enhanced version of the Raytheon ERAAM. It can be argued that conflicting voices within the FMOD are of little consequence, as the German defence budget cannot afford either missile. However, it must not be forgotten that the British decision to buy European transport aircraft and missiles is conditional on commitments by other European governments and on suppliers meeting cost and performance targets. These are no small caveats and if strictly adhered to, could well force the Europeans to face harsh budgetary realities.

A setback to the *Eurofighter* project was the decision by the Norwegian government in May 2000 to delay a plan to replace the Norwegian Air Force's F-5 fleet. The F-5s will retire on schedule before 2004, leaving Norway's fighter capability at 58 F-16s until 2010–12, when a replacement will be sought. *Eurofighter* had been in a good position to win the now-postponed order for 20 aircraft. This new decision puts the US Joint Strike Fighter (JSF) in the frame and aligns Norway's plans with those of Belgium, Denmark and the Netherlands, thus opening the prospect of a similar deal to one in the 1970s, when the four agreed a combined order for F-16s. The Norwegian action could well trigger other countries to put back fighter replacement programmes in order to save money, while holding out the prospect of acquiring the JSF and skipping the *Eurofighter* generation of capability. This could harm *Eurofighter's* export potential, which is already suffering from the project running ten years late and well over budget. Ironically, the financial success of the *Eurofighter* could now hinge on whether the JSF itself is cancelled or unduly delayed. However, an order from Greece for 60 *Eurofighters*, with an option for 30 more, is a welcome boost to the project, provided that the order is sustained.

Naval capabilities In general, very few confirmed orders have been made that will increase NATO members' naval capabilities. Most orders are for equipment to replace and update existing capabilities. With the exception of the UK, there have been no new orders for either power-projection platforms or dedicated sealift ships, despite the fact that these were the two most glaring deficiencies in Europe's naval contribution to the 1999 Kosovo campaign. The UK is slowly increasing its naval-power projection and sealift capabilities. Initial assessment phase contracts were awarded to Thomson-CSF and British Aerospace Systems (BAe Systems) in December 1999 for the two 35,000–40,000-tonne aircraft carriers, which have planned in-service dates of 2012 and 2015. The carriers' principal aircraft remains undecided: JSF is still the favourite but *Eurofighter*, *Rafale* and the F/A 18 are being considered in case of difficulties in the US with JSF's funding. The conversion of the 12 nuclear-fuelled submarines (SSN) to launch *Tomahawk* land-attack missiles is proceeding slowly: so far, only three SSNs have been converted. An order for six roll-on roll-off ferries is expected in late 2000. To reduce running costs, the ships will operate commercially but will be ready for recall to government service. They have an in-service date of 2005 and will represent a significant expansion of the UK's sealift capability. Replacements for current capabilities are as follows:

• In July 2000, the government announced its intention to order a first batch of three (of an intended 12) Type 45 air-defence destroyers, to enter service from 2007. They will replace the existing 11 Type 42 destroyers; a contract is expected to be awarded in late 2000.

- The fourth and last *Vanguard* nuclear-fuelled ballistic-missile submarine (SSBN) was commissioned in late 1999; these replace the four Polaris SSBNs.
- Two landing platform docks (LPD) are still in construction to replace the two *Fearless*-class LPDs, only one of which is still in service. The in-service dates have slipped to 2003.
- A single joint carrier airborne group has been formed, Joint Force 2000, which unites RAF GR-7 Harriers and the RN's FA-2 Harriers.

France, unless a second carrier is ordered, is not expanding its naval capabilities. The nuclear-powered 40,000-tonne *Charles de Gaulle* aircraft carrier will be commissioned in October 2000 to replace the *Foch*. Defence Minister Alain Richard said in June 2000 that the government would make a decision on a second carrier in autumn 2000 or early 2001. The second *Le Triomphant*-class SSBN has been commissioned; two more are under construction with in-service dates of 2004 and 2008, to replace the four *L'Inflexible*-class. France is still building the *Horizon* air-defence destroyer based on a modified *La Fayette* design with Italy, and is planning to procure two, as is Italy, with in-service dates of 2005–08. Instead of upgrading the present escort force, plans have been made to buy 17 new multi-purpose frigates with an in-service date of 2010.

Belgium has announced that, as part of its defence reform programme for 2000–15, it will replace its three *Wielingen* anti-surface-unit warfare (ASUW) frigates with two general-purpose frigates. It will upgrade its seven *Flower* minehunters and has cancelled an order for four KMV coastal minehunters.

Germany is updating its current naval capabilities with the Type 124 *Sachsen*-class frigate programme, a collaborative design with the Netherlands. Three are being built with a first in-service date of 2002 and will replace the three *Lutjens* destroyers. Germany signed a Memorandum of Understanding with Italy in late 1999 for the combined acquisition of 26 maritime-patrol aircraft (MPA), of which Germany will buy 10–12 to replace its 14 ageing *Atlantiques*. These should be delivered in 2007. Two *Berlin*-class AOs (tankers) are being built for delivery in 2001–02. Delivery of four Type 212 diesel submarines with air-independent propulsion should commence in 2002. A contract is expected to be signed in early 2001 for five new Type K130 corvettes, whose main role will be anti-surface-unit warfare (ASUW) and reconnaissance in coastal areas. Delivery is currently planned for between 2005–08. Two further batches of five are planned with deliveries finishing in 2015; the equipment for each of the ships will include two unmanned aerial vehicles (UAV).

Greece has ordered three German Type 214 diesel submarines to be built under licence with an option on a fourth; these should be in service by the end of the decade. Four *Zubr* air-cushion vehicles have been ordered for delivery in 2001; two from Russia and two from Ukraine. Italy's orders for naval equipment are being maintained, in particular, the delivery of two Type 212 diesel submarines, license-built from a German design and expected in 2005–06.

The Netherlands has started construction of four *De Seven Provincien* destroyers for delivery in 2002–05, to replace existing escorts. A second *Rotterdam*-class landing platform dock (LPD) is planned to come into service in 2007–08.

Norway has signed a contract with Spain's Bazan for five new *Fridtjof Nansen*-class frigates with an *Aegis* combat system. As Norway has only three elderly frigates in service, this order represents an expansion of current capabilities. This may be the last naval order for several years as, according to Norway's *Defence Analysis 2000* plan, its 14 *Hauk* surface-to-surface-missile-bearing fast patrol craft (PFM/SSM) will be retired from service, along with the newer *Skjold*-class PFMs and the ageing *Kobben* diesel submarines. Under *Defence Analysis 2000*, a decision to replace the six *Ula* diesel submarines will not be made until 2002. Poland is updating its escort force and

Table 13 Defence R&D and procurement spending in NATO and non-NATO Western Europe, 1996–2000

Constant 1999 US$m	Defence Budget					Research and Development (R&D)					Equipment Procurement				
	1996	1997	1998	1999	2000	1996	1997	1998	1999	2000	1996	1997	1998	1999	2000
NATO															
Be	3,314	2,920	2,879	2,547	2,402	3	2	1	2	1	226	200	211	191	234
Da	3,224	2,836	2,760	2,552	2,283	5	5	5	5	1	400	353	365	335	333
Fr	39,388	34,031	31,942	29,497	26,538	5,131	3,975	3,385	3,025	3,053	7,894	6,726	5,847	5,902	5,317
Ge	34,066	28,444	27,052	25,423	22,871	1,924	1,547	1,467	1,313	1,299	3,854	3,075	3,594	3,865	3,413
Gr	3,743	3,750	4,037	3,426	3,195	10	19	24	22	24	1,192	1,193	1,339	1,324	1,351
It	21,514	18,973	18,201	16,239	15,704	787	781	555	310	333	2,108	2,185	2,491	1,982	2,276
Lu	128	113	110	102	99	0	0	0	0	0	7	6	6	5	6
Nl	8,567	7,251	7,248	6,535	6,047	126	107	103	66	66	1,642	1,378	1,645	1,435	1,369
No	3,974	3,551	3,422	3,303	2,820	37	23	22	22	23	873	943	805	719	788
Por	1,826	1,767	1,617	1,626	1,524	4	4	4	4	4	274	366	379	416	371
Sp	7,297	6,179	6,123	7,358	6,857	293	252	206	177	175	1,294	1,053	813	774	1,065
Tu	5,383	4,180	7,903	8,901	7,577	15	41	47	44	47	2,402	2,672	3,051	3,150	3,121
UK	36,599	35,603	38,090	35,945	33,890	3,560	3,632	3,938	4,067	4,026	8,520	8,808	9,732	8,596	8,537
Sub-total	**169,025**	**149,599**	**151,384**	**143,453**	**131,808**	**11,895**	**10,387**	**9,757**	**9,058**	**9,052**	**30,686**	**28,957**	**30,276**	**28,695**	**28,182**
Cz	1,194	1,028	1,165	1,164	1,131	29	25	20	16	21	148	140	155	183	204
Hu	742	692	673	745	776	5	2	12	12	12	86	135	186	186	235
Pl	3,109	3,119	3,429	3,219	3,104	63	57	96	80	83	581	494	526	486	697
Sub-total	**5,045**	**4,840**	**5,267**	**5,128**	**5,010**	**97**	**84**	**129**	**108**	**116**	**816**	**769**	**867**	**855**	**1,135**
Total	**174,070**	**154,439**	**156,651**	**148,581**	**136,817**	**11,992**	**10,471**	**9,885**	**9,165**	**9,169**	**31,501**	**29,726**	**31,143**	**29,551**	**29,317**
Ca	8,227	7,451	6,448	6,996	7,456	97	78	117	114	118	2,252	1,954	1,499	1,308	1,282
US (DoD)	282,699	281,243	276,618	292,147	287,466	37,163	37,873	37,824	36,635	33,692	45,081	44,662	45,657	48,951	51,970
US and Ca	290,926	288,694	283,066	299,143	294,922	37,260	37,951	37,941	36,749	33,810	47,333	46,616	47,157	50,259	53,252
Total NATO	**464,996**	**443,133**	**439,717**	**447,724**	**431,740**	**49,252**	**48,422**	**47,826**	**45,915**	**42,979**	**78,835**	**76,342**	**78,300**	**79,810**	**82,569**
Non-NATO															
A	2,082	1,858	1,835	1,664	1,497	10	10	10	10	10	386	322	416	300	312
SF	2,056	1,908	1,929	1,695	1,583	9	9	10	14	8	772	697	901	615	624
Irl	787	797	811	745	711	0	0	0	0	0	26	28	35	42	47
Swe	6,506	5,197	4,885	4,525	4,405	167	165	167	98	104	2,021	1,739	1,972	2,294	2,179
CH	4,720	3,878	3,700	3,169	2,893	101	80	72	64	64	1,914	1,714	1,580	1,368	1,300
Total	**16,151**	**13,639**	**13,160**	**11,799**	**11,089**	**287**	**264**	**258**	**187**	**186**	**5,119**	**4,500**	**4,902**	**4,618**	**4,462**

in 2000 commissioned a US *Oliver Hazard Perry*-class frigate. A second is planned for 2001. An order for four German *Meko* 100 corvettes may be made in late 2000.

Turkey has announced its long-term naval procurement plans. They include six TF 2000 frigates; four new German diesel submarines, probably the Type 214, for delivery in 2006; 12 *Milgem Kormet* patrol craft; eight *Seahawk* search and rescue/anti-surface-unit warfare (SAR/ASUW) helicopters; and nine CN-235 MPAs from Spain.

Non-NATO Europe

Defence spending in 1999 by the European countries outside NATO fell by 9.4% in real terms to $20.2bn (measured in constant 1999 US dollars). The combined defence budgets of non-NATO EU countries and Switzerland for 2000 were broadly unchanged from 1999 in local currency terms; however, the weakness of European currencies resulted in a 4% fall in dollar terms. In the Transcaucasus, Armenia increased its defence budget by around 30% in dollar terms to $96m; the budgets of Azerbaijan and Georgia increased in local currency terms but were little changed in dollar terms. In the Balkan states, combined budgets in 2000 were unchanged from 1999. In Ukraine, the defence budget increased from h1.5bn ($385m) to h2.4bn ($441m). In reality, only h1.4bn was actually available; the balance was to be 'found' through economies. In January 2000, however, an additional $350m was allocated to acquire five AN-70 transport aircraft being developed jointly with Russia in order for to acquire their certificate of airworthiness. A total of 229 aircraft are on order under the 1999 bilateral agreement. Romania has embarked on a radical restructuring of its armed forces, largely for financial reasons. Commands in the Army, Navy and Air Force will be merged and Ministry of Defence staff will be radically reduced. The restructuring plan is due to be completed by 2004. Bulgaria announced similar plans in late 1999, but has made little progress to date with the reform programme.

European Defence Industry

On 10 July 2000, the European Aeronautic and Defence and Space Company (EADS) came into being. It has a combined annual revenue of about $22bn, which puts it in third place in the rankings of international defence companies behind Boeing ($55bn) and Lockheed Martin ($28bn), and just ahead of BAe Systems ($21bn). One of the main challenges for the new company is its complex management and ownership structure. Top management is shared between two chief executive officers, one French and one German, and its style is predicated on the devolution of decision making to the lowest level possible within its five principal divisions (Airbus, Military Transport Aircraft, Aeronautics, Space and Defence and Civil Systems). The divisions are not well-balanced: Airbus is by far the biggest revenue-earner and EADS lacks a strong position in defence electronics, essential for integrating and producing advanced military systems. Mainly because of the latter weakness, EADS is heavily dependent on joint ventures with BAe Systems for over two-thirds of its revenue. The process of industrial integration is far more complex than other major mergers in the same field, such as that between Boeing and McDonnell Douglas and between Lockheed and Martin Marietta and others within this merger. Lockheed Martin is still faltering, having never properly resolved all its integration problems.

EADS involves a much greater number of companies and the situation is further complicated by its labyrinthine ownership arrangements. French and Spanish government equity stakes are involved, although somewhat diluted through a series of holding and pooling companies. The immediate ownership of the company is split between the public, which has a 34% stake, and a holding company controlling the remaining 66%. This company is itself owned by Daimler Chrysler (46%), the Spanish government (8.5%) and another holding company, French this time, with 44%. It is this company through which the French government has its share in the EADS action. It owns half the shares. The remainder are held by the French defence and aerospace company Lagardère, with 37% and a number of French private institutions which hold 13% between them.

In a merger operation, economies of scale should be one of the main benefits, particularly in saving on production costs, joint ordering of components, research and development, and winning orders that the individual companies would have difficulty in winning on their own. There will be concerns that, with a significant government stake, there will be more than the usual government interference with the integration process on issues such as plant closures.

[1] Armenia, Azerbaijan, Bulgaria, Georgia, Moldova, Romania, Russia, Turkey and Ukraine in the south-east and Iceland, Norway and Russia in the north-west.

Table 14 **Arms orders and deliveries, NATO Europe and Canada, 1998–2000**

Country	Country supplier	Classification	Designation	Quantity	Order date	Delivery date	Comment
Belgium	US	FGA	**F-16**	110	1993	1998	Upgrade; 88 AMRAAM on order
	Sgp	tpt	**A-310**	2	1996	1998	2nd aircraft delivered May 1998
	A	APC	***Pandur***	54	1997	1998	
	Il	UAV	***Hunter***	18	1998	2000	
	US	FGA	**F-16**	18	1999	2000	Upgrade; option on 18 exercised
Canada	dom	MCMV	***Kingston***	12	1992	1996	Deliveries complete by 1999
	US	hel	**B-412EP**	100	1992	1994	Deliveries to 1998 at 3 per month
	dom	LAV	**LAV-25**	240	1996	1998	Deliveries continue
	US	APC	**M-113**	400	1997	1998	Upgrade; deliveries continue
	Ge	MBT	***Leopard* 1**	114	1997	1999	*Leopard* C1A5 Upgrade
	dom	LAV	**LAV-25**	120	1998	2001	Follow-on order of 240
	UK	SSK	***Upholder***	4	1998	2000	Deliveries to 2001
	col	hel	**EH-101**	15	1998	2001	Deliveries to 2002
Czech Republic							
	dom	MBT	**T-72**	250	1995	2000	Upgrade to 1999
	dom	trg	**L-39**	27	1997	1999	Originally for Nga
	dom	FGA	**L-159**	72	1997	1999	
	col	UAV	***Sojka* 3**	8	1998	2000	Upgraded Sojka III, with Hu
	RF	cbt hel	**Mi-24**	7	1999	1999	Exchange for debt repayment
Denmark	US	FGA	**F-16**	63	1993	1998	Mid-life update; deliveries continue 1999
	dom	AFV	***Hydrema***	12	1996	1997	Deliveries to 1998
	Ge	MBT	***Leopard* 2A4**	51	1998	2000	Ex-Ge Army
	Ca	tpt	***Challenger* 604**	3	1998		
	UK	hel	***Lynx***	8	1998	2000	Upgrade to *Super Lynx* standard
	CH	APC	***Piranha* III**	2	1998	1999	Option for 20 more
	Ge	APC	**M-113**	100	1999	2000	Upgrade. Deliveries until 2001
	Fr	UAV	***Sperwer***	2	1999		
France	col	hel	***Tiger***	215	1984	2003	With Ge
	dom	FGA	***Rafale***	60	1984	1999	Deliveries of first ten 1999–2001

Country	Country supplier	Classification	Designation	Quantity	Order date	Delivery date	Comment
	dom	FGA	***Rafale***	234	1984	1999	1st 3 1999. ISD 2005
	dom	MBT	***Leclerc***	406	1985	1992	153 delivered by end 1998; to 2002
	col	radar	***Cobra***	10	1986	2002	With UK, Ge
	dom	LSD	***Foudre***	2	1986	1990	2nd of class delivered 1998
	dom	SSBN	***Le Triomphant***	3	1986	1996	Deliveries to 2001
	dom	CVN	***Charles de Gaulle***	1	1986	1999	Sea trials 1998; commission 2000
	col	hel	**NH-90**	27	1987	2004	With Ge, It, Nl
	col	ATGW	***Trigat***		1988	2004	With Ge
	col	hel	**AS-555**	44	1988	1990	Deliveries through 1990s
	dom	FFG	***Lafayette***	5	1990	1996	Deliveries to 2003
	col	SAM	**FSAF**		1990	2006	Future SAM family, with It, UK
	col	torp	**MU-90**	150	1991	2000	With It and Ge. Deliveries 2000–2
	dom	FGA	***Mirage* 2000-D**	86	1991	1994	Deliveries to 2000
	col	hel	**AS-532**	4	1992	1996	Deliveries to 1998
	dom	FGA	***Mirage* 2000-5F**	37	1993	1998	Upgrade, deliveries to 2002
	col	UAV	***Eagle***				Devpt with UK
	col	sat	***Helios* 1A**	2	1994	1999	With Ge, It, Sp; *Helios* 1B
	col	sat	***Helios* 2**	1	1994	2002	Devpt with Ge
	col	sat	***Horus***		1994	2005	Fr has withdrawn funding
	US	AEW	**E2-D**	3	1994	1999	1st delivered in Jan 1999
	col	ALCM	**SCALP**	600	1994	2000	2 orders for delivery over 11 years
	col	hel	**AS-532**	4	1995	1999	Combat SAR, requirement for 6
	Sp	tpt	**CN 235**	7	1996	1998	Offset for Sp AS-532 purchase
	dom	SLBM	**M-51**		1996	2008	To replace M-45; devpt continues
	col	hel	**AS 565**	8	1996	1997	7 delivered 1997; 1 delivered 1998
	dom	APC	**VBL**	120	1996	1998	20 delivered in 1998
	dom	SAM	***Mistral***	1,130	1996	1997	To 2002
	dom	recce	***Falcon*-50**	4	1997	1998	Deliveries to 2000
	col	hel	**BK-117**	32	1997	1999	
	dom	ATGW	***Eryx***	6,400	1997	1997	Missiles, to 2002
	dom	ATGW	**LAW**	30,800	1997	1997	Missiles, for delivery 1997–2002
	col	ASM	***Vesta***		1997	2005	In development
	col	sat	***Skynet*** 5	4	1998	2005	Comms; dev in 1998; with Ge, UK
	dom	SSN	**SSN**	6	1998	2010	Design studies approved Oct 1998
	col	AAM	***Mica***	225	1998	1999	
	col	APC	**VBCI**	50	1998	2004	Dev with Ge (GTK), UK (MRAV)
	dom	AIFV	**AMX-10**	300	1999	2001	Upgrade
	Swe	APC	**Bv 206S**	12	1999	1999	For units serving in Kososvo
	col	FFG	**Modified *Horizon***	2	1999	2005	Joint Fr/It project
	dom	LSD	**NTCD**	2	2000	2005	2 on order
Germany	col	hel	***Tiger***	212	1984	2003	With Fr
	col	FGA	**EF-2000**	180	1985	2001	With UK, It, Sp; 44 ordered
	dom	SP arty	**PzH 2000**	186	1986	1998	Req 594 units; 4 delivered 1998
	col	hel	**NH-90**	80	1987	2004	With Fr, It, Nl
	dom	MHC	**Type 332**	12	1988	1992	Deliveries completed in 1998

Country	Country supplier	Classification	Designation	Quantity	Order date	Delivery date	Comment
	col	ATGW	***Trigat***		1988	2004	With Fr
	col	tpt	**FLA**	75	1989	2008	Development
	dom	SSK	**Type 212**	4	1994	2003	Deliveries to 2006
	col	recce	***Fennek***	164	1994	2000	With Nl. Prod in 2000
	col	sat	***Helios* 2**	1	1994	2001	Dev with Fr, It
	col	sat	***Horus***	1	1994	2005	Dev with Fr
	dom	FFG	**Type F 124**	3	1996	2002	Deliveries 2002–05.
	dom	AOE	**Type 702**	2	1996	2000	1st delivered in 2000
	UK	hel	***Lynx***	7	1996	1999	
	dom	AAA	***Gepard***	147	1996	1999	Upgrade, deliveries started 1999
	col	sat	***Skynet* 5**	4	1997	2005	With UK, Fr
	col	AAM	**IRIS-T**		1997	2003	Devpt with It, Swe, Gr, Ca, No
	col	hel	**EC-135**	15	1997	1998	For *Tiger* hel trg
	col	hel	**AS-365**	13	1997	1998	Delivery from 1998–2001
	col	APC	**GTK**	200	1998	2004	With Fr (VBCI), UK (MRAV)
	dom	SAM	***Wiesel* 2**	50	1998	1999	
	US	SAM	***Patriot***	7	1998		Upgrade to PAC-3
	US	SAM	***Patriot***	12	1998		*Roland/Patriot* cost total $2.1bn
	US	SAM	***Roland***	21	1998		Air defence system
	dom	APC	**TPz KRK**	50	1998	1999	
	col	radar	**COBRA**	12	1998		
	UK	hel	***Lynx***	17	1998	2000	Upgrade to *Super Lynx* standard
	col	torp	**MU-90**	600	1998	2000	
	col	ASM	***Taurus***		1998	2001	Devpt with Swe (KEPD-350)
	dom	FFG	**Type F 125**	8	1999	2010	Feasibility study stage
	dom	AG	**Type 751**	1	1999	2002	Defence research and test ship
	dom	AFV	**ATF-2K**	56	1999		
Greece	US	FGA	**F-16**	80	1985	1988	2nd batch of 40 1997–99
	Ge	FFG	***Meko***	4	1988	1992	Deliveries to 1998; 2 built in Gr
	dom	AIFV	***Kentaurus***		1994	2000	In devpt and trials in late 1998
	US	hel	**CH-47D**	7	1995	2001	In addition to 9 in inventory
	US	FGA	**F-4**	38	1996	1999	Upgrade in Ge; deliveries to 2000
	US	MRL	**ATACM**	81	1996	1998	Deliveries completed by 1998
	Swe	PCI	**CB 90**	3	1996	1998	
	US	AAM	**AIM-120B**	90	1997	1999	
	US	SP arty	**M-109A5**	12	1997	1999	
	Ge	MBT	***Leopard* 1A5**	170	1997	1998	In addition to prev delivery of 75
	US	SAM	***Stinger***	188	1998	2000	
	US	trg	**T-6A**	45	1998	2000	Completion of deliveries 2003
	US	SAM	***Patriot* PAC-3**	5	1998	2001	5 batteries, option for 1 more
	RF	SAM	**SA-15**	32	1998	1999	Aka Tor-M1; for Cy
	Br	AEW	**RJ-145**	4	1998	2000	Interim lease from Swe of Saab 350 *Argus*
	Ge	SSK	**Type 214**	3	1998	2006	1 built in Ge; 2 in Gr
	UK	MCMV		2	1998	2000	1 in 2000. 1 in 2001

Country	Country supplier	Classification	Designation	Quantity	Order date	Delivery date	Comment
	It	AK	***Hunt***	1	1999	2002	
	Fr	hel	**AS-532**	4	1999	1999	
	US	MRL	**MLRS**	18	1999	2002	
	US	FGA	**F-16**	58	1999	2000	
	Fr	FGA	***Mirage* 2000-5**	15	1999	2001	Delivery to 2003
	Fr	FGA	***Mirage* 2000**	10	1999		Upgrade 10 of existing 35
	Fr	SAM	***Crotale* NG**	11	1999	2001	9 for air force; 2 for navy
	US	hel	**S-70B**	2	2000		
	col	FGA	**EF-2000**	60	2000	2005	May increase to 90
	dom	PFM	***Super Vita***	3	2000	2003	Option on further 4
	dom	PCO		4	2000		
	dom	AO		1	2000		
	RF	LCAC	***Zubr***	4	2000	2001	
	US	AAM	**AMRAAM**	560	2000		
	US	recce	**C-12**	2	2000		
Hungary	RF	APC	**BTR-80**	555	1993	1996	Delivery completed in 1999
	Bel	MBT	**T-72**	100	1994	1996	Ex-Bel, 31 delivered in 1996
	Fr	SAM	***Mistral***	45	1996	1998	27 launchers
Italy	dom	hel	**A-129**	66	1978	1990	Deliveries through 1998
	dom	MBT	**C1 *Ariete***	200	1982	1995	Deliveries to 2001
	dom	AIFV	**VCC-80**	200	1982	1999	First ordered 1998. Aka *Dardo*
	col	FGA	**EF-2000**	121	1985	2002	With UK, Ge, Sp; 29 ordered
	col	hel	**NH 90**	117	1987	2004	With Fr, Ge, Nl
	dom	APC	***Puma***	600	1988	1999	Deliveries to 2004
	col	tpt	**FLA**	44	1989	2008	With Fr, Ge, Sp, Be, Por, Tu, UK
	col	SAM	**FSAF**		1990	2006	With Fr, UK
	col	hel	**EH101**	16	1993	1999	With UK. Navy require 38
	dom	PCO	***Esploratore***	4	1993	1997	Deliveries to 2000
	dom	AO	***Etna***	1	1994	1998	
	dom	CV	**Project NUM**	1	1996	2004	Partly funded 1996 budget
	US	tpt	**C-130J**	18	1997	1999	Options for 14 more
	Fr	tpt	***Falcon* 900EX**	2	1997	1999	
	It	tpt	**P-180**	12	1997	1998	
	Ge	SSK	**Type 212**	2	1997	2005	Licence
	dom	AGI	**A-5353**	2	1998	2000	2nd for delivery 2001
	dom	PCO	***Aliscarfi***	4	1999	2001	
	dom	LPD	***San Giorgio***	2	1999	2001	Upgrade to carry 4 hel
	Ge	SP arty	**PzH 2000**	70	1999	2001	
	Col	FFG	**Modified *Horizon***	2	1999	2005	With Fr and It
	dom	AT	**C-27J**	12	1999	2001	
	US	UAV	***Predator***	6	2000		
	US	tpt	**C-130J**	2	2000		
	US	AAM	***Stinger***	30	2000		For use on A-129

Country	Country supplier	Classification	Designation	Quantity	Order date	Delivery date	Comment
Netherlands							
	col	hel	**NH-90**	20	1987	2004	With Fr, Ge, It
	dom	LPD	***Rotterdam***	1	1993	1998	Hangar space for 4 hel
	US	FGA	**F-16**	136	1993	1997	Upgrade continues to 2001
	Ge	MBT	***Leopard* 2A5**	330	1994	1996	Upgrade programme
	US	hel	**AH-64A**	12	1995	1997	Lease until AH-64D
	US	hel	**AH-64D**	30	1995	1998	4 delivered in 1998
	US	hel	**CH-47C**	7	1995	1999	
	dom	FFG	***De Zeven***	4	1995	2001	2 ordered 1995; 2 more 1997
	SF	APC	**XA-188**	90	1996	1998	24 delivered in 1998
	US	MPA	**P-3C**	7	1999	2001	Upgrade
Norway	Swe	AIFV	**CV-90**	104	1990	1996	Option for 70 more; deliveries 1996–99
	US	FGA	**F-16**	58	1993	1997	Mid-Life upgrade to 2001
	US	AAM	**AMRAAM**	500	1993	1995	Deliveries to 2000
	US	MRL	**MLRS**	12	1995	1997	Deliveries to 1998
	Swe	LCA	**90-H**	16	1995	1997	Deliveries to 1998
	dom	FAC	***Skjold***	8	1996	1999	Deliveries to 2004
	UK	arty	**105mm**	21	1997	1998	
	US	MPA	**PC-3**	4	1997	1999	Upgrade
	US	FGA	**F-16**	30	1998	2001	
	Ge	AFV	***Leopard* 1**	73	1998	1999	Deliveries to 2000
	Sp	FFG	***Fridtjof***	5	2000	2004	
Poland	dom	hel	**W-3**	11	1994	1998	1 for Navy. First in 1998
	dom	SAR	**PLZ M-28**	3	1998	1999	
	UK	SP arty	**AS-90**	80	1999	2001	Turret system only
	Ge	FGA	**MiG-29**	22	1999	2002	Upgrade
	US	FFG	***Perry***	1	1999	2000	
Spain	col	tpt	**FLA**	36	1989	2008	With Fr, Ge, It, Be, Por, Tu, UK
	col	MHC	***Segura***	4	1989	1999	Deliveries to 2000
	dom	LPD	***Galicia***	2	1991	1998	2nd delivered 2000
	dom	FFG	**F-100**	4	1992	2002	Deliveries to 2006
	US	FGA	**F/A-18A**	30	1994	1995	Deliveries continue to 1998
	col	FGA	**EF-2000**	87	1994	2001	With Ge, It, UK
	Fr	hel	**AS-532**	18	1995	1996	1st in 1996. Deliveries to 2003
	US	tpt	**C-130**	12	1995	1999	Upgrade programme
	A	AIFV	***Pizarro***	144	1996	1998	Licence. Full order for 463
	It	SAM	***Spada* 2000**	2	1996	1998	First of 2 btys delivered
	dom	arty	**SBT-1**		1997	2000	Development
	dom	MPA	**P-3**	7	1997	2002	Upgrade
	US	AAM	**AIM-120B**	100	1998	1999	
	Ge	MBT	***Leopard* 2**	235	1998	2002	Built in Sp. Includes 16 ARVs
	Fr	trg	**EC120B**	12	2000	2000	Deliveries July 2000 – July 01
	dom	AT	**C295**	9	2000		To be delivered by 2004

Country	Country supplier	Classification	Designation	Quantity	Order date	Delivery date	Comment
Turkey	US	FGA	**F-16**	240	1984	1987	All but 8 assembled in Tu; to 1999
	Ge	FFH	***Meko*-200**	8	1985	1987	7 by 1999; final delivery 2000
	Ge	SSK	***Preveze***	8	1987	1994	Delivery of first 5 to 2003
	US	APC	**M-113**	1698	1988	1992	Final deliveries in 1999
	Sp	tpt	**CN-235**	43	1990	1992	41 delivered by 1998
	Ge	PCM	**P-330**	3	1993	1998	1st built Ge; 2nd & 3rd Tu; to 1999
	US	TKR AC	**KC-135R**	9	1994	1995	Deliveries in 1995 and 1998
	US	tpt hel	**CH-47**	4	1996	1999	
	UK	SAM	***Rapier***	78	1996	1998	Upgrade programme
	US	FFG	***Perry***	6	1996	1998	5 in 1998–99. Last in 2000
	Il	FGA	**F-4**	54	1996	1999	Upgrade; deliveries to 2002
	US	MRL	**ATACM**	72	1996	1998	36 msl delivered in 1998
	Fr	hel	**AS-532**	30	1996	2000	To be completed by 2003
	US	AAM	**AIM-120B**	138	1997	2000	
	US	asw hel	**SH-60B**	8	1997	2000	
	dom	APC	**RN-94**	5	1997		Development
	Il	AGM	***Popeye* 1**	50	1997	1999	For upgraded F-4
	Sp	MPA	**CN-235**	9	1997	2000	
	Fr	MHC	***Circe***	5	1997	1998	3 in 1998, 2 in 1999
	It	SAR hel	**AB-412**	5	1998	2000	
	Il	FGA	**F-5**	48	1998	2001	IAI to upgrade
	US	hel	**CH-53E**	8	1998	2003	
	US	SAM	***Stinger***	208	1999	2001	
	US	hel	**S-70 *Blackhawk***	50	1999	1999	Deliveries to 2001
	dom	PCC		10	1999	2000	For coastguard
	UK	SAM	***Rapier* MK 2**	840	1999	2000	Licence; 80 a year for 10 years
	US	FGA	**F-16**	32	1999	2002	License
	Ge	SSK	**Type 214**	4	2000	2006	
	Ge	MHC	**Type 332**	6	2000	2004	1st built in Ge, remainder in Tu
	US	hel	**S-80E**	8	2000		Heavy lift
	US	hel	**S-70 *Seahawk***	6	2000		Heavy lift
United Kingdom							
	Col	hel	**EH 101**	44	1979	1999	With It; for RN
	dom	AAM	**ASRAAM**	1300	1981	1998	
	dom	SSBN	***Vanguard***	4	1982	1993	Deliveries to 1999
	US	SLBM	***Trident* D-5**	48	1982	1994	Deliveries to 1999
	col	FGA	**EF-2000**	232	1984	2002	1st 55 ordered 1998
	dom	FFG	**Type 23**	16	1984	1997	Delivery completed 1999
	dom	UAV	***Phoenix***	50	1985	1998	Upgrade programme planned
	dom	MHC	***Sandown***	12	1985	1989	11 delivered by 2000
	dom	FGA	***Sea Harrier***	35	1985	1994	Upgrade; deliveries to 1999
	col	radar	***Cobra***		1986	1999	Counter-battery radar with Fr, Ge
	col	hel	**EH 101**	22	1987	2000	With It; for RAF
	dom	SSN	***Swiftsure***	5	1988	1999	Upgrade to carry TLAM
	dom	SSN	***Trafalgar***	7	1988	2000	Upgrade to carry TLAM

Country	Country supplier	Classification	Designation	Quantity	Order date	Delivery date	Comment
	dom	FGA	***Sea Harrier***	18	1990	1995	Deliveries to 1999
	dom	SSN	***Astute***	3	1991	2005	
	dom	LPD	***Albion***	2	1991	2003	Expected in 2003
	dom	hel	***Lynx***	50	1992	1995	Upgrade to 1999
	dom	MBT	***Challenger* 2**	386	1993	1998	78 delivered 1998
	col	sat	***Skynet* 5**	4	1993	2005	With Fr and Ge
	dom	LPH	***Ocean***	1	1993	1998	Delivered in 1998
	col	SAM	**PAAMS**		1994	2003	Dev with Fr, It; part of FSAF prog
	US	AAM	**AIM-120**		1994	1998	
	US	tpt	**C-130J**	25	1994	1999	Option for 20 more
	dom	FGA	***Tornado* GR4 ID**	142	1994	1998	Upgrade; deliveries to 2003
	US	hel	**CH-47**	14	1995	1997	Deliveries to 2000
	US	SLCM	***Tomahawk***	65	1995	1998	Delivered. 20 fired in FRY conflict
	dom	ASM	***Brimstone***		1996	2001	1st 12 to be delivered 2001
	col	ASM	***Storm Shadow***	900	1996	2001	
	dom	FGA	***Tornado* F-3**	100	1996	1998	Upgrade
	col	FGA	**JSF**	150	1996	2012	Devpt with US
	dom	MPA	***Nimrod* 2000**	21	1996	2003	Upgrade
	US	hel	**AH-64D**	67	1996	2000	Deliveries to 2003
	dom	APC	***Stormer***	18	1996	1998	
	dom	AO	***Wave Knight***	2	1997	2001	
	dom	PCI	***Archer***	2	1997	1998	
	dom	AK	***Sea Chieftain***	1	1997	1998	18 month lease. Heavy sealift
	col	AEW	**ASTOR**	5	1997	2005	Delivery slipped from 2003
	Ge	trg	***Grob*-115D**	85	1998	2000	
	dom	CV	**CV**	2	1998	2012	
	col	lt tk	**TRACER**	200	1998	2007	With US; in feasibility phase
	col	APC	**MRAV**	200	1998	2002	With Fr, Ge
	dom	DDG	**Type 45**	3	2000	2005	First 3 of intended 12
	Swe	APC	**BvS 10**	125	1999	2001	
	col	UAV	***Sender***		1999		Devpt with US
	US	SLCM	***Tomahawk***	30	1999	2002	
	dom	AGHS	**ECHO**	2	2000	2002	Deliveries in 2002 & 2003
	US	tpt	**C-17**	4	2000	2001	Lease to 2010 interim to A-400M
	col	tpt	**A-400M**	25	2000	2010	With Be, Fr, Ge, It, Sp, Tu
	US	AAM	**AIM-120 B/C**		2000	2002	Interim to *Meteor*
	col	AAM	***Meteor***		2000	2010	With Fr, Ge, It

Table 15 **Arms orders and deliveries, non-NATO Europe, 1998–2000**

Country	Country supplier	Classification	Designation	Quantity	Order date	Delivery date	Comment
Armenia	PRC	AAA	***Typhoon***	8	1998	1999	
Azerbaijan	Kaz	FGA	**MiG-25**	8	1996	1998	
	Tu	PCC	**AB-34**	1	2000	2000	
Austria	US	SPA	**M109A2**	46	1996	1998	27 delivered in 1998
	dom	APC	***Pandur***	269	1997	1999	

Country	Country supplier	Classification	Designation	Quantity	Order date	Delivery date	Comment
	Ge	ATGW	***Jaguar***	90	1997	1998	
	Nl	MBT	***Leopard* 2A4**	114	1997	1998	79 delivered in 1998
	Swe	FGA	**J-35**	5	1999	1999	
	col	lt tk	**ULAN**	112	1999	2002	Delivery to 2004
Belarus	RF	trg	**MiG-29UB**	8	1999	1999	
Bosnia-Herzegovina							
	US	hel	**UH-1**	15	1996	1998	US Train and Equip programme
	UAE	arty	**105mm**	36	1996	1998	
	Et	arty	**122mm**	12	1996	1998	
	Et	arty	**130mm**	12	1996	1998	
	Et	AD	**23mm**	18	1996	1998	
	R	arty	**122mm**	18	1996	1998	
	R	arty	**130mm**	8	1996	1998	
Bulgaria	US	hel	**B-206**	6	1998	1999	2 delivered
Croatia	dom	MBT	**M-84**		1992	1996	In production
	dom	MBT	***Degman***		1995	2001	
	dom	MHC	***Rhino***	1	1995	1999	
	dom	PCI		1	1996	1998	Deliveries to 2002
	US	FGA	**F-16**	18	1999	2001	Ex-US inventory
	Il	FGA	**MiG-21**	40	1999		Upgrade
Cyprus	It	SAM	***Aspide***	44	1996	1998	12 more ordered
	RF	SAM	**S-300**	48	1997	1999	Delivered to Gr, based in Crete
	Gr	MBT	**AMX-30**	37	1997	1997	Last ten delivered in 1998
Estonia	SF	arty	**105mm**	18	1996	1997	105mm. Deliveries 1997–98
	SF	ML		2	1998	1999	Free transfer
	Ge	MCMV	***Lindau***	1	1999	1999	Free Transfer
Finland	dom	APC	**XA-185**	450	1982	1983	Deliveries to 1999
	US	FGA	**F/A-18C/D**	64	1992	1995	To 2000, 57 assembled in SF
	SF	arty	**K-98**	24	1996	1998	
	dom	ACV	**RA-140**	10	1997	1998	Mine-clearing vehicle
	dom	PFM	***Hamina***	2	1997	1998	
	dom	AIFV	**TA 2000**	150	1998		Development
	dom	APC	**XA-200**	48	1999	1999	Deliveries to 2001
	CH	UAV	***Ranger***	9	1999	1999	Under licence from Il
	US	ATGW	***Javelin***	242	2000		3,190 Missiles
Georgia	Tu	PCI	**SG-48**	1	1997	1998	Free transfer
	Ge	MSC	***Lindau***	2	1997	1998	Free transfer; deliveries to 1999
	UK	PFC		2	1998	1999	Free transfer
	Ukr	PFM	***Konotop***	1	1999	1999	
	US	hel	**UH-1**	10	1999	1999	
Ireland	UK	arty	**105mm**	12	1996	1998	
	UK	OPV	***Guardian***	1	1997	1999	Delivered in 1999

Country	Country supplier	Classification	Designation	Quantity	Order date	Delivery date	Comment
	CH	APC	***Piranha* III**	40	1999	2002	
Latvia	Ge	MCMV	***Lindau***	1	1999	1999	Free Transfer
Lithuania	Cz	trg	**L-39**	2	1997	1998	
	Ge	MCMV	***Lindau***	1	1999	1999	Free Transfer
Macedonia	Kaz	APC	**BTR-80**	12	1997	1998	
	Ge	APC	**BTR-70**	60	1998	1998	Free transfer
	Bg	arty	**152mm**	10	1998	1998	Free transfer
	Bg	arty	**76mm**	72	1998	1998	Aka ZIS-3. Free transfer
	US	ACV	**HMMWV**	41	1998	1998	Free transfer
	Bg	MBT	**T-55**	150	1998	1999	36 type T-55AM2
	Bg	arty	**122mm**	142	1998	1999	Free transfer
	US	arty	**105mm**	18	1998	1999	Free transfer
	Tu	FGA	**F-5A/B**	20	1998	1999	Free transfer
Malta	UK	MPA	**BN2B**	2	1997	1998	
Romania	dom	FGA	**MiG-21**	110	1994	1997	Upgrade programme, with Il
	US	tpt	**C-130**	5	1995	1998	
	dom	hel	**IAR-330L**	26	1995	1998	Upgrade
	Il	UAV	***Shadow***	6	1995	1998	
	US	cbt hel	**AH-1RO**	96	1997		Licence; delayed
	Ge	AAA	**35mm**	43	1997	1999	
	R	trg	**IAR-99**	24	1998	2000	6 delivered in 2000
Slovakia	dom	MBT	**M-2 Moderna**		1995	2000	T-72 upgrade programme
	Cz	APC	**OT-64**	100	1997	1998	Also 2 BVP-2 from Ukr for Indo
	col	hel	**EC-135**	12	1997	1999	
	col	hel	**AS-532**	5	1997	1999	
	col	hel	**AS-550**	2	1997	1999	
	dom	arty	***Zuzana* 2000**	8	1997	1998	155mm. Deliveries 1998
	Il	mor	**120mm**	56	1996	1998	
	Il	arty	**M845**	18	1996	1998	155mm 45 cal. towed arty
	Il	trg	**PC-9**	9	1997	1998	Upgrade
	CH	trg	**PC-9**	2	1997	1998	
	dom	MBT	**M-55**	30	1998	1999	T-55 upgrade
	A	APC	***Pandur***	70	1998	1999	
	dom	MBT	**T-84**	40	1999	2002	Upgrade
Sweden	dom	FGA	**JAS-39**	204	1981	1995	Deliveries to 2007, 18 in 1998
	dom	AIFV	**CV-90**	600	1984	1993	200 delivered by 1998; to 2004
	dom	LCA	**90**	199	1988	1989	To 2001. 100 delivered by 1997
	dom	PCI	***Tapper***	12	1992	1993	Deliveries to 1999. Coastal arty
	dom	AEW	***Saab* 340**	6	1993	1997	Deliveries 1997–98.
	US	AAM	**AMRAAM**	110	1994	1998	Option for a further 700
	Ge	MBT	***Leopard* 2**	120	1994	1998	New *Leopard* 2A5; to 2002
	Ge	MBT	***Leopard* 2**	160	1994	1997	Ex-Ge Army. Upgrade
	dom	MCM	**YSB**	4	1994	1996	Deliveries to 1998

Country	Country supplier	Classification	Designation	Quantity	Order date	Delivery date	Comment
	Ge	APC	**MT-LB**	584	1994	1995	Deliveries to 1998. Last 34 delivered 1998
	dom	FSG	***Visby***	4	1995	2001	Deliveries to 2003
	CH	APC	***Pirahna***	8	1996	1998	Deliveries in 2000
	col	AAM	**IRIS-T**		1997	2003	Devpt with Ge
	col	ASM	**KEPD 350**		1997	2003	With Ge to 2002, also KEPD 150
	dom	LCA	***Transportbat***	14	1997	1999	
	dom	ACV	**M 10**	6	1997	1998	
	Fr	UAV	***Ugglan***	3	1997		
	Fr	hel	**AS532**	12	1998	2001	Deliveries 2002
	dom	SP arty	***Karelin***	50	1998		155mm. Development
	dom	PCI	**KBV 201**	2	1999	2002	
	Ge	ARV	***Buffel***	10	1999	2002	
	dom	FSG	***Visby***	2	1999	2008	
Switzerland							
	US	FGA	**F/A-18C/D**	34	1993	1997	Licence; deliveries to 1999
	US	AAM	**AIM-120**	100	1993	1998	
	Il	UAV	***Ranger***	28	1995	1998	Licensed,deliveries to 1999
	dom	AD	***Skyguard***	100	1997	1999	Upgrade
	US	SP arty	**M-109**	456	1997	1998	Upgrade, deliveries to 2000
	Fr	hel	**AS-532**	12	1997	2000	Deliveries to 2002
	dom	APC	***Eagle 2***	205	1997	1999	Deliveries 2003
	US	AD	***Florako***	1	1999	2007	Upgrade
Ukraine	dom	CGH	***Ukraina***	1	1990	2000	
	dol	tpt	AN-70	5	1991	2003	Up to 65 required

Belgium Be

franc fr		1998	1999	2000	2001
GDP	fr	9.1tr	9.4tr		
	US$	250bn	237bn		
per capita	US$	24,200	25,000		
Growth	%	2.9	1.8		
Inflation	%	1.0	1.1		
Publ debt	%	117.4	114.3		
Def exp	fr	133bn	136bn		
	US$	3.6bn	3.4bn		
Def bdgt	fr	102.5bn	100.8bn	101bn	102bn
	US$	2.8bn	2.5bn	2.5bn	2.4bn
US$1=fr		36.3	39.6	41.2	

Population			10,126,000
Age	13–17	18–22	23–32
Men	307,000	313,000	697,000
Women	293,000	301,000	676,000

Total Armed Forces

ACTIVE 39,250
(incl 1,250 Medical Service; 2,570 women)

RESERVES 152,050 (to be 62,000)
Army 105,200 **Navy** 6,250 **Air Force** 20,700 **Medical Service** 19,900

Army 26,800

(incl 1,500 women)
1 joint service territorial comd (incl 2 engr, 2 sigs bn)
1 op comd HQ
1 mech inf div with 3 mech inf bde (each 1 tk, 2 armd inf, 1 SP arty bn, 1 engr coy) (2 bde at 70%, 1 bde at 50% cbt str), 1 AD arty bn, 2 recce coy (Eurocorps); 1 recce bn (MNDC)
1 cbt spt div (11 mil schools forming, 1 arty, 1 engr bn – augment mech inf div, plus 1 inf, 1 tk bn for bde at 50% cbt str)
1 para-cdo bde (2 para, 1 cdo, 1 ATK/recce bn, 1 arty, 1 AD bty, 1 engr coy)
1 lt avn gp (2 ATK, 1 obs bn)

RESERVES

Territorial Defence 11 lt inf bn (9 province, 1 gd, 1 reserve)

EQUIPMENT

MBT 140 *Leopard* 1A5
RECCE 141 *Scimitar* (29 in store)
AIFV 230 YPR-765 (plus 53 'look-alikes')
APC 185 M-113 (plus 144 'look-alikes'), 115 *Spartan* (plus 20 'look-alikes'), 4 YPR-765, 54 *Pandur* incl 'look-alikes'
TOTAL ARTY 242
TOWED 19: **105mm**: 5 M-101, 14 LG Mk II
SP 132: **105mm**: 18 M-108 (trg); **155mm**: 114 M-109A2
MOR 107mm: 89 M-30; **120mm**: 2 (for sale), plus **81mm**: 100
ATGW 476: 420 *Milan* (incl 217 YPR-765 (24 in store), 56 M-113 (4 in store))
RL 66mm: LAW
AD GUNS 35mm: 51 *Gepard* SP (all for sale)
SAM 118 *Mistral*
AC 10 BN-2A *Islander*
HELICOPTERS 76
ASLT 28 A-109BA
OBS 18 A-109A
SPT 30 SA-318 (5 in store)
UAV 28 *Epervier*

Navy 2,600

(incl 270 women)
BASES Ostend, Zeebrugge. Be and Nl navies under joint op comd based at Den Helder (Nl)

FRIGATES 3

FFG 3 *Wielingen*, 4 MM-38 *Exocet* SSM, 8 *Sea Sparrow* SAM, 1 × 100mm gun, 2 × TT (Fr L5 HWT), 1 × 6 ASW rkt

MINE COUNTERMEASURES 11

MCMV 4 *Van Haverbeke* (US *Aggressive* MSO) (incl 1 used for trials)
MHC 7 *Aster* (tripartite)

SUPPORT AND MISCELLANEOUS 12

2 log spt/comd with hel deck, 1 PCR, 1 sail trg, 7 AT; 1 AGOR

HEL 3 SA-316B

Air Force 8,600

(incl 800 women)
Flying hours 165
FGA 5 sqn with F-16A/B
FGA/RECCE 1 sqn with F-16A(R)/B
OCU with 8 F-16B
TPT 2 sqn
1 with 11 C-130H
1 with 2 Airbus A310-200, 3 HS-748, 5 *Merlin* IIIA, 2 *Falcon* 20, 1 *Falcon* 900
TRG 4 sqn
2 with *Alpha Jet*
1 with SF-260
1 with CM-170
SAR 1 sqn with *Sea King* Mk 48

EQUIPMENT

90 cbt ac (plus 59 in store), no armed hel
AC 129 **F-16** (72 **-A**, 18 **-B**, plus 39 in store (110 to receive mid-life update)) • 11 **C-130** (tpt) • 2 **Airbus A310-200** (tpt) • 3 **HS-748** (tpt) • 2 ***Falcon* 20** (VIP) • 1 ***Falcon* 900B** • 5 **SW 111 *Merlin*** (VIP, photo, cal) • 13 **CM-170** (trg, liaison) • 33 **SF-260** (trg) • 29 ***Alpha Jet*** (trg)

HEL 5 (SAR) *Sea King*
IN STORE 6 *Mirage* 5
MISSILES
AAM AIM-9 *Sidewinder*, AIM-120 AMRAAM
ASM AGM-65G *Maverick*
SAM 24 *Mistral*

Forces Abroad

GERMANY 2,000; 1 mech inf bde (1 inf, 1 arty bn, 1 recce coy)
STANAVFORLANT/STANAVFORMED
1 FFG (part time basis)
1 MHC
MCMFORMED
1 MCMV (part time basis)
UN AND PEACEKEEPING
BOSNIA/CROATIA (SFOR II): up to 550 (UNMOP): 1 obs **DROC** (MONUC): 1 obs **INDIA/PAKISTAN** (UNMOGIP): 2 obs **ITALY** (DELIBERATE FORGE): 4 F-16A **MIDDLE EAST** (UNTSO): 6 obs **WESTERN SAHARA** (MINURSO): 1 obs **YUGOSLAVIA** (KFOR): 900

Foreign Forces

NATO HQ NATO Brussels; HQ SHAPE Mons
WEU Military Planning Cell
US 790: **Army** 170 **Navy** 100 **Air Force** 520

Canada — Ca

Canadian dollar C$		1998	1999	2000	2001
GDP	C$	896bn	949bn		
	US$	604bn	644bn		
per capita	US$	23,700	24,800		
Growth	%	3.2	3.7		
Inflation	%	1.0	1.7		
Publ debt	%	97	93		
Def exp	C$	11.2bn	11.5bn		
	US$	7.5bn	7.8bn		
Def bdgt	C$	9.4bn	10.3bn	11.2bn	
	US$	6.3bn	7.0bn	7.6bn	
US$1=C$		1.48	1.47	1.47	

Population			29,512,000
Age	13–17	18–22	23–32
Men	1,000,000	980,000	2,055,000
Women	955,000	946,000	2,009,000

Canadian Armed Forces are unified and org in functional comds. Land Force Comd has op control of TAG. Maritime Comd has op control of maritime air. This entry is set out in traditional single-service manner.

Total Armed Forces

ACTIVE 59,100
(incl 6,100 women). Some 15,700 are not identified by service

RESERVES 43,300
Primary 28,600 **Army** (Militia) (incl comms) 22,200 **Navy** 4,000 **Air Force** 2,100 **Primary Reserve List** 300 *Supplementary* **Ready Reserve** 14,700

Army (Land Forces) 20,900

(incl 1,600 women)
1 Task Force HQ • 3 mech inf bde gp, each with 1 armd regt, 3 inf bn (1 lt), 1 arty, 1 engr regt, 1 AD bty • 1 indep AD regt • 1 indep engr spt regt

RESERVES
Militia 20,100 (excl comms); 18 armd, 51 inf, 19 arty, 12 engr, 20 log bn level units, 14 med coy
Canadian Rangers 3,250; 127 patrols
EQUIPMENT
MBT 114 *Leopard* C-1/C-2
RECCE 5 *Lynx* (in store), 195 *Cougar*, 203 *Coyote*
APC 1,790: 1,214 M-113 A2 (341 to be upgraded, 82 in store), 61 M-577, 269 *Grizzly*, 199 *Bison*, ε47 *Kodiak* (more to be delivered)
TOWED ARTY 213: **105mm**: 185 C1/C3 (M-101), 28 LG1 Mk II
SP ARTY 155mm: 58 M-109A4 (18 in store)
MOR 81mm: 167
ATGW 150 TOW (incl 72 TUA M-113 SP), 425 *Eryx*
RL 66mm: M-72
RCL 84mm: 1,040 *Carl Gustav*; **106mm**: 111
AD GUNS 35mm: 34 GDF-005 with *Skyguard*; **40mm**: 57 L40/60 (in store)
SAM 22 ADATS, 96 *Javelin*, *Starburst*

Navy (Maritime Command) 9,000

(incl 2,800 women)
SUBMARINES 0
but 1 *Victoria* SSK (UK *Upholder*) to commission late 2000
PRINCIPAL SURFACE COMBATANTS 16
DESTROYERS 4
DDG 4 modified *Iroquois* with 1 Mk-41 VLS for 29 SM-2 MR SAM, 1 × 76mm gun, 6 ASTT, 2 CH-124 *Sea King* ASW hel (Mk 46 LWT)
FRIGATES 12
FFG 12 *Halifax* with 8 *Harpoon* SSM, 16 *Sea Sparrow* SAM, 2 × ASTT, 1 CH-124A *Sea King* hel (Mk 46 LWT)
PATROL AND COASTAL COMBATANTS 14
12 *Kingston* MCDV, 2 *Fundy* PCC (trg)

MINE COUNTERMEASURES 2

2 *Anticosti* MSA (converted offshore spt vessels)

SUPPORT AND MISCELLANEOUS 6

2 *Protecteur* AO with 3 *Sea King* hel, 1 AOT; 1 diving spt; 2 AGOR

DEPLOYMENT AND BASES

NATIONAL Ottawa (Chief of Maritime Staff)

ATLANTIC Halifax (National and Marlant HQ; Commander Marlant is also OMCANLANT): 1 SSK, 2 DDG, 7 FFG, 1 AO, 1 AK, 6 MCDV; 2 MR plus 1 MR (trg) sqn with CP-140 and 3 CP-140A, 1 ASW and 1 ASW (trg) hel sqn with 26 CH-125 hel

PACIFIC Esquimalt (HQ): 2 DDG, 5 FFG, 1 AO, 6 MCDV, 1 MSA; 1 MR sqn with 4 CP-140 and 1 ASW hel sqn with 6 CH-124 hel

RESERVES

HQ Quebec

4,000 in 24 div: patrol craft, coastal def, MCM, Naval Control of Shipping, augmentation of regular units

Air Force (Air Command) 13,500

(incl 1,700 women)

Flying hours 210

1 Air Div with 13 wg responsible for operational readiness, combat air-support, air tpt, SAR, MR and trg

EARLY WARNING Canadian NORAD Regional HQ at North Bay: 47 North Warning radar sites: 11 long-range, 36 short-range; Regional Op Control Centre (ROCC) (2 Sector Op Control Centres (SOCC)): 4 Coastal Radars and 2 Transportable Radars. Canadian Component – NATO Airborne Early Warning (NAEW)

EQUIPMENT

140 (incl 18 MR) cbt **ac**, 30 armed **hel**

AC 122 **CF-18** (83 **-A**, 39 **-B**) - 60 operational (5sqns) and 62 fighter trg, testing and rotation • 4 sqns with 18 **CP-140** (MR) and 3 **CP-140A** (environmental patrol) • 4 sqns with 32 **CC-130E/H** (tpt) and 5 **KCC-130** (tkr) • 1 sqn with 5 **CC-150** (Airbus A-310) and 5 Boeing CC-137 • 1 sqn with 7 **CC-109** (tpt) and 16 11 **CC-144** (EW trg, coastal patrol, VIP/tpt) • 4 sqns with 7 **CC-138** (SAR/tpt), 10 **CC-115** (SAR/tpt), 45 **CT-133** (EW trg/tpt plus 9 in store)

HEL 15 **CH-113** (SAR/tpt) • 3 sqns of 30 **CH-124** (ASW, afloat) • 99 **CH-146** (tpt, SAR) of which 30 armed • 8 **CH-118** (EW) • first of 15 **CH-149** to be delivered early 2001

TRG 2 Flying Schools 130 **CT-114** *Turor*, 6 **CT-142 hel** 9 **CH-139** *Jet Ranger*

NATO FLIGHT TRAINING CANADA 12 T-6A/CT-156 (primary), (another 12 to be delivered by end 2000). First of 18 Hawk 115 (advanced/wpns/tactics trg) delivered

AAM AIM-7M *Sparrow*, AIM-9L *Sidewinder*

Forces Abroad

UN AND PEACEKEEPING

BOSNIA (SFOR II): 1,350: 1 inf bn, 1 armd recce, 1 engr sqn **CROATIA** (UNMOP): 1 obs **CYPRUS** (UNFICYP): 2 **DROC** (MONUC): 3 obs **EAST TIMOR** (UNTAET): 3 **EGYPT** (MFO): 28 **IRAQ/KUWAIT** (UNIKOM): 6 obs **ITALY** (DELIBERATE FORGE): 6 CF-18 **MIDDLE EAST** (UNTSO): 10 obs **SIERRA LEONE** (UNAMSIL): 5 obs **SYRIA/ISRAEL** (UNDOF): 186: log unit **YUGOSLAVIA** (KFOR): 800

Paramilitary 9,350

Canadian Coast Guard has merged with **Department of Fisheries and Oceans**. Both are civilian-manned.

CANADIAN COAST GUARD (CCG) 4,700

some 103 vessels incl 31 navaids/tender, 12 survey/research, 6 icebreaker, 22 cutter, 5 PCO, 12 PCI, 8 utility, 4 ACV, 3 trg; plus **hel** 1 S-61, 6 Bell-206L, 5 Bell-212, 16 BO-105

DEPARTMENT OF FISHERIES AND OCEANS (DFO) 4,650

some 90 vessels incl 35 AGOR/AGHS, 38 patrol, 17 icebreakers

Foreign Forces

UK 355: Army 200; Air Force 155

Czech Republic Cz

koruna Kc		1998	1999	2000	2001
GDP	Kc	1.8tr	1.8tr		
	US$	55bn	51.5bn		
per capita	US$	13,000	13,200		
Growth	%	-2.7	-0.2		
Inflation	%	10.7	2.1		
Debt	US$	23.1bn	25.3bn		
Def exp	Kc	37.3bn	41.7bn		
	US$	1,155m	1,169m		
Def bdgt	Kc	36.9bn	41.4bn	44.0bn	
	US$	1,141m	1,163m	1,153m	
FMA (US)	US$	1.4m	1.4m	1.6m	1.7m
US$1=Kc		32.2	35.7	38.1	

Population 10,290,000

(Slovak 3%, Polish 0.6%, German 0.5%)

Age	13–17	18–22	23–32
Men	347,000	412,000	796,000
Women	330,000	396,000	768,000

Total Armed Forces

ACTIVE 57,700

(incl 25,000 conscripts; 19,200 MoD, centrally controlled formations and HQ units)

NATO

NATO and Non-NATO Europe

Terms of service 12 months

Army 25,100

(incl 15,500 conscripts)
1 rapid-reaction bde (2 mech, 1 AB, 1 recce, 1 arty, 1 engr bn)
2 mech bde (each with 3 mech, 1 recce, 1 arty, 1 ATK, 1 AD, 1 engr bn)
1 combined msl regt
7 trg and mob base (incl arty, AD, engr)

RESERVES

14–15 territorial def bde

EQUIPMENT

MBT 792 (176 in store): 123 T-54, 128 T-55, 541 T-72M (250 to be upgraded)
RECCE some 182 BRDM, OT-65
AIFV 801: 612 BMP-1, 174 BMP-2, 15 BRM-1K
APC 403 OT-90, 7 OT-64 plus 570 'look-alikes'
TOTAL ARTY 740
TOWED 122mm: 148 D-30
SP 364: **122mm**: 91 2S1; **152mm**: 273 *Dana* (M-77)
MRL 122mm: 135 RM-70
MOR 93: **120mm**: 85 M-1982, 8 MSP-85
SSM FROG-7, SS-21
ATGW 721 AT-3 *Sagger* (incl 621 on BMP-1, 100 on BRDM-2), 21 AT-5 *Spandrel*
AD GUNS 30mm: M-53/-59
SAM SA-7, ε140 SA-9/-13
SURV GS-13 (veh), *Small Fred/Small Yawn* (veh, arty)

Air Force 13,400

(incl AD and 8,500 conscripts); 110 cbt ac (incl 17 MiG-23 in store), 34 attack hel
Organised into two main structures – Tactical Air Force and Air Defence
Flying hours 60
FGA/RECCE 2 sqn with 32 Su-22MK/UM3K, 24 Su-25Bk/UBk, 2 L-159 (a further 19 to be delivered by end 2000, 26 by end 2001, 25 in 2002)
FTR 1 sqn with 37 MiG-21
TPT 2 sqn with 14 L-410, 8 An-24/26/30, 2 Tu-154, 1 Challenger CL-600 **hel** 2 Mi-2, 4 Mi-8, 1 Mi-9, 10 Mi-17
HEL 2 sqn (aslt/tpt/attack) with 24 Mi-2, 9 Mi-8/20, 32 Mi-17, 34* Mi-24, 11 PZL W-3
TRG 1 regt with **ac** 24 L-29, 14 L-39C, 17 L-39ZO, 3 L-39MS, 8 Z-142C **hel** 8 Mi-2
AAM AA-2 *Atoll*, AA-7 *Apex*, AA-8 *Aphid*
SAM SA-2, SA-3, SA-6

Forces Abroad

UN AND PEACEKEEPING

BOSNIA (SFOR II): up to 560; 1 mech inf bn **CROATIA** (UNMOP): 1 obs (SFOR): 7 **DROC** (MONUC): 4 obs **GEORGIA** (UNOMIG): 4 obs **SIERRA LEONE** (UNAMSIL): 5 obs **YUGOSLAVIA** (KFOR): 160

Paramilitary 5,600

BORDER GUARDS 4,000
(1,000 conscripts)
INTERNAL SECURITY FORCES 1,600
(1,500 conscripts)

Denmark Da

kroner kr		1998	1999	2000	2001
GDP	Kr	1,167bn	1,213bn		
	US$	174bn	166bn		
per capita	US$	24,100	24,800		
Growth	%	2.7	1.3		
Inflation	%	1.8	2.4		
Publ debt	%	59.5	55.4		
Def exp	Kr	19.1bn	19.5bn		
	US$	2.9bn	2.7bn		
Def bdgt	Kr	18.1bn	18.6bn	18.4bn	18.5bn
	US$	2.7bn	2.6bn	2.3bn	2.3bn
US$1=kr		6.7	7.3	7.9	

Population			**5,267,000**
Age	13–17	18–22	23–32
Men	144,000	154,000	380,000
Women	138,000	149,000	368,000

Total Armed Forces

ACTIVE 21,810
(excluding civilians but incl some 460 central staff; about 5,025 conscripts; 685 women)
Terms of service 4–12 months (up to 24 months in certain ranks)

RESERVES 64,900
Army 46,000 **Navy** 7,300 **Air Force** 11,600
Home Guard (*Hjemmevaernet*) (volunteers to age 50) 58,680 **Army** 48,700 **Navy** 4,380 **Air Force** 5,600

Army 12,850

(incl 4,400 conscripts, 350 women)
1 op comd • 1 mech inf div with 3 mech inf bde (each 2 mech inf, 1 tk, 1 SP arty bn), 1 regt cbt gp (1 mech inf, 1 mot inf bn, 1 engr coy), 1 recce, 1 tk, 1 AD, 1 engr bn; div arty • 1 rapid reaction bde with 2 mech inf, 1 tk, 1 SP arty bn (20% active cbt str) • 1 recce, 1 tk, 1 AD, 1 engr bn, 1 MLRS coy • Army avn (1 attack hel coy, 1 recce hel det) • 1 SF unit

RESERVES

5 local def region (1–2 mot inf bn), 2 regt cbt gp (3 mot inf, 1 arty bn)

EQUIPMENT

MBT 248: 230 *Leopard* 1A5 (58 in store), 18 *Leopard* 2
RECCE 36 Mowag *Eagle*
APC 315 M-113 (plus 340 'look-alikes' incl 55 SP mor)
TOTAL ARTY 475
TOWED 105mm: 134 M-101; **155mm**: 97 M-114/39
SP 155mm: 76 M-109
MRL 227mm: 8 MLRS
MOR 120mm: 160 Brandt; **81mm**: 338 (incl 53 SP)
ATGW 140 TOW (incl 56 SP)
RL 84mm: AT-4
RCL 1,151: **84mm**: 1,131 *Carl Gustav*; **106mm**: 20 M-40
SAM *Stinger*
SURV *Green Archer*
ATTACK HEL 12 AS-550C2
SPT HEL 13 Hughes 500M/OH-6

Navy 4,060

(incl 500 conscripts, 150 women)
BASES Korsør, Frederikshavn, Vaerlose (naval air)

SUBMARINES 3

SSK 3
2 *Tumleren* (mod No *Kobben*) with Swe Type 61 HWT
1 *Narhvalen*, with Type 61 HWT

CORVETTES 3

FSG 3 *Niels Juel* with 8 *Harpoon* SSM, 8 *Sea Sparrow* SAM, 1 × 76mm gun

PATROL AND COASTAL COMBATANTS 27

MISSILE CRAFT 5 *Flyvefisken* (Stanflex 300) PFM with 2 × 4 *Harpoon* SSM, 6 *Seasparrow* SAM, 1 × 76mm gun, 2 × 533mm TT
PATROL CRAFT 22
OFFSHORE 4
4 *Thetis* PCO with 1 × 76mm gun, 1 *Lynx* hel
COASTAL 18
6 *Flyvefisken* (Stanflex 300) PFC, 3 *Agdlek* PCC, 9 *Barsøe* PCC

MINE WARFARE 7

MINELAYERS 4
(All units of *Flyvefisken* class can also lay up to 60 mines)
2 *Falster* (400 mines), 2 *Lindormen* (50 mines)
MINE COUNTERMEASURES 3
3 *Flyvefisken* (SF300) MHC

SUPPORT AND MISCELLANEOUS 13

1AE, 1 tpt; 4 icebreakers, 6 environmental protection, 1 Royal Yacht
HEL 8 *Lynx* (up to 4 embarked)

COASTAL DEFENCE

1 coastal fortress; **150mm** guns, coastal radar
2 mobile coastal missile batteries: 2 × 8 *Harpoon*

RESERVES (Home Guard)

40 inshore patrol craft/boats

Air Force 4,900

(incl 125 conscripts, 185 women)
Flying hours 180

TACTICAL AIR COMD

FGA/FTR 3 sqn with 69 F-16A/B (60 operational, 9 attritional reserve)
TPT 1 sqn with 3 C-130H, 1 *Challenger*-604 (2 more on order for MR/VIP), 2 *Gulfstream* G-III
SAR 1 sqn with 8 S-61A hel
TRG 1 flying school with SAAB T-17

AIR DEFENCE GROUP

2 SAM bn: 8 bty with 36 I HAWK, 32 **40mm**/L70

CONTROL/REPORTING GROUP

5 radar stations, one in the Faroe Islands

EQUIPMENT

69 cbt ac, no armed hel
AC 69 **F-16A/B** (FGA/ftr) • 3 **C-130H** (tpt) • 1 *Challenger*-**604** (tpt) • 28 **SAAB T-17** • 2 *Gulfstream* G-III
HEL 8 **S-61** (SAR)

MISSILES

ASM AGM-12 *Bullpup*
AAM AIM-9 *Sidewinder*, AIM-120A AMRAAM
SAM 36 I HAWK

Forces Abroad

UN AND PEACEKEEPING

BOSNIA (SFOR II): 425. 40% of Nordic-Polish BG incl 1 tk sqn (10 *Leopard* MBT); aircrew with NATO E-3A operations; Air Force personnel in tac air-control parties (TACP) **CROATIA** (UNMOP): 1 obs **DROC** (MONUC): 2 obs **EAST TIMOR** (UNTAET): 2 obs **GEORGIA** (UNOMIG): 5 obs **INDIA/PAKISTAN** (UNMOGIP): 6 obs **IRAQ/KUWAIT** (UNIKOM): 5 obs **ITALY** (DELIBERATE FORGE): 3 F-16 **MIDDLE EAST** (UNTSO): 10 obs **SIERRA LEONE** (UNAMSIL): 2 obs **YUGOSLAVIA** (KFOR): 900: 1 inf bn gp incl 1 tk sqn

Foreign Forces

NATO HQ Joint Command North-East

France Fr

franc fr		1998	1999	2000	2001
GDP	fr	8.5tr	8.8tr		
	US$	1.4tr	1.4tr		
per capita	US$	23,100	24,000		
Growth	%	3.2	2.9		
Inflation	%	0.7	0.6		
Publ debt	%	64.7	65.0		
Def exp	fr	240bn	239bn		
	US$	40.0bn	37.1bn		
Def bdgt	fr	184.8bn	190.0bn	187.9bn	
	US$	31.3bn	29.5bn	27.0bn	
US$1=fr		5.90	6.44	6.94	

Population			59,425,000
Age	13–17	18–22	23–32
Men	1,974,000	1,943,000	4,298,000
Women	1,885,000	1,860,000	4,122,000

Total Armed Forces

ACTIVE 294,430

(incl 58,710 conscripts, 18,920 women; incl 5,200 **Central Staff**, 8,600 (750 conscripts) ***Service de santé***, 1,340 ***Service des essences*** not listed)

Terms of service 10 months (can be voluntarily extended to 12–24 months)

RESERVES 419,000 (1999 figures – re-org in process)

Army 242,500 **Navy** 97,000 **Air Force** 79,500

Potential 1,058,500 **Army** 782,000 **Navy** 97,000 **Air Force** 179,500

Strategic Nuclear Forces (8,400)

(**Navy** 4,700 **Air Force** 3,100 ***Gendarmerie*** 600)

NAVY 64 SLBM in 4 SSBN

SSBN 4

2 *L'Inflexible* with 16 M-4/TN-70 or -71, SM-39 *Exocet* USGW and 4 × 533mm HWT

2 *Le Triomphant* with 16 M-45/TN-75 SLBM, SM-39 *Exocet* USGW and 4 × 533mm HWT

36 *Super Etendard* strike **ac** (ASMP); plus 16 in store

AIR FORCE

3 sqn with 60 *Mirage* 2000 N(ASMP)

TKR 1 sqn with 11 C-135FR, 3 KC-135

RECCE 1 sqn with 5 *Mirage* IV P

AIRBORNE RELAY 4 C-160H *Astarte*

CBT TRG 6 *Mystere* 20, 6 *Jaguar* E

Army 169,300

(incl 9,150 women, 47,000 conscripts) regt normally bn size

1 Land Comd HQ

5 Regional, 4 Task Force HQ

2 armd bde (each 2 armd, 2 armd inf, 1 SP arty, 1 engr regt)

2 mech inf bde (each 1 armd, 1 armd inf, 1 APC inf, 1 SP arty, 1 engr regt)

2 lt armd bde (each 2 armd cav, 2 APC inf, 1 arty, 1 engr regt)

1 mtn inf bde with 1 armd cav, 3 APC inf, 1 arty, 1 engr bde)

1 AB bde with 1 armd cav, 4 para inf, 1 arty, 1 engr regt

1 air mobile bde with 3 cbt hel, 1 spt hel regt

1 arty bde with 1 MLRS, 3 *Roland* SAM, 1 *Hawk* SAM regt

1 arty, 1 engr, 1 sigs, 1 Int and EW bde

1 Fr/Ge bde (2,500): Fr units incl 1 armd cav, 1 APC inf regt, 1 recce coy

FOREIGN LEGION (8,000)

1 armd, 1 para, 5 inf, 2 engr regt (incl in units listed above)

MARINES (16,500)

(incl conscripts, mainly overseas enlisted)

11 regt in France (incl in units listed above), 13 regt overseas

SPECIAL OPERATIONS FORCES

2 para regt, 2 hel units (EW, special ops) (incl in units listed above)

RESERVES

Territorial def forces: 75 coy (all arms)

EQUIPMENT

MBT 834 (CFE: 1,234): 635 AMX-30B2, 199 *Leclerc*

RECCE 337 AMX-10RC, 192 ERC-90F4 *Sagaie*, 899 VBL M-11

AIFV 713 AMX-10P/PC

APC 3,900 VAB (incl variants)

TOTAL ARTY 802

TOWED 155mm: 105 TR-F-1

SP 155mm: 273 AU-F-1

MRL 227mm: 61 MLRS

MOR 120mm: 363 RT-F1

ATGW 780 *Eryx*, 1,348 *Milan*, HOT (incl 135 VAB SP)

RL 89mm: 9,850; **112mm**: 9,690 APILAS

AD GUNS 20mm: 774 53T2

SAM 69 HAWK, 113 *Roland* I/II, 331 *Mistral*

SURV RASIT-B/-E (veh, arty), RATAC (veh, arty)

AC 2 Cessna *Caravan* II , 5 PC-6, 2 TBM-700

HELICOPTERS 498

ATTACK 339: 154 SA-341F, 155 SA-342M, 30 SA-342AATCP

RECCE 4 AS-532 *Horizon*

SPT 155: 27 AS-532, 128 SA-330

UAV 6 CL-289 (AN/USD-502), 2 *Crecerelle*, 4 *Hunter*

Navy 49,490

(incl 2,000 Marines, 3,500 Naval Air, 3,470 women, 5,020 conscripts)

COMMANDS SSBN (ALFOST) HQ Brest **Atlantic** (CECLANT) HQ Brest **North Sea/Channel** (COMAR CHERBOURG) HQ Cherbourg **Mediterranean** (CECMED) HQ Toulon **Indian Ocean** (ALINDIEN) HQ afloat **Pacific Ocean** (ALPACI) HQ Papeete
ORGANIC COMMANDS ALFAN (Surface Ships) **ALFAN/Brest** (Surface Ships ASW) **ALFAN/Mines** (mine warfare) **ALAVIA** (naval aviation) **COFUSCO** (Marines) **ALFOST** (Submarines)
BASES France Cherbourg, Brest (HQ), Lorient, Toulon (HQ) **Overseas** Papeete (HQ) (Tahiti), La Réunion, Noumea (New Caledonia), Fort de France (Martinique), Cayenne (French Guiana)

SUBMARINES 11

STRATEGIC SUBMARINES 4 SSBN (see **Strategic Nuclear Forces**)
TACTICAL SUBMARINES 7
SSN 6 *Rubis* ASW/ASUW with F-17 HWT, L-5 LWT and SM-39 *Exocet* USGW
SSK 1 *Agosta* with F-17 HWT and L-5 LWT, SM-39 *Exocet* USGW

PRINCIPAL SURFACE COMBATANTS 35

CARRIERS 1 *Clémenceau* CVS (32,700t), capacity 40 ac (typically 2 flt with 25 *Super Etendard*, 1 with 6 *Alizé*; 1 det with 4 *Etendard* IVP, 2 *Super Frelon*, 2 *Dauphin* hel)
(1 *Charles de Gaulle* CVN to commission late 2000)
CRUISERS 1 *Jeanne d'Arc* CG (trg/ASW) with 6 MM-38 *Exocet* SSM, 4 × 100mm guns, capacity 8 SA-319B hel
DESTROYERS 4
DDG 4
2 *Cassard* with 8 MM-40 *Exocet* SSM, 1 × 2 SM-1MR SAM, 1 × 100mm gun, 2 ASTT, 1 *Panther* hel
2 *Suffren* with 4 MM-38 *Exocet* SSM, 1 × 2 *Masurca* SAM, 2 × 100mm gun, 1 *Malafon* SUGW
FRIGATES 29
FFG 29
6 *Floréal* with 2 MM-38 *Exocet* SSM, 1 × 100mm gun, 1 AS-365 hel
7 *Georges Leygues* with *Croatale* SAM, 1 × 100mm gun, 2 ASTT, 2 *Lynx* hel (Mk 46 LWT); 5 with 8 MM-40 *Exocet* SSM, 2 with 4 MM-38 *Exocet* SSM
2 *Tourville* with MM-38 *Exocet* SSM, *Croatale* SAM, 2 × 100mm gun, 1 *Malafon* SUGW, 2 ASTT
10 *D'Estienne d'Orves* with 1 × 100mm gun, 4 ASTT, 6 ASW mor; 4 with 2 MM-38 *Exocet* SSM, 6 with 4 MM-40 *Exocet* SSM
4 *La Fayette* with 8 MM-40 *Exocet* SSM, *Croatale* SAM, 1 × 100mm gun, 1 *Panther* hel

PATROL AND COASTAL COMBATANTS 40

PATROL, OFFSHORE 1 *Albatross* PCO (Public Service Force)
PATROL, COASTAL 23
10 *L'Audacieuse* PCC, 8 *Léopard* PCC (instruction), 3 *Flamant* PCC, 1 *Sterne* PCC, 1 *Grebe* PCC (Public Service Force)
PATROL, INSHORE 16
2 *Athos* PCI<, 2 *Patra* PCI<, 1 *La Combattante* PCI<, 6 *Stellis* PCI<, 5 PCI< (manned by *Gendarmarie Maritime*)

MINE WARFARE 21

COMMAND AND SUPPORT 1 Loire MCCS
MINELAYERS 0, but submarines and *Thetis* (trials ship) have capability
MINE COUNTERMEASURES 20
13 *Eridan* MHC, 4 *Vulcain* MCM diver spt, 3 *Antares* (route survey/trg)

AMPHIBIOUS 9

2 *Foudre* LPD, capacity 450 tps, 30 tk, 4 *Cougar* hel, 2 CDIC LCT or 10 LCM
2 *Ouragan* LPD: capacity 350 tps, 25 tk, 2 *Super Frelon* hel or 4 *Puma* hel
5 *Champlain* LSM: capacity 140 tps, tk
Plus craft: 5 LCT, 21 LCM

SUPPORT AND MISCELLANEOUS 30

UNDER WAY SUPPORT 4 *Durance* AO with 1 SA-319 hel
MAINTENANCE AND LOGISTIC 7
4 AOT, 1 *Jules Verne* AR with 2 SA-319 hel, 2 *Rhin* depot/spt, with hel
SPECIAL PURPOSES 14
8 trial ships, 2 *Glycine* trg, 4 AT/F (3 civil charter)
SURVEY/RESEARCH 5
4 AGHS, 1 AGOR

DEPLOYMENT

CECLAND (HQ, Brest): 4 SSBN, 2 SS, 1 CG, 9 DDG/FFG, 3 MCMV, 1 MCCS, 8 MHC, 1 diver spt, 3 AGS, 1 AGOR
COMAR CHERBOURG (HQ, Cherbourg): 1 clearance diving ship
CECMED (HQ, Toulon): 6 SSN, 1 CV, 19 DDG/FFG, 3 LSD, 3 AO, 1 LSM, 4 LCT, 2 diver spt, 3 MHC, 1 AR

NAVAL AIR (3,500)

Flying hours *Etendard*: 180–220 (night qualified pilots)
NUCLEAR STRIKE 2 flt with *Super Etendard* (ASMP nuc ASM)
RECCE 1 sqn with *Etendard* IV P
AEW 2 flt with *Alizé*
MR 4 flt with N-262 *Frégate*
MP 2 sqn with *Atlantique*
ASW 2 sqn with *Lynx*
AEW 1 E-2C
TRG 3 units with N-262 *Frégate*, *Rallye* 880, CAP 10, 1 unit with SA-316B
COMMANDOS 1 aslt sqn with SA-321
MISC 1 SAR unit with SA-321, SA-365, 1 unit with SA-365F, SA-319, 2 liaison units with EMB-121

EQUIPMENT

52 cbt ac (plus 30 in store); 32 armed hel (plus 8 in store)
AC 29* ***Super Etendard*** plus 23 in store (52 to be mod

for ASMP) • 5* *Etendard* IV P • 6 *Alizé* (AEW) plus 2 in store • 18* ***Atlantique* 2** (MP) plus 7 in store • 13 ***Nord* 262** (MR/trg) • 8 ***Xingu*** (misc) • 7 ***Rallye* 880** • 8 **CAP-10** (trg) • 5 ***Falcon* 10MER** (trg)

HEL 32 ***Lynx*** (ASW) plus 8 in store • 13 **AS-565SA** (SAR, trg) plus 2 in store

MISSILES

ASM *Exocet* AM-39

AAM R-550 *Magic* 2

MARINES (2,000)

COMMANDO UNITS (400) 4 aslt gp

1 attack swimmer unit

FUSILIERS-MARIN (1,600) 14 naval-base protection gp

PUBLIC SERVICE FORCE naval personnel performing general coast guard, fishery, SAR, anti-pollution and traffic surv duties: 1 *Albatross*, 1 *Sterne*, 1 *Grebe*, 3 *Flamant* PCC; **ac** 4 N-262 **hel** 4 SA-365 (ships incl in naval patrol and coastal totals). Comd exercised through ***Maritime Préfectures*** (Premar): ***Manche*** (Cherbourg), ***Atlantique*** (Brest), ***Méditerranée*** (Toulon)

Air Force 60,500

(incl 6,300 women, 5,940 conscripts, 5,500 civilians; incl strategic nuc forces)

Flying hours 180

AIR SIGNALS AND GROUND ENVIRONMENT COMMAND

CONTROL automatic *STRIDA* II, 6 radar stations, 1 sqn with 4 E3F

SAM 11 sqn (1 trg) with *Crotale*, *Aspic*, SATCP and AA gun bty **(20mm)**

AIR COMBAT COMMAND

FTR 6 sqn with *Mirage* 2000C/B/5F

FGA 7 sqn

3 with *Mirage* 2000D • 2 with *Jaguar* A • 2 with *Mirage* F1-CT

RECCE 2 sqn with *Mirage* F-1CR

TRG 3 OCU sqn

1 with *Jaguar* A/E • 1 with F1-C/B • 1 with *Mirage* 2000/BC

EW 1 sqn with C-160 ELINT/ESM

AIR MOBILITY COMMAND (CFAP)

TPT 14 sqn

1 hy with DC-8F, A310-300

6 tac with C-160/-160NG, C-130H

7 lt tpt/trg/SAR/misc with C-160, DHC-6, CN235, *Falcon* 20, *Falcon* 50, *Falcon* 900, TBM-700, N-262, AS-555

EW 1 sqn with DC-8 ELINT

HEL 5 sqn with AS-332, SA-330, AS-555, AS-355, SA-319

TRG 1 OCU with C-160, N-262, 1 OCU with SA-319, AS-555, SA-330

AIR TRAINING COMMAND

TRG *Alpha Jet*, EMB-121, TB-30, EMB-312, CAP-10/-20/-231, CR-100, N262

EQUIPMENT

517 cbt ac, no armed hel

AC 352 ***Mirage*** (10 **F-1B** (OCU), 23 **F-1C** (OCU plus 6 in Djibouti), 40 **F-1CR** (recce), 40 **F-1CT** (FGA), 5 **MIVP** (recce), 114 **-M-2000B/C/5F** (64 -C (ftr), 30 -5F (upgraded C), 20 -B (OCU)), 60 **-M-2000N** (strike, FGA), 60 **-M-2000D**) • 66 ***Jaguar*** (FGA) (plus 54 in store) • 99* ***Alpha Jet*** (trg, plus 29 in store) • 4 **E-3F** (AEW) • 2 **A 310-300** (tpt) • 2 **DC-8F** (tpt) • 1 **DC-8E** • 14 **C-130** (**5 -H** (tpt), 9 **-H-30** (tpt)) • 11 **C-135FR** (tkr) • 77 **C-160** (13 **-AG**, 60 **-NG** (tpt/14 tkr) 4 **-H**) • 3 **KC-135** • 14 **CN-235M** (tpt) • 19 **N-262** • 17 ***Falcon*** (7 **-20**), 4 **-50** (VIP), 2 **-900** (VIP)) • 17 **TBM-700** (tpt) • 6 **DHC-6** (tpt) • 32 **EMB-121** (trg) • 92 **TB-30** (trg plus 50 in store) • 9 **CAP-10/20/231** (trg) • 48 **EMB-312** (trg) • 2 **CR-100** (trg)

HEL 3 **SA-319** (*Alouette* III) • 29 **SA-330** (26 tpt, SAR, 3 OCU) (*Puma*) • 7 **AS-332** (tpt/VIP) (*Super Puma*) • **3 AS-532** (tpt) (*Cougar*) • 4 **AS-355** (*Ecureuil*) • 43 **AS-555** (34 tpt, 9 OCU) (*Fennec*)

MISSILES

ASM ASMP, AS-30/-30L

AAM *Super* 530F/D, R-550 *Magic* 1/II, AIM-9 *Sidewinder*, *Mica*

Forces Abroad

GERMANY 2,700: incl elm Eurocorps

ANTILLES (HQ Fort de France): 3,800: 3 marine inf regt (incl 2 SMA), 1 marine inf bn, 1 air tpt unit **ac** 2 C-160 **hel** 2 SA-330, 2 AS-555, 1 FFG (1 AS-365 hel), 2 PCI, 1 LSM, 1 spt ***Gendarmerie*** 860

FRENCH GUIANA (HQ Cayenne): 3,250: 2 marine inf (incl 1 SMA), 1 Foreign Legion regt, 2 PCI 1 *Atlantic* **ac**, 1 air tpt unit **hel** 2 SA-330, 3 AS-555 ***Gendarmerie*** 600

INDIAN OCEAN (Mayotte, La Réunion): 4,200: 2 Marine inf (incl 1 SMA) regt, 1 spt bn, 1 air tpt unit **ac** 2 C-160 **hel** 2 AS 555, 1 LSM, 1 spt ***Gendarmerie*** 850 **Navy** Indian Ocean Squadron, Comd ALINDIEN (HQ afloat): 1 FFG (2 AS-365 hel), 2 PCI, 1 AOR (comd), reinforcement 2 FFG, 1 *Atlantic* ac

NEW CALEDONIA (HQ Nouméa): 3,100: 1 Marine inf regt; some 12 AML recce, 5 **120mm** mor; 1 air tpt unit, det **ac** 3 CN-235 **hel** 2 AS-555, 5 SA-330 **Navy** 2 FFG (2 AS-365 hel), 2 PCI, 1 LSM, 1 AGS, 1 spt **ac** 2 *Guardian* MR ***Gendarmerie*** 1,050

POLYNESIA (HQ Papeete) 3,100 (incl *Centre d'Expérimentation du Pacifique*): 1 Marine inf regt, 1 Foreign Legion bn, 1 air tpt unit; 2 CN-235, 3 AS-332 ***Gendarmerie*** 600 **Navy** 1 FFG, 3 patrol combatants, 1 amph, 1 AGHS, 5 spt **ac** 3 *Guardian* MR

CHAD 990: 2 inf coy, 1 AML sqn (-) **ac** 1 C-160, 1 C-130,

3 *Mirage* FICT, 2 *Mirage* FICR **hel** 3 SA-330
CÔTE D'IVOIRE 500: 1 marine inf bn (18 AML-60/-90) **hel** 1 AS-555
DJIBOUTI 3,200: 2 inf coy, 2 AMX sqn, 1 engr unit; 1 sqn with **ac** 6 *Mirage* F-1C (plus 4 in store), 1 C-160 **hel** 2 SA-330, 1 AS-555
GABON 680: 1 marine inf bn (4 AML-60) **ac** 2 C-160 **hel** 1 AS-555, 13 AS-532
SENEGAL 1,170: 1 marine inf bn (14 AML-60/-90) **ac** 1 *Atlantic* MR, 1 C-160 tpt **hel** 1 SA-319

UN AND PEACEKEEPING

BOSNIA (SFOR II): 3,200 (UNMIBH): 1 **CROATIA**: **SFOR Air Component** 11 *Jaguar*, 10 Mirage 2000C/D, 1 E-3F, 1 KC-135, 1 N-262 **DROC** (MONUC): 3 obs **EAST TIMOR** (UNTAET): 3 obs **EGYPT** (MFO): 17; 1 DHC-6 **GEORGIA** (UNOMIG): 3 obs **IRAQ/KUWAIT** (UNIKOM): 11 obs **ITALY** (DELIBERATE FORGE): 6 *Mirage* 2000C/D, 3 *Jaguar* **LEBANON** (UNIFIL): 251: elm 1 log bn **MIDDLE EAST** (UNTSO): 4 obs **SAUDI ARABIA** (*Southern Watch*): 170; 5 *Mirage* 2000C, 3 F-1CR, 1 C-135 **SIERRA LEONE** (UNAMSIL): 3 obs **WESTERN SAHARA** (MINURSO): 25 obs (*Gendarmerie*) **YUGOSLAVIA** (KFOR): 5,080

Paramilitary 94,950

GENDARMERIE 94,950
(incl 4,970 women, 15,650 conscripts, 1,870 civilians)
Territorial 62,050 **Mobile** 17,000 **Schools** 5,200 **Overseas** 3,200 **Maritime, Air** (personnel drawn from other dept.) **Republican Guard, Air tpt, Arsenals** 5,000 **Administration** 2,500 **Reserves** 50,000
EQPT 28 VBC-90 armd cars; 155 VBRG-170 APC; 826 **60mm, 81mm** mor; 5 PCIs (listed under Navy), plus 35 other patrol craft and 4 AT **hel** 12 SA-319, 30 AS-350

Foreign Forces

SINGAPORE AIR FORCE 200; 18 TA-4SU *Skyhawks* (Cazaux AFB)

Germany Ge

deutschmark DM		1998	1999	2000	2001
GDP	DM	3.8tr	3.7tr		
	US$	2.1tr	1.9tr		
per capita	US$	23,000	23,500		
Growth	%	2.2	1.3		
Inflation	%	1.0	0.6		
Publ debt	%	63.1	63.5		
Def exp	DM	58.1bn	59.7bn		
	US$	33.0bn	31.1bn		
Def bdgt	DM	46.7bn	48.8bn	48.3bn	47.8bn
	US$	26.5bn	25.4bn	23.3bn	23.1bn
US$1=DM		1.76	1.92	2.07	

Population			82,112,000
Age	13–17	18–22	23–32
Men	2,366,000	2,263,000	6,265,000
Women	2,231,000	2,157,000	5,883,000

Total Armed Forces

ACTIVE some 321,000
(incl 128,400 conscripts, 4,500 women)
Terms of service 10 months; 12–23 months voluntary

RESERVES 364,300
(men to age 45, officers/NCO to 60) **Army** 295,400 **Navy** 9,600 **Air Force** 59,300

Army 221,100

(incl 102,100 conscripts, 3,000 women)

ARMY FORCES COMMAND
1 air-mobile force comd (div HQ) with 2 AB (1 Crisis Reaction Force (CRF)) • 1 cdo SF bde • 1 army avn bde with 5 regt • 1 SIGINT/ELINT bde • 1 spt bde

ARMY SUPPORT COMMAND
3 log, 1 medical bde

CORPS COMMANDS
I Ge/Nl Corps 2 MDC/armd div
II Corps 2 MDC/armd div; 1 MDC/mtn div
IV Corps 1 MDC/armd inf div; 1 armd inf div; 1 MDC
Corps Units 2 spt bde and Ge elm of Ge/Nl Corps, 1 air mech bde (CRF), 1 ATGW hel regt
Military District Commands (MDC)/Divisions
6 MDC/div; 1 div; 1 MDC comd and control 9 armd bde, 7 armd inf and the Ge elm of the Ge/Fr bde, 2 armd (not active), 2 armd inf (not active), 1 inf, 1 mtn bde. Bde differ in their basic org, peacetime str, eqpt and mob capability; 4 (2 armd, 1 inf and Ge/Fr bde) are allocated to the CRF, the remainder to the Main Defence Forces (MDF). The MDC also comd and control 27 Military Region Commands (MRC). One armd div earmarked for Eurocorps, another for Allied Rapid Reaction Corps (ARRC). 7 recce bn, 7 arty regt, 7 engr bde and 7 AD regt available for cbt spt

EQUIPMENT
MBT 2,815: 1,033 *Leopard* 1A1/A3/A4/A5 (249 to be destroyed), 1,782 *Leopard* 2 (250 to be upgraded to A5)
RECCE 523: 409 SPz-2 *Luchs*, 114 TPz-1 *Fuchs* (NBC)
AIFV 2,120 *Marder* A2/A3, 133 *Wiesel* (with **20mm** gun)
APC 917 TPz-1 *Fuchs* (incl variants), 2,109 M-113 (incl variants, 320 arty obs)
TOTAL ARTY 2,115
TOWED 353: **105mm**: 18 Geb H, 143 M-101; **155mm**: 192 FH-70
SP 155mm 621: 572 M-109A3G, 49 PzH 2000
MRL 232: **110mm**: 78 LARS; **227mm**: 154 MLRS

NATO
NATO and Non-NATO Europe

MOR 909: **120mm**: 394 Brandt, 515 Tampella
ATGW 1,973: 1,606 *Milan*, 157 RJPz-(HOT) *Jaguar* 1, 210 *Wiesel* (TOW)
AD GUNS 1,525: **20mm**: 1,145 Rh 202 towed; **35mm**: 380 *Gepard* SP (147 being upgraded)
SAM 143 *Roland* SP, *Stinger*
SURV 19 *Green Archer* (mor), 110 RASIT (veh, arty), 65 RATAC (veh, arty)
HELICOPTERS 592
ATTACK 204 PAH-1 (BO-105 with HOT)
SPT 388: 145 UH-1D, 108 CH-53G, 95 BO-105M, 40 *Alouette* II
UAV CL-289 (AN/USD-502)
MARINE (River Engineers) 13 LCM

Navy 26,600

(incl 4,200 Naval Air; 5,500 conscripts, 500 women)
FLEET COMMAND Type comds Frigate, Patrol Boat, MCMV, Submarine, Naval Air **Spt comds** Naval Comms, Electronics
BASES Glücksburg (Maritime HQ), Wilhelmshaven, Kiel, Olpenitz, Eckernförde, Warnemünde

SUBMARINES 14
SSK 12 Type 206/206A SSC with *Seeaal* DM2 A3 HWT
SSC 2 Type 205

PRINCIPAL SURFACE COMBATANTS 14
DESTROYERS 2
DDG 2 *Lütjens* (mod US *Adams*) with 1 × 1 SM-1 MR SAM/*Harpoon* SSM launcher, 2 × 127mm guns, 8 ASROC (Mk 46 LWT), 6 ASTT
FRIGATES 12
FFG 12
8 *Bremen* with 8 *Harpoon* SSM, 1 × 76mm gun, 2 × 2 ASTT, 2 *Lynx* hel
4 *Brandenburg* with 4 MM-38 *Exocet* SSM, 1 VLS Mk-41 SAM, 1 × 76mm gun, 4 × 324mm TT, 2 *Lynx* hel

PATROL AND COASTAL COMBATANTS 28
MISSILE CRAFT 28
10 *Albatross* (Type 143) PFM with 4 *Exocet* SSM, and 2 533mm TT
10 *Gepard* (T-143A) PFM with 4 *Exocet* SSM
8 *Tiger* (Type 148) PFM with 4 *Exocet* SSM

MINE WARFARE
MINE COUNTERMEASURES 35
10 *Hameln* (T-343) comb ML/MCC (5 being converted to MHC, 5 to MSC)
3 converted *Lindau* (T-331) MHC
12 *Frankenthal* (T-332) MHC
4 HL 351 *Troika* MSC control and guidance, each with 3 unmanned sweep craft
5 *Frauenlob* MSI
1 MCM/T-742A diver spt ship

AMPHIBIOUS craft only
5 LCU/LCM

SUPPORT AND MISCELLANEOUS 45
UNDER WAY SUPPORT 2 *Spessart* AO
MAINTENANCE AND LOGISTIC 15
6 *Elbe* spt, 4 small (2,000t) AOT, 3 *Lüneburg* log spt, 2 AE
SPECIAL PURPOSE 24
3 AGI, 2 trials, 8 multi-purpose (T-748/745), 1 trg, 8 AT, 2 icebreakers (civil)
RESEARCH AND SURVEY 4
1 AGOR, 3 AGHS (civil-manned for Ministry of Transport)

NAVAL AIR (4,200)
Flying hours *Tornado*: 180
3 wg, 8 sqn, plus 2 sqn GBAD *Roland*
1 wg with *Tornado*, 2 sqn FGA/recce, 1 sqn trg, 1 sqn GBAD *Roland*
1 wg with MPA/SIGINT/SAR/pollution control/tpt, 2 sqn with *Atlantic* (MPA/SIGINT), Do-228 (pollution control/tpt), 1 sqn with *Lynx* (ASW/ASUW), 1 sqn trg with *Atlantic*, Do-228, *Lynx*
1 SAR/ASUW/tpt wg with 1 sqn *Sea King* Mk 41 hel

EQUIPMENT
50 cbt ac, 40 armed hel
AC 50 ***Tornado*** • 17 ***Atlantic*** (13 MR, 4 ELINT) • 2 **Do-228** (pollution control) • 2 **Do-228** (tpt)
HEL 15 ***Sea Lynx* Mk 88** (ASW/ASUW) • 21 ***Sea King* Mk 41** (SAR/tpt/ASUW), 4 *Lynx* Mk 88A (ASW/ASUW)

MISSILES
ASM *Kormoran*, *Sea Skua*, HARM
AAM AIM-9 *Sidewinder*, *Roland*

Air Force 73,300

(incl 20,800 conscripts, 1,000 women)
Flying hours 150

AIR FORCE COMMAND
2 air cmds (North and South), 4 air div
FGA 5 wg with 10 sqn *Tornado*; 1 wg operates ECR *Tornado* in SEAD role
FTR 4 wg (with F-4F (7 sqn); MiG-29 1 sqn)
RECCE 1 wg with 2 sqn *Tornado*
SAM 6 mixed wg (each 1 gp *Patriot* (6 sqn) plus 1 gp HAWK (4 sqn plus 2 reserve sqn)); 14 sqn *Roland*
RADAR 2 tac Air Control regts, 8 sites; 11 remote radar posts

TRANSPORT COMMAND (GAFTC)
TPT 3 wg, 4 sqn with *Transall* C-160, incl 1 (OCU) with C-160, 4 sqn (incl 1 OCU) with Bell UH-1D, 1 special air mission wg with Boeing 707-320C, Tu-154, Airbus A-310, VFW-614, CL-601, L-410S (VIP), 3 AS-532U2 (VIP)
TRAINING
FGA OCU with 27 *Tornado*
FTR OCU with 23 F-4F
NATO joint jet pilot trg (Sheppard AFB, TX) with 35 T-

37B, 40 T-38A; primary trg sqn with Beech *Bonanza* (Goodyear AFB, AZ), GAF Air Defence School (Fort Bliss TX)

EQUIPMENT

457 cbt ac (50 trg (overseas)) (plus 102 in store); no attack hel

AC 154 **F-4** *Phantom* II (incl 7 in store), 8 **F-104** (2 in store), 267 ***Tornado*** (189 FGA, 35* ECR, 41 Recce, 2 in store), 1 **MiG-21**, 3 **MiG-23** (2 in store) • 23 **MiG-29** (19 (ftr), 4* **-UB** (trg)) • 1 **Su-22** • 92 ***Alpha Jet*** (89 in store) • 84 ***Transall* C-160** (tpt, trg) • 7 **A-310** (VIP, tpt) • 7 **CL-601** (VIP) • 4 **L-410-S** (VIP) • 35 **T-37B** • 40 **T-38A**

HEL 99 **UH-1D** (95 SAR, tpt, liaison; 4 VIP) • 3 **AS-532U2** (VIP)

MISSILES

ASM AGM-65 *Maverick*, AGM-88A HARM

AAM AIM-9 *Sidewinder*, AA-8 *Aphid*, AA-10 *Alamo*, AA-11 *Archer*

SAM HAWK, *Roland*, *Patriot*

Forces Abroad

NAVY 1 DDG/FFG with STANAVFORLANT, 1 DDG/FFG with STANAVFORMED, 1 MCMV with MCMFORNORTH, 1 MCMV with MCMFORMED, 3 MPA in ELMAS/Sardinia

US Army trg area with 40 *Leopard* 2 MBT, 32 *Marder* AIFV, 12 M-109A3G **155mm** SP arty **Air Force** 812 flying trg at Goodyear, Sheppard, Holloman AFBs, NAS Pensacola, Fort Rucker AFBs with 35 T-37, 40 T-38, 23 F-4F; 27 *Tornado*, missile trg at Fort Bliss

UN AND PEACEKEEPING

BOSNIA (SFOR II): 2,369; 56 SPz-2 *Luchs* recce, 29 TPz-1 *Fuchs* APC, hel 4 CH-53, 4 UH-1D, UAV 1 CL-289 **GEORGIA** (UNOMIG): 10 obs **IRAQ/KUWAIT** (UNIKOM): 14 **YUGOSLAVIA** (KFOR): 5,300; 37 *Leopard* 2 MBT, 32 *Marder* AIFV, 54 TPz-1 *Fuchs* APC, 12 *Wiesel* TOW ATGW; 3 CH-53, 8 UH-1D hel

Foreign Forces

NATO HQ Allied Rapid Reaction Corps (ARRC), HQ Allied Air Forces North (AIRNORTH), HQ Joint Command Centre (JCCENT), HQ Multi-National Division (Central) (MND(C)), Airborne Early Warning Force: 17 E-3A *Sentry*, 2 Boeing-707 (trg)
BELGIUM 2,000: 1 mech inf bde(-)
FRANCE 2,700: incl elm Eurocorps
NETHERLANDS 3,000: 1 lt bde
UK 20,600: **Army** 18,300: 1 corps HQ (multinational), 1 armd div **Air Force** 2,300: 1 air base, 3 sqn with **ac** 39 *Tornado* GR1, 1 RAF regt sqn with *Rapier* SAM
US 57,580: **Army** 42,200: 1 army HQ, 1 corps HQ; 1 armd (-), 1 mech inf div (-) **Navy** 300 **USMC** 200 **Air Force** 14,880: HQ USAFE, (HQ 17th Air Force), 1 tac ftr wg with 4 sqn FGA/ftr, 1 cbt spt wg, 1 air-control wg, 1 tac airlift wg; 1 air base wg, 54 F-16C/D, 12 A-10, 6 OA-10, 16 C-130E, 9 C-9A, 9 C-21, 2 C-20, 1 CT-43

Greece Gr

drachma dr		1998	1999	2000	2001
GDP	dr	35.9tr	38.2tr		
	US$	121bn	107.4bn		
per capita	US$	13,100	13,700		
Growth	%	3.7	3.3		
Inflation	%	4.7	2.7		
Publ debt	%	105.4	104.4		
Def exp	dr	1.7tr	1.9tr		
	US$	5.8bn	5.3bn		
Def bdgt	dr	1,170bn	1,220bn	1,160bn	
	US$	4.0bn	3.4bn	3.3bn	
FMA (US)	US$	15m	0.025m	0.035m	0.025m
US$1=dr		296	319	356	

Population[a] **10,692,000** (Muslim 1%)

Age	13–17	18–22	23–32
Men	336,000	382,000	825,000
Women	317,000	362,000	789,000

[a] Excl ε350–400,000 Albanians working in Greece in 1999

Total Armed Forces

ACTIVE 159,170

(incl 98,321 conscripts, 5,520 women)

Terms of service **Army** up to 18 months **Navy** up to 21 months **Air Force** up to 21 months

RESERVES some 291,000

(to age 50) **Army** some 235,000 (Field Army 200,000, Territorial Army/National Guard 35,000) **Navy** about 24,000 **Air Force** about 32,000

Army 110,000

(incl 81,000 conscripts, 2,700 women)

FIELD ARMY

3 Mil Regions • 1 Army, 2 comd, 5 corps HQ (incl 1 RRF) • 5 div HQ (1 armd, 3 mech, 1 inf) • 5 inf div (3 inf, 1 arty regt, 1 armd bn) • 5 indep armd bde (each 2 armd, 1 mech inf, 1 SP arty bn) • 7 mech bde (2 mech, 1 armd, 1 SP arty bn) • 5 inf bde • 1 army avn bde with 5 avn bn (incl 1 ATK, 1 tpt hel) • 1 indep avn coy • 1 marine bde (1 marine, 1 tk, 1 arty bn), 1 cdo regt, 1 AB regt • 4 recce bn • 5 fd arty bn • 10 AD arty bn • 2 SAM bn with I HAWK

Units are manned at 3 different levels

Cat A 85% fully ready **Cat B** 60% ready in 24 hours **Cat C** 20% ready in 48 hours

RESERVES 34,000

National Guard internal security role

EQUIPMENT

MBT 1,735: 714 M-48 (15 A3, 699 A5), 669 M-60 (357 A1, 312 A3), 352 *Leopard* (105 -1GR, 170 -1V, 77 - 1A5)
RECCE 130 M-8, ε50 VBL
AIFV 500 BMP-1
APC 308 *Leonidas* Mk1/Mk2, 1,669 M-113A1/A2
TOTAL ARTY 1,894
TOWED 729: **105mm**: 18 M-56, 445 M-101; **155mm**: 266 M-114
SP 407: **105mm**: 73 M-52A1; **155mm**: 141 M-109A1/A2/A5, **175mm**: 12 M-107; **203mm**: 181 M-110A2
MRL 122mm: 116 RM-70; **227mm**: 18 MLRS (incl ATACMS)
MOR 107mm: 624 M-30 (incl 191 SP); plus **81mm**: 2,800
ATGW 290 *Milan*, 336 TOW (incl 320 M-901), 262 AT-4 *Spigot*
RL 64mm: 18,520 RPG-18; **66mm**: 10,700 M-72
RCL 84mm: 2000 *Carl Gustav*; **90mm**: 1,314 EM-67; **106mm**: 1,291 M-40A1
AD GUNS 23mm: 506 ZU-23-2
SAM 1,000 *Stinger*, 42 I HAWK, 21 SA-15, 20 SA-8B, SA-10 (S-300) in Crete, originally intended for Cyprus
SURV 10 AN/TPQ-36 (arty, mor), 2 AN/TPQ-37(V)3
AC 43 U-17A
HELICOPTERS
ATTACK 20 AH-64A
SPT 9 CH-47D (1 in store), 76 UH-1H, 31 AB-205A, 14 AB-206

Navy 19,000

(incl 9,800 conscripts, 1,300 women)
BASES Salamis, Patras, Soudha Bay

SUBMARINES 8

SSK 8
4 *Glavkos* (Ge T-209/1100) with 533mm TT, and *Harpoon* USGW (1 in refit)
4 *Poseidon* (Ge T-209/1200) with 533mm TT and *Harpoon* USGW

PRINCIPAL SURFACE COMBATANTS 16

DESTROYERS 4
DDG 4 *Kimon* (US *Adams*) with 6 *Harpoon* SSM, 1 × 1 SM-1 SAM, 2 × 127mm gun, 2 × 3 ASTT, 1 × 8 *ASROC* SUGW
FRIGATES 12
FFG 12
4 *Hydra* (Ge MEKO 200) with 8 *Harpoon* SSM, 1 × 127mm gun, 6 ASTT, 1 SH-60 hel
2 *Elli* (Nl *Kortenaer* Batch 2) with 8 *Harpoon* SSM, *Sea Sparrow* SAM, 1 × 76mm gun, 4 ASTT, 2 AB-212 hel
4 *Aegean* (Nl *Kortenaer* Batch 1) with 8 *Harpoon* SSM, *Sea Sparrow* SAM, 1 × 76mm gun ASTT, 2 AB-212 hel
2 *Makedonia* (ex-US *Knox*) (US lease) with *Harpoon* SSM (from ASROC launcher), 1 × 127mm gun, 4 ASTT, 8 *ASROC* SUGW

PATROL AND COASTAL COMBATANTS 42

CORVETTES 5 *Niki* (ex-Ge *Thetis*) FS with 4 ASW RL, 4 × 533mm TT
MISSILE CRAFT 19
13 *Laskos* (Fr *La Combattante* II, III, IIIB) PFM, all with 2 × 533mm TT; 8 with 4 MM-38 *Exocet* SSM, 5 with 6 *Penguin* SSM
4 *Votis* (Fr *La Combattante* IIA) PFM 2 with 4 MM-38 *Exocet* SSM, 2 with *Harpoon* SSM
2 *Stamou* with 4 SS-12 SSM
TORPEDO CRAFT 8
4 *Hesperos* (Ge *Jaguar*) PFT with 4 533mm TT
4 *Andromeda* (No *Nasty*) PFT with 4 533mm TT
PATROL CRAFT 10
OFFSHORE 4
2 *Armatolos* (Dk *Osprey*) PCO, 2 *Pirpolitis* PCO
COASTAL/INSHORE 6
2 *Tolmi* PCC, 4 PCI<

MINE WARFARE 17

MINELAYERS 2 *Aktion* (US LSM-1) (100–130 mines)
MINE COUNTERMEASURES 15
1 MHC (UK *Hunt*)
8 *Alkyon* (US MSC-294) MSC
6 *Atalanti* (US *Adjutant*) MSC, plus 4 MSR

AMPHIBIOUS 7

5 *Chios* LST with hel deck: capacity 300 tps, 4 LCVP plus veh
2 *Inouse* (US *County*) LST: capacity 400 tps, 18 tk
Plus about 57 craft: 2 LCT, 6 LCU, 11 LCM, some 31 LCVP, 7 LCA

SUPPORT AND MISCELLANEOUS 14

2 AOT, 4 AOT (small), 1 *Axios* (ex-Ge *Lüneburg*) log spt, 1 AE, 5 AGHS, 1 trg

NAVAL AIR (250)

0 cbt ac, 20 armed hel
AC 6 P-3B (MR) (Air Force operated)
HEL 2 sqn with 8 AB-212 (ASW), 2 AB-212 (EW), 2 SA-319 (ASW), 8 S-70B (ASW)

Air Force 30,170

(incl 7,521 conscripts, 1,520 women)

TACTICAL AIR FORCE

8 cbt wg, 1 tpt wg
FGA 11 sqn
2 with A-7H, 2 with A-7E, 2 with F-16CG/DG, 2 with F-4E, 1 with F-5A/B, 2 with *Mirage* F-1CG
FTR 6 sqn
2 with F-16 CG/DG, 2 with *Mirage* 2000 EG/BG, 6 surplus French Air Force 2000-5 used for conversion trg pending arrival of 10 upgraded 2000 EG, 2 with F-4E
RECCE 1 sqn with RF-4E

TPT 3 sqn with C-130H/B, YS-11, C-47, Do-28, *Gulfstream*
HEL 1 sqn with AB-205A, AB-212, Bell 47G
AD 1 bn with *Nike Hercules* SAM (36 launchers), 12 bty with *Skyguard/Sparrow* SAM, twin **35mm** guns

AIR TRAINING COMMAND

TRG 4 sqn
1 with T-41A, 1 with T-37B/C, 2 with T-2E (first of 45 T-6A in service to replace T-41 and T-37)

EQUIPMENT

458 cbt ac, no armed hel
AC 90 **A-7 H** (FGA), 4 **TA-7H** (FGA) • 87 **F-5A/B**, 10 **NF-5A**, 1 **NF-5B** • 95 **F-4E/RF-4E**, of which 39 being upgraded • 75 **F-16CG** (FGA)/**DG** (trg) • 27 *Mirage* **F-1 CG** (ftr) • 34 *Mirage* **2000** (**EG** (FGA)/ **BG*** (trg))-10 EG to be upgraded to 2000-5 from 2001 • (94 F-TF-104Gs in storage) • 4 **C-47** (tpt) • 10 **C-130H** (tpt) • 5 **C-130B** (tpt) • 10 **CL-215** (tpt, fire-fighting) • 2 **CL-415** (fire-fighting) - 8 more to follow by late 2001. 13 **Do-28** (lt tpt) • 1 *Gulfstream* **I** (VIP tpt) • 35* **T-2E** (trg) • 34 **T-37B/C** (trg) • 20 **T-41D** (trg) • 1 **YS-11-200** (tpt)
HEL 13 **AB-205A** (SAR) • 1 **AB-206** • 4 **AB-212** (VIP, tpt) • 7 **Bell 47G** (liaison)

MISSILES

ASM AGM-65 *Maverick*, AGM-88 HARM
AAM AIM-7 *Sparrow*, AIM-9 *Sidewinder* L/P, R-550 *Magic* 2, AIM 120 AMRAAM, *Super* 530D
SAM 1 bn with 36 *Nike Hercules*, 3 *Patriot* PAC-2 for training, prior to delivery of 4 PAC-3 bty from 2001, 12 bty with *Skyguard*, 40 *Sparrow*, 4 SA-15, 9 *Crotale*, **35mm** guns

Forces Abroad

CYPRUS 1,250: incl 1 mech bde and officers/NCO seconded to Greek-Cypriot forces

UN AND PEACEKEEPING

ADRIATIC (*Sharp Guard* if re-implemented): 2 MSC **BOSNIA** (SFOR II): 250 **SFOR Air Component** 1 C-130 **GEORGIA** (UNOMIG): 4 obs **IRAQ/KUWAIT** (UNIKOM): 5 obs **WESTERN SAHARA** (MINURSO): 1 obs **YUGOSLAVIA** (KFOR): 430

Paramilitary 4,000

COAST GUARD AND CUSTOMS 4,000

some 100 patrol craft, **ac** 2 Cessna *Cutlass*, 2 TB-20 *Trinidad*

Foreign Forces

NATO HQ Joint Command South-Centre (SOUTHCENT)
US 420: **Navy** 240; facilities at Soudha Bay **Air Force** 180; air base gp; facilities at Iraklion

Hungary Hu

forint f		1998	1999	2000	2001
GDP	f	10.0tr	11.4tr		
	US$	47bn	47bn		
per capita	US$	7,600	8,000		
Growth	%	5.0	3.8		
Inflation	%	14.4	10.3		
Debt	US$	28.5bn	29.2bn		
Def exp	f	142bn	182bn		
	US$	660m	745m		
Def bdgt	f	142bn	182bn	217bn	
	US$	660m	745m	791m	
FMA (US)	US$	1.5m	1.5m	1.6m	1.7m
US$1=f		214	244	275	

Population **10,005,000**
(Romany 4%, German 3%, Serb 2%, Romanian 1%, Slovak 1%)

Age	13–17	18–22	23–32
Men	320,000	388,000	712,000
Women	303,000	364,000	667,000

Total Armed Forces

ACTIVE 43,790
(incl 22,900 conscripts; 8,790 HQ staff and centrally controlled formations/units)
Terms of service 9 months

RESERVES 90,300
Army 74,900 **Air Force** 15,400 (to age 50)

Land Forces 23,500

(incl 17,800 conscripts)
Land Forces HQ • 1 Mil District HQ
2 mech div
1 with 3 mech inf, 1 engr bde, 1 arty, 1 ATK, 1 recce bn
1 with 2 trg centres
Corps tps
1 army maritime wing, 1 counter mine bn
MoD tps (Budapest): 1 MP regt

RESERVES

4 mech inf bde

EQUIPMENT

MBT 806: 568 T-55 (209 in store), 238 T-72
RECCE 104 FUG D-442
AIFV 490 BMP-1, 12 BRM-1K, 70 BTR-80A
APC 420 BTR-80, 336 PSZH D-944 (83 in store), 30 MT-LB (plus some 369 'look-alike' types)
TOTAL ARTY 839
TOWED 532: **122mm**: 230 M-1938 (M-30) (24 in store); **152mm**: 302 D-20 (218 in store)
SP 122mm: 151 2S1 (18 in store)
MRL 122mm: 56 BM-21
MOR 120mm: 100 M-120 (2 in store)

ATGW 369: 115 AT-3 *Sagger*, 30 AT-4 *Spigot* (incl BRDM-2 SP), 224 AT-5 *Spandrel*
ATK GUNS 85mm: 162 D-44 (all in store); **100mm**: 106 MT-12
AD GUNS 57mm: 186 S-60 (43 in store)
SAM 243 SA-7, 60 SA-14, 45 *Mistral*
SURV PSZNR-5B, SZNAR-10

Army Maritime Wing (290)

BASE Budapest

RIVER CRAFT 50

6 *Nestin* MSI (riverine), some 44 An-2 mine warfare/patrol boats

Air Force 11,500

(incl 5,100 conscripts)

AIR DEFENCE COMMAND

68 cbt ac, 24 attack hel
Flying hours 50
FGA 2 tac ftr wgs, 1 with 22 MiG-21bis/MF/UM (all to be withdrawn by 2001), 1 with 27 MiG-29A/UB (only 12 available)
ATTACK HEL 1 wg with 24 Mi-24 (plus 15 in store), 38 Mi-8/17 (tpt/assault), 1 Mi-9 (Cmd Post), 2 Mi-17PP (EW)
TPT 1 mixed tpt wg, 1 mixed tpt sqn, ac 9 An-26, 4 Z-43, hel 16 Mi-2, some Mi-8
TRG 19 L-39*, 12 Yak-52
AAM AA-2 *Atoll*, AA-8 *Aphid*, AA-10 *Alamo*, AA-11 *Archer*
ASM AT-2 *Swatter*, AT-6 *Spiral*
SAM 2 regt with 66 SA-2/-3/-5, 12 SA-4, 20 SA-6

Forces Abroad

UN AND PEACEKEEPING

BOSNIA (SFOR II): 4 obs **CROATIA** (SFOR II): 310; 1 engr bn **CYPRUS** (UNFICYP): 111 **EGYPT** (MFO): 41 mil pol **GEORGIA** (UNOMIG): 7 obs **IRAQ/KUWAIT** (UNIKOM): 5 obs **WESTERN SAHARA** (MINURSO): 3 obs **YUGOSLAVIA** (KFOR): 325

Paramilitary 14,000

BORDER GUARDS (Ministry of Interior) 12,000 (to reduce)

11 districts/regts plus 1 Budapest district (incl 7 rapid-reaction coy; 33 PSZH, 68 BTR-80 APC)

INTERNAL SECURITY FORCES (Police) 2,000

33 PSZH, 40 BTR-80APC

Iceland — Icl

kronur K		1998	1999	2000	2001
GDP	K	586bn	638bn		
	US$	8.3bn	8.5bn		
per capita	US$	23,900	25,500		
Growth	%	5.0	6.0		
Inflation	%	1.7	3.2		
Publ debt	%	47.7	43.6		
Sy exp[a]	K	1.3bn			
	US$	18m			
Sy bdgt[a]	K		1.4bn	1.4bn	
	US$		19m	19m	
US$1=K		71.0	75.5	75.3	

[a] Iceland has no Armed Forces. Sy bdgt is mainly for Coast Guard

Population			**283,000**
Age	13–17	18–22	23–32
Men	11,000	11,000	22,000
Women	10,000	10,000	20,000

Total Armed Forces

ACTIVE Nil

Paramilitary 120

COAST GUARD 120

BASE Reykjavik
PATROL CRAFT 4
2 *Aegir* PCO with hel, 1 *Odinn* PCO with hel deck, 1 PCI<
AVN ac 1 F-27, **hel** 1 SA-365N, 1 SA-332, 1 AS-350B

Foreign Forces

NATO Island Commander Iceland (ISCOMICE, responsible to CINCEASTLANT)
US 1,640: **Navy** 960; MR: 1 sqn with 4 P-3C **Marines** 80 **Air Force** 600; 4 F-15C/D, 1 HC-130, 1 KC-135, 4 HH-60G
NETHERLANDS 16: **Navy** 1 P-3C

Italy — It

lira L		1998	1999	2000	2001
GDP	L	2,017tr	2,125tr		
	US$	1.2tr	1.1tr		
per capita	US$	21,500	22,000		
Growth	%	1.4	1.0		
Inflation	%	2.0	1.7		
Publ debt	%	117.7	116.6		
Def exp	L	40.1tr	41.8tr		
	US$	23.1bn	22.0bn		
Def bdgt	L	31.0tr	30.9tr	32.8tr	
	US$	17.8bn	16.2bn	16.0bn	
US$1=L		1,736	1,900	2,050	

Population			57,930,000
Age	13–17	18–22	23–32
Men	1,578,000	1,751,000	4,456,000
Women	1,451,000	1,670,000	4,305,000

Total Armed Forces

ACTIVE 250,600
(incl 111,800 conscripts)
Terms of service all services 10 months

RESERVES 65,200 (immediate mobilisation)
Army 11,900 (500,000 obligation to age 45) **Navy** 23,000 (to age 39 for men, variable for officers to 73) **Air Force** 30,300 (to age 25 or 45 (specialists))

Army 153,000

(incl 83,000 conscripts)
1 Op Comd HQ, 3 mil region HQ
1 Projection Force with 1 mech, 1 airmobile, 1 AB bde, 1 amph, 1 engr regt
1 mtn force with 3 mtn bde, 1 engr, 1 avn regt, 1 alpine AB bn
2 div defence force
1 with 1 armd, 1 mech, 1 armd cav bde, 1 engr regt
1 with 3 mech, 1 armd bde, 1 engr, 1 avn regt
1 spt comd with
1 AD div: 3 HAWK SAM, 2 AAA regt
1 arty bde: 1 hy arty, 3 arty, 1 NBC regt
1 avn div: 2 avn regt, 2 avn bn

EQUIPMENT
MBT 699 (CFE: 1,301): 133 *Leopard* 1A1, 98 *Leopard* 1A5, 363 *Centauro* B-1, 105 *Ariete*
AIFV 20 *Dardo*
APC 882 M-113 (incl variants), 1,638 VCC1/-2, 26 Fiat 6614, 101 *Puma*
AAV 15 LVTP-7
TOTAL ARTY 895 (CFE: 1,390)
TOWED 222: **105mm**: 72 Model 56 pack; **155mm**: 96 FH-70, 54 M-114 (in store)
SP 155mm: 192 M-109G/-L
MRL 227mm: 22 MLRS
MOR 120mm: 459; **81mm**: 1,200 (386 in store)
ATGW 426 TOW 2B, 432 I-TOW, 752 *Milan*
RL 1,860 *Panzerfaust* 3
RCL 80mm: 434 *Folgore*
AD GUNS 25mm: 208 SIDAM SP
SAM 60 HAWK, 112 *Stinger*, 32 *Skyguard*/*Aspide*
AC 6 SM-1019, 3 Do-228, 3 P-180
HELICOPTERS
ATTACK 45 A-129
ASLT 27 A-109, 62 AB-206
SPT 86 AB-205A, 68 AB-206 (obs), 14 AB-212, 23 AB-412, 36 CH-47C
UAV 5 *Mirach* 20

Navy 38,000

(incl 2,500 Naval Air, 1,000 Marines and 11,000 conscripts)
COMMANDS 1 Fleet Commander CINCNAV (also NATO COMEDCENT) **Area Commands** 4 Upper Tyrrhenian, Ionian and Strait of Otranto, Sicily, Sardinia
BASES La Spezia (HQ), Taranto (HQ), Brindisi, Augusta
SUBMARINES 7
SSK 7
4 *Pelosi* (imp *Sauro*) with Type 184 HWT
3 *Sauro* with Type 184 HWT
PRINCIPAL SURFACE COMBATANTS 30
CARRIERS 1 *G. Garibaldi* CV with total ac capacity 16 AV-8B *Harrier* V/STOL or 18 SH-3 *Sea King* hel; usually a mix
CRUISERS 1 *Vittorio Veneto* CG with 4 *Teseo* SSM, 1 × 2 SM-1 ER SAM, 8 × 76mm gun, 2 × 3 ASTT, 6 AB-212 ASW hel (Mk 46 LWT)
DESTROYERS 4
DDG 4
2 *Luigi Durand de la Penne* (ex-*Animoso*) with 8 *Teseo* SSM, 1 SM-1 MR SAM, 1 × 127mm gun, 6 ASTT, 2 AB-312 hel
2 *Audace* with 4 *Teseo* SSM, 1 SM-1 MR SAM, 1 × 127mm gun, 6 ASTT, 2 AB-212 hel
FRIGATES 24
FFG 16
8 *Maestrale* with 4 *Teseo* SSM, *Aspide* SAM, 1 × 127mm gun, 2 × 533mm TT, 2 AB-212 hel
4 *Lupo* with 8 *Teseo* SSM, *Sea Sparrow* SAM, 1 × 127mm gun, 2 × 3 ASTT, 1 AB-212 hel
4 *Artigliere* with 8 *Teseo* SSM, 8 *Aspide* SAM, 1 × 127mm gun, 1 AB-212 hel
FF 8 *Minerva* with *Aspide* SAM, 1 × 76mm gun, 6 × ASTT
PATROL AND COASTAL COMBATANTS 9
PATROL, OFFSHORE 5
4 *Cassiopea* PCO with 1 × 76mm gun, 1 AB-212 hel
1 *Storione* (US *Aggressive*) PCO
PATROL, COASTAL 4
4 *Esplatore* PCC
MINE COUNTERMEASURES 13
1 MCCS (ex *Alpino*)
4 *Lerici* MHC/MSC
8 *Gaeta* MHC/MSC
plus 2 MCD
AMPHIBIOUS 3
2 *San Giorgio* LPD: capacity 350 tps, 30 trucks, 2 SH-3D or CH-47 hel, 7 craft
1 *San Giusto* LPD: capacity as above
Plus some 33 craft: about 3 LCU, 10 LCM and 20 LCVP

SUPPORT AND MISCELLANEOUS 32

2 *Stromboli* AO, 1 *Etna* AO; 7 AWT, 2 AR; 2 ARS, 7 sail trg, 8 AT (plus 44 coastal AT); 3 AGOR

SPECIAL FORCES (Special Forces Command – COMSUBIN)

3 gp; 1 underwater ops; 1 school; 1 research

MARINES (San Marco gp) (1,000)

1 bn gp, 1 trg gp, 1 log gp

EQUIPMENT

30 VCC-1 APC, 10 LVTP-7 AAV, 16 **81mm** mor, 8 **106mm** RCL, 6 *Milan* ATGW

NAVAL AIR (2,500)

18 cbt ac, 80 armed hel
FGA 1 sqn with 16 AV-8B plus and 2*TAV-8B plus
ASW 5 hel sqn with 21 SH-3D, 45 AB-212
HEL 8* SH-3D (amph aslt), 6* AB-212 (amph aslt)
AAM AIM-9L *Sidewinder*
ASM *Marte* Mk 2, AS-12
AGM 65 *Maverik*

Air Force 59,600

(incl 17,800 conscripts)
AFHO 3 Inspectorates (Naval Aviation, Flight Safety, Operational), 1 Op Cmd (responsible for 5 op bde), 1 Force Cmd, 1 Logs Cmd, 1 Trg Cmd
FGA 8 sqn
4 with *Tornado* IDS (50% of 1 sqn devoted to recce) • 4 with AMX (50% of 1 sqn devoted to recce)
FTR 5 sqn
3 with F-104 ASA • 2 with *Tornado* ADV
RECCE 1 sqn with AMX
MR 2 sqn with *Atlantic* (OPCON to Navy)
EW 1 ECM/recce sqn with G-222VS, PD-808, P-180, P-166DL-3
TPT/TAC 3 sqn
2 with G-222 • 1 with C-130H/C-130J
TKR/CAL 1 sqn with B707-320, G-222 RM
TPT/VIP 2 sqn with **ac** *Gulfstream* III, *Falcon* 50, *Falcon* 900, DC-9 **hel** SH-3D
TRG 7 sqn with **ac** TF-104G, *Tornado*, AMX-T, PD-808, MB-339A (incl aerobatic team), SF-260M **hel** 2 sqn with NH-500, AB-212, HH-3F
SAR 4 sqn with HH-3F, 5 det with AB-212
AD 15 bty: 3 HSAM bty with *Nike Hercules*, 12 SAM bty with *Spada*

EQUIPMENT

336 cbt ac, no armed hel
AC 116 ***Tornado*** (76 IDS, 24 ADV, 16 ECR) • 91 **F-104** (70 ASA, 21 TF-104G) • 104 **AMX** (60 (FGA), 20 **-T** (trg)) (plus 22 FGA and 2 T in store) • 11* **MB-339CD** • 14* ***Atlantic*** (MR) • 4 **Boeing-707-320** (tkr/tpt) • 15 **C-130H/J** (tpt/tac) • 38 **G-222** (tpt/tac/calibration) • 2 **DC9-32** (VIP), being replaced by 2 Airbus A319CJ • 2 ***Gulfstream*** **III** (VIP) • 4 ***Falcon*** **50** (VIP), 2 ***Falcon*** **900** (VIP) • 7 **P-166** (2 **-M**, 5 **-DL3** (liaison/trg)) • 18 **P-180** (liaison) • 1 **PD-808** (ECM, cal, tpt) • 26 **SF-260M** (trg) • 75 MB-339 (62 trg, 13 aerobatic team) • 30 **SIAI-208** (liaison)
HEL 21 **HH-3F** (SAR/CSAR) • 2 **SH-3D** (liaison/VIP) • 27 **AB-212** (SAR) • 51 **NH-500D** (trg)
UAV 6 *Predator* being delivered

MISSILES

ASM AGM-88 HARM, *Kormoran*
AAM AIM-9L *Sidewinder*, *Aspide*
SAM *Nike Hercules*, *Aspide*

Forces Abroad

GERMANY 92: **Air Force, NAEW Force**
MALTA 16: **Air Force** with 2 AB-212
US 33: **Air Force** flying trg

UN AND PEACEKEEPING

BOSNIA (SFOR II): 1,640: 1 mech inf bde gp **EGYPT** (MFO): 77 **INDIA/PAKISTAN** (UNMOGIP): 7 obs **IRAQ/KUWAIT** (UNIKOM): 6 obs **LEBANON** (UNIFIL): 59; hel unit **MIDDLE EAST** (UNTSO): 9 obs **WESTERN SAHARA** (MINURSO): 5 obs **YUGOSLAVIA** (KFOR): 6,400

Paramilitary 252,500

CARABINIERI 110,000 (Ministry of Interior)

Territorial 5 bde, 18 regt, 94 gp **Trg** 1 bde **Mobile def** 1 div, 2 bde, 1 cav regt, 1 special ops gp, 13 mobile bn, 1 AB bn, avn and naval units
EQPT 40 Fiat 6616 armd cars; 40 VCC2, 91 M-113 APC **hel** 24 A-109, 4 AB-205, 39 AB-206, 24 AB-412

PUBLIC SECURITY GUARD 79,000 (Ministry of Interior)

11 mobile units; 40 Fiat 6614 APC **ac** 5 P-68 **hel** 12 A-109, 20 AB-206, 9 AB-212

FINANCE GUARDS 63,500 (Treasury Department)

14 Zones, 20 Legions, 128 gp **ac** 5 P-166-DL3 **hel** 15 A-109, 65 Breda-Nardi NH-500M/MC/MD; 3 PCI; plus about 300 boats

HARBOUR CONTROL (*Capitanerie di Porto*)

(subordinated to Navy in emergencies): 12 PCI, 130+ boats and 4 AB-412 (SAR)

Foreign Forces

NATO HQ Allied Forces South Europe, HQ Allied Air Forces South (AIRSOUTH), HQ Allied Naval Forces South (NAVSOUTH), HQ Joint Command South (JCSOUTH), HQ 5 Allied Tactical Air Force (5 ATAF)
US 10,300: **Army** 1,700; 1 inf bn gp **Navy** 4,400 **Air Force** 4,200 **USMC** 200
DELIBERATE FORGE COMPONENTS Be 4 F-16A **Ca** 6 CF-18 **Da** 3 F-16A **Fr** 6 *Mirage* 2000C/D, 3 *Jaguar* **Nl** 4 F-16A **Sp** 5 EF-18, 1 KC-130 **Tu** 4 F-16C **UK** 4 *Harrier* GR-

7, 1 *Nimrod*, 1 K-1 *Tristar*, 2 E-3D *Sentry* **US** 32 F-16C/D, 1 AC-130, 1 KC-135, 6 UH-60, 2 U-2, 10 P-3C, 5 C-12, 2 C-21
SUPPORT COMPONENTS (for NATO ops in Kosovo) **Sp** 1 CASA 212, **US** 4 C-12, 1 LJ-35, 1 BE-20, 4 C-130, 3 KC-135, 4 H-53, 2 H-3, 1 C-5, 3 P-3, 1 C-9, 2 C-2

Luxembourg Lu

franc fr		1998	1999	2000	2001
GDP	fr	665bn	708bn		
	US$	18.3bn	17.9bn		
per capita	US$	27,300	29,000		
Growth	%	5.7	5.1		
Inflation	%	1.0	1.0		
Publ debt	%	6.5	6.5		
Def exp	fr	5.1bn	5.5bn		
	US$	142m	138m		
Def bdgt	fr	3.9bn	4.0bn	4.3bn	
	US$	107m	102m	100m	
US$1=fr		36.3	39.6	42.7	

Population	**420,000** (ε124,000 foreign citizens)		
Age	13–17	18–22	23–32
Men	12,000	12,000	27,000
Women	12,000	12,200	28,000

Total Armed Forces

ACTIVE 899

Army 899

1 lt inf bn, 2 recce coy, 1 to Eurocorps/BE div, 1 to AMF(L)

EQUIPMENT

MOR 81mm: 6
ATGW 6 TOW
RL LAW

Air Force

(none, but for legal purposes NATO's E-3A AEW ac have Lu registration)
1 sqn with 17 E-3A *Sentry* (NATO standard), 2 Boeing 707 (trg)

Forces Abroad

UN AND PEACEKEEPING

BOSNIA (SFOR II): 23 **Deliberate Forge Air Component** 5 E-3A **YUGOSLAVIA** (KFOR): some

Paramilitary 612

GENDARMERIE 612

Netherlands Nl

guilder gld		1998	1999	2000	2001
GDP	gld	776bn	813bn		
	US$	391bn	375bn		
per capita	US$	22,800	23,800		
Growth	%	3.8	3.0		
Inflation	%	1.9	2.2		
Publ debt	%	67	63.7		
Def exp	gld	13.4bn	15.0bn		
	US$	6.8bn	6.9bn		
Def bdgt	gld	14.1bn	14.1bn	14.2bn	
	US$	7.1bn	6.5bn	6.2bn	
US$1=gld		1.98	2.16	2.30	

Population			**15,794,000**
Age	13–17	18–22	23–32
Men	447,000	451,000	1,141,000
Women	428,000	431,000	1,081,000

Total Armed Forces

ACTIVE 51,940

(incl 5,200 Royal Military Constabulary, 4,155 women; excl 19,600 civilians)

RESERVES 32,200

(men to age 35, NCOs to 40, officers to 45) **Army** 22,200 **Navy** some 5,000 **Air Force** 5,000 (immediate recall)

Army 23,100

(incl 1,630 women)
1 Corps HQ (Ge/Nl), 1 mech div HQ • 3 mech inf bde (2 cadre) • 1 lt bde • 1 air-mobile bde (3 inf bn) • 1 fd arty gp, 1 AD bn • 1 engr gp
Summary of cbt arm units
6 tk bn • 6 armd inf bn • 3 air-mobile bn • 3 recce bn • 6 arty bn • 1 AD bn • 1 SF bn • 2 MLRS bty

RESERVES

(cadre bde and corps tps completed by call-up of reservists)
National Command (incl Territorial Comd): 6 inf bn, could be mob for territorial defence
Home Guard 3 sectors; lt inf weapons

EQUIPMENT

MBT 330 *Leopard* 2 (180 to be A5; 136 for sale)
AIFV 383 YPR-765, 65 M-113C/-R all with **25mm**
APC 269 YPR-765 (plus 491 look-a-likes), 70 XA-188 *Sisu*
TOTAL ARTY 397
TOWED 155mm: 20 M-114, 80 M-114/39, 15 FH-70 (trg); **203mm**: 1 M-110
SP 155mm: 126 M-109A3
MRL 227mm: 22 MLRS
MOR 81mm: 40; **120mm**: 133
ATGW 753 (incl 135 in store): 427 *Dragon*, 326 (incl 90

YPR-765) TOW
RL 84mm: *Carl Gustav*, AT-4
AD GUNS 35mm: 77 *Gepard* SP (60 to be upgraded); **40mm**: 60 L/70 towed
SAM 312 *Stinger*
SURV AN/TPQ-36 (arty, mor)
UAV *Sperwer* (reported)
MARINE 1 tk tpt, 3 coastal, 3 river patrol boats

Navy 12,340

(incl 950 Naval Air, 3,100 Marines, 1,150 women)
BASES Netherlands Den Helder (HQ). Nl and Be Navies under joint op comd based Den Helder. Valkenburg (MPA) De Kooy (hel) **Overseas** Willemstad (Curaçao)

SUBMARINES 4
SSK 4 *Walrus* with Mk 48 HWT; plus provision for *Harpoon* USGW

PRINCIPAL SURFACE COMBATANTS 15
DESTROYERS 3
DDG (Nl desig = FFG) 3
1 *Tromp* with 8 *Harpoon* SSM, 1 SM-1 MR SAM, 2 × 120mm gun, 6 ASTT (Mk 46 LWT), 1 *Lynx* hel
2 *Van Heemskerck* with 8 *Harpoon* SSM, 1 SM-1 MR SAM, 4 ASTT
FRIGATES 12
FFG 12
8 *Karel Doorman* with 8 *Harpoon* SSM, *Sea Sparrow* SAM, 1 × 76mm gun, 4 ASTT, 1 *Lynx* hel
4 *Kortenaer* with 8 × *Harpoon* SSM, 8 × *Sea Sparrow* SAM, 1 × 76mm gun, 4 ASTT, 2 *Lynx* hel

MINE WARFARE 14
MINELAYERS none, but *Mercuur*, listed under spt and misc, has capability
MINE COUNTERMEASURES 14
14 *Alkmaar* (tripartite) MHC
plus 4 diving vessels

AMPHIBIOUS 1
1 *Rotterdam* LPD: capacity 4 LCU or 6 LCA, 600 troops, 6 *Lynx* hel or 4 NH-90
plus craft: 5 LCU, 6 LCA

SUPPORT AND MISCELLANEOUS 8
1 *Amsterdam* AO (4 *Lynx* or 2 NH-90), 1 *Zuideruis* AO (2 *Lynx* or 2 NH-90), 1 *Pelikaan* spt; 1 *Mercuur* torpedo tender, 2 trg; 1 AGOR, 1 AGHS

NAVAL AIR (950)
MR/ASW 2 sqn with P-3C
ASW/SAR 2 sqn with *Lynx* hel
EQPT 13 cbt ac, 21 armed hel
AC 13 **P-3C** (MR)
HEL 21 ***Lynx*** (ASW, SAR)

MARINES (3,100)
3 Marine bn (1 cadre); 1 spt bn
(1 bn integrated with UK 3rd Cdo Bde to form UK/NL Amph Landing Force)
EQUIPMENT
APC 20 XA-188 *Sisu*
TOWED ARTY 105mm: 8 lt
MOR 81mm: 18; **120mm**: 14
ATGW *Dragon*
RL AT-4
SAM *Stinger*

Air Force 11,300

(incl 975 women)
Flying hours 180
FTR/FGA 5 sqn: 3 F-16AM (MLU), 2 F-16A/B
OCU 1 F-16 conversion unit
FTR/RECCE 1 sqn with F-16A(R). Disbanding early 2001; recce ac to be divided among other sqns
TPT 1 sqn with F-50, F-60, C-130H-30, KDC-10 (tkr/tpt), *Gulfstream* IV
TRG 1 sqn with PC-7
HEL
1 sqn with AH-64A
1 sqn with AH-64D (of which 8 in US for trg purposes)
1 sqn with BO-105
1 sqn with AS-532U2, SA-316
1 sqn with CH-47D
SAR 1 sqn with AB-412 SP
AD 4 sqns, each with 2 HAWK SAM bty; 1 *Patriot* SAM bty; 7 *Stinger*

EQUIPMENT
157 cbt ac, 42 armed hel
AC 157 **F-16A/B**: 92 F-16A, 21 F-16A(R) and 25 F-16B being converted under European mid-life update programme • 2 **F-50** • 4 **F-60** • 2 **C-130H-30** • 2 **KDC-10** (tkr/tpt) • 1 *Gulfstream* IV • 10 **PC-7** (trg)
HEL 3 **AB-412 SP** (SAR) • 4 **SA-316** • 15 **BO-105** • 12* **AH-64A** (leased from US Army pending delivery of AH-64D) • 30* **AH-64D** (deliveries to be completed by end 2002) • 13 **CH-47D** • 17 **AS-532U2**

MISSILES
AAM AIM-9/L/N *Sidewinder*, AIM-120B AMRAAM
ASM AGM-65G *Maverick*, AGM-114K *Hellfire*
SAM 48 HAWK, 5 *Patriot*, 100 *Stinger*
AD GUNS 25 VL 4/41 *Flycatcher* radar, 75 L/70 **40mm** systems

Forces Abroad

GERMANY 3,000: 1 lt bde (1 armd inf, 1 tk bn), plus spt elms
ICELAND 16: **Navy** 1 P-3C
NETHERLANDS ANTILLES Netherlands, Aruba and the Netherlands Antilles operate a Coast Guard Force to combat org crime and drug smuggling. Comd by Netherlands Commander Caribbean. HQ Curaçao,

bases Aruba and St Maarten **Navy** 20 (to expand); 1 FFG, 1 amph cbt det, 3 P-3C, 1 Marine bn

UN AND PEACEKEEPING

BOSNIA (SFOR II): 1,267; 1 mech inf bn gp (UNMIBH): 1 **CYPRUS** (UNFICYP): 99 **ETHIOPIA/ ERITREA** (UNMEE): 1 obs **ITALY**: 155 (DELIBERATE FORGE) 4 F-16 **MIDDLE EAST** (UNTSO): 10 obs **YUGOSLAVIA** (KFOR): 1,450

Paramilitary 5,200

ROYAL MILITARY CONSTABULARY (*Koninklijke Marechaussee*) 5,200 (incl 400 women)
6 districts with 60 'bde'

Foreign Forces

NATO HQ Allied Forces North Europe
US 370: **Army** 60 **Air Force** 290 **Navy** 10 **USMC** 10

Norway No

kroner kr		1998	1999	2000	2001
GDP	kr	1,101bn	1,192bn		
	US$	146bn	150bn		
per capita	US$	25,000	25,500		
Growth	%	2.0	0.6		
Inflation	%	2.2	2.4		
Publ debt	%	33.7	34.6		
Def exp	kr	25.1bn	26.1bn		
	US$	3.3bn	3.3bn		
Def bdgt	kr	25.3bn	26.3bn	25.3bn	25.8bn
	US$	3.4bn	3.3bn	2.9bn	2.9bn
US$1=kr		7.55	7.96	8.81	

Population			**4,443,000**
Age	13–17	18–22	23–32
Men	138,000	138,000	331,000
Women	131,000	130,000	312,000

Total Armed Forces

ACTIVE 26,700
(incl 15,200 conscripts; 400 Joint Services org, 500 Home Guard permanent staff)
Terms of service **Army, Navy, Air Force**, 12 months, plus 4–5 refresher trg periods

RESERVES

222,000 mobilisable in 24–72 hours; obligation to 44 (conscripts remain with fd army units to age 35, officers to age 55, regulars to age 60)
Army 89,000 **Navy** 25,000 **Air Force** 25,000 **Home Guard** some 83,000 on mob

Army 14,700

(incl 8,700 conscripts)
2 Comd, 4 district comd, 14 territorial regt
North Norway 1 inf/ranger bn, border gd, cadre and trg units for 1 div (1 armd, 2 inf bde) and 1 indep mech inf bde
South Norway 2 inf bn (incl Royal Guard), indep units plus cadre units for 1 mech inf and 1 armd bde

RESERVES

17 inf, 3 ranger plus some indep coy and spt units. 1 arty bn; 19 inf coy, engr coy, sigs units

LAND HOME GUARD 77,000

18 districts each divided into 2–6 sub-districts and some 465 sub-units (pl)

EQUIPMENT

MBT 170 *Leopard* (111 -1A5NO, 59 -1A1NO)
AIFV 53 NM-135 (M-113/**20mm**), ε104 CV 9030N
APC 109 M-113 (incl variants), 48 XA-186 *Sisu*
TOTAL ARTY 184
TOWED 155mm: 46 M-114
SP 155mm: 126 M-109A3GN
MRL 227mm: 12 MLRS
MOR 81mm: 450 (40 SP incl 24 M-106A1, 12 M-125A2)
ATGW 320 TOW-1/-2 incl 97 NM-142 (M-901), 424 *Eryx*
RCL 84mm: 2,517 *Carl Gustav*
AD GUNS 20mm: 252 Rh-202 (192 in store)
SAM 300 RBS-70 (120 in store)
SURV *Cymberline* (mor)

Navy 6,100

(incl 160 Coastal Defence, 270 Coast Guard, 3,300 conscripts)
OPERATIONAL COMMANDS: 2 JOINT OPERATIONAL COMMANDS, COMNAVSONOR and COMNAVNON with regional naval commanders and 7 regional Naval districts
BASES Horten, Haakonsvern (Bergen), Olavsvern (Tromsø)

SUBMARINES 10

SSK 6 *Ula* with DM 2 A3 HWT
SSC 4 *Kobben* with TP 613 HWT plus Mk 37 LWT

FRIGATES 4

FFG 4 *Oslo* with 4 *Penguin 1* SSM, *Sea Sparrow* SAM, 1 × twin 76mm gun, 6 *Terne* ASW RL, *Stingray* LWT (1 in reserve)

PATROL AND COASTAL COMBATANTS 15

MISSILE CRAFT 15
14 *Hauk* PFM with 6 × *Penguin* 2 SSM, 2 × *Mistral* SAM, 2 (Swe TP-613) HWT
1 *Skjold* PFM fitted for 8 *Kongsberg* SSM

MINE WARFARE 12

MINELAYERS 3

2 *Vidar*, coastal (300–400 mines), 1 *Tyr* (amph craft also fitted for minelaying)

MINE COUNTERMEASURES 9

4 *Oskøy* MHC, 5 *Alta* MSC, plus 2 diver spt

AMPHIBIOUS craft only

5 LCT, 22 S90N LCA

SUPPORT AND MISCELLANEOUS 6

1 *Horten* sub/patrol craft depot ship; 1 *Valkyrien* TRV, 1 Royal Yacht, 2 *Hessa* trg, 1 *Mariata* AGI

NAVAL HOME GUARD 4,900

on mob assigned to 10 HQ sectors incl 31 areas; 235 vessels plus 77 boats

COASTAL DEFENCE

FORTRESS 6 **75mm**: 3; **120mm**: 3; **127mm**: 6; **150mm**: 2 guns; 3 cable mine and 3 torpedo bty

COAST GUARD (270)

PATROL AND COASTAL COMBATANTS 18

PATROL, OFFSHORE 11

3 *Nordkapp* with 1 *Lynx* hel (SAR/recce), fitted for 6 *Penguin* Mk 2 SSM, 1 *Nornen*, 7 chartered (partly civ manned)

PATROL INSHSORE 7 PCI plus 7 cutters

AVN ac 2 P-3N *Orion* **hel** 6 *Lynx* Mk 86 (Air Force-manned)

Air Force 5,000

(incl 3,200 conscripts, 185 women)

Flying hours 180

OPERATIONAL COMMANDS 2 joint with COMSONOR and COMNON

FGA 4 sqn with F-16A/B

FTR 1 trg sqn with F-5A/B

MR 1 sqn with 4 P-3C *Orion* and 2 P-3N *Orion* (assigned to Coast Guard)

TPT 2 sqn: 1 with C-130, 1 with DHC-6

CAL/ECM 1 sqn with 2 *Falcon* 20C (EW) and 1 *Falcon* 20C (Flight Inspection Service)

TRG MFI-15

SAR 1 sqn with *Sea King* Mk 43B

TAC HEL 2 sqn with Bell-412SP

SAM 6 bty NASAMS, 10 bty RB-70

AAA 8 bty L70 (with Fire-Control System 2000) org into 5 gps

EQUIPMENT

79 cbt ac (incl 4 MR), no armed hel

AC 15 **F-5A/B** (ftr/trg) (to be phased out in near future) • 58 **F-16A/B** • 6* **P-3** (4 **-D** (MR), 2 **-N** (Coast Guard)) • 6 **C-130H** (tpt) • 3 ***Falcon* 20C** (EW/FIS) • 3 **DHC-6** (tpt) • 15 **MFI-15** (trg)

HEL 18 **Bell 412 SP** (tpt) • 12 ***Sea King* Mk 43B** (SAR) • 6 ***Lynx* Mk 86** (Coast Guard)

MISSILES

ASM CRV-7, *Penguin* Mk-3

AAM AIM-9L/N *Sidewinder*, AIM 120 AMRAAM

AA HOME GUARD

(on mob under comd of Air Force): 2,500; 2 bn (9 bty) AA **20mm** NM45

Forces Abroad

UN AND PEACEKEEPING

BOSNIA (SFOR II): 125 **CROATIA** (UNMOP): 1 obs **EAST TIMOR** (UNTAET): 6 **EGYPT** (MFO): 5 Staff Officers **MIDDLE EAST** (UNTSO): 11 obs **SIERRA LEONE** (UNAMSIL): 5 obs **YUGOSLAVIA** (KFOR): 1,200 (incl 200 in FYROM)

Foreign Forces

Prepositioned eqpt for **Marines**: 1 MEB **Army**: 1 arty bn **Air Force**: ground handling eqpt

Ge prepositioned eqpt for 1 arty bn

NATO HQ Joint Command North Europe (JC North)

Poland Pl

zloty z		1998	1999	2000	2001
GDP	z	551bn	614bn		
	US$	159bn	157bn		
per capita	US$	7,500	7,400		
Growth	%	4.8	4.0		
Inflation	%	11.7	7.3		
Debt	US$	47bn	49bn		
Def exp	z	11.9bn	12.7bn		
	US$	3.4bn	3.2bn		
Def bdgt	z	11.7bn	12.6bn	13.7bn	14.8bn
	US$	3.4bn	3.2bn	3.2bn	3.4bn
FMA (US)	US$	1.6m	1.6m	1.6m	1.7m
US$1=z		3.48	3.91	4.32	

Population			**38,648,000**
(German 1.3%, Ukrainian 0.6%, Belarussian 0.5%)			
Age	13–17	18–22	23–32
Men	1,647,000	1,658,000	2,764,000
Women	1,564,000	1,581,000	2,646,000

Total Armed Forces

ACTIVE 217,290

(incl 111,950 conscripts; 21,480 centrally controlled staffs, units/formations)

Terms of service 12 months

RESERVES 406,000

Army 343,000 **Navy** 14,000 (to age 50) **Air Force** 49,000 (to age 60)

Army 132,750

(incl 82,750 conscripts)
To reorg:
2 Mil Districts/Army HQ
1 Multi-national Corps HQ (Pl/Ge/Da)
1 Air-Mechanised Corps HQ
6 mech div (incl 1 coastal)
1 armd cav div
1 air cav div HQ
5 bde (incl 1 armd, 1 mech, 1 air aslt, 2 mtn inf)
5 arty bde (incl 1 AD)
2 engr, 4 territorial def bde
1 recce regt
2 SSM regt
3 AD regt
2 cbt hel regt
1 special ops, 1 gd regt

EQUIPMENT

MBT 1,704: 812 T-55, 706 T-72, 186 PT-91
RECCE 510 BRDM-2
AIFV 1,405: 1,367 BMP-1, 38 BRM-1
APC 33 OT-64 plus some 693 'look-alike' types
TOTAL ARTY 1,558
TOWED 412: **122mm**: 277 M-1938 (M-30); **152mm**: 135 M-1938 (ML-20)
SP 658: **122mm**: 539 2S1; **152mm**: 111 *Dana* (M-77); **203mm**: 8 2S7
MRL 258: **122mm**: 228 BM-21, 30 RM-70
MOR 230: **120mm**: 214 M-120, 16 2B11/2S12
SSM launchers: 32 FROG, SS-C-2B
ATGW 395: 263 AT-3 *Sagger*, 108 AT-4 *Spigot*, 18 AT-5 *Spandrel*, 6 AT-6 *Spiral*
ATK GUNS 85mm: 711 D-44
AD GUNS 871: **23mm**: 395 ZU-23-2, 70 ZSU-23-4 SP; **57mm**: 406 S-60
SAM 979: 115 SA-6, 628 SA-7, 64 SA-8, 168 SA-9 (*Grom*), 4 SA-13
HELICOPTERS
ATTACK 46 Mi-24, 22 Mi-2URP, 28 Pzlw-3W
SPT 8 Mi-2URN
TPT 29 Mi-8, 3 Mi-17, 42 Mi-2
SURV GS-13 (arty), 1 L219/200 PARK-1 (arty), *Long Trough* ((SNAR-1) arty), *Pork Trough* ((SNAR-2/-6) veh, arty), *Small Fred/Small Yawn* (veh, arty), *Big Fred* ((SNAR-10) veh, arty)

Navy 16,860

(incl 2,500 Naval Aviation, 9,500 conscripts)
BASES Gdynia, Swinoujscie, Kolobrzeg, Hel, Gydnia-Babie Doly (Naval Air Brigade)

SUBMARINES 3

SSK 3
1 *Orzel* SS (RF *Kilo*) with 533mm TT
2 *Wilk* (RF *Foxtrot*) with 533mm TT

PRINCIPAL SURFACE COMBATANTS 3

DESTROYERS 1
DDG 1 *Warszawa* (Sov mod *Kashin*) with 4 SS-N-2C *Styx* SSM, 2 × 2 SA-N-1 *Goa* SAM, 5 × 533mm TT, 2 ASW RL
FRIGATES 2
FFG 1 *Pulawski* (US *Perry*) with *Harpoon* SSM, SM-1MR SAM, 1 × 76mm gun, 2 × 3 324mm ASTT
FF 1 *Kaszub* with SA-N-5 *Grail* SAM, 1 × 76mm gun, 2 × 2 533mm ASTT, 2 ASW RL

PATROL AND COASTAL COMBATANTS 25

CORVETTES 4 *Gornik* (Sov *Tarantul* I) FSG with 2 × 2 SS-N-2C *Styx* SSM, 1 × 4 SA-N-5 *Grail* SAM, 1 × 76mm gun
MISSILE CRAFT 7 Sov *Osa* I PFM with 4 SS-N-2A SSM
PATROL CRAFT 14
COASTAL 3 *Sassnitz* PCC with 1 × SA-N-5 *Grail* SAM, 1 × 76mm gun, 8 *Obluze* PCC
INSHORE 11
11 *Pilica* PCI<

MINE WARFARE 24

MINELAYERS none, but SS, *Krogulec* MSC and *Lublin* LSM have minelaying capability
MINE COUNTERMEASURES 24
5 *Krogulec* MSC, 13 *Goplo* (*Notec*) MSC, 4 *Mamry* (*Notec*) MHC/MSC, 2 *Leniwka* MSI

AMPHIBIOUS 5

5 *Lublin* LSM, capacity 135 tps, 9 tk
Plus craft: 3 *Deba* LCU (none employed in amph role)

SUPPORT AND MISCELLANEOUS 19

1 AOT; 1 *Polochny B* AK, 5 ARS; 1 *Polochny C* AGF, 5 trg, 1 sail trg, 2 mod *Moma* AGI; 3 AGHS

NAVAL AVIATION (2,460)

28 cbt ac, 11 armed hel
Flying hours (MiG-21) 60
7 sqn
2 with 28 MiG-21 BIS/UM
1 with 5 PZL-3RM (SAR), 2 PZL-W3T (tpt), 3 An-28 (tpt), 25 Mi-2 (tpt)
1 ASW with 11 Mi-14PL
1 SAR with 3 Mi-14 PS, 3 Mi-2RM
1 SAR with 3 PZL An-28RM, 6 PZL An-2
1 Recce with 17 PZL TS-11 *Iskra*

Air Force 46,200

(incl 19,700 conscripts); 267 cbt ac, 10 attack hel
Flying hours 60–120
2 AD Corps - North and South
FTR 2 sqn
1 with 22 MiG-29 (18 -29U, 4 -29UB)
1 with 25 MiG-23 (20 -23MF, 5 23U)
FGA 4 sqn with 99 Su-22 (81 -22M4, 18 -22UM3K)
5 sqn with 114 MiG-21 (29 -21 bis, 27 -21MF, 15 -21M, 10 -21R, 33 -21UM)
RECCE 1 sqn with 7 Su-22M4
TPT 2 regt with 10 An-26, 2 An-28, 13 Yak-40, 2 Tu-154,

23 An-2
HEL 35 Mi-2, 5 Mi-8, 10* W-3 *Sokol*
TRG 125 TS-11 *Iskra*, 11 PZL I-22 *Iryda*, 34 PZL-130 *Orlik*
AAM AA-2 *Atoll*, AA-3 *Anab*, AA-8 *Aphid*, AA-11 *Archer*
ASM AS-7 *Kerry*
SAM 5 bde; 1 indep regt with 2 rdrs S-200 WEGA, about 200 SA-2/-3/-4/-5

Forces Abroad

UN AND PEACEKEEPING

BOSNIA (SFOR II): 290; 1 AB bn; (UNMIBH): 1 obs **CROATIA** (UNMOP): 1 obs **DROC** (MONUC): 2 obs **GEORGIA** (UNOMIG): 4 obs **IRAQ/KUWAIT** (UNIKOM): 6 obs **LEBANON** (UNIFIL): 629: 1 inf bn, mil hospital **SYRIA** (UNDOF): 358: 1 inf bn **WESTERN SAHARA** (MINURSO): 4 obs **YUGOSLAVIA** (KFOR): 763

Paramilitary 21,500

BORDER GUARDS (Ministry of Interior and Administration) 14,500

11 district units, 2 trg centres

MARITIME BORDER GUARD

about 23 patrol craft: 2 PCC, 11 PCI and 10 PCI<

PREVENTION UNITS OF POLICE (OPP-Ministry of Interior) 7,000

(1,000 conscripts)

Portugal Por

escudo esc		1998	1999	2000	2001
GDP	esc	19.2tr	20.6tr		
	US$	106bn	104bn		
per capita	US$	14,800	15,500		
Growth	%	3.9	3.1		
Inflation	%	2.8	2.3		
Publ debt	%	57.7	58.3		
Def exp	esc	420bn	448bn		
	US$	2.3bn	2.3bn		
Def bdgt	esc	286bn	320bn	ε339bn	
	US$	1.6bn	1.6bn	1.6bn	
FMA (US)	US$	0.8m	0.7m	0.7m	0.75m
US$1=esc		180	197	212	

Population			**9,875,000**
Age	13–17	18–22	23–32
Men	327,000	372,000	810,000
Women	308,000	355,000	789,000

Total Armed Forces

ACTIVE 44,650

(5,860 conscripts, 2,300 women)

Terms of service **Army** 4–8 months **Navy** and **Air Force** 4–12 months

RESERVES 210,930

(all services) (obligation to age 35) **Army** 210,000 **Navy** 930

Army 25,650

(incl 5,500 conscripts)

5 Territorial Comd (1 mil governance, 2 mil zone, 2 mil region) • 1 mech inf bde (2 mech, 1 tk, 1 fd arty bn) • 3 inf bde (on mob), 1 AB bde • 1 lt intervention bde • 3 mech, 1 tk, 1 composite regt (3 inf bn, 2 AA bty) • 3 armd cav regt • 8 inf regt • 2 fd, 1 AD, regt • 2 engr regt • 1 MP regt • 1 special ops centre

EQUIPMENT

MBT 187: 86 M-48A5, 101 M-60 (8 -A4, 86 -A3)
RECCE 15 V-150 *Chaimite*, 25 ULTRAV M-11
APC 249 M-113, 44 M-557, 81 V-200 *Chaimite*
TOTAL ARTY 318 (excl coastal)
TOWED 134: **105mm**: 51 M-101, 24 M-56, 21 L119; **155mm**: 38 M-114A1
SP 155mm: 6 M-109A2
MOR 107mm: 62 M-30 (14 SP); **120mm**: 116 *Tampella*; **81mm**: incl 21 SP
COASTAL 21: **150mm**: 9; **152mm**: 6; **234mm**: 6 (inactive)
RCL 84mm: 162 *Carl Gustav*; **90mm**: 112; **106mm**: 128 M-40
ATGW 131 TOW (incl 18 M-113, 4 M-901), 83 *Milan* (incl 6 ULTRAV-11)
AD GUNS 95, incl **20mm**: Rh202; **40mm**: L/60
SAM 15 *Stinger*, 37 *Chaparral*

DEPLOYMENT

AZORES AND MADEIRA 2,250; 1 composite regt (3 inf bn, 2 AA bty)

Navy 11,600

(incl 1,460 Marines, 360 conscripts, 130 recalled reserves)

COMMANDS Naval Area Comd, 4 Subordinate Comds Azores, Madeira, North Continental, South Continental

BASES Lisbon (Alfeite), 4 spt bases Leca da Palmeira (North), Portimao (South), Funchal (Madeira), Ponta Delgada (Azores), Montido (naval air)

SUBMARINES 3

SSK 3 *Albacora* (Fr *Daphné*) with 12 × 550mm TT

FRIGATES 6

FFG 3 *Vasco Da Gama* (MEKO 200) with 8 *Harpoon* SSM, 8 *Sea Sparrow* SAM, 1 × 100mm gun, 6 ASTT, some with 2 *Super Lynx* hel
FF 3 *Commandante João Belo* (Fr *Cdt Rivière*) with 2 × 100mm gun, 6 ASTT

PATROL AND COASTAL COMBATANTS 30

PATROL, OFFSHORE 10

6 *João Coutinho* PCO with 2 × 76mm gun, hel deck
4 *Baptista de Andrade* PCO with 1 × 100mm gun, hel deck

PATROL, COASTAL 9 *Cacine* PCC

PATROL, INSHORE 10

5 *Argos* PCI<
5 *Albatros* PCI<

RIVERINE 1 *Rio Minho* PCR

AMPHIBIOUS craft only

1 LCU

SUPPORT AND MISCELLANEOUS 13

1 *Berrio* (UK *Green Rover*) AO; 2 trg, 1 ocean trg, 1 div spt; 8 AGHS

NAVAL AIR 5 *Super Lynx*-Mk 95 hel

MARINES (1,460)

2 bn, 1 police, 1 special ops det
1 fire spt coy

EQUIPMENT

MOR 120mm: 36

Air Force 7,400

Flying hours F-16: 180; A-7P: 160
1 op air com (COFA), 5 op gps
FGA 2 sqn
1 with F-16A/B, 1 with *Alpha Jet*
SURVEY 1 sqn with C-212
MR 1 sqn with P-3P
TPT 3 sqn
1 with C-130H, 1 with C-212, 1 with *Falcon* 20 and *Falcon* 50
SAR 2 sqn
1 with SA-330 hel, 1 with SA-330 hel and C-212
LIAISON/UTILITY 1 sqn with Cessna FTB-337G, hel 1 sqn with SA-330
TRG 2 sqn
1 with *Socata* TB-30 *Epsilon*, 1 with *Alpha Jet*
hel and multi-engine trg provided by SA-316 and one of C-212 sqns

EQUIPMENT

51 cbt ac (plus 15 in store), no attack hel
AC 25 ***Alpha Jet*** (FGA/trg) (plus 15 in store) • 20 **F-16A/B** (17 -A, 3 -B) • 6* **P-3P** (MR) • 6 **C-130H** (tpt/SAR) • 24 **C-212** (20 **-A** (12 tpt/SAR, 1 Nav trg, 2 ECM trg, 5 fisheries protection), 4 **-B** (survey)) • 12 **Cessna 337** (utility) • 1 ***Falcon* 20** (tpt, cal) • 3 ***Falcon* 50** (tpt) • 16 ***Epsilon*** (trg)
HEL 10 **SA-330** (SAR/tpt) • 18 **SA-316** (trg, utl)
MISSILES
ASM AGM-65B/G *Maverick*, AGM-84A *Harpoon*
AAM AIM-9Li *Sidewinder*

Forces Abroad

SAO TOME & PRINCIPE

5 Air Force, 1 C-212

UN AND PEACEKEEPING

BOSNIA (SFOR II): 335; 1 inf bn(-) **CROATIA** (UNMOP): 1 obs **EAST TIMOR** (UNTAET): 761, 24 Air Force, 1 C-130H **WESTERN SAHARA** (MINURSO): 5 obs **YUGOSLAVIA** (KFOR): 340

Paramilitary 45,800

NATIONAL REPUBLICAN GUARD 25,300

Commando Mk III APC **hel** 7 SA-315

PUBLIC SECURITY POLICE 20,500

Foreign Forces

NATO HQ South Atlantic at Lisbon (Oeiras)
US 980: **Navy** 50 **Air Force** 930

Spain Sp

peseta pts		1998	1999	2000	2001
GDP	pts	86.7tr	93tr		
	US$	582bn	569bn		
per capita	US$	16,900	17,900		
Growth	%	4.0	3.7		
Inflation	%	1.8	2.3		
Publ debt	%	69	67.6		
Def exp	pts	1.1tr	1.2tr		
	US$	7.4bn	7.3bn		
Def bdgt	pts	897bn	1,201bn	1,231bn	
	US$	5.9bn	7.4bn	7.0bn	
US$1=pts		149	163	176	

Population			39,237,000
Age	13–17	18–22	23–32
Men	1,210,000	1,472,000	3,328,000
Women	1,136,000	1,390,000	3,174,000

Total Armed Forces

ACTIVE 166,050

(incl 51,700 conscripts (to be reduced), some 5,515 women)
Terms of service 9 months

RESERVES 447,900

Army 436,000 **Navy** 3,900 **Air Force** 8,000

Army 100,000

(incl 30,000 conscripts, 4,000 women)
8 Regional Op Comd incl 2 overseas
1 mech inf div with 3 mech inf bde, 1 armd cav, 1 arty, 1

engr regt
1 Rapid Action Force (FAR) with
1 Spanish Legion bde (3 lt inf, 1 arty, 1 engr bn, 1 ATK coy)
1 AB bde
1 air-portable bde
1 lt armd cav regt
2 armd bde
1 mtn bde
3 lt inf bde (cadre)
2 Spanish Legion regt (each 1 mech, 1 mot bn, 1 ATK coy)
3 island garrison: Ceuta and Melilla, Balearic, Canary
1 arty bde
1 AD regt
1 engr bde
1 Army Avn bde (1 attack, 1 tpt hel bn, 4 utl units)
1 AD Comd: 5 AD regt incl 1 HAWK SAM, 1 composite *Aspide*/**35mm**, 1 *Roland* bn
1 Coast Arty Comd (2 coast arty regt)
3 special ops bn

EQUIPMENT

MBT 665: 209 AMX-30 ER1/EM2, 164 M-48A5E, 184 M-60A3TTS, 108 *Leopard* 2 A4 (Ge tempy transfer)
RECCE 340 BMR-VEC (100 **90mm**, 208 **25mm**, 32 **20mm** gun)
AIFV 14 *Pizarro*
APC 1,624: 1,311 M-113 (incl variants), 313 BMR-600 (plus 220 variants)
TOTAL ARTY 1,082 (excluding coastal)
TOWED 457: **105mm**: 223 M-26, 158 M-56 pack, 56 L 118; **155mm**: 20 M-114
SP 202: **105mm**: 48 M-108; **155mm**: 90 M-109A1; **203mm**: 64 M-110A2
COASTAL ARTY 53: **6in**: 44; **305mm**: 6; **381mm**: 3
MRL 140mm: 14 *Teruel*
MOR 120mm: 409 (incl 169 SP); plus **81mm**: 1,314 (incl 187 SP)
ATGW 442 *Milan* (incl 106 SP), 28 HOT, 200 TOW (incl 68 SP)
RCL 106mm: 638
AD GUNS 20mm: 329 GAI-BO1; **35mm**: 92 GDF-002 twin; **40mm**: 183 L/70
SAM 24 I HAWK, 18 *Roland*, 13 *Skyguard/Aspide*, 108 *Mistral*
HELICOPTERS 174 (28 attack)
23 AS-532UL, 48 HU-10B, 69 HA/HR-15 (31 with **20mm** guns, 28 with HOT, 9 trg), 6 HU-18, 11 HR-12B, 17 HT-17 (incl 12-D models)
SURV 2 AN/TPQ-36 (arty, mor)

DEPLOYMENT

CEUTA AND MELILLA 8,100; 2 armd cav, 2 Spanish Legion, 2 mot inf, 2 engr, 2 arty regt; 2 lt AD bn, 1 coast arty bn
BALEARIC ISLANDS 4,500; 1 mot inf regt: 3 mot inf bn; 1 mixed arty regt: 1 fd arty, 1 AD; 1 engr bn
CANARY ISLANDS 8,600; 3 mot inf regt each 2 mot inf bn; 1 mot inf bn, 2 mixed arty regt each: 1 fd arty, 1 AD bn; 2 engr bn

Navy 36,950

(incl 700 Naval Air, 6,900 Marines, 10,700 conscripts and 830 women)
NAVAL ZONES Cantabrian, Strait (of Gibraltar), Mediterranean, Canary (Islands)
BASES El Ferrol (La Coruña) (Cantabrian HQ), San Fernando (Cadiz) (Strait HQ), Rota (Cadiz) (Fleet HQ), Cartagena (Murcia) (Mediterranean HQ), Las Palmas (Canary Islands HQ), Palma de Mallorca and Mahón (Menorca)

SUBMARINES 8

SSK 8
4 *Galerna* (Fr *Agosta*) with 20 *L-5* HWT
4 *Delfin* (Fr *Daphné*) with 12 *L-5* HWT

PRINCIPAL SURFACE COMBATANTS 16

CARRIERS 1 (CVV) *Príncipe de Asturias* (16,200t); air gp: typically 6 to 10 AV-8/AV-8B FGA, 4 to 6 SH-3D ASW hel, 2 SH-3D AEW hel, 2 utl hel
FRIGATES 15
FFG 15
6 *Santa Maria* (US *Perry*) with 1 × 1 SM-1 MR SAM/*Harpoon* SSM launcher, 1 × 76mm gun, 2 × 3 ASTT, 2 SH-60B hel
5 *Baleares* with 8 *Harpoon* SSM, 1 × 1 SM-1 MR SAM, 1 × 127mm gun, 2 × 2 ASTT, 8 ASROC SUGW
4 *Descubierta* with 8 *Harpoon* SSM, *Sea Sparrow* SAM, 1 × 76mm gun, 6 ASTT, 1 × 2 ASW RL

PATROL AND COASTAL COMBATANTS 33

PATROL, OFFSHORE 8
4 *Serviola* PCO with 1 × 76mm gun, 1 *Chilreu* PCO, 2 *Alboran* PCI, 1 *Descubierta* PCO
PATROL, COASTAL 10 *Anaga* PCC
PATROL, INSHORE 15
6 *Barceló* PFI<, 9 PCI<

MINE COUNTERMEASURES 12

1 *Descubierta* MCCS
1 *Guadalete* (US *Aggressive*) MHO
4 *Segura* MHO
6 *Júcar* (US *Adjutant*) MSC

AMPHIBIOUS 4

2 *Hernán Cortés* (US *Newport*) LST, capacity: 400 tps, 500t vehicles, 3 LCVPs, 1 LCPL
2 *Galicia* LPD, capacity 620 tps, 6 LCVP
Plus 13 craft: 3 LCT, 2 LCU, 8 LCM

SUPPORT AND MISCELLANEOUS 32

1 AOE; 1 AOT, 3 AWT, 3 AK; 5 AT, 3 diver spt, 5 trg, 4 sail trg; 6 AGHS, 1 AGOR

NAVAL AIR (700)

(incl 290 conscripts)
Flying hours 160
FGA 1 sqn with 9 AV-8B/8 AV-8B plus

LIAISON 1 sqn with 3 *Citation* II
HELICOPTERS 5 sqn
ASW 2 sqn
1 with SH-3D/G *Sea King* (mod to SH-3H standard)
1 with SH-60B (LAMPS-III fit)
COMD/TPT 1 sqn with 10 AB-212
TRG 1 sqn with 10 Hughes 500
AEW 1 flt with SH-3D (*Searchwater* radar)

EQUIPMENT

17 cbt ac, 27 armed hel
Flying hours 160
AC 9 **AV-8B**, 8 **AV-8B** plus (trg) • 3 ***Citation*** **II** (liaison)
HEL 10 **AB-212** (Amph asslt) • 11 **SH-3D** (8 **-H** ASW, 3 **-D** AEW) •10 **Hughes 500** (trg) • 6 **SH-60B** (ASW)

MARINES (6,900)

(incl 2,800 conscripts)
1 marine bde (3,000); 2 inf, 1 spt bn; 3 arty bty
5 marine garrison gp

EQUIPMENT

MBT 16 M-60A3
AFV 17 *Scorpion* lt tk, 16 LVTP-7 AAV, 4 BLR
TOWED ARTY 105mm: 12 M-56 pack
SP ARTY 155mm: 6 M-109A
ATGW 12 TOW, 18 *Dragon*
RL 90mm: C-90C
RCL 106mm: 54
SAM 12 *Mistral*

Air Force 29,100

(incl 11,000 conscripts, 685 women)
Flying hours EF-18: 180; F-5: 220; *Mirage* F-1: 180

CENTRAL AIR COMMAND (Torrejon) 4 wg

FTR 2 sqn with EF-18 (F-18 *Hornet*)
RECCE 1 sqn with RF-4C
TPT 7 sqn
2 with C-212, 2 with CN-235, 1 with *Falcon* (20, 50, 900), 1 with Boeing 707 (tkr/tpt), 1 with AS-332 (tpt)
SPT 5 sqn
1 with CL-215, 1 with C-212 (EW) and *Falcon* 20, 1 with C-212, AS-332 (SAR), 1 with C-212 and Cessna *Citation*, 1 with Boeing 707
TRG 3 sqn
1 with C-212, 1 with C-101, 1 with Beech (*Bonanza*)

EASTERN AIR COMMAND (Zaragosa) 2 wg

FTR 3 sqn
2 with EF-18, 1 OCU with EF-18
TPT 2 sqn
1 with C-130H, 1 tkr/tpt with KC-130H
SPT 1 sqn with **ac** C-212 (SAR) **hel** AS-330

STRAIT AIR COMMAND (Seville) 4 wg

FTR 3 sqn
2 with *Mirage* F-1 CE/BE
1 with EF/A-18
LEAD-IN TRG 2 sqn with F-5B
MP 1 sqn with P-3A/B
TRG 6 sqn
2 hel with *Hughes* 300C, S-76C, 1 with C-212, 1 with E-26 (*Tamiz*), 1 with C-101, 1 with C-212

CANARY ISLANDS AIR COMMAND (Gando) 1 wg

FGA 1 sqn with EF-18
TPT 1 sqn with C-212
SAR 1 sqn with **ac** F-27 **hel** AS-332 (SAR)

LOGISTIC SUPPORT COMMAND (MALOG)

1 trials sqn with C-101, C-212 and F-5A, EF/A-18, F-1

EQUIPMENT

211 cbt ac, no armed hel
AC 90 **EF/A-18 A/B** (ftr, OCU) • 35 **F-5B** (FGA) • 65 ***Mirage*** **F-1CF/-BE/-EE** • 14* **RF-4C** (recce) 7* **P-3** (2 **-A** (MR), 5 **-B** (MR)) • 4 **Boeing 707** (tkr/tpt) • 7 **C-130H/H-30** (tpt), 5 **KC-130H** (tkr) • 78 **C-212** (34 tpt, 9 SAR, 6 recce, 26 trg, 2 EW, 1 trials) • 2 **Cessna 560** ***Citation*** (recce) • 74 **C-101** (trg) • 15 **CL-215** (spt) • 5 ***Falcon*** **20** (3 VIP tpt, 2 EW) • 1 ***Falcon*** **50** (VIP tpt) • 2 ***Falcon*** **900** (VIP tpt) • 21 Do-27 (U-9, liaison/trg) • 3 **F-27** (SAR) • 37 **E-26** (trg) • 20 **CN-235** (18 tpt, 2 VIP tpt) • 25 **E-24** (*Bonanza*) trg
HEL 5 **SA-330** (SAR) • 16 **AS-332** (10 SAR, 6 tpt) • 13 **Hughes 300C** (trg) • 8 **S-76C** (trg) • 15 **EC 120B** *Colibri* being delivered between July 2000 and July 2001

MISSILES

AAM AIM-7 *Sparrow*, AIM-9 *Sidewinder*, AIM-120 AMRAAM, R-530
ASM AGM-65G *Maverick*, AGM-84D *Harpoon*, AGM-88A HARM
SAM *Mistral*, *Skyguard/Aspide*

Forces Abroad

UN AND PEACEKEEPING

BOSNIA (SFOR II): 1,600; 1 inf bn gp, 12 obs, 2 TACP
ITALY (Deliberate Forge) 5 F/A-18, 1 KC-130
YUGOSLAVIA (KFOR): 900

Paramilitary 75,760

GUARDIA CIVIL **75,000**

(incl 2,200 conscripts); 9 regions, 19 inf *tercios* (regt) with 56 rural bn, 6 traffic security gp, 6 rural special ops gp, 1 special sy bn; 22 BLR APC, 18 Bo-105, 5 BK-117 hel

GUARDIA CIVIL DEL MAR 760

32 PCI

Foreign Forces

NATO HQ Joint Command South-West (JCSOUTHWEST)
US 2,130: **Navy** 1,760 **Air Force** 250 **USMC** 120

Turkey Tu

lira L		1998	1999	2000	2001
GDP	L	52,224tr	79,814tr		
	US$	200bn	186bn		
per capita	US$	6,300	6,000		
Growth	%	2.8	-2.3		
Inflation	%	69.7	64.8		
Debt	US$	102bn	105bn		
Def exp	L	2,179tr	4,367tr		
	US$	8.4bn	10.1bn		
Def bdgt	L	2,021tr	3,818tr	4,742tr	
	US$	7.8bn	8.9bn	7.7bn	
FMA (US)	US$	22m	1.5m	1.5m	1.6m
US$1=L		260,724	428,920	613,390	

Population		**66,130,000** (Kurds ε20%)	
Age	13–17	18–22	23–32
Men	3,266,000	3,254,000	6,098,000
Women	3,180,000	3,090,000	5,765,000

Total Armed Forces

ACTIVE 609,700

(incl ε528,000 conscripts) *Terms of service* 18 months

RESERVES 378,700

(all to age 41) **Army** 258,700 **Navy** 55,000 **Air Force** 65,000

Army ε495,000

(incl ε462,000 conscripts)

4 army HQ: 9 corps HQ • 1 mech div (1 mech, 1 armd bde) • 1 mech div HQ • 1 inf div • 14 armd bde (each 2 armd, 2 mech inf, 2 arty bn) • 17 mech bde (each 2 armd, 2 mech inf, 1 arty bn) • 9 inf bde (each 4 inf, 1 arty bn) • 4 cdo bde (each 4 cdo bn) • 1 inf regt • 1 Presidential Guard regt • 5 border def regt • 26 border def bn

RESERVES

4 coastal def regt • 23 coastal def bn

EQUIPMENT

Figures in ( ) were reported to CFE on 1 Jan 2000

MBT 4,205 (2,464): 2,876 M-48 A5T1/T2 (1,300 to be stored), 932 M-60 (658 -A3, 274-A1), 397 *Leopard* (170-1A1, 227-1A3)

RECCE some *Akrep*, some ARSV (*Cobra*)

TOTAL AIFV/APC (2,616)

AIFV 650 AIFV

APC 830 AAPC, 2,813 M-113/-A1/-A2

TOTAL ARTY (2,883)

TOWED 105mm: M-101A1; **155mm**: 517 M-114A1\A2; **203mm**: 162 M-115

SP 105mm: 365 M-52T, 26 M-108T; **155mm**: 222 M-44T1; **175mm**: 36 M-107; **203mm**: 219 M-110A2

MRL 60: **107mm**: 48; **227mm**: 12 MLRS (incl ATACMS)

MOR 2,021: **107mm**: 1,264 M-30 (some SP); **120mm**: 757 (some 179 SP); plus **81mm**: 3,792 incl SP

ATGW 943: 186 *Cobra*, 365 TOW SP, 392 *Milan*

RL M-72

RCL 57mm: 923 M-18; **75mm**: 617; **106mm**: 2,329 M-40A1

AD GUNS 1,664: **20mm**: 439 GAI-DO1; **35mm**: 120 GDF-001/-003; **40mm**: 803 L60/70, 40 T-1, 262 M-42A1

SAM 108 *Stinger*, 789 *Redeye* (being withdrawn)

SURV AN/TPQ-36 (arty, mor)

AC 168: 3 Cessna 421, 34 *Citabria*, 4 B-200, 4 T-42A, 98 U-17B, 25 T-41D

ATTACK HEL 37 (26) AH-1W/P

SPT HEL 262: 20 S-70A, 19 AS-532UL, 12 AB-204B, 64 AB-205A, 20 AB-206, 2 AB-212, 28 H-300C, 3 OH-58B, 94 UH-1H

UAV CL-89 (AN/USD-501), *Gnat* 750, *Falcon* 600

Navy 54,600

(incl 3,100 Marines, 1,050 Coast Guard, 34,500 conscripts)

BASES Ankara (Navy HQ and COMEDNOREAST), Gölcük (HQ Fleet), Istanbul (HQ Northern area and Bosphorus), Izmir (HQ Southern area and Aegean), Eregli (HQ Black Sea), Iskenderun, Aksaz Bay, Mersin (HQ Mediterranean)

SUBMARINES 14

SSK 10

6 *Atilay* (Ge Type 209/1200) with 8 × 533mm TT (SST 4 HWT)

4 *Preveze* (Ge Type 209/1400) with *Harpoon* SSM, 8 × 533mm TT

SSC 4

2 *Canakkale* (US *Guppy*)† with 10 × 533mm TT

2 *Hizirreis* (US *Tang*) with 8 × 533mm TT (Mk 37 HWT)

PRINCIPAL SURFACE COMBATANTS 22

FRIGATES 22

FFG 21

5 *Gaziantep* (US *Perry*) with 4 *Harpoon* SSM, 36 SM-1 MR SAM, 1 × 76mm gun, 2 × 3 ASTT

4 *Yavuz* (Ge MEKO 200) with 8 *Harpoon* SSM, *Sea Sparrow* SAM, 1 × 127mm gun, 2 × 3 ASTT, 1 AB-212 hel

8 *Muavenet* (US *Knox*-class) with *Harpoon* SSM (from ASROC launcher), 1 × 127mm gun, 4 ASTT, 8 ASROC SUGW, 1 AB 212 hel

4 *Barbaros* (MOD Ge MEKO 200) with 8 *Harpoon* SSM, 8 *Sea Sparrow* SAM, 1 × 127mm gun, 6 × 324mm TT, 1 AB-212 hel

FF 1 *Berk* with 4 × 76mm guns, 6 ASTT, 2 Mk 11 *Hedgehog*

PATROL AND COASTAL COMBATANTS 49

MISSILE CRAFT 21

3 *Kilic* PFM with 8 × *Harpoon* SSM, 1 × 76mm gun
8 *Dogan* (Ge Lürssen-57) PFM with 8 *Harpoon* SSM, 1 × 76mm gun
8 *Kartal* (Ge *Jaguar*) PFM with 4 *Penguin* 2 SSM, 2 × 533mm TT
2 *Yildiz* PFM with 8 *Harpoon* SSM, 1 × 76mm gun

PATROL CRAFT 28

COASTAL 28

1 *Girne* PFC, 6 *Sultanhisar* PCC, 2 *Trabzon* PCC, 4 PGM-71 PCC, 1 *Bora* (US *Asheville*) PFC, 10 AB-25 PCC, 4 AB-21 PCC

MINE WARFARE 24

MINELAYERS 1

1 *Nusret* (400 mines) plus 3 ML tenders (*Bayraktar*, *Sarucabey* and *Çakabey* LST have minelaying capability)

MINE COUNTERMEASURES 23

5 *Edineik* (Fr *Circe*) MHC
8 *Samsun* (US *Adjutant*) MSC
6 *Karamürsel* (Ge *Vegesack*) MSC
4 *Foça* (US *Cape*) MSI (plus 8 MCM tenders)

AMPHIBIOUS 8

1 *Osman Gazi* LST: capacity 980 tps, 17 tk, 4 LCVP
2 *Ertugru* LST (US *Terrebonne Parish*): capacity 400 tps, 18 tk
2 *Bayraktar* LST (US LST-512): capacity 200 tps, 16 tk
2 *Sarucabey* LST: capacity 600 tps, 11 tk
1 *Çakabey* LSM: capacity 400 tps, 9 tk
Plus about 59 craft: 35 LCT, 2 LCU, 22 LCM

SUPPORT AND MISCELLANEOUS 27

1 *Akar* AO, 5 spt tkr, 2 Ge *Rhein* plus 3 other depot ships, 3 tpt, 2 AR; 3 ARS, 5 AT, 1 div spt; 2 AGHS

NAVAL AVIATION

16 armed hel
ASW 3 AB-204AS, 13* AB-212
TRG 7 TB-20

MARINES (3,100)

1 regt, HQ, 3 bn, 1 arty bn (18 guns), spt units

Air Force 60,100

(incl 31,500 conscripts) 2 tac air forces (divided between east and west), 1 tpt comd, 1 air trg comd, 1 air log comd
Flying hours 180
FGA 11 sqn
1 OCU with F-5A/B, 4 (1 OCU) with F-4E, 6 (1 OCU) with F-16C/D
FTR 7 sqn
2 with F-5A/B, 2 with F-4E, 3 with F-16C/D
RECCE 2 sqn with RF-4E
TPT 5 sqn
1 with C-130B/E, 1 with C-160D, 2 with CN-235, 1 VIP tpt unit with *Gulfstream*, *Citation* and CN 235
TKR 2 KC-135R
LIAISON 10 base flts with **ac** T-33 **hel** UH-1H
SAR hel AS-532
TRG 3 sqn
1 with T-41, 1 with SF-260D, 1 with T-37B/C and T-38A. Each base has a stn flt with **hel** UH-1H and in some cases, **ac** CN-235
SAM 4 sqn with 92 *Nike Hercules*, 2 sqn with 86 *Rapier*

EQUIPMENT

505 cbt ac, no attack hel
AC 240 **F-16C/D** (210 **-C**, 30 **-D**); further package of 32, including 20 recce configuration, to be delivered by 2002) • 87 **F/NF-5A/B** (FGA) (48 being upgraded as lead-in trainers) • 178 **F-4E** (92 FGA, 47 ftr, 39 **RF-4E** (recce)) (54 being upgraded to *Phantom* 2000) • 13 **C-130B/E** (tpt) • **7 KC-135R** • 19 **C-160D** (tpt) • 2 ***Citation*** **VII** (VIP) • 50 **CN-235** (tpt/EW) • 38 **SF-260D** (trg) • 34 **T-33** (trg) • 60 **T-37** trg • 70 **T-38** (trg) • 28 **T-41** (trg)
HEL 20 **UH-1H** (tpt, liaison, base flt, trg schools), 20 **AS-532** (14 SAR/6 CSAR) being delivered

MISSILES

AAM AIM-7E *Sparrow*, AIM 9 S *Sidewinder*, AIM-120 AMRAAM
ASM AGM-65 *Maverick*, AGM-88 HARM, AGM-142, *Popeye* 1

Forces Abroad

CYPRUS ε36,000; 1 corps; 386 M-48A5 MBT; 265 M-113, 211 AAPC APC; 72 **105mm**, 18 **155mm**, 12 **203mm** towed; 60 **155mm** SP; 127 **120mm**, 148 **107mm**, 175 **81mm** mor; **20mm**, 16 **35mm**; 48 **40mm** AA guns; **ac** 3 **hel** 4 **Navy** 1 PCI

UN AND PEACEKEEPING

BOSNIA (SFOR II): 1,300; 1 inf bn gp **EAST TIMOR** (UNTAET): 2 obs **GEORGIA** (UNOMIG): 5 obs **IRAQ/KUWAIT** (UNIKOM): 6 obs **ITALY** (Deliberate Forge): 4 F-16 C **YUGOSLAVIA** (KFOR): 950

Paramilitary

***GENDARMERIE*/NATIONAL GUARD** 218,000 (Ministry of Interior, Ministry of Defence in war)

50,000 reserve; some *Akrep* recce, 535 BTR-60/-80, 25 *Condor* APC **ac** 2 Dornier 28D, 0-1E **hel** 19 Mi-17, 8 AB-240B, 6 AB-205A, 8 AB-206A, 1 AB-212, 14 S-70A

COAST GUARD 2,200

(incl 1,400 conscripts); 48 PCI, 16 PCI<, plus boats, 2 tpt

Foreign Forces

NATO HQ Joint Command South-East (JCSOUTHEAST), HQ 6 Allied Tactical Air Force (6 ATAF)

OPERATION NORTHERN WATCH

UK Air Force 160; 4 *Jaguar* GR-3A/-B, 2 VC-10 (tkr)
US 2,040: **Navy** 20 **Air Force** 1,800; 1 wg (**ac** on det only), numbers vary (incl F-16, F-15C, KC-135, E-3B/C, C-12, HC-130, HH-60) **USMC** 220
US Installations for seismic monitoring
ISRAEL Periodic det of F-16 at Akinci

United Kingdom UK

pound £		1998	1999	2000	2001
GDP	£	847bn	890bn		
	US$	1.4tr	1.4tr		
per capita	US$	21,600	22,300		
Growth	%	2.1	1.7		
Inflation	%	3.4	1.6		
Publ debt	%	56.2	53		
Def exp	£	22.5bn	22.9bn		
	US$	37.4bn	36.9bn		
Def bdgt	£	22.5bn	22.8bn	22.8bn	23.0bn
	US$	37.4bn	35.9bn	34.5bn	34.8bn
US$1=£		0.60	0.64	0.66	
Population					**58,882,000**

(*Northern Ireland* 1,600,000: Protestant 56%, Roman Catholic 41%)

Age	13–17	18–22	23–32
Men	1,905,000	1,795,000	4,096,000
Women	1,818,000	1,711,000	3,926,000

Total Armed Forces

ACTIVE 212,450

(incl 15,060 women, and 4,060 locally enlisted personnel)

RESERVES 302,850

Army 187,200 (Regular 142,800) **Territorial Army** (TA) 43,300 (to be 41,200) **Navy/Marines** 28,500 (Regular 24,450, Volunteer Reserves 4,050) **Air Force** 43,850 (Regular 42,000, Volunteer Reserves 1,850)

Strategic Forces (1,900)

SLBM 58 msl in 4 SSBN, fewer than 200 op available warheads
SSBN 4
4 *Vanguard* SSBN each capable of carrying 16 *Trident* (D5); will not deploy with more than 48 warheads per boat, but each msl could carry up to 12 MIRV (some *Trident* D5 missiles loaded with single warheads for sub-strategic role)
EARLY WARNING
Ballistic-Missile Early-Warning System (BMEWS) station at Fylingdales

Army 113,950

(incl 6,380 women, 4,060 enlisted outside the UK, of whom 3,670 are Gurkhas)
regt normally bn size
1 Land Comd HQ • 3 (regenerative) div HQ (former mil districts) and UK Spt Comd (Germany) • 1 armd div with 3 armd bde, 3 arty, 4 engr, 1 avn, 1 AD regt • 1 mech div with 3 mech bde (*Warrior/Saxon*), 3 arty, 3 engr, 1 avn, 1 AD regt • ARRC Corps tps: 3 armd recce, 2 MLRS, 2 AD, 1 engr regt (EOD) • 1 joint hel comd (RN/Army/RAF) with 1 air aslt bde • 1 AD bde HQ • 14 inf bde HQ (3 control ops in N. Ireland, remainder mixed regular and TA for trg/administrative purposes only)
1 joint NBC regt (RAF/Army)
Summary of combat arm units
7 armd regt • 3 armd recce regt • 4 mech inf bn (*Saxon*) • 8 armd inf bn (*Warrior*) • 25 lt inf bn (incl 2 air mobile, 2 Gurkha) • 3 AB bn (1 only in para role) • 1 SF (SAS) regt • 11 arty regt (2 MLRS, 5 SP, 3 fd (1 cdo, 1 AB, 1 air-mobile), 1 trg) • 4 AD regt (2 *Rapier*, 2 *Javelin*) • 10 engr regt • 5 army avn regt (2 ATK, 2 air mobile, 1 general)

HOME SERVICE FORCES

N. Ireland 4,200: 6 inf bn (2,400 full-time)
Gibraltar 360: 1 regt (170 full-time)
Falkland Island Defence Force 60

RESERVES

Territorial Army 4 lt recce, 14 inf bn, 1 AB (not in role), 2 SF (SAS), 3 arty (1 MLRS, 1 fd, 1 obs), 4 AD, 5 engr, 1 avn regt

EQUIPMENT

Figures in ( ) were reported to CFE on 1 Jan 2000
MBT 616: 192 *Challenger* 2, 410 *Challenger*, 14 *Chieftain*
LT TK 11 *Scorpion*
RECCE 332 *Scimitar*, 138 *Sabre*, 11 *Fuchs*
TOTAL AIFV/APC 2,426 (2,330)
AIFV 527 *Warrior* (plus 199 'look-alikes'), 11 AFV 432 *Rarden*
APC 748 AFV 432, 529 FV 103 *Spartan*, 609 *Saxon*, 2 *Saracen*, (plus 1,390 'look-alikes')
TOTAL ARTY 457 (424)
TOWED 214: **105mm**: 166 L-118/-119, 3 M-56; **140mm**: 1 5.5in; **155mm**: 44 FH-70
SP 105mm: 1 *Abbot*; **155mm**: 179 AS-90
MRL 227mm: 63 MLRS
MOR 81mm: 543 (incl 110 SP)
ATGW 755 *Milan*, 60 *Swingfire* (FV 102 *Striker* SP), TOW
RL 94mm: LAW-80
SAM 135 HVM (SP), 147 *Starstreak* (LML), 374 *Javelin*, 105 *Rapier* (some 24 SP)
SURV 33 *Cymbeline* (mor)
AC 6 BN-2
ATTACK HEL 269 (249): 144 SA-341, 110 *Lynx* AH-1/-7/-9

UAV *Phoenix*
LANDING CRAFT 2 LCL, 6 RCL, 4 LCVP, 4 workboats

Navy (RN) 43,770

(incl 6,740 Fleet Air Arm, 6,740 Royal Marines Command, 3,330 women)

ROYAL FLEET AUXILIARY (RFA)

(2,400 civilians) mans major spt vessels

MARINE SERVICES

(280 MoD civilians and 780 commercial contractors) 203 craft, provides harbour/coastal services

BASES UK Northwood (HQ Fleet, CINCEASTLANT), Devonport (HQ), Faslane, Portsmouth; Culdrose, Prestwick, Yeovilton (Fleet Air Arm); **Overseas** Gibraltar

SUBMARINES 16

STRATEGIC SUBMARINES 4 SSBN
TACTICAL SUBMARINES 12
SSN 12
5 *Swiftsure* with *Spearfish* or *Tigerfish* HWT and *Sub-Harpoon* SSM (2 in refit); one (*Splendid*) with 12 *Tomahawk* Block III LAM
7 *Trafalgar* with *Spearfish* and *Tigerfish* HWT and *Sub-Harpoon* SSM (3 in refit); two (*Triumph* and *Trafalgar*) with 12 *Tomahawk* Block III LAM

PRINCIPAL SURFACE COMBATANTS 34

CARRIERS 3: 2 mod *Invincible* CVS each with **ac** 8 FA-2 *Sea Harrier* V/STOL **hel** 12 *Sea King*, up to 9 ASW, 3 AEW; plus 1 *Invincible* in extended refit
Full 'expeditionary air group' comprises 8 *Sea Harrier* FA-2, 8 RAF *Harrier* GR-7, 2 *Sea King* ASW, 4 *Sea King* AEW
DESTROYERS 11
DDG 11
7 Type 42 Batch 1/2 with 2 × *Sea Dart* SAM, 1 × 114mm gun, 6 × 324mm ASTT, 1 *Lynx* hel
4 Type 42 Batch 3 with wpns as above
FRIGATES 20
FFG 20
4 *Cornwall* (Type 22 Batch 3) with 8 *Harpoon* SSM, *Seawolf* SAM, 1 × 114mm gun, 6 × 324mm ASTT (*Stingray* LWT)
2 *Broadsword* (Type 22 Batch 2) with 4 × MM 38 *Exocet* SSM, *Seawolf* SAM, 6 × 324mm ASTT (*Stingray* LWT), 2 *Lynx* or 1 *Sea King* hel
14 *Norfolk* (Type 23) with 8 *Harpoon* SSM, *Seawolf* VL SAM, 1 × 114mm gun, 4 × 324mm ASTT (*Stingray* LWT)

PATROL AND COASTAL COMBATANTS 23

PATROL, OFFSHORE 7
2 *Castle* PCO, 5 *Island* PCO
PATROL, INSHORE 16
16 *Archer* (incl 8 trg)

MINE WARFARE 21

MINELAYER no dedicated minelayer, but all submarines have limited minelaying capability
MINE COUNTERMEASURES 21
13 *Hunt* MCC (4 mod *Hunt* MCC/PCC), 10 *Sandown* MHO (5 batch 1, 5 batch 2)

AMPHIBIOUS 6

1 *Fearless* LPD with 4 LCU, 4 LCVP; capacity 350 tps, 15 tk, 3 hel
1 *Ocean* LPH with 4 LVCP, capacity 800 tps, 18 hel
4 *Sir Bedivere* LSL; capacity 340 tps, 16 tk, 1 hel (RFA manned)
Plus 23 craft: 9 LCU, 14 LCVP
(see Army for additional amph lift capability)

SUPPORT AND MISCELLANEOUS 24

UNDER WAY SUPPORT 9
2 *Fort Victoria* AO, 2 *Olwen* AO, 3 *Rover* AO, 2 *Fort Grange* AF (all RFA manned)
MAINTENANCE AND LOGISTIC 7
1 *Diligence* AR, 1 *Sea Crusader* AK, 1 *Sea Centurion* AK, 4 AOT (all RFA manned)
SPECIAL PURPOSE 2
1 *Argus* AVB (RFA manned), 1 *Endurance* (ice patrol)
SURVEY 6
1 *Scott* AGHS, 2 *Bulldog* AGHS, 1 *Roebuck* AGHS, 1 *Gleaner* AGHS, 1 *Herald* AGOS

FLEET AIR ARM (6,740)

(incl 330 women)
Flying hours *Harrier*: 275
A typical CVS air group consists of 8 FA-2 *Harrier*, 7 *Sea King* (ASW), 3 *Sea King* (AEW) (can carry 8 RAF GR-7 *Harrier* instead of 4 *Sea King*)
FTR/ATK 2 ac sqn with *Sea Harrier* F/A2 plus 1 trg sqn with *Harrier* T-4/-8
ASW 5 hel sqn with *Sea King* Mk-5/6
ASW/ATK 2 sqn with *Lynx* HAS-3 HMA8 (in indep flt)
AEW 1 hel sqn with *Sea King* AEW-2
COMMANDO SPT 3 hel sqn with *Sea King* HC-4
SAR 1 hel sqn with *Sea King* MK-5
TRG 1 sqn with *Jetstream*
FLEET SPT 13 *Mystère-Falcon* (civil registration), 1 Cessna *Conquest* (civil registration), 1 Beech *Baron* (civil registration) 5 GROB 115 (op under contract)
TPT *Jetstream*

EQUIPMENT

34 cbt ac (plus 21 in store), 120 armed hel
AC 29 *Sea Harrier* FA-2 (plus 19 in store) • 5* T-4/T-8 (trg) plus 2 in store • 15 *Hawk* (spt) • 10 *Jetstream* • 7 T-2 (trg) • 3 T-3 (spt)
HEL 92 *Sea King* (49 HAS-5/6, 33 HC-4, 10 AEW-2) • 36 *Lynx* HAS-3 • 23 *Lynx* HAS-8, 12 EH-101 *Merlin*

MISSILES

ASM *Sea Skua*
AAM AIM-9 *Sidewinder*, AIM-120C AMRAAM

ROYAL MARINES COMMAND (6,740, incl RN and Army)

1 cdo bde: 3 cdo; 1 cdo arty regt (Army); 1 cdo AD bty

(Army), 2 cdo engr (1 Army, 1 TA), 1 LCA sqn. Serving with RN/Other comd: 1 sy gp, Special Boat Service, 1 cdo lt hel sqn, 2 LCA sqn, 3 dets/naval parties

EQUIPMENT

MOR 81mm
ATGW *Milan*
SAM *Javelin*
HEL 9 SA-341 (*Gazelle*); plus 3 in store, 6 *Lynx* AH-7
AMPH 24 RRC, 4 LACV

RESERVES

About 1,000

Air Force (RAF) 54,730

(incl 5,350 women)
Flying hours *Tornado* GRI/4: 196, F3: 192; *Harrier*: 195; *Jaguar*: 199
FGA/BBR 5 sqn with *Tornado* GRI/4
FGA 5 sqn
3 with *Harrier* GR-7, 2 with *Jaguar* GR-1A/GR-3/3A
FTR 5 sqn with *Tornado* F-3 plus 1 flt in the Falklands
RECCE 4 sqn
2 with *Tornado* GR-1A/4A, 1 with *Canberra* PR-9, 1 with *Jaguar* GR-1A/GR-3/3A
MR 3 sqn with *Nimrod* MR-2
AEW 2 sqn with E-3D *Sentry*
ELINT 1 sqn with *Nimrod* R-1
TPT/TKR 3 sqn
1 with VC-10 C1K, VC-10 K-3/-4, and 1 with *Tristar* K-1/KC-2A, plus 1 VC-10 flt in the Falklands
TPT 4 sqn with *Hercules* C-130K/J, 1 comms sqn with **ac** BAe-125, BAe-146 **hel** AS-355 (*Twin Squirrel*)
TARGET FACILITY/CAL 1 sqn with *Hawk* T-1/T-1A
OCU 6: *Tornado* GR-1/4, *Tornado* F-3, *Jaguar* GR-3/3A/T2A, *Harrier* GR-7/-T10, *Hercules* C-130K/J, *Nimrod* MR-2
TRG *Hawk* T-1/-1A/-1W, *Jetstream* T-1, *Bulldog* T-1, G.115E *Tutor*, HS-125 *Dominie* T-1, *Tucano* T-1, T-67 *Firefly*
TAC HEL 9 sqn
1 with CH-47 (*Chinook*) and SA-341 (*Gazelle* HT3), 1 with *Wessex* HC-2, 2 with SA-330 (*Puma*), 1 with CH-47 and *Sea King* HAR-3, 2 with CH-47, 1 with *Wessex* HC-2 and SA-330 (*Puma*), 1 with *Merlin* HC3
SAR 2 hel sqn with *Sea King* HAR-3/3A
TRG *Sea King* (including postgraduate training on 203(R) sqn), Tri-Service Defence Helicopter School with AS-350 (*Single Squirrel*) and Bell-412

EQUIPMENT

429 cbt ac (plus 137 in store), no armed hel
AC 214 ***Tornado*** (63 **GR-4/4A**, 58 **GR-1/1A/1B**), 93 **F-3** (plus 63 **GR** and 26 **F-3** in store) • 53 ***Jaguar*** (43 **GR-1A/3/3A**, 10 **T-2A/B** (plus 26 in store)) • 64 ***Harrier*** (53 **GR-7**, 11 **T-10** (plus 22 **GR-7** in store)) • 125 ***Hawk*** **T-1/1-A-W** (incl 75* (T1-A)) (plus 16 in store) • 7 ***Canberra*** (2 **T-4**, 5 **PR-9**) • 26 ***Nimrod*** (3 **R-1** (ECM), 23* **MR-2** (MR) • 7 ***Sentry*** **(E-3D)** (AEW) • 9 ***Tristar*** (2 **K-1** (tkr/pax), 4 **KC-1** (tkr/pax/cgo), 2 **C-2** (pax), 1 **C-2A** (pax) • 21 **VC-10** (12 **C-1K** (tkr/cgo), 4 **K-3** (tkr), 5 **K-4** (tkr)) • 51 ***Hercules*** **C-130** (26 **-K**, 25 **-J** by mid 2001) • 6 **BAe-125 CC-3** (comms) • 2 ***Islander*** **CC-MK2** • 2 **BAe-146** Mk 2 (VIP tpt) • 1 **BAe-748** 'Open Skies' • 84 ***Tucano*** (trg) (plus 44 in store) • 11 ***Jetstream*** (trg) • 80 ***Bulldog*** (trg)(plus 25 in store) • 10 ***Dominie*** (trg) • 40 ***Tutor*** (trg) • 43 *Firefly* (trg)
HEL 15 ***Wessex*** • 38 **CH-47** (*Chinook*) • 6 *Merlin* HC3 (22 on order) • 39 **SA-330 (*Puma*)** • 25 ***Sea King*** • 38 **AS-350B** (***Single Squirrel***) • 3 **AS-355** (***Twin Squirrel***) • 9 **Bell-412EP**

MISSILES

ASM AGM-65G2 *Maverick*, AGM-84D-1 *Harpoon*
AAM ASRAAM, AIM-9L/M *Sidewinder*, *Sky Flash* AMRAAM
ARM ALARM

ROYAL AIR FORCE REGIMENT

5 fd sqn, 4 gd based air defence sqns with 24 *Rapier* field standard C fire units; joint *Rapier* trg unit (with Army), 3 tactical Survival To Operate (STO) HQs
VOLUNTEER RESERVE AIR FORCES (Royal Auxiliary Air Force/RAF Reserve): 3 field sqns, 1 gd based AD sqn, 1 air movements sqn, 2 medical sqns, 2 intelligence sqns, 2 training and standardisation sqns, 6 op support sqns covering STO duties, 1 HQ augmentaion sqn, 1 mobile meteorological unit

Deployment

ARMY

LAND COMMAND
Assigned to ACE Rapid Reaction Corps **Germany** 1 armd div plus Corps cbt spt tps **UK** 1 mech inf div, 1 air aslt bde (assigned to MND(C)); additional TA units incl 8 inf bn, 2 SAS, 3 AD regt **Allied Command Europe Mobile Force** (*Land*) (AMF(L)): UK contribution 1 inf BG (incl 1 inf bn, 1 arty bty, 1 sigs sqn)
HQ NORTHERN IRELAND
(some 8,600 (incl 200 RN, 1,100 RAF), plus 4,200 Home Service); 3 inf bde HQ, up to 12 major units in inf role (5 in province, 1 committed reserve, up to 6 roulement inf bn), 1 engr, 1 avn regt, 6 Home Service inf bn.
The roles of the remainder of Army regular and TA units incl Home Defence and the defence of Dependent Territories, the Cyprus Sovereign Base Areas and Brunei.

NAVY

FLEET (CinC is also CINCEASTLANT and COMNAVNORTHWEST): almost all regular RN forces are declared to NATO, split between SACLANT and SACEUR
MARINES 1 cdo bde (declared to SACLANT)

AIR FORCE

STRIKE COMMAND responsible for all RAF front-line forces. Day-to-day control delegated to 3 Groups **No. 1** (All RAF front-line fast jet ac, excl *Harrier*) **No. 2** (AT, AAR, airborne C3I support and RAF regt) **No. 3** (Joint Force *Harrier* (all *Harrier* GR7s and RN *Sea Harrier*), maritime assets (*Nimrod* MR-2 and SAR hel force) and 1 HQ Augmentation sqn)

Forces Abroad

ANTARCTICA 1 ice patrol ship (in summer only)
ASCENSION ISLAND RAF 37
BELGIUM RAF 196
BELIZE Army 180
BRUNEI Army some 1,050: 1 Gurkha inf bn, 1 hel flt (3 hel)
CANADA Army 200 trg and liaison unit **RAF** 143; routine training deployment of **ac** *Tornado, Harrier, Jaguar*
CYPRUS 3,200: **Army** 2,000; 2 inf bn, 1 engr spt sqn, 1 hel flt **RAF** 1,200; 1 hel sqn (5 *Wessex* HC-2), plus **ac** on det
FALKLAND ISLANDS 1,650: **Army** 1 inf coy gp, 1 engr sqn (fd, plant) **RN** 1 DDG/FFG, 1 PCO, 1 spt, 1 AR **RAF**, 4 *Tornado* F-3, 1 *Hercules* C-1, 1 VC-10 K (tkr), 2 *Sea King* HAR-3, 2 CH-47 hel, 1 sqn RAF regt (*Rapier* SAM)
GERMANY about 20,610: **Army** 18,300; 1 corps HQ (multinational), 1 armd div **RAF** 2,300; 3 sqn with 39 *Tornado*, 1 RAF regt sqn with *Rapier* SAM
GIBRALTAR 330: **Army** 60; Gibraltar regt (350) **RN/Marines** 270; 2 PCI; Marine det, 2 twin *Exocet* launchers (coastal defence), base unit **RAF** some 116; periodic ac det
INDIAN OCEAN (*Armilla Patrol*): 2 DDG/FFG, 1 spt **Diego Garcia** 1 Marine/naval party
NEPAL Army 90 (Gurkha trg org)
NETHERLANDS RAF 137
OMAN & MUSCAT RAF 33
USA RAF 136
WEST INDIES 1 DDG/FFG, 1 spt

UN AND PEACEKEEPING

BAHRAIN (*Southern Watch*): RAF 50 1 VC-10 (tkr) **BOSNIA** (SFOR II): 2,700 (incl log and spt tps in Croatia); 1 Augmented Brigade HQ (multinational) with 2 recce sqn, 1 armd inf bn, 1 tk sqn, 2 arty bty, 1 engr sqn, 1 hel det **hel** 2 *Sea King* MK4 (RN), 3 *Lynx* AH-7 (Army), 2 *Gazelle* (Army), 3 CH-47 *Chinook* (RAF) **CYPRUS** (UNFICYP): 312: 1 bn in inf role, 1 hel flt, engr spt (incl spt for UNIFIL) **DROC** (MONUC): 6 obs **EAST TIMOR** (UNTAET): 4 obs **GEORGIA** (UNOMIG): 7 obs **IRAQ/KUWAIT** (*Southern Watch*): RAF 300; 8 *Tornado* GRI; (UNIKOM): 11 obs **ITALY** (Deliberate Forge): 350; 4 *Harrier* GR-7, 1 K-1 *Tristar* (tkr), 2 E-3D *Sentry* (periodic) **SAUDI ARABIA** (*Southern Watch*): RAF 569; 6 *Tornado* F3 **SIERRA LEONE** (UNOMSIL): 18 incl 3 obs **TURKEY** (*Northern Watch*): RAF 185; 4 *Jaguar* GR-3/3A, 2 VC-10 (tkr) **YUGOSLAVIA** (KFOR): 3,500; 1 armd bde with 1 armd, 1 armd inf, 1 inf bn, 1 arty, 1 engr regt; hel 2 *Puma*
MILITARY ADVISERS 455 in 30 countries

Foreign Forces

US 11,340: **Army** 450 **Navy** 1,220 **Air Force** 9,500; 1 Air Force HQ (3rd Air Force) 1 ftr wg, 81 cbt ac, 2 sqn with 27 F-15E, 1 sqn with 27 F-15C/D, 1 air refuelling wg with 15 KC-135, 1 Special Ops Gp with 5 MC-130P, 5 MC-130H, 1 C-130E, 8 MH-53J, 1 Recce sqn with 2 RC-135Js (ac not permanently assigned), 1 naval air flt with 2 C-12 **USMC** 170
NATO HQ Allied Naval Forces North (HQNAVNORTH), HQ East Atlantic (HQEASTLANT)

Albania Alb

leke		1998	1999	2000	2001
GDP	leke	460bn	523bn		
	US$	3.1bn	3.9bn		
per capita	US$	3,700	4,000		
Growth	%	8.0	8.0		
Inflation	%	20.1	0.4		
Debt	US$	820m	975m		
Def exp	leke	ε15.0bn	ε18.7bn		
	US$	100m	139m		
Def bdgt	leke	10.5bn	5.8bn		
	US$	70m	43m		
FMA[a] (US)	US$	0.6m	0.6m	0.6m	
FMA (Tu)	US$		5m		
US$1=leke		1.51	1.34	1.42	

Population 3,792,000

(Muslim 70%, Albanian Orthodox 20%, Roman Catholic 10%; Greek ε3–8%)

Age	13–17	18–22	23–32
Men	189,000	175,000	327,000
Women	172,000	161,000	303,000

Total Armed Forces

The Albanian armed forces are in the process of being reconstituted under Plan 2000. The army is to consist of 7 divs, plus a cdo bde of 3 bn. Restructuring was planned for completion by the beginning of year 2000, but appears not to have met this target. Eqpt details are primarily those reported prior to the unrest and should be treated with caution.

Army some 40,000

EQUIPMENT

MBT ε600: incl T-34 (in store), T-59
LT TK 35 Type-62
RECCE 15 BRDM-1
APC 103 PRC Type-531
TOWED ARTY 122mm: 425 M-1931/37, M-30, 208 PRC Type-60; **130mm**: 100 PRC Type-59-1; **152mm**: 90 PRC Type-66
MRL 107mm: 270 PRC Type-63
MOR 82mm: 259; **120mm**: 550 M-120; **160mm**: 100 M-43
RCL 82mm: T-21
ATK GUNS 45mm: M-1942; **57mm**: M-1943; **85mm**: 61 D-44 PRC Type-56; **100mm**: 50 Type-86
AD GUNS 23mm: 12 ZU-23-2/ZPU-1; **37mm**: 100 M-1939; **57mm**: 82 S-60; **85mm**: 30 KS-12; **100mm**: 56 KS-19

Navy ε2,500

BASES Durrës, Sarandë, Shëngjin, Vlorë
PATROL AND COASTAL COMBATANTS† 21
TORPEDO CRAFT 11 PRC *Huchuan* PHT with 2 533mm TT
PATROL CRAFT 10
1 Sov *Kronstadt* PCC, 1 PRC *Shanghai* II PCC, 3 Sov Po-2 PFI<, 5 (US) PB Mk3 (for Coast Guard use)<
MINE COUNTERMEASURES† 3
3 Sov T-301 MSC, (plus 3 Sov T-43 MSO in reserve)
SUPPORT 2
1 AGOR, 1 AT†

Air Force 4,500

98 cbt ac†, no armed hel
Flying hours 10–15
FGA 1 air regt with 10 J-2 (MiG-15), 14 J-6 (MiG-17), 23 J-6 (MiG-19)
FTR 2 air regt
1 with 20 J-6 (MiG-19), 10 J-7 (MiG-21)
1 with 21 J-6 (MiG-19)
TPT 1 sqn with 10 C-5 (An-2), 3 Il-14M, 6 Li-2 (C-47)
HEL 1 regt with 20 Z-5 (Mi-4), 4 SA-316, 1 Bell 222
TRG 8 CJ-5, 15 MiG-15UTI, 6 Yak-11
SAM† some 4 SA-2 sites, 22 launchers

Forces Abroad

UN AND PEACEKEEPING

BOSNIA (SFOR II): 100 **GEORGIA** (UNOMIG): 1 obs

Paramilitary

INTERNAL SECURITY FORCE 'SPECIAL POLICE': 1 bn (Tirana) plus pl sized units in major towns
BORDER POLICE (Ministry of Public Order): ε500

Foreign Forces

NATO (COMMZW): ε2,400 spt tps for KFOR

Armenia Arm

dram d		1998	1999	2000	2001
GDP	d	952bn	992bn		
	US$	1.78bn	1.85bn		
per capita	US$	2,800	2,900		
Growth	%	7.2	4.0		
Inflation	%	8.9	0.7		
Debt	US$	799m	870m		
Def exp	d	ε75bn	ε85bn		
	US$	149m	159m		
Def bdgt	d	33bn	40bn	50bn	
	US$	66m	75m	96m	
US$1=d		505	536	528	

Population			3,803,000
(Armenian Orthodox 94%, Russian 2%, Kurd 1%)			
Age	13–17	18–22	23–32
Men	185,000	176,000	294,000
Women	181,000	173,000	285,000

Total Armed Forces

incl 780 MoD and comd staff
Terms of service conscription, 24 months

RESERVES

some mob reported, possibly 210,000 with mil service within 15 years

Army 41,300

(incl 3,200 Air and AD Component; conscripts)
5 Army Corps HQ
1 with 1 mot rifle bde, 1 MRR, 1 indep recce bn
1 with 2 MRR
1 with 1 mot rifle bde, 3 MRR, 1 tk bn
1 with 3 MRR, 1 tk bn, 1 SP arty regt
1 with 2 MRR
1 mot rifle trg bde
2 arty regt (1 SP), 1 ATK regt
1 SAM bde, 2 SAM regt
1 mixed avn regt, 1 avn sqn
1 SF, 1 engr regt

EQUIPMENT (CFE-declared totals as at 1 Jan 2000)

MBT 102 T-72
AIFV 133 BMP-1, 5 BMP-2, 20 BRM-1K, 10 BMD-1
APC 11 BTR-60, 21 BTR-70, 4 BTR-80
TOTAL ARTY 229
TOWED 121: **122mm**: 59 D-30; **152mm**: 2 D-1, 34 D-20, 26 2A36
SP 38: **122mm**: 10 2S1; **152mm**: 28 2S3
MRL 51: **122mm**: 47 BM-21, 4 WM-80
MOR 120mm: 19 M-120
SSM 8 *Scud* (reported)
ATK GUNS 105: **85mm**: D-44; **100mm**: T-12
ATGW 18 AT-3 *Sagger*, 27 AT-6 *Spiral*
SAM 25 SA-2/-3, 27 SA-4, 20 SA-8, SA-13
SURV GS-13 (veh), *Long Trough* ((SNAR-1) arty), *Pork Trough* ((SNAR-2/-6) arty), *Small Fred/Small Yawn* (arty), *Big Fred* ((SNAR-10) veh/arty)

AIR AND AIR DEFENCE AVIATION FORCES (3,200 incl AD)
6 cbt ac, 12 armed hel
FGA 1 sqn with 5* Su-25, 1* MiG-25, 2 L-39
HEL 1 sqn with 7 Mi-24P* (attack), 3 Mi-24K*, 2 Mi-24R*, 7 Mi-8MT (combat support), 9 Mi-2 (utility)
TPT 1 An-24, 1 An-32
TRG CENTRE 6 An-2, 10 Yak-52, 6 Yak-55/Yak-18T

Paramilitary 1,000

MINISTRY OF INTERIOR ε1,000
4 bn: 34 BMP-1, 47 BTR-60/-70/-152

Foreign Forces

RUSSIA 3,100: **Army** 1 mil base (div) with 74 MBT, 17 APC, 148 ACV, 84 arty/MRL/mor **Air Defence** 1 sqn MiG-29, 1 SA-10 (S-300) bty

Austria

A

Austrian schilling OS		1998	1999	2000	2001
GDP	OS	2.6tr	2.7tr		
	US$	212bn	198bn		
per capita	US$	22,500	23,400		
Growth	%	3.3	2.2		
Inflation	%	0.9	1.1		
Publ Debt	%	63.5	64.9		
Def exp	OS	22.3bn	22.5bn		
	US$	1.8bn	1.7bn		
Def bdgt	OS		22.5bn	22.3bn	22.5bn
	US$		1.7bn	1.5bn	1.5bn
US$1=OS		12.4	13.5	14.6	

Population			8,138,000
Age	13–17	18–22	23–32
Men	244,000	241,000	592,000
Women	232,000	232,000	570,000

Total Armed Forces

(Air Service forms part of the Army)

ACTIVE some 35,500

(incl ε18,000 active and short term; ε17,500 conscripts; excl ε9,000 civilians; some 66,000 reservists a year undergo refresher trg, a proportion at a time)
Terms of service 7 months recruit trg, 30 days reservist refresher trg during 8 years (or 8 months trg, no refresher); 60–90 days additional for officers, NCO and specialists

RESERVES

75,000 ready (72 hrs) reserves; 990,000 with reserve trg, but no commitment. Officers, NCO and specialists to age 65, remainder to age 50

Army 35,500

(incl ε17,500 conscripts)
2 corps
1 with 2 inf bde (each 3 inf bn), 1 mech inf bde (2 mech inf, 1 tk, 1 recce, 1 SP arty bn), 1 SP arty regt, 1 recce, 2 engr, 1 ATK bn
1 with 1 inf bde (3 inf bn), 1 mech inf bde (1 mech inf, 2 tk, 1 SP arty bn), 1 SP arty regt, 1 recce, 1 engr bn

1 Provincial mil comd with 1 inf regt (plus 5 inf bn on mob)
8 Provincial mil comd (15 inf bn on mob)

EQUIPMENT

MBT 169 M-60A3 (being withdrawn), 114 *Leopard* 2A4
LT TK 152 *Kuerassier* JPz SK (plus 133 in store)
APC 465 Saurer 4K4E/F, 68 *Pandur*
TOWED ARTY 105mm: 85 IFH (M-101 - in store); **155mm**: 20 M-2A1 (in store)
SP ARTY 155mm: 189 M-109/-A2/-A5Ö
FORTRESS ARTY 155mm: 24 SFK M-2 (deactivated)
MOR 81mm: 498; **120mm**: 241 M-43
ATGW 378 RBS-56 *Bill*, 89 RJPz-(HOT) *Jaguar* 1
RCL 84mm: 2,196 *Carl Gustav*; **106mm**: 374 M-40A1 (in store)
ANTI-TANK GUNS
STATIC 105mm: some 227 L7A1 (*Centurion* tk – being deactivated)
AD GUNS 20mm: 145 (plus 323 in store)

MARINE WING

(under School of Military Engineering)
2 river patrol craft<; 10 unarmed boats

Air Force (6,500)

(ε3,400 conscripts); 52 cbt ac, 11 armed hel
Flying hours 130
1 air div HQ, 3 air regt, 3 AD regt, 1 air surv regt
FTR/FGA 1 wg with 23 SAAB J-35Oe
LIAISON 12 PC-6B
HEL
LIAISON/RECCE 11 OH-58B*
TPT 22 AB-212
UTILITY/SAR 23 SA-319 *Alouette* III
TPT 2 *Skyvan* 3M
TRG 16 PC-7, 29* SAAB 105Oe hel 11 AB-206A
MISSILES
AAM AiM-9P3
AD 76 *Mistral* with Thomson RAC 3D radars; 89 **20mm** AA guns: 74 Twin **35mm** AA towed guns with *Skyguard* radars; air surv *Goldhaube* with *Selenia* MRS-403 3D radars and Thomson RAC 3D

Forces Abroad

UN AND PEACEKEEPING

BOSNIA (SFOR II): 56 **CYPRUS** (UNFICYP): 238; 1 inf bn **GEORGIA** (UNOMIG): 5 obs **IRAQ/KUWAIT** (UNIKOM): 5 obs **MIDDLE EAST** (UNTSO): 6 obs **SYRIA** (UNDOF): 367; 1 inf bn **WESTERN SAHARA** (MINURSO): 4 obs **YUGOSLAVIA** (KFOR): 480

Azerbaijan — Az

manat m		1998	1999	2000	2001
GDP	m	16.3tr	18.0tr		
	US$	4.2bn	4.5bn		
per capita	US$	1,800	1,950		
Growth	%	10	7.4		
Inflation	%	-0.8	-8.5		
Debt	US$	693m			
Def exp	m	ε750bn	ε800bn		
	US$	193m	203m		
Def bdgt	m	463bn	472bn	520bn	
	US$	119m	120m	119m	
FMA (Tu)	US$		3m		
US$1=m		3,890	3,950	4,378	

Population 7,284,000

(Daghestani 3%, Russian 2%, Armenian 2–3% (mostly in Nagorno-Karabakh))

Age	13–17	18–22	23–32
Men	407,000	364,000	616,000
Women	386,000	340,000	621,000

Total Armed Forces

ACTIVE 72,100

Terms of service 17 months, but can be extended for ground forces

RESERVES

some mob 575,700 with mil service within 15 years

Army 61,800

3 Army Corps HQ • 1 MRD • 19 MR bde • 2 arty bde, 1 ATK regt

EQUIPMENT (CFE declared totals as at 1 Jan 2000)

MBT 220: 120 T-72, 100 T-55
AIFV 135: 44 BMP-1, 41 BMP-2, 1 BMP-3, 28 BMD-1, 21 BRM-1
APC 25 BTR-60, 28 BTR-70, 11 BTR-80, 11 BTR-D, 280 MT-LB
TOTAL ARTY 282
TOWED 144: **122mm**: 92 D-30; **152mm**: 30 D-20, 22 2A36
SP 122mm: 12 2S1
COMBINED GUN/MOR 120mm: 26 2S9
MRL 122mm: 53 BM-21
MOR 120mm: 47 PM-38
SAM 60+ SA-4/-8/-13
SURV GS-13 (veh); *Long Trough* ((SNAR-1) arty), *Pork Trough* ((SNAR-2/-6) arty), *Small Fred/Small Yawn* (veh, arty), *Big Fred* ((SNAR-10) veh, arty)

Navy 2,200†

BASE Baku

PATROL AND COASTAL COMBATANTS 9

MISSILE CRAFT 2 *Osa* II PFM with 4 SS-N-2B *Styx* SSM

PATROL, INSHORE 7

5 *Stenka* PFI<, 1 *Zhuk* PCI<, 1 *Svetlyak* PCI<

MINE COUNTERMEASURES 5

3 *Sonya* MSC, 2 *Yevgenya* MSI

AMPHIBIOUS 4

4 *Polnochny* LSM capacity 180 tps

SUPPORT AND MISCELLANEOUS 3

1 *Vadim Popov* (research), 2 *Balerian Uryvayev* (research)

Air Force and Air Defence 8,100

50 cbt ac, 15 attack hel

FGA regt with 4 Su-17, 5 Su-24, 2 Su-25, 5 MiG-21

FTR sqn with 29* MiG-25, 3* MiG-25UB

RECCE sqn with 2* MiG-25

TPT 4 ac (1 An-12, 3 Yak-40)

TRG 28 L-29, 12 L-39

HEL 1 regt with 7 Mi-2, 13 Mi-8, 15* Mi-24

IN STORE 33 **ac** MiG-25, MiG-21, Su-24

SAM 100 SA-2/-3/-5

Forces Abroad

UN AND PEACEKEEPING

YUGOSLAVIA (KFOR II): 34

Paramilitary ε15,000+

MILITIA (Ministry of Internal Affairs) 10,000+

EQPT incl 3 T-55 MBT; 17 ACV incl BMP-1, BMD, BTR-60/-70/-80; 2 122mm D-30 arty; 2 120mm mor

BORDER GUARD (Ministry of Internal Affairs) ε5,000

EQPT incl 100 BMP-2 AIFV, 19 BTR-60/-70/-80 APC

Opposition

ARMENIAN ARMED GROUPS

ε15–20,000 in Nagorno-Karabakh, perhaps 40,000 on mob

(incl ε8,000 personnel from Armenia)

EQPT (reported) 316 incl T-72, T-55 MBT; 324 ACV incl BTR-70/-80, BMP-1/-2; 322 arty incl D-44, 102 D-30, 53 D-20, 99 2A36, 44 BM-21, KS-19

Belarus — Bel

rubel r		1998	1999	2000	2001
GDP	r	675tr	2.9tr		
	US$	11.7bn	9.3bn		
per capita	US$	6,800	7,100		
Growth	%	8.3	3.0		
Inflation	%	73	293		
Debt	US$	1,120m	1,312m		
Def exp	r	ε26.7tr	ε14.5tr		
	US$	462m	466m		
Def bdgt	r	9.9tr	29bn	74bn	
	US$	170m	94m	75m	
FMA[a] (US)	US$	0.1m			
US$1=r		57,850	311	976	

[a] Excl US Cooperative Threat Reduction programme: **1992–96** US$119m budget, of which US$44m spent by Sept 1996. Programme continues through 1999

Population 10,045,000

(Russian 13%, Polish 4%, Ukrainian 3%)

Age	13–17	18–22	23–32
Men	406,000	394,000	711,000
Women	392,000	384,000	708,000

Total Armed Forces

ACTIVE 83,100

(incl 17,100 in centrally controlled units and MoD staff, 2,100 women; 40,000 conscripts)

Terms of service 18 months

RESERVES some 289,500

with mil service within last 5 years

Army 43,500

MoD tps: 1 MRD (1 trg), 3 indep mob bde, 1 arty div (5 'bde')

2 SSM, 1 ATK, 1 *Spetsnaz*

3 Corps

1 with 3 indep mech, 1 SAM bde, 1 arty, 1 MRL, 1 ATK regt

1 with 1 SAM bde, 1 arty, 1 MRL regt

1 with 1 SAM bde, 1 arty, 1 ATK, 1 MRL regt

EQUIPMENT (CFE declared totals as at 1 Jan 2000)

MBT 1,724 (238 in store): 60 T-55, 1,569 T-72, 95 T-80

AIFV 1,560 (53 in store): 81 BMP-1, 1,164 BMP-2, 161 BRM, 154 BMD-1

APC 918 (306 in store): 188 BTR-60, 445 BTR-70, 193 BTR-80, 22 BTR-D, 70 MT-LB

TOTAL ARTY 1,465 (153 in store) incl

TOWED 428: **122mm**: 178 D-30; **152mm**: 6 M-1943 (D-1), 58 D-20, 136 2A65, 50 2A36

SP 572: **122mm**: 235 2S1; **152mm**: 168 2S3, 120 2S5; **152mm**: 13 2S19; **203mm**: 36 2S7

COMBINED GUN/MOR 120mm: 54 2S9

MRL 334: **122mm**: 208 BM-21, 1 9P138; **130mm**: 1

BM-13; **220mm**: 84 9P140; **300mm**: 40 9A52
MOR 120mm: 77 2S12
ATGW 480: AT-4 *Spigot*, AT-5 *Spandrel* (some SP), AT-6 *Spiral* (some SP), AT-7 *Saxhorn*
SSM 60 *Scud*, 36 FROG/SS-21
SAM 350 SA-8/-11/-12/-13
SURV GS-13 (arty), *Long Trough* ((SNAR-1) arty), *Pork Trough* ((SNAR-2/-6) arty), *Small Fred/Small Yawn* (veh, arty), *Big Fred* ((SNAR-10) veh, arty)

Air Force 22,500

(incl 10,200 Air Defence); 230 cbt ac, 60 attack hel
Flying hours 28
FGA 36 Su-24, 80 Su-25
FTR 35 MiG-23, 50 MiG-29, 23 Su-27
RECCE 6* Su-24
HELICOPTERS
ATTACK 55 Mi-24, 4 Mi-24R, 1 Mi-24K
CBT SPT 29 Mi-6, 131 Mi-8, 8 Mi-24K, 4 Mi-24R
TPT ac 16 Il-76, 3 An-12, 1 An-24, 6 An-26, 1 Tu-134 **hel** 14 Mi-26
AWAITING DISPOSAL 6 MiG-23, 2 MiG-25, 30 Su-17, 2 Su-25, 2 Mi-24
MISSILES
AAM AA-7, AA-8, AA-10, AA-11
ASM AS-10, AS-11, AS-14

AIR DEFENCE FORCE (10,000)
SAM 175 SA-3/-5/-10

Paramilitary 110,000

BORDER GUARDS (Ministry of Interior) 12,000
MINISTRY OF INTERIOR TROOPS 11,000
MILITIA (Ministry of Interior) 87,000

Bosnia-Herzegovina BiH

convertible mark		1998	1999	2000	2001
GDP	US$	ε3.9bn	ε4.4bn		
per capita	US$	ε6,300	ε7,000		
Growth	%	18	8		
Inflation	%	10	5		
Debt	US$	2.9bn	8.1bn		
Def exp[a]	US$	ε397m	ε365m		
Def bdgt[a]	US$	227m	318m	163m	
FMA[bc] (US)	US$	0.6m	0.6m	0.6m	0.8m
$1=convertible mark		1.5	1.85	1.86	

[a] Excl Bosnian Serb def exp
[b] Eqpt and trg valued at εUS$450m from US, Sau, Kwt, UAE, Et and Tu in 1996–99
[c] UNMIBH **1997** US$190m **1998** US$190m; SFOR **1997** εUS$4bn **1998** US$4bn

Population			ε**3,600,000**
(Bosnian Muslim 44%, Serb 33%, Croat 17%)			
Age	*13–17*	*18–22*	*23–32*
Men	194,000	186,000	338,000
Women	184,000	176,000	318,000

Total Armed Forces

ACTIVE see individual entries below
BiH and HVO forces are to merge and form the armed forces of a Muslim–Croat Federation with a probable structure of 4 Corps (3 Muslim, 1 Croat), 14 bde, 1 rapid-reaction force (bde), and an arty div. It is reported that this force will be equipped with 250 MBT (200), 240 ACV (200), 2,000 arty (1,000). Figures in ( ) denote eqpt delivered in country, but under US control

Army (BiH) some 30,000

1 'Army' HQ • 3 'Corps' HQ • 8 div HQ • 1 armd, 10 mot inf, 4 arty bde

RESERVES 150,000: 59 inf, 1 arty bde
EQUIPMENT
MBT 170: T-34, T-55, AMX-30, M-60A3
RECCE 44 AML-90
APC 150 incl M-113A2
TOTAL ARTY 1,500
ARTY incl **105mm:** 36 L-118; **122mm:** 12 D-30; **130mm:** M-59; **155mm:** 126 M-114; **203mm**
MRL incl **262mm**: M-87 *Orkan*
MOR 82mm; 120mm
ATGW 250 AT-3 *Sagger*, *Red Arrow* (TF-8) reported
AD GUNS 20mm; 23mm: 19 ZU-23; **30mm**
SAM SA-7/-14
HEL 10 Mi-8/-17, 15 UH-1H
AC 3 UTVA-75

Other Forces

CROAT (Croatian Defence Council (HVO)) 10,000
4 MD • 4 gd, 2 arty bde, 1 inf bn, 1 MP bn
RESERVES 40,000: 12 Home Guard inf regt, 6 Home Guard inf bn
EQUIPMENT
MBT 80, incl T-34, T-55, M-84/T-72M, M-47
AFV 90 M-60, M-80
TOTAL ARTY some 500 incl
TOWED incl **76mm**: M-48; **105mm**: M-56; **122mm**: D-30; **130mm**: M-46
MRL 122mm: BM-21; **128mm**: M-63, M-77
MOR 82mm
ATGW 50 AT-3 *Sagger*, AT-4 *Fagot*, AT-6 (reported)
RL ε100 *Armbrust*, M-79, RPG-7/-22
RCL 84mm: 30 *Carl Gustav*

AD GUNS 20mm: M-55, Bov-3; **30mm**: M-53; **57mm**: S-60
SAM SA-7/-9/-14/-16
HEL Mi-8

SERB (Army of the Serbian Republic of Bosnia and Herzegovina–VRS) 30,000

4 'Corps' HQ • 38 inf/armd/mot inf bde • 1 SF 'bde' • 12 arty/ATK/AD regt

RESERVES 80,000

EQUIPMENT

MBT 250+ incl T-55, M-84, T-72
APC 350
TOTAL ARTY ε750 incl
TOWED 122mm: D-30, M-1938 (M-30); **130mm**: M-46; **152mm**: D-20
SP 122mm: 2S1
MRL 128mm: M-63; **262mm**: M-87 *Orkan*
MOR 120mm
SSM FROG-7
ATGW ε500
AD GUNS 975: incl **20mm**, **23mm** incl ZSU 23-4; **30mm**: M53/59SP; **57mm**: ZSU-57-2; **90mm**
SAM SA-2, some SA-6/-7B/-9
AC 6 *Orao*, 13 *Jastreb*, 1 *Super Galeb*
HEL 20 SA-341, 10 Mi-8

Foreign Forces

NATO (SFOR II): about 20,000: Be, Ca, Da, Fr, Ge, Gr, Hu, It, Nl, No, Pl, Por, Sp, Tu, UK, US **Non-NATO** Alb, A, Cz, Ea, Lat, L, Mor, R, RF

Bulgaria Bg

leva L		1998	1999	2000	2001
GDP	L	22bn	23bn		
	US$	12.2bn	12.0bn		
per capita	US$	4,200	4,400		
Growth	%	3.5	2.5		
Inflation	%	18.8	2.6		
Debt	US$	9.9bn	9.9bn		
Def exp	L	700m	750m		
	US$	398m	392m		
Def bdgt	L	487m	561m	723m	802m
	US$	277m	293m	351m	389m)
FMA (US)	US$	1.0m	1.0m	$1.0m	
US$1=L		1.76	1.91	2.06	

Population			8,231,000
(Turkish 9%, Macedonian 3%, Romany 3%)			
Age	13–17	18–22	23–32
Men	283,000	306,000	594,000
Women	268,000	290,000	568,000

Total Armed Forces

ACTIVE 79,760

(incl about 12,500 centrally controlled staff, 1,300 MoD staff, but excl some 10,000 construction tps; perhaps 49,000 conscripts)
Terms of service 12 months

RESERVES 303,000

Army 250,500 **Navy** (to age 55, officers 60 or 65) 7,500 **Air Force** (to age 60) 45,000

Army 42,400

(incl ε33,300 conscripts)
3 Mil Districts/Corps HQ
1 with 1 MRD, 1 tk, 2 mech bde • 1 with 1 MRD, 1 Regional Training Centre (RTC), 1 tk bde • 1 with 2 MRD, 2 tk, 1 mech bde
Army tps: 4 *Scud*, 1 SS-23, 1 SAM bde, 2 arty, 1 MRL, 3 ATK, 3 AD arty, 1 SAM regt
1 AB bde

EQUIPMENT

MBT 1,475: 1,042 T-55, 433 T-72
ASLT GUN 68 SU-100
RECCE 58 BRDM-1/-2
AIFV 100 BMP-1, 114 BMP-23, BMP-30
APC 1,750: 737 BTR-60, 1,013 MT-LB (plus 1,270 'look-alikes')
TOTAL ARTY 1,750 (CFE total as at 1 Jan 2000)
TOWED 100mm: M-1944 (BS-3); **122mm**: 195 M-30, M-1931/37 (A-19); **130mm**: 72 M-46; **152mm**: M-1937 (ML-20), 206 D-20
SP 122mm: 692 2S1
MRL 122mm: 222 BM-21
MOR 120mm: M-38, 2S11, B-24, 359 *Tundzha* SP
SSM launchers: 28 FROG-7, 36 *Scud*, 8 SS-23
ATGW 200 AT-3 *Sagger*
ATK GUNS 85mm: 150 D-44; **100mm**: 200 T-12
AD GUNS 400: **23mm**: ZU-23, ZSU-23-4 SP; **57mm**: S-60; **85mm**: KS-12; **100mm**: KS-19
SAM 20 SA-3, 27 SA-4, 20 SA-6
SURV GS-13 (veh), *Long Trough* ((SNAR-1) arty), *Pork Trough* ((SNAR-2/-6) arty), *Small Fred/Small Yawn* (veh, arty), *Big Fred* ((SNAR-10) veh, arty)

Navy ε5,260

(incl ε2,000 conscripts)
BASES Coastal Varna (HQ), Atya **Danube** Vidin (HQ), Balchik, Sozopol. Zones of operational control at Varna and Burgas

SUBMARINES 1

SSK 1 *Pobeda* (Sov *Romeo*)-class with 533mm TT†

FRIGATES 1

FF 1 *Smeli* (Sov *Koni*) with 1 × 2 SA-N-4 *Gecko* SAM, 2 × twin 76mm guns, 2 × 12 ASW RL

PATROL AND COASTAL COMBATANTS 23

CORVETTES 7

1 *Tarantul* II FSG with 2 × 2 SS-N-2C *Styx* SSM, 2 × 4 SA-N-5 *Grail* SAM, 1 × 76mm gun
4 *Poti* FS with 2 ASW RL, 4 ASTT
2 *Pauk* I FS with 1 SA-N-5 *Grail* SAM, 2 × 5 ASW RL, 4 × 406mm TT, 2 × 5 ASW RL

MISSILE CRAFT 6 *Osa* I/II PFM with 4 SS-N-2A/B *Styx* SSM

PATROL, INSHORE 10
10 *Zhuk* PFI<

MINE WARFARE 20

MINE COUNTERMEASURES 20
4 *Sonya* MSC, 4 *Vanya* MSC, 4 *Yevgenya* MSI<, 6 *Olya* MSI<, 2 PO-2 MSI<

AMPHIBIOUS 2 Sov *Polnocny A* LSM, capacity 150 tps, 6 tk
Plus 6 LCU

SUPPORT AND MISCELLANEOUS 16
3 AO, 1 diving tender, 1 degaussing, 1 AT, 7 AG; 3 AGHS

NAVAL AVIATION

9 armed hel
HEL 1 ASW sqn with 9 Mi-14
COASTAL ARTY 2 regt, 20 bty
GUNS 100mm: ε150; **130mm**: 4 SM-4-1
SSM SS-C-1B *Sepal*, SSC-3 *Styx*

NAVAL GUARD

3 coy

Air Force 18,300

181 cbt ac, 43 attack hel, 1 Tactical Aviation corps, 1 AD corps
Flying hours 30–40
FGA 3 regt
1 with 39 Su-25 (35 -A, 4 -UB)
FTR 3 regt with some 30 MiG-23 (being progressively withdrawn), 60 MiG-21 bis, 21 MiG-29 (17 -A, 4 -UB)
RECCE 1 regt with 21 Su-22* (18 -M4, 3 -UM3), 10 MiG-21MF/UM*
TARGET FACILITIES 12 L-29 operated by front-line sqns
TPT 1 regt with 2 Tu-134, 2 An-24, 5 An-26, 6 L-410, 1 Yak-40 (VIP)
SURVEY 1 An-30 (*Open Skies*)
HEL 2 regt
1 with 43 Mi-24 (attack)
1 with 8 Mi-8, 31 Mi-17, 6 Bell-206
TRG 2 trg schools with 12 L-29 (basic), 30 L-39ZA (advanced)
MISSILES
ASM AS-7 *Kerry*, AS-14 *Kedge*
AAM AA-2 *Atoll*, AA-7 *Apex*, AA-8 *Aphid*, AA-11 *Archer*
SAM SA-2/-3/-5/-10 (20 sites, some 110 launchers)

Forces Abroad

UN AND PEACEKEEPING

BOSNIA (SFOR II): 1 pl

Paramilitary 34,000

BORDER GUARDS (Ministry of Interior) 12,000
12 regt; some 50 craft incl about 12 Sov PO2 PCI<
SECURITY POLICE 4,000
RAILWAY AND CONSTRUCTION TROOPS 18,000

Croatia Cr

kuna k		1998	1999	2000	2001
GDP	k	138bn	142bn		
	US$	18.5bn	19.0bn		
per capita	US$	6,700	6,700		
Growth	%	3.0	-2.0		
Inflation	%	5.7	3.5		
Debt	US$	8.3bn	9.1bn		
Def exp	k	6.8bn	5.8bn		
	US$	1.1bn	776m		
Def bdgt	k	7.5bn	6.1bn	4.8bn	
	US$	1.2bn	814m	590m	
FMA[a] (US)	US$	0.4m	0.4m	0.6m	0.5m
US$1=k		6.36	7.47	8.1	

[a] UNTAES **1997** US$266m; UNMOP (UNMIBH) **1997** US$190m **1998** US$190m

Population	ε**4,500,000** (Serb 3%, Slovene 1%)		
Age	13–17	18–22	23–32
Men	164,000	169,000	330,000
Women	155,000	159,000	316,000

Total Armed Forces

ACTIVE 61,000
(incl ε21,320 conscripts)
Terms of service 9 months

RESERVES 220,000
Army 150,000 **Home Defence** 70,000

Army 53,000

(incl ε20,000 conscripts)
6 Mil Districts • 8 Guard bde (each 3 mech, 1 tk, 1 arty bn) • 3 inf 'bde' (each 3 inf bn, 1 tk, 1 arty unit) • 1 mixed arty/MRL bde • 1 ATK bde • 4 AD bde • 1 engr bde

RESERVES

26 inf 'bde' (incl 1 trg), 39 Home Def regt

EQUIPMENT

MBT 305: T-34, T-55, M-84/T-72M

LT TK 1 PT-76
RECCE 7 BRDM-2
AIFV 109 M-80
APC 18 BTR-50, 20 M-60PB plus 7 'look-alikes'
TOTAL ARTY some 1,004
TOWED 423: **76mm**: ZIS-3; **85mm; 105mm**: 100 M-56, 50 M-2A1; **122mm**: 50 M-1938, 40 D-30; **130mm**: 83 M-46; **152mm**: 20 D-20, 21 M-84; **155mm**: 37 M-1; **203mm**: 22
SP 122mm: 8 2S1
MRL 122mm: 40 BM-21; **128mm**: 200 M-63/-91; **262mm**: 2 M-87 *Orkan*
MOR 1,000 incl: **82mm; 120mm**: 325 M-75, 6 UBM-52; **240mm**: reported
ATGW AT-3 *Sagger* (9 on BRDM-2), AT-4 *Spigot*, AT-7 *Saxhorn*, *Milan* reported
RL 73mm: RPG-7/-22. **90mm**: M-79
ATK GUNS 100mm: 133 T-12
AD GUNS 600+: **14.5mm**: ZPU-2/-4; **20mm**: BOV-1 SP, M-55; **30mm**: M-53/59, BOV-3SP

Navy 3,000

BASES Split, Pula, Sibenik, Ploce, Dubrovnik **Minor facilities** Lastovo, Vis
SUBMARINES 1
SSI 1 *Velebit* (Mod *Una*) for SF ops (4 SDV or 4 mines)
PATROL AND COASTAL COMBATANTS 8
MISSILE CRAFT 2
1 *Kralj Petar* PFM with 4 or 8 RBS-15 SSM
1 *Rade Koncar* PFM with 4 RBS-15 SSM
PATROL, COASTAL/INSHORE 6
1 *Dubrovnik* (Mod Sov *Osa* 1) PFC, can lay mines
4 *Mirna* PCC, 1 RLM-301 PCI< plus 5 PCR
MINE WARFARE 0
but *Silba* LCT has minelaying capability
AMPHIBIOUS craft only
2 *Silba* LCT, 1 DTM LCT/ML, 1 DSM-501 LCT/ML, 7 LCU
SUPPORT AND MISCELLANEOUS 4
2 AT, 1 *Spasilac* ARS, 1 Sov *Moma* AGHS

MARINES
2 indep inf coy

COASTAL DEFENCE
some 10 coast arty bty, 2 RBS-15 SSM bty (reported)

Air Force 5,000

(incl 1,320 conscripts)
41+ cbt ac, 10 armed hel
Flying hours 50
FGA/FTR 2 sqn with 20 MiG-21 bis/4 MiG-21 UM, some *Kraguj* and *Jastreb*
TPT 4 An-2, 2 An-26, 2 An-32, 5 UTVA, 2 Do-28
HEL 16 Mi-8/17, 10* Mi-24, 2 MD-500, 1 UH-1, 9 Bell-206B
TRG 17* PC-9, 1 *Galeb*
AAM AA-2 *Atoll*, AA-8 *Aphid*
SAM SA-7, SA-9, SA-10 (reported to be returned this year), SA-14/-16

Forces Abroad

UN AND PEACEKEEPING
SIERRA LEONE (UNAMSIL): 10 obs

Paramilitary 40,000

POLICE 40,000 armed
COAST GUARD boats only

Foreign Forces

UN (UNMOP): 27 obs from 23 countries; (SFOR II): n.k.

Cyprus Cy

pound C£		1998	1999	2000	2001
GDP	C£	4.7bn	4.9bn		
	US$	9.1bn	8.7bn		
per capita	US$	12,300	13,000		
Growth	%	5.0	4.5		
Inflation	%	2.2	1.5		
Debt	US$	7.2bn			
Def exp	C£	256m	300m		
	US$	499m	530m		
Def bdgt	C£	265m	300m	330m	
	US$	515m	530m	573m	
US$1=C£		0.51	0.57	0.6	

UNFICYP **1997** US$46m **1998** US$45m

Population		**880,000** (Turkish 23%)	
Age	13–17	18–22	23–32
Men	33,000	29,000	54,000
Women	31,000	27,000	51,000

Total Armed Forces

ACTIVE 10,000
(incl 8,700 conscripts; 423 women)
Terms of service conscription, 26 months, then reserve to age 50 (officers 65)

RESERVES
60,000 all services

National Guard 10,000

(incl 8,700 conscripts) (all units classified non-active under Vienna Document)

1 Corps HQ, 1 air comd, 1 naval comd • 2 lt inf div HQ • 2 lt inf bde HQ • 1 armd bde (3 bn) • 1 arty comd (regt) • 1 Home Guard comd • 1 SF comd (regt of 3 bn)

EQUIPMENT

MBT 104 AMX-30 (incl 52 -B2), 41 T-80U
RECCE 124 EE-9 *Cascavel*, 15 EE-3 *Jararaca*
AIFV 27 VAB-VCI, 43 BMP-3
APC 268 *Leonidas*, 118 VAB (incl variants), 16 AMX-VCI
TOWED ARTY 75mm: 4 M-116A1 pack; **88mm**: 36 25-pdr (in store); **100mm**: 20 M-1944; **105mm**: 72 M-56; **155mm**: 12 TR F1
SP ARTY 155mm: 12 F3
MRL 128mm: 18 FRY M-63
MOR 376+: **81mm**: 170 E-44, 70+ M1/M29 (in store); **107mm**: 20 M-30/M-2; **120mm**: 116 RT61
ATGW 45 *Milan* (15 on EE-3 *Jararaca*), 22 HOT (18 on VAB)
RL 66mm: M-72 LAW; **73mm**: 850 RPG-7; **112mm**: 1,000 *Apilas*
RCL 90mm: 40 EM-67; **106mm**: 144 M-40A1
AD GUNS 20mm: 36 M-55; **35mm**: 24 GDF-003 with *Skyguard*; **40mm**: 20 M-1 (in store)
SAM 60 *Mistral* (some SP), 24 *Aspide*, 6 SA-15, SA-10 (S-300 deployed on Greek island of Crete)

MARITIME WING

1 *Kyrenia* (Gr *Dilos*) PCC
1 *Salamis* PCC< (plus 11 boats)
1 coastal def SSM bty with 3 MM-40 *Exocet*

AIR WING

AC 1 BN-2 *Islander*, 2 PC-9
HEL 3 Bell 206C, 4 SA-342 *Gazelle* (with HOT), 2 Mi-2 (in store)

Paramilitary some 750

ARMED POLICE about 500

1 mech rapid-reaction unit (350), 2 VAB/VTT APC, 1 BN-2A *Maritime Defender* ac, 2 Bell 412 hel

MARITIME POLICE 250

2 *Evagoras* PFI, 1 *Shaltag* PFI, 5 SAB-12 PCC

Foreign Forces

GREECE 1,250: 1 mech inf bde incl 950 (ELDYK) (Army); 2 mech inf, 1 armd, 1 arty bn, plus ε200 officers/NCO seconded to Greek-Cypriot National Guard
Eqpt: 61 M-48A5 MOLF MBT, 80 *Leonidas* APC (from National Guard), 12 M-114 155mm towed arty, 6 M-110A2 203mm SP arty
UK (in Sovereign Base Areas) 3,200: **Army** 2,000; 2 inf bn, 1 eng spt sqn, 1 hel flt **Air Force** 1,200; 1 hel sqn, plus ac on det
UN (UNFICYP) some 1,209; 3 inf bn (Arg, A, UK), tps from Ca, SF, Hu, Irl, N, Nl, Slvn, plus 33 civ pol from 2 countries

'Turkish Republic of Northern Cyprus'

Data presented here represent the *de facto* situation on the island. This in no way implies international recognition as a sovereign state.

(Turkish) lira L		1998	1999	2000	2001
GNP	US$	ε950m			
per capita	US$	ε5,100			
Def exp (Tu)	US$	ε700m			
Def bdgt	L		17.3tr		
	US$		41m		
Population					ε215,000

Total Armed Forces

ACTIVE ε5,000

Terms of service conscription, 24 months, then reserve to age 50

RESERVES 26,000

11,000 **first-line** 10,000 **second-line** 5,000 **third-line**

Army ε5,000

7 inf bn

EQUIPMENT

MOR 120mm: 73
ATGW 6 *Milan*
RCL 106mm: 36

Paramilitary

ARMED POLICE ε150

1 Police SF unit

COAST GUARD

(operated by TRNC Security Forces)
1 *Raif Denktash* PCC • 2 ex-US Mk5 PCC • 2 SG45/SG46 PCC • 1 PCI

Foreign Forces

TURKEY

ARMY ε36,000 (mainly conscripts)

1 Corps HQ, 2 Inf div, 1 armd bde, 1 indep mech inf bde

EQUIPMENT

MBT 386 M-48A5 T1/T2, 8 M-48A2 (trg)
APC 211 AAPC, 265 M-113
TOWED ARTY 105mm: 72 M-101A1; **155mm**: 18 M-114A2; **203mm**: 12 M-115
SP ARTY 105mm: 36 M-52A1; **155mm**: 24 M-44T
MOR 81mm: 175; **107mm**: 148 M-30; **120mm**: 54 HY-12
ATGW 66 *Milan*, 48 TOW
RL 66mm: M-72 LAW

RCL 90mm: M-67; **106mm**: 156 M-40A1
AD GUNS 20mm: Rh 202; **35mm**: 16 GDF-003; **40mm**: 48 M-1
SAM 50+ *Stinger*
SURV AN/TPQ-36
AC 3 U-17. Periodic det of F-16C/D, F-4E
HEL 4 UH-1H. Periodic det of S-70A, AS-532UL, AH-1P

NAVY

1 *Caner Goyneli* PCI

Estonia Ea

kroon kn		1998	1999	2000	2001
GDP	kn	73.2bn	75.4bn		
	US$	4.7bn	4.5bn		
per capita	US$	8,600	8,600		
Growth	%	4.0	-1.3		
Inflation	%	10.7	1.0		
Debt	US$	781m			
Def exp[a]	kn	843m	1,083m		
	US$	60m	71m		
Def bdgt	kn	843m	1,134m	1,328m	1,460m
	US$	60m	74m	80m	88m
FMA (US)	US$	0.7m	0.7m	0.8m	0.8m
US$1=kn		14.1	15.4	16.5	

[a] Incl exp on paramilitary forces

Population			1,450,000
(Russian 28%, Ukrainian 3%, Belarussian 2%)			
Age	13–17	18–22	23–32
Men	59,000	57,000	104,000
Women	56,000	55,000	101,000

Total Armed Forces

ACTIVE some 4,800
(incl 2,870 conscripts)
Terms of service 12 months

RESERVES some 14,000
Militia

Army 4,320

(incl 2,600 conscripts)
4 Defence Regions, 14 Defence Districts, 5 inf, 1 arty • 1 guard, 1 recce bn • 1 peace ops centre

RESERVES

Militia 7,500, 15 *Kaitseliit* (Defence League) units

EQUIPMENT

RECCE 7 BRDM-2
APC 32 BTR-60/-70/-80
TOWED ARTY 105mm: 19 M 61-37
MOR 81mm: 44; **120mm**: 14
ATGW 10 *Mapats*, 3 RB-56 *Bill*
RL 82mm: 200 B-300
RCL 84mm: 109 *Carl Gustav*; **90mm**: 100 PV-1110; **106mm**: 30 M-40A1
AD GUNS 23mm: 100 ZU-23-2

Navy 250

(incl 140 conscripts)
Lat, Ea and L have set up a joint Naval unit BALTRON
BASES Tallinn (HQ BALTRON), Miinisadam (Navy and BALTRON)

PATROL CRAFT 3

PATROL, COASTAL/INSHORE 3
1 *Kondor* PCC, 1 *Ahti* (*Da Maagen*) PCC, 1 *Grif* (RF *Zhuk*) PCI<

MINE WARFARE 5

MINELAYERS 2

2 *Rihtiniemi* ML

MINE COUNTERMEASURES 3

1 *Lindau* (Ge) MHC
2 *Kalev* (Ge *Frauenlob*) MSI

SUPPORT AND MISCELLANEOUS 1

1 *Laine* (Ru *Mayak*) AK

Air Force 140

1 air base and 1 air surveillance div
(incl 50 conscripts)
Flying hours 70
ac 2 An-2 (another expected this year), 1 PZL-140 *Wilga* **hel** 3 Mi-2, 4 Robinson R-44

Forces Abroad

UN AND PEACEKEEPING

BOSNIA (SFOR II): 46 **MIDDLE EAST** (UNTSO): 1 obs

Paramilitary 2,800

BORDER GUARD (Ministry of Internal Affairs) 2,800 (360 conscripts); 1 regt, 3 rescue coy; maritime elm of Border Guard also fulfils task of Coast Guard

BASES Tallinn
PATROL CRAFT 20
PATROL, OFFSHORE 3
1 *Kou* (*Silma*), 1 *Linda* (*Kemio*), 1 *Valvas* (US *Bittersweet*)
PATROL, COASTAL 6
3 PVL-100 (*Koskelo*), 1 *Pikker*, 1 *Torm* (*Arg*), 1 *Maru* (*Viima*)
PATROL, INSHORE 11
3 PVK-001 (Type 257 KBV), 1 PVK-025 (Type 275 KBV), 5 PVK-006 (RV), 1 PVK-010 (RV 90), 1 PVK-017, plus 2 LCU
SPT AND MISC 1 *Linda* (SF *Kemio*) PCI (trg)
AVN 2 L-410 UVP-1 *Turbolet*, 5 Mi-8 (In war, subordinated to Air Force staff)

Finland SF

markka m		1998	1999	2000	2001
GDP	m	686bn	718bn		
	US$	128bn	123bn		
per capita	US$	21,000	22,200		
Growth	%	5.0	3.6		
Inflation	%	1.4	1.2		
Publ debt	%	61.5	63.4		
Def exp	m	10.1bn	9.9bn		
	US$	1.9bn	1.7bn		
Def bdgt[a]	m	10.0bn	9.0bn	10.2bn	10.3bn
	US$	1.9bn	1.7bn	1.6bn	1.6bn
US$1=m		5.34	5.84	6.29	

[a] Excl supplementary multi-year budget for procurement of m6.1bn (US$1.1bn) approved in April 1998

Population			5,183,000
Age	13–17	18–22	23–32
Men	167,000	170,000	336,000
Women	158,000	162,000	322,000

Total Armed Forces

ACTIVE 31,700

(incl 23,100 conscripts, some 500 women)
Terms of service 6–12 months (12 months for officers, NCO and soldiers with special duties)

RESERVES some 485,000 (to be 430,000)

Total str on mob some 485,000 (all services), with 100,000 operational forces, 27,000 territorial forces and 75,000 in local forces. Some 35,000 reservists a year do refresher trg: total obligation 40 days (75 for NCO, 100 for officers) between conscript service and age 50 (NCO and officers to age 60)

Army 24,000 (to be 315,000 on mob)

(incl 19,000 conscripts)
(all bdes reserve, some with peacetime trg role)
3 Mil Comd
- 1 with 6 mil provinces, 2 armd (1 trg), 3 *Jaeger* (trg), 9 inf bde
- 1 with 2 mil provinces, 3 *Jaeger* (trg) bde
- 1 with 4 mil provinces, 4 *Jaeger* (trg), 5 inf bde

Other units
- 3 AD regt, 4 engr bn

RESERVES

some 150 local bn and coy

EQUIPMENT

MBT 70 T-55M, 160 T-72
AIFV 163 BMP-1PS, 110 BMP-2 (incl 'look-alikes')
APC 120 BTR-60, 450 XA-180/185/200 *Sisu*, 220 MT-LBV (incl 'look-alikes')
TOWED ARTY 122mm: 486 H 63 (D-30); **130mm**: 36 K 54, **152mm**: 324 incl: H 55 (D-20), H 88-40, H 88-37 (ML-20), H 38 (M-10); **155mm**: 108 M-74 (K-83), 24 K 98
SP ARTY 122mm: 72 PsH 74 (2S1); **152mm**: 18 *Telak* 91 (2S5)
MRL 122mm: 24 Rak H 76 (BM-21), 36 Rak H 89 (RM-70)
MOR 81mm: 800; **120mm**: 789: KRH 40, KRH 92
ATGW 100: incl 24 M-82 (AT-4 *Spigot*), 12 M-83 (BGM-71D TOW 2), M-82M (AT-5 *Spandrel)*
RL 112mm: APILAS
RCL 66mm: 66 KES-75, 66 KES-88; **95mm**: 100 SM-58-61
AD GUNS 23mm: 400 ZU-23; **30mm; 35mm**: GDF-005, *Marksman* GDF-005 SP; **57mm**: 12 S-60 towed, 12 ZSU-57-2 SP
SAM SAM-86M (SA-18), SAM-86 (SA-16), 20 SAM-90 (*Crotale* NG), 18 SAM-96 (SA-11)
SURV *Cymbeline* (mor)
HEL 4 Hughes 500D/E, 7 Mi-8

Navy 5,000

(incl 2,600 conscripts)
BASES Upinniemi (Helsinki), Turku
4 functional sqn (2 missile, 2 mine warfare). Approx 50% of units kept fully manned; others in short-notice storage, rotated regularly

PATROL AND COASTAL COMBATANTS 10

CORVETTES 1 *Turunmaa* FS with 1 × 120mm gun, 2 × 5 ASW RL

MISSILE CRAFT 9
- 4 *Helsinki* PFM with 4 × 2 MTO-85 (Swe RBS-15SF) SSM
- 4 *Rauma* PFM with 2 × 2 and 2 × 1 MTO-85 (Swe RBS-15SF) SSM, 1 × 6 *Mistral* SAM
- 1 *Hamina* PFM with 6 RBS 15 SF SSM, 1 × 6 *Mistal* SAM

MINE WARFARE 23

MINELAYERS 10
- 2 *Hämeenmaa*, 150–200 mines, plus 1 × 6 Matra *Mistral* SAM
- 1 *Pohjanmaa*, 100–150 mines; 2 × 5 ASW RL
- 3 *Pansio* aux minelayer, 50 mines
- 4 *Tuima* (ex-PFM), 20 mines

MINE COUNTERMEASURES 13
- 6 *Kuha* MSI, 7 *Kiiski* MSI

AMPHIBIOUS craft only
- 3 *Kampela* LCU tpt, 2 *Kala* LCU

SUPPORT AND MISCELLANEOUS 37
- 1 *Kustaanmiekka* command ship, 5 *Valas* tpt, 6 *Hauki* tpt, 4 *Hila* tpt, 2 *Lohi* tpt, 1 *Aranda* AGOR (Ministry of Trade control), 9 *Prisma* AGS, 9 icebreakers (Board of Navigation control)

COASTAL DEFENCE
100mm: D-10T (tank turrets); **130mm**: 195 K-54 (static) arty
COASTAL SSM 5 RBS-15

Air Force 2,700

(incl 1,500 conscripts) wartime strength 35,000; 64 cbt ac, no armed hel, 3 Air Comds: Satakunta (West), Karelia (East), Lapland (North). Each Air Comd assigned to one of the 3 AD areas into which Finland is divided. 3 ftr wgs, one in each AD area.

Flying hours 120

FGA 3 wg with 57 F/A-18C, 7 F/A-18D

Advanced AD/Attack Trg/Recce

20 *Hawk* 50/51A. One F-27 ESM/*Elint*

All remaining SAAB 35 *Drakens* being retired during 2000

SURVEY 3 *Learjet* 35A (survey, ECM trg, target-towing)

TPT 1 **ac** sqn with 2 F-27, 3 Learjet-35A

TRG 22 *Hawk* Mk 51, 28 L-70 *Vinka*

LIAISON 14 Piper (8 *Cherokee Arrow*, 6 *Chieftain*), 9 L-90 *Redigo*

UAV Tactical (6 *Ranger* systems to be delivered)

AAM AA-8 *Aphid*, AIM-9 *Sidewinder*, RB-27, RB-28 (*Falcon*), AIM-120 AMRAAM

Forces Abroad

UN AND PEACEKEEPING

BOSNIA (SFOR II): 480; 1 mech inf bn **CROATIA** (UNMOP): 1 obs **CYPRUS** (UNFICYP): 11 **INDIA/PAKISTAN** (UNMOGIP): 6 obs **IRAQ/KUWAIT** (UNIKOM): 5 obs **LEBANON** (UNIFIL): 624; 1 inf bn **MIDDLE EAST** (UNTSO): 11 obs **YUGOSLAVIA** (KFOR): 800

Paramilitary 3,400

FRONTIER GUARD (Ministry of Interior) 3,400

(on mob 22,000); 4 frontier, 3 Coast Guard districts, 1 air comd; 5 offshore, 2 coastal, 4 inshore patrol craft (plus boats and ACVs); air patrol sqn with **hel** 3 AS-332, 4 AB-206L, 4 AB-412 **ac** 2 Do-228 (Maritime Surv)

Georgia Ga

lari		1998	1999	2000	2001
GDP	lari	6.8bn	5.6bn		
	US$	4.4bn	2.5bn		
per capita	US$	4,600	4,800		
Growth	%	2.9	3.0		
Inflation	%	3.5	19.1		
Debt	US$	1.7bn			
Def exp	lari	170m	250m		
	US$	110m	111m		
Def bdgt[a]	lari	82m	55m	43.7m	
	US$	53m	24m	22m	
FMA[b] (US)	US$	0.4m	0.4m	0.4m	0.4m
FMA (Tu)	US$		3.8m		
US$1=lari		1.55	2.25	1.98	

[a] Abkhazia def bdgt 1997 US$5m

[b] UNOMIG **1997** US$18m **1998** US$19m

Population			5,472,000
(Armenian 8%, Azeri 6%, Russian 6%, Ossetian 3%, Abkhaz 2%)			
Age	13–17	18–22	23–32
Men	214,000	208,000	381,000
Women	205,000	200,000	359,000

Total Armed Forces

ACTIVE 26,900

(incl 430 MoD staff)

Terms of service conscription, 2 years

RESERVES up to 250,000

with mil service in last 15 years

Army 23,800

2 comd HQ

2 MR 'bde', 1 national gd bde plus trg centre • 1 arty 'bde' (bn) • 1 recce bn, 1 marine inf bn, 1 peacekeeping bn

EQUIPMENT

MBT 79: 48 T-55, 31 T-72

AIFV/APC 185: 68 BMP-1, 13 BMP-2, 11 BRM-1K, 18 BTR-70, 3 BTR-80, 72 MT-LB

TOWED ARTY 85mm: D-44; **100mm**: KS-19 (ground role); **122mm**: 60 D-30; **152mm**: 3 2A36, 10 2A65

SP ARTY 152mm: 1 2S3, 1 2S19; **203mm**: 1 2S7

MRL 122mm: 16 BM-21

MOR 120mm: 17 M-120

SAM some SA-13

Navy 800

BASES Tbilisi (HQ), Poti

PATROL AND COASTAL COMBATANTS 10

PATROL CRAFT 10

1 *Matka* PHM, 1 *Lindau* PCC, 2 *Dilos* PCC, 1 *Stenka* PCC, 1 *Zhuk* PCI<, plus 4 other PCI<

AMPHIBIOUS craft only

2 LCT, 4 LCM

Air Force 1,870

7 cbt ac, 3 armed hel

ATTACK 7 Su-25 (1 -25, 5 - 25K, 1 -25UB), 5 Su-17 (non-operational)

TPT 6 An-2, 2 Yak-40, 1 Tu-134A (VIP)

HEL 3 Mi-24 (attack), 4 Mi-8/17

TRG ac 10 L-29 hel 2 Mi-2

AIR DEFENCE

SAM 75 SA-2/-3/-4/-5/-7

Forces Abroad

UN AND PEACEKEEPING

YUGOSLAVIA (KFOR): 34

Opposition

ABKHAZIA ε5,000

50+ T-72, T-55 MBT, 80+ AIFV/APC, 80+ arty

SOUTH OSSETIA ε2,000

5–10 MBT, 30 AIFV/APC, 25 arty incl BM-21

Paramilitary ε6,500

BORDER GUARD ε6,500

COAST GUARD

3 *Zhuk* PCI

Foreign Forces

RUSSIA 5,000: **Army** 3 mil bases (each = bde+); 140 T-72 MBT, 500 ACV, 173 arty incl **122mm** D-30, 2S1; **152mm** 2S3; **122mm** BM-21 MRL; **120mm** mor plus 118 ACV and some arty deployed in Abkhazia **Air Force** 1 composite regt, some 35 tpt **ac** and **hel** incl An-12, An-26 and Mi-8

PEACEKEEPING

Abkhazia ε1,500; 2 MR, 1 AB bn (Russia) **South Ossetia** 1,700; 1 MR bn
UN (UNOMIG): 102 obs from 22 countries

Ireland Irl

pound I£		1998	1999	2000	2001
GDP	I£	59.6bn	67.3bn		
	US$	85bn	87bn		
per capita	US$	20,400	22,400		
Growth	%	8.9	8.6		
Inflation	%	2.4	1.6		
Publ debt	%	55.6	51.9		
Def exp	I£	506m	576m		
	US$	722m	745m		
Def bdgt	I£	558m	576m	601m	
	US$	796m	745m	725m	
US$1=I£		0.71	0.77	0.83	

Population			3,723,000
Age	13–17	18–22	23–32
Men	154,000	168,000	332,000
Women	145,000	159,000	314,000

Total Armed Forces

ACTIVE 11,460

(incl 200 women)

RESERVES 14,800

(obligation to age 60, officers 57–65) **Army** first-line 500, second-line 14,000 **Navy** 300

Army 9,300

3 inf bde each 3 inf bn, 1 arty regt, 1 cav recce sqn, 1 engr coy
Army tps: 1 lt tk sqn, 1 AD regt, 1 Ranger coy
Total units: 9 inf bn • 1 UNIFIL bn *ad hoc* with elm from other bn, 1 lt tk sqn, 3 recce sqn, 3 fd arty regt (each of 2 bty) • 1 indep bty, 1 AD regt (1 regular, 3 reserve bty), 4 fd engr coy, 1 Ranger coy

RESERVES

4 Army Gp (garrisons), 18 inf bn, 6 fd arty regt, 3 cav sqn, 3 engr sqn, 3 AD bty

EQUIPMENT

LT TK 14 *Scorpion*
RECCE 15 AML-90, 32 AML-60
APC 47 Panhard VTT/M3, 5 *Timoney* Mk 6, 2 A-180 *Sisu*
TOWED ARTY 88mm: 42 25-pdr; **105mm**: 24 L-118
MOR 81mm: 400; **120mm**: 68
ATGW 21 *Milan*
RL 84mm: AT-4
RCL 84mm: 444 *Carl Gustav*; **90mm**: 96 PV-1110
AD GUNS 40mm: 24 L/60, 2 L/70
SAM 7 RBS-70

Naval Service 1,100

BASE Cork, Haulbowline

PATROL AND COASTAL COMBATANTS 8

PATROL OFFSHORE 8

1 *Eithne* with 1 *Dauphin* hel PCO, 3 *Emer* PCO, 1 *Deirdre* PCO, 2 *Orla* (UK *Peacock*) PCO with 1 × 76mm gun, 1 *Roisin* PCO with 1 × 76mm gun

Air Corps 1,060

7 cbt ac, 15 armed hel; 3 wg (1 trg)
CCT 1 sqn
1 with 7 SF-260WE,
MR 2 CN-235MP
TPT 1 *Super King Air* 200, 1 *Gulfstream* IV
LIAISON 1 sqn with 6 Cessna Reims FR-172H, 1 FR-172K
HEL 4 sqn
1 Army spt with 8 SA-316B (*Alouette* III)
1 Navy spt with 2 SA-365FI (*Dauphin*)
1 SAR with 3 SA-365FI (*Dauphin*)
1 trg with 2 SA-342L (*Gazelle*)

Forces Abroad

UN AND PEACEKEEPING

BOSNIA (SFOR II): 50 **CROATIA** (UNMOP): 1 obs

CYPRUS (UNFICYP): 4 **EAST TIMOR** (UNTAET): 44 incl 2 obs **IRAQ/KUWAIT** (UNIKOM): 7 obs **LEBANON** (UNIFIL): 660; 1 bn; 4 AML-90 armd cars, 10 *Sisu* APC, 4 **120mm** mor **MIDDLE EAST** (UNTSO): 9 obs **WESTERN SAHARA** (MINURSO): 6 obs **YUGOSLAVIA** (KFOR): 104

Latvia Lat

lats L		1998	1999	2000	2001
GDP	L	3.8bn	3.7bn		
	US$	6.4bn	6.0bn		
per capita	US$	6,000	6,300		
Growth	%	3.6	0.8		
Inflation	%	4.5	3.2		
Debt	US$	755m			
Def exp[a]	L	ε23m	ε35m		
	US$	39m	58m		
Def bdgt	L	23.2m	35.4m	43m	L43m
	US$	39m	58m	72m	72m
FMA (US)	US$	0.8m	0.7m	0.7m	0.8m
FMA (Swe)	US$	1.5m			
US$1=L		0.59	0.61	0.60	

[a] Incl exp on paramilitary forces.

Population 2,420,000

(Russian 34%, Belarussian 5%, Ukrainian 3%, Polish 2%)

Age	13–17	18–22	23–32
Men	97,000	94,000	165,000
Women	94,000	91,000	162,000

Total Armed Forces

ACTIVE 5,050

(incl 1,600 National Guard; 1,690 conscripts)

Terms of service 12 months

RESERVES 14,500

National Guard

Army 2,400

(incl 1,330 conscripts)

1 mobile rifle bde with 1 inf bn • 1 recce bn • 1 HQ bn • 1 arty unit • 1 peacekeeping coy • 1 SF team

RESERVES

National Guard 5 bde each of 5–7 bn

EQUIPMENT

MBT 3 T-55
RECCE 2 BRDM-2
APC 13 *Pskbil* m/42
TOWED ARTY 100mm: 26 K-53
MOR 82mm: 5; **120mm**: 26
AD GUNS 14.5mm: 12 ZPU-4

Navy 840

(incl 360 conscripts, 250 Coastal Defence)

Lat, Ea and L have set up a joint Naval unit BALTRON with bases at Liepaja, Riga, Ventspils, Tallin (Ea), Klaipeda (L)

BASES Liepaja, Riga, Ventspils

PATROL CRAFT 9

1 *Osa* PFM (unarmed), 1 *Storm* PCC (unarmed), 2 *Ribnadzor* PCC, 5 KBV 236 PCI<, plus 3 boats

MINE COUNTERMEASURES 3

2 *Kondor* II MCC, 1 *Namejs* (Ge *Lindau*) MHC

SUPPORT AND MISCELLANEOUS 3

1 *Nyrat* AT, 1 *Goliat* AT, 1 diving vessel

COASTAL DEFENCE (250)

1 coastal def bn
10 patrol craft: 2 *Ribnadzor* PCC, 5 KBV 236 PC, 3 PCH

Air Force 210

AC 13 An-2, 1 L-410, 5 PZL Wilga
HEL 3 Mi-2, 1 Mi-8

Forces Abroad

UN AND PEACEKEEPING

BOSNIA (SFOR II): 40 **YUGOSLAVIA** (KFOR): 10

Paramilitary 3,500

BORDER GUARD (Ministry of Internal Affairs) 3,500

1 bde (7 bn)

Lithuania L

litas L		1998	1999	2000	2001
GDP	L	43bn	42.6bn		
	US$	10.8bn	10.7bn		
per capita	US$	5,700	5,500		
Growth	%	5.1	-3.3		
Inflation	%	5.0	0.8		
Debt	US$	1.9bn			
Def exp	L	545m	425m		
	US$	136m	106m		
Def bdgt	L	541m	716m	629m	843m
	US$	135m	179m	157m	210m
FMA (US)	US$	0.7m	0.7m	0.8m	
US$1=L		4.0	4.0	4.0	

Population 3,700,000

(Russian 9%, Polish 8%, Belarussian 2%)

Age	13–17	18–22	23–32
Men	142,000	138,000	257,000
Women	137,000	133,000	252,000

Total Armed Forces

ACTIVE 12,700

(incl 2,000 Voluntary National Defence Force; 4,000 conscripts) *Terms of service* 12 months

RESERVES 355,650

27,700 **first line** (ready 72 hrs, incl 10,200 Voluntary National Defence Service), 327,950 **second line** (age up to 59)

Army 9,340

(incl 3,720 conscripts)

3 mil region, 1 motor rifle bde (4 bn), 1 motor rifle bde (2 bn plus 1 coast def bn) • 1 Jaeger, 1 trg regt (3 bn), 1 engr bn • 1 peacekeeping coy

EQUIPMENT

RECCE 11 BRDM-2
APC 14 BTR-60, 13 *Pskbil* m/42, 10 MT-LB
MOR 120mm: 42 M-43
RL 82mm: 170 RPG-2
RCL 84mm: 119 *Carl Gustav*

RESERVES

Voluntary National Defence Service: 10 Territorial Defence regt, 38 territorial def bn with 200 territorial def coy, 2 air sqn

Navy 560

(incl 280 conscripts)

Lat, Ea and L have set up a joint Naval unit BALTRON with bases at Liepaja, Riga, Ventspils, Tallin (Ea), Klaipeda (L)

BASES Klaipeda, HQ BALTRON Tallinn (Ea)

PATROL AND COASTAL COMBATANTS 6

CORVETTES 2

2 Sov *Grisha III* FS, with 4 × 533mm TT, 2 × 12 ASW RL

PATROL COASTAL/INSHORE 4

1 *Storm* PCC, 1 SK-21 PCI<, 1 SK-23 PCI<, 1 SK-24 PCI<

MINE WARFARE 1

1 *Suduvis* (Ge *Lindau*) MHC

SUPPORT AND MISCELLANEOUS 1

1 *Valerian Uryvayev* AGOR/AG

Air Force 800

no cbt ac

Air Surveillance and Control Command, 2 air bases

Flying hours 60

AC

TPT 2 L-410, 3 An-26, 22 An-2
TRG 6 L-39

HEL 8 Mi-8 (tpt/SAR), 5 Mi-2

AIRFIELD DEFENCE 1 AD bn

Forces Abroad

UN AND PEACEKEEPING

BOSNIA (SFOR II): 41 **YUGOSLAVIA** (KFOR): 30

Paramilitary 3,900

BORDER POLICE 3,500

COAST GUARD 400

Macedonia, Former Yugoslav Republic of FYROM

dinar d		1998	1999	2000	2001
GDP	US$	3.5bn	3.4bn		
per capita	US$	3,700	3,900		
Growth	%	2.9	2.5		
Inflation	%	0.6	-1.1		
Debt	US$	2.4bn	1.9bn		
Def exp	d	ε4.0bn	3.8bn		
	US$	70m	67m		
Def bdgt	d	3.7bn	3.8bn	4.6bn	
	US$	68m	66m	77m	
FMA[ab] (US)	US$	3.4m	0.5m	0.5m	0.5m
US$1=d		54.5	57.0	59.8	

[a] UNPREDEP **1997** US$45m **1998** US$21m

[b] UNPREDEP figures exclude US costs paid as voluntary contributions

Population			**2,322,000**
(Albanian 22%, Turkish 4%, Romany 3%, Serb 2%)			
Age	13–17	18–22	23–32
Men	96,000	94,000	175,000
Women	87,000	85,000	160,000

Total Armed Forces

ACTIVE ε16,000

(incl about 1,000 HQ staff; 8,000 conscripts) *Terms of service* 9 months

RESERVES 60,000

Army ε15,000

2 Corps HQ (cadre), 3 bde incl 1 border gd bde

EQUIPMENT

MBT 4 T-34, 94 T-55
RECCE 10 BRDM-2, 41 HMMWV
APC 60 BTR-70, 12 BTR-80, 30 M-113A, 10 *Leonidas*
TOWED ARTY 76mm: M-48, M-1942; **105mm**: 18 M-56, 18 M-2A1; **122mm**: M-30
MRL 128mm: 73 M-71 (single barrel), 12 M-77

MOR 60mm; 82mm; 120mm
ATGW AT-3 *Sagger*
RCL 57mm; 82mm

MARINE WING (400)
9 river patrol craft

Air Force (under 700)

ac 4 *Zlin*-242 (trg), 10 UTVA-75 **hel** 4 Mi-17
AD GUNS 50: **20mm; 40mm**
SAM 30 SA-7

Paramilitary 7,500

POLICE 7,500
(some 4,500 armed)

Foreign Forces

UN (KFOR) about 5,000 providing logistic spt for tps deployed in the Yugoslav province of Kosovo

Malta M

lira ML		1998	1999	2000	2001
GDP	ML	1.3bn	1.4bn		
	US$	3.4bn	3.5bn		
per capita	US$	8,700	9,100		
Growth	%	3.1	3.5		
Inflation	%	2.4	2.1		
Debt	US$	990m			
Def exp	ML	11.5m	11.3m		
	US$	30m	27m		
Def bdgt	ML	11.5m	11.2m	11.3m	
	US$	30m	27.3m	27.6m	
FMA (US)	US$	0.1m	0.1m	0.1m	0.1m
US$1=ML		0.39	0.41	0.43	

Population 380,000

Age	13–17	18–22	23–32
Men	14,000	15,000	26,000
Women	14,000	14,000	25,000

Total Armed Forces

ACTIVE 2,140

Armed Forces of Malta 2,140

Comd HQ, spt tps
No. 1 Regt (inf bn): 3 rifle, 1 spt coy
No. 2 Regt (composite regt)
1 air wg (76) with **ac** 4 0-1 *Bird Dog*, 2 BN-2B *Islander* **hel** 5 SA-316B, 2 NH-369M Hughes, 2 AB-47G2
1 maritime sqn (210) with 3 ex-GDR *Kondor* 1 PCC, 4 PCI, 3 harbour craft, 1 LCVP
1 AD bty; **14.5mm**: 50 ZPU-4; **40mm**: 40 Bofors
No. 3 Regt (Depot Regt): 1 engr sqn, 1 workshop, 1 ordnance, 1 airport coy

Foreign Forces

ITALY 47 **Air Force** 2 AB-212 **hel**

Moldova Mol

leu L		1998	1999	2000	2001
GDP	L	9.1bn	12.2bn		
	US$	1.7bn	1.1bn		
per capita	US$	3,400	3,200		
Growth	%	-8.6	-5.0		
Inflation	%	6.6	46		
Debt	US$	1.0bn	972m		
Def exp[a]	L		63m		
	US$		6m		
Def bdgt	L	96m	63m	64m	70m
	US$	18m	7m	5m	5.5m
FMA (US)	US$	0.5m	0.5m	0.6m	0.6m
US$1=L		5.37	11.11	12.67	

[a] Incl exp on paramilitary forces

Population 4,428,000

(Moldovan/Romanian 65%, Ukrainian 14%, Russian 13%, Gaguaz 4%, Bulgarian 2%, Jewish <1.5%)

Age	13–17	18–22	23–32
Men	204,000	187,000	301,000
Women	187,000	184,000	299,000

Total Armed Forces

ACTIVE 9,500
(incl ε5,200 conscripts) *Terms of service* up to 18 months

RESERVES some 66,000

Army 8,500

(incl ε5,200 conscripts)
3 MR bde • 1 arty bde • 1 SF, 1 indep engr bn

EQUIPMENT
AIFV 54 BMD-1
APC 11 BTR-80, 11 BTR-D, 2 BTR-60PB, 131 TAB-71, plus 135 'look-alikes'
TOTAL ARTY 153
TOWED ARTY 122mm: 17 M-30; **152mm**: 32 D-20, 21 2A36
COMBINED GUN/MOR 120mm: 9 2S9
MRL 220mm: 14 9P140 *Uragan*
MOR 82mm: 54; **120mm**: 60 M-120
ATGW 70 AT-4 *Spigot*, 19 AT-5 *Spandral*, 27 AT-6 *Spiral*
RCL 73mm: SPG-9
ATK GUNS 100mm: 36 MT-12

Non-NATO Europe
NATO and Non-NATO Europe

AD GUNS 23mm: 30 ZU-23; **57mm**: 12 S-60
SURV GS-13 (arty), 1 L219/200 PARK-1 (arty), *Long Trough* ((SNAR-1) arty), *Pork Trough* ((SNAR-2/-6) veh, arty), *Small Fred/Small Yawn* (veh, arty), *Big Fred* ((SNAR-10) veh, arty)

Air Force 1,000

(incl Defence Aviation)
TPT 1 mixed sqn **ac** 3 An-72, 1 Tu-134, (6 MiG-29 in store) **hel** 11 Mi-8
SAM 1 bde with 25 SA-3/-5

Paramilitary 3,400

INTERNAL TROOPS (Ministry of Interior) 2,500
OPON (Ministry of Interior) 900 (riot police)

Opposition

DNIESTR 5,000 (up to 10,000 reported)
incl Republican Guard (Dniestr bn), Delta bn, ε1,000 Cossacks

Foreign Forces

RUSSIA 2,600: 1 op gp
PEACEKEEPING
Russia (500) 1 MR bn

Romania R

lei		1998	1999	2000	2001
GDP	lei	339tr	521tr		
	US$	38bn	33bn		
per capita	US$	4,400	4,400		
Growth	%	-5.4	-3.9		
Inflation	%	59	45		
Debt	US$	9.5bn	8.6bn		
Def exp	lei	7.9tr	9.6tr		
	US$	887m	607m		
Def bdgt	lei	8.0tr	9.6tr		
	US$	900m	607m		
FMA (US)	US$	1.0m	1.3m	1.2m	1.3m
US$1=lei		8,876	15,835	20,750	

Population		**22,500,000** (Hungarian 9%)	
Age	13–17	18–22	23–32
Men	860,000	918,000	1,860,000
Women	830,000	885,000	1,798,000

Total Armed Forces

ACTIVE 207,000
(incl 36,700 in centrally controlled units; 108,600 conscripts)
Terms of service All services 12 months

RESERVES 470,000
Army 400,000 **Navy** 30,000 **Air Force** 40,000

Army 106,000

(incl 71,000 conscripts)
3 Army HQ, 7 Corps HQ each with 2–3 mech 1 tk, 1 mtn, 1 arty, 1 ATK bde
Army tps: 1 arty, 1 ATK, 1 SAM bde, 1 engr regt
Defence Staff tps: 2 AB (Air Force), 1 gd bde
Land Force tps: 1 SAM, 2 engr regt
Determining the manning state of units is difficult. The following is based on the latest available information: one-third at 100%, one-third at 50–70%, one-third at 10–20%.
EQUIPMENT
MBT 1,253: 821 T-55, 30 T-72, 314 TR-85 M1, 88 TR-580
ASLT GUN 84 SU-100
RECCE 121 BRDM-2, 49 ABC-M
AIFV 177 MLI-84
APC 1,619: 167 TAB-77, 384 TABC-79, 920 TAB-71, 88 MLVM, 60 TAB ZIMBRU, plus 1,015 'look-alikes'
TOTAL ARTY 1,276
TOWED 748: **122mm**: 204 M-1938 (M-30) (A-19); **130mm**: 70 Gun 82; **150mm**: 12 Skoda (Model 1934); **152mm**: 114 Gun-how 85, 294 Model 81, 54 M-1937 (ML-20)
SP 48: **122mm**: 6 2S1, 42 Model 89
MRL 122mm: 148 APR-40
MOR 120mm: 332 M-1982
SSM launchers: 9 FROG
ATGW 174 AT-3 *Sagger* (incl BRDM-2), 54 AT-5 *Spandrel*
ATK GUNS 57mm: M-1943; **76mm**: M-1942; **85mm**: D-44; **100mm**: 877 Gun 77, 75 Gun 75
AD GUNS 1,093: **30mm**; **35mm**: 32 *Gepard*, GDF-003; **37mm**; **57mm**; **85mm**; **100mm**
SAM 62 SA-6/-7/-8
SURV GS-13 (arty), 1 L219/200 PARK-1 (arty), *Long Trough* ((SNAR-1) arty), *Pork Trough* ((SNAR-2/-6) veh, arty), *Small Fred/Small Yawn* (veh, arty), *Big Fred* ((SNAR-10) veh, arty)
UAV 6 *Shadow*-600

Navy 20,800

(incl 12,600 conscripts; incl 10,200 Naval Inf and Coastal Defence)
Navy HQ with 1 Naval fleet, 1 Danube flotilla, 1 Naval inf corps
BASES Coastal Mangalia, Constanta **Danube** Braila, Giurgiu, Tulcea, Galati

SUBMARINES 1
SSK 1 Sov *Kilo* with 6 × 533mm TT
PRINCIPAL SURFACE COMBATANTS 7
DESTROYERS 1
DDG 1 *Muntena* with 4 × 2 SS-N-2C *Styx* SSM, SA-N-5 *Grail* SAM, 4 × 76mm guns, 2 × 3 533mm ASTT, 2 IAR 316 hel
FRIGATES 6
FF 6
4 *Tetal* 1 with 4 × 76mm guns, 4 ASTT, 2 ASW RL
2 *Tetal* II with 1 × 76mm gun, 4 ASTT, 2 ASW RL
PATROL AND COASTAL COMBATANTS 65
MISSILE CRAFT 6
3 *Zborul* PFM (Sov *Tarantul* I) with 2 × 2 SS-N-2C *Styx* SSM, 1 × 76mm gun
3 Sov *Osa* I PFM with 4 SS-N-2A *Styx* SSM
TORPEDO CRAFT 28
12 *Epitrop* PFT with 4 × 533mm TT
16 PRC *Huchuan* PHT with 2 533mm TT†
PATROL CRAFT 31
OFFSHORE 4 *Democratia* (GDR M-40) PCO
RIVERINE 27
some 6 *Brutar* with 1 × 100mm gun, 1 × 122mm RL, 3 *Kogalniceanu* with 2 × 100mm gun, 18 VB 76 PCI
MINE WARFARE 37
MINELAYERS 2 *Cosar*, capacity 100 mines
MINE COUNTERMEASURES 35
4 *Musca* MSO, 6 T-301 MSI, 25 VD141 MSI
SUPPORT AND MISCELLANEOUS 11
2 *Constanta* log spt with 1 *Alouette* hel, 1 AK, 3 AOT; 1 trg, 2 AT; 2 AGOR
HELICOPTERS 7
3 1AR-316, 4 Mi-14 PL

NAVAL INFANTRY (10,200)
1 Corps HQ
2 mech, 1 mot inf, 1 arty bde, 1 ATK, 1 marine bn
EQUIPMENT
MBT 120 TR-580
APC 208: 172 TAB-71, 36 TABC-79 plus 101 'look-alikes'
TOTAL ARTY 138
TOWED 90: **122mm**: 54 M-1938 (M-30); **152mm**: 36 Model 81
MRL 122mm: 12 APR-40
MOR 120mm: 36 Model 1982
ATK GUNS 100mm: 57 Gun 77

COASTAL DEFENCE
4 coastal arty bty with 32 **130mm**

Air Force 43,500

(incl 5,500 AB; 25,000 conscripts); 323 cbt ac, 16 attack hel
Flying hours 40
Air Force comd: 2 Air and Air Defence Corps, 6–7 air bases, 4–6 air defence artillery bde or rgt
FGA 4 regt with 73 IAR-93, 5 regt with 180 MiG-21 (110 being upgraded to Lancer standard: 75 Lancer A (air-to-gd), 25 Lancer C (AD), 10 Lancer B (two-seat trainers))
FTR 1 regt with 40 MiG-23, 1 regt with 18 MiG-29
RECCE 1 sqn with 12* H-5 (recce/ECM/trg towing)
TPT ac 6 An-24, 11 An-26, 2 Boeing 707, 4 C-130B **hel** 5 IAR-330, 9 Mi-8, 4 SA-365
SURVEY 3 An-30
HELICOPTERS
ATTACK 16 IAR 316A
CBT SPT 74 IAR-330, 80 IAR-316, 15 Mi-8, 2 Mi-17
TRG ac 45 L-29, 32 L-39, 14 IAR-99 (reportedly increasing to 24), 36 IAR-823, 23 Yak-52
AAM AA-2 *Atoll*, AA-3 *Anab*, AA-7 *Apex*, AA-11 *Archer*
ASM AS-7 *Kerry*
AD 2 div bde
20 SAM sites with 120 SA-2, SA-3

Forces Abroad

UN AND PEACEKEEPING
BOSNIA (SFOR II): 200; 1 engr bn **DROC** (MONUC): 12 obs **IRAQ/KUWAIT** (UNIKOM): 3 obs

Paramilitary 75,900

BORDER GUARDS (Ministry of Interior) 22,900 (incl conscripts) 9 regional formations, 3 regional maritime dets
33 TAB-71 APC, 18 SU-100 aslt gun, 12 M-1931/37 (A19) **122mm** how, 18 M-38 **120mm** mor, 7 PRC *Shanghai* II PFI
GENDARMERIE (Ministry of Interior) 53,000

Slovakia Slvk

koruna Ks		1998	1999	2000	2001
GDP	Ks	717bn	779bn		
	US$	20.4bn	17.5bn		
per capita	US$	7,600	7,800		
Growth	%	4.4	1.0		
Inflation	%	6.7	10.6		
Debt	US$	9.9bn	10.4bn		
Def exp	Ks	14.6bn	14.6bn		
	US$	415m	329m		
Def bdgt	Ks	14.6bn	13.8bn	15.8bn	
	US$	416m	311m	348m	
FMA (US)	US$	0.6m	0.6m	0.7m	0.7m
US$1=Ks		35.2	44.4	45.3	

Population **5,400,000**
(Hungarian 11%, Romany ε5%, Czech 1%)

Age	13–17	18–22	23–32
Men	222,000	236,000	420,000
Women	214,000	229,000	411,000

Total Armed Forces

ACTIVE 38,600
(incl 3,300 centrally controlled staffs, log and spt tps; 13,600 conscripts)
Terms of service 12 months

RESERVES ε20,000 on mob
National Guard Force

Army 23,800

(incl 13,600 conscripts)
1 Corps HQ
1 tk bde (2 tk, 1 mech, 1 recce, 1 arty bn)
1 mech inf bde (2 mech inf, 1 tk, 1 recce, 1 arty bn)
1 arty bde
1 Rapid Reaction bn

RESERVES
1 Corps HQ, 2 mech bde, 1 arty bde
National Guard Force

EQUIPMENT
MBT 275: 272 T-72M, 3 T-55
RECCE 129 BRDM, 90 OT-65, 72 BPVZ
AIFV 311 BMP-1, 93 BMP-2
APC 207 OT-90, 11 OT-64
TOTAL ARTY 383 (68 in store)
TOWED 122mm: 76 D-30
SP 191: **122mm**: 49 2S1; **152mm**: 134 *Dana* (M-77); **155mm**: 8 *Zuzana* 2000
MRL 122mm: 87 RM-70
MOR 120mm: 29 M-1982
SSM 9 FROG-7, SS-21, *Scud*
ATGW 538 (incl BMP-1/-2 and BRDM mounted): AT-3 *Sagger*, AT-5 *Spandrel*
AD GUNS ε200: **30mm**: M-53/-59, *Strop* SP; **57mm**: S-60
SAM SA-7, ε48 SA-9/-13
SURV GS-13 (veh), *Long Trough* (SNAR-1), *Pork Trough* ((SNAR-2/-6) arty), *Small Fred/Small Yawn* (veh, arty), *Big Fred* ((SNAR-10) veh, arty)

Air Force 11,500

84 cbt ac, 19 attack hel
Flying hours 45
FGA 10 Su-22, 12 Su-25
FTR 30 MiG-21, 24 MiG-29
RECCE 8* MiG-21 RF
TPT 1 An-12, 2 An-24, 2 An-26, 4 L410M
TRG 14 L-29, 20 L-39
ATTACK HEL 19 Mi-24
ASLT TPT 13 Mi-2, 7 Mi-8, 17 Mi-17
AAM AA-2 *Atoll*, AA-7 *Apex*, AA-8 *Aphid*, AA-10 *Alamo*, AA-11 *Archer*
AD SA-2, SA-3, SA-6, SA-10B

Forces Abroad

UN AND PEACEKEEPING
MIDDLE EAST (UNTSO): 2 obs **SIERRA LEONE** (UNAMSIL): 2 obs **SYRIA** (UNDOF): 93 **YUGOSLAVIA** (KFOR): 40

Paramilitary 2,600

INTERNAL SECURITY FORCES 1,400
CIVIL DEFENCE TROOPS 1,200

Slovenia — Slvn

tolar t		1998	1999	2000	2001
GDP	t	3.2tr	3.6tr		
	US$	19.5bn	18.5bn		
per capita	US$	10,900	11,500		
Growth	%	3.9	3.8		
Inflation	%	6.5	6.6		
Debt	US$	4.9bn	5.5bn		
Def exp	t	52bn	65bn		
	US$	317m	337m		
Def bdgt	t	41.4bn	50bn	59bn	81bn
	US$	241m	259m	273m	373m
FMA (US)	US$	0.6m	0.7m	0.7m	0.7m
US$1=t		166	192	218	

Population 2,020,000
(Croat 3%, Serb 2%, Muslim 1%)

Age	13–17	18–22	23–32
Men	69,000	75,000	148,000
Women	64,000	71,000	144,000

Total Armed Forces

ACTIVE ε9,000
(incl ε4–5,000 conscripts) *Terms of service* 7 months

RESERVES 61,000
Army (incl 300 maritime)

Army ε9,000

7 Mil Regions, 27 Mil Districts • 7 inf bde (each 1 active, 3 reserve inf bn) • 1 SF 'bde' • 1 SAM 'bde' (bn) • 2 indep mech bn • 1 avn 'bde' • 1 arty bn

RESERVES
2 indep mech, 1 arty, 1 coast def, 1 ATK bn

EQUIPMENT
MBT 46 M-84, 54 T-55M
RECCE 16 BRDM-2
AIFV 52 M-80
APC ε70 *Valuk* (*Pandur*)
TOWED ARTY 105mm: 18 M-2; **155mm**: 18 Model 845

MRL 128mm: 56 M-71 (single tube), 4 M-63
MOR 120mm: 120 M-52
ATGW AT-3 *Sagger* (incl 14 BOV-1SP), AT-4 *Spigot*

MARITIME ELEMENT (100)
(effectively police)
BASE Koper
1 PCI

AIR ELEMENT (120)
8 armed hel
AC 10 PC-9, 3 *Zlin*-242, 1 LET L-410, 3 UTVA-75, 2 PC-6
HEL 1 AB-109, 3 B-206, 8* B-412
SAM 9 SA-9
AD GUNS 20mm: 9 SP; **30mm**: 9 SP; **57mm**: 21 SP

Forces Abroad

UN AND PEACEKEEPING
BOSNIA (SFOR II) some **CYPRUS** (UNFICYP): 29
MIDDLE EAST (UNTSO): 2 obs

Paramilitary 4,500

POLICE 4,500
armed (plus 5,000 reserve) **hel** 2 AB-206 *Jet Ranger*, 1 AB-109A, 1 AB-212, 1 AB-412

Sweden Swe

kronor Skr		1998	1999	2000	2001
GDP	Skr	1.9tr	2.0tr		
	US$	237bn	230bn		
per capita	US$	21,700	22,900		
Growth	%	3.0	3.8		
Inflation	%	0.4	0		
Publ Debt	%	73.7	68.3		
Def exp	Skr	44.9bn	44.9bn		
	US$	5.6bn	5.2bn		
Def bdgt	Skr	38.1bn	38.7bn	39.6bn	42.7bn
	US$	4.8bn	4.5bn	4.5bn	4.8bn
US$1=Skr		7.95	8.56	8.8	
Population					**8,947,000**
Age		13–17	18–22		23–32
Men		267,000	260,000		601,000
Women		251,000	246,000		573,000

Total Armed Forces

ACTIVE 52,700
(incl 32,800 conscripts and active reservists)
Terms of service **Army, Navy** 7–15 months **Air Force** 8–12 months

RESERVES[a] 570,000
(obligation to age 47) **Army** (incl Local Defence and Home Guard) 450,000 **Navy** 50,000 **Air Force** 70,000

[a] About 48,000 reservists carry out refresher trg each year; length of trg depends on rank (officers up to 31 days, NCO and specialists, 24 days, others 17 days). Commitment is five exercises during reserve service period, plus mob call-outs

Army 35,100

(incl 24,200 conscripts and active reservists)
3 joint (tri-service) comd each with: Army div and def districts, Naval Comd (2 in Central Joint Comd)
Air Comd, logistics regt
No active units (as defined by Vienna Document)
3 div with total of 6 mech, 4 inf, 3 arctic bde, 3 arty regt HQ, 12 arty bn
15 def districts

EQUIPMENT
MBT 60 *Centurion*, 239 Strv-103B (in store), 160 Strv-121 (*Leopard* 2), 78 Strv-122 (*Leopard* 2 (S))
LT TK 211 Ikv-91
AIFV 647 Pbv-302, 202 Strf-9040, 361 Pbv-501 (BMP-1)
APC 550 Pbv 401A (MT-LB), 96 *Pskbil* M/42 (plus 323 ACV 'look-alikes')
TOWED ARTY 105mm: 127 m/40; **155mm**: 206 FH-77A, 168 Type F
SP ARTY 155mm: 24 BK-1C
MOR 81mm: 160; **120mm**: 525
ATGW 55 TOW (Pvrbv 551 SP), RB-55, RB-56 *Bill*
RL 84mm: AT-4
RCL 84mm: *Carl Gustav*; **90mm**: PV-1110
AD GUNS 40mm: 600
SAM RBS-70 (incl Lvrbv SP), RB-77 (I HAWK), RBS-90
SURV *Green Archer* (mor)
AC 1 C-212
HEL see under Air Force 'Armed Forces Helicopter Wing'

Navy 9,200

(incl 1,100 Coastal Defence, 320 Naval Air, 4,200 conscripts)
BASES Muskö, Karlskrona, Härnösand, Göteborg (spt only)

SUBMARINES 9
SSK 9
3 *Gotland* with 4 × 533mm TT, TP-613 HWT and TP-43/45 LWT (AIP powered)
4 *Västergötland* with 6 × 533mm TT, TP-613 HWT and TP-43/45 LWT (2 being fitted with AIP)
2 *Näcken* with 6 × 533mm TT, *TP-613* HWT and *TP-421* LWT

PATROL AND COASTAL COMBATANTS 45
MISSILE CRAFT 20 PFM
4 *Göteborg* with 4 × 2 RBS-15 SSM, 4 ASW torpedoes, 4 ASW mor

2 *Stockholm* with 4 × 2 RBS-15 SSM, 2 Type 613 HWT, 4 ASW torpedoes, 4 ASW mor (in refit until 2002)
8 *Kaparen* with 6 RBS-12 *Penguin* SSM, ASW mor
6 *Norrköping* with 4 × 2 RBS-15 SSM, 2–6 Type 613 HWT

PATROL CRAFT 25
About 25 PCI<

MINE WARFARE 22

MINELAYERS 2
1 *Carlskrona* (200 mines) trg, 1 *Visborg* (200 mines)
(Mines can be laid by all submarine classes)

MINE COUNTERMEASURES 20
4 *Styrsö* MCMV, 1 *Utö* MCMV spt, 1 *Skredsvic* MCM/ diver support, 7 *Landsort* MHO, 2 *Gassten* MSO, 1 *Vicksten* MSO, *4 *Hisingen* diver support

AMPHIBIOUS
craft only about 120 LCU

SUPPORT AND MISCELLANEOUS 23
1 AK, 1 AR; 1 AGI, 1 ARS, 2 TRV, 8 AT, 7 icebreakers, 2 sail trg

COASTAL DEFENCE (1,100)

2 mobile coastal arty bde: 5 naval bde, 6 amph, 3 mobile arty (**120mm**), 12 specialist protection (incl inf, static arty (**75mm, 105mm, 120mm**), SSM and mor units)

EQUIPMENT
APC 3 *Piranha*
GUNS 40mm, incl L/70 AA; **75mm, 105mm, 120mm** 24 CD-80 *Karin* (mobile); **120mm** *Ersta* (static)
MOR 81mm, 120mm: 70
SSM 90 RBS-17 *Hellfire*, 6 RBS-15KA
SAM RBS-70
MINELAYERS 5 inshore
PATROL CRAFT 12 PCI
AMPH 16 LCM, 52 LCU, 123 LCA

Air Force 8,400

(incl 2,600 conscripts and 1,800 active reservists); 250 cbt ac, no armed hel
Flying hours 110–140
1 Air Comd, 8 sqns
FGA/RECCE 1 sqn with 30 SAAB AJS-37/AJSH-37/ AJSF-37, 1 OCU/EW trg with 14 SAAB SK-37
MULTI-ROLE (FTR/FGA/RECCE) 5 sqn with 90 SAAB JAS-39
FTR 2 sqn with 130 SAAB JA-37
SIGINT 2 S-102B *Korpen* (*Gulfstream* IV)
AEW 6 S-100B *Argus* (SAAB-340B)
TPT 8 C-130, 3 *King Air* 200, 1 Tp-100A (SAAB 340B) (VIP), 1 Tp-102A (*Gulfstream* IV) (VIP)
TRG 106 Sk-60, 38 SK-61 (*Bulldog*)
AAM RB-71 (*Skyflash*), RB-74 AIM 9L (*Sidewinder*), RB-99, AIM 120 (AMRAAM)
ASM RB-15F, RB-75 (*Maverick*), BK-39
AD semi-automatic control and surv system, *Stric*, coordinates all AD components

ARMED FORCES HELICOPTER WING
(1,000 personnel from all three services)
HEL 15 Hkp-3c tpt/SAR, 14 Hkp-4 ASW/tpt/SAR, 25 Hkp-5b (*Hughes* 300c) trg, 19 Hkp-6a (Bell-206) utl, 10 Hkp-6b, 20 Hkp-9a, 12 Hkp-10 (*Super Puma*) SAR, 5 Hkp-11 SAR, 1 C-212 ASW, MP

Forces Abroad

UN AND PEACEKEEPING

BOSNIA (SFOR II): 510 **CROATIA** (UNMOP): 1 obs (SFOR): 1 **DROC** (MONUC): 2 obs **EAST TIMOR** (UNTAET): 2 obs **GEORGIA** (UNOMIG): 5 obs **INDIA/PAKISTAN** (UNMOGIP): 8 obs **IRAQ/ KUWAIT** (UNIKOM): 6 obs **LEBANON** (UNIFIL): 44 **MIDDLE EAST** (UNTSO): 11 obs **SIERRA LEONE** (UNAMSIL): 3 obs **SYRIA** (UNDOF): 1 **YUGOSLAVIA** (KFOR): 760

Paramilitary 600

COAST GUARD 600
1 *Gotland* PCO and 1 KBV-171 PCC (fishery protection), some 65 PCI
AIR ARM 2 C-212 MR

CIVIL DEFENCE shelters for 6,300,000
All between ages 16–25 liable for civil defence duty

VOLUNTARY AUXILIARY ORGANISATIONS some 35,000

Switzerland CH

franc fr		1998	1999	2000	2001
GDP	fr	380bn	389bn		
	US$	262bn	246bn		
per capita	US$	28,600	29,600		
Growth	%	2.1	1.7		
Inflation	%	0	0.9		
Publ Debt	%	53.8	54.0		
Def exp	fr	5.3bn	4.9bn		
	US$	3.6bn	3.1bn		
Def bdgt	fr	5.3bn	5.0bn	4.9bn	4.7bn
	US$	3.6bn	3.2bn	2.9bn	2.8bn
US$1=fr		1.45	1.58	1.66	

Population			**7,090,000**
Age	13–17	18–22	23–32
Men	206,000	204,000	560,000
Women	198,000	197,000	566,000

Total Armed Forces

ACTIVE about 3,470
plus recruits (2 intakes in 1999 (total 24,500) each for 15 weeks only)

Terms of service 15 weeks compulsory recruit trg at age 19–20, followed by 10 refresher trg courses of 3 weeks over a 22-year period between ages 20–42. Some 200,000 attended trg in 1999

RESERVES 351,200

Army 321,000 (to be mobilised)

Armed Forces Comd (All units non-active/Reserve status)
Comd tps: 2 armd bde, 2 inf, 1 arty, 1 airport, 2 engr regt
3 fd Army Corps, each 2 fd div (3 inf, 1 arty regt), 1 armd bde, 1 engr, 1 cyclist, 1 fortress regt, 1 territorial div (5/6 regt)
1 mtn Army corps with 3 mtn div (2 mtn inf, 1 arty regt), 3 fortress bde (each 1 mtn inf regt), 2 mtn inf, 2 fortress, 1 engr regt, 1 territorial div (6 regt), 2 territorial bde (1 regt)

EQUIPMENT

MBT 556: 186 Pz-68/88, 370 Pz-87 (*Leopard* 2)
RECCE 233 *Eagle*/II
AIFV 435 (incl 6 in store): 120 M-63/73, 315 M-63/89 (all M-113 with **20mm**)
APC 812 M-63/73 (M-113) incl variants, 291 *Piranha*
SP ARTY 155mm: 558 PzHb 66/74/-74/-79/-88 (M-109U)
MOR 81mm: 1,469 M-33, M-72; **120mm**: 534: 402 M-87, 132 M-64 (M-113)
ATGW 3,012 *Dragon*, 303 TOW-2 SP (MOWAG) *Piranha*
RL 13,484 incl: **60mm**: *Panzerfaust*; **83mm**: M-80
SAM *Stinger*
HEL 60 *Alouette* III

MARINE

10 *Aquarius* patrol boats

Air Force 30,200 (to be mobilised)

(incl AD units, mil airfield guard units); 154 cbt ac, no armed hel
1 Air Force bde, 1 AD bde, 1 Air-Base bde, 1 C³I bde, AF Maintenance Service
Flying hours: 150–200; reserves approx 50
FTR 8 sqn
6 with 89 *Tiger* II/F-5E, 12 *Tiger* II/F-5F
2 with 26 F/A-18 C and 7 F/A-18D
RECCE 2 sqn with 16* *Mirage* IIIRS 2, 4* *Mirage* IIIDS
TPT 1 sqn with 16 PC-6, 1 *Learjet* 35A, 2 Do-27, 1 *Falcon*-50
HEL 3 sqn with 15 AS-332 M-1 (*Super Puma*), 10 SA-316 (*Alouette* III)
TRG 19 *Hawk* Mk 66, 38 PC-7, 11 PC-9 (tgt towing)
UAV *Ranger*/ADS 95 (2 systems to be deployed in 2000). 1 UAV bn operational in 2001
AAM AIM-9 *Sidewinder*, AIM-120 AMRAAM

AIR DEFENCE

1 AD bde with
1 SAM regt (3 bn, each with 2 or 3 bty; 59 B/L-84 *Rapier*)
5 AD Regt (each with 2 or 3 bn; each bn of 3 bty; 35mm guns, Skyguard fire control radar)

Forces Abroad

UN AND PEACEKEEPING

BOSNIA (OSCE): 50 **CROATIA** (UNMOP): 1 obs **DROC** (MONUC): 1 obs **GEORGIA** (UNOMIG): 4 obs **KOREA** (NNSC): 5 Staff **MIDDLE EAST** (UNTSO): 9 obs **YUGOSLAVIA** (KFOR): 130; 1 coy

Paramilitary

CIVIL DEFENCE 280,000 (not part of Armed Forces)

Ukraine Ukr

hryvnia h		1998	1999	2000	2001
GDP	h	104bn	127bn		
	US$	42bn	49bn		
per capita	US$	4,450	4,550		
Growth	%	-1.7	-0.4		
Inflation	%	10.6	22.7		
Debt	US$	12.7bn	12.6bn		
Def exp[a]	h	3.4bn	5.7bn		
	US$	1.4bn	1.4bn		
Def bdgt	h	1.3bn	1.5bn	2.4bn	
	US$	547m	385m	441m	
FMA[b] (US)	US$	1.3m	2.2m	1.3m	1.5m
US$1=h		2.45	3.97	5.44	

[a] Incl exp on paramilitary forces
[b] Excl US Cooperative Threat Reduction programme: **1992–96** US$395m, of which US$171m spent by Sept 1996. Programme continues through 2000

Population			**49,980,000**
(Russian 22%, Polish ε4%, Jewish 1%)			
Age	13–17	18–22	23–32
Men	1,898,000	1,869,000	3,576,000
Women	1,830,000	1,823,000	3,554,000

Total Armed Forces

ACTIVE ε303,800

(excl Strategic Nuclear Forces and Black Sea Fleet; incl 43,600 in central staffs and units not covered below)
Terms of service **Army, Air Force** 18 months **Navy** 2 years

RESERVES some 1,000,000

mil service within 5 years

Strategic Nuclear Forces

Elimination of Ukraine's nuclear wpns on schedule to be completed by December 2001

ICBM 27 SS-24 *Scalpel* (RS-22); silo-based (without warheads)

BBR 25

17 Tu-95H16 (with AS-15 ALCM) (START accountable)

7 Tu-160 (with AS-15 ALCM) (START accountable)

Ground Forces 151,200

3 Op Comd (North, South, West)

MoD tps: 1 air mobile bde, 1 SSM bde (SS-21), 1 arty (trg), 1 engr bde

WESTERN OP COMD

Comd tps 1 arty div (1 arty, 1 MRL, 1 ATK bde), 3 SSM (SS-21) bde, 1 air mobile regt, 1 engr bde, 2 army avn bde

2 Army Corps

1 with 2 mech div (each 3 mech, 1 tk, 1 SP arty regt), 2 mech bde, 1 arty regt, 1 MRL regt, 1 ATK bde

1 with 2 mech div (each 3 mech, 1 tk, 1 SP arty regt), 1 mech bde, 1 arty regt, 1 MRL regt, 1 ATK bde

SOUTHERN OP COMD

Comd tps 1 mech div (2 mech bde), 1 air mobile div (1 air aslt, 1 airmobile bde, 1 arty regt), 1 arty div (1 arty, 1 MRL, 1 ATK bde), 1 air mobile, 1 SSM (*Scud*), 1 avn bde

2 Army Corps

1 with 1 tank div (3 tk, 1 SP arty regt), 2 mech div (each 2 mech, 1 tk, 1 SP arty regt), 1 arty bde, 1 MRL regt, 1 ATK bde, 1 engr regt

1 with 2 mech, 1 arty bde, 1 MRL, 1 ATK, 1 engr regt

NORTHERN OP COMD

Comd tps 1 mech div (3 mech, 1 SP arty regt), 1 tk trg centre, 1 tank, 2 SSM bde (1 *Scud*, 1 SS-21), 1 army avn bde, 1 engr regt

1 Army Corps with 1 tank div (3 tk, 1 SP arty regt), 1 mech div (2 mech, 1 SP arty regt), 1 mech trg centre, 1 arty bde, 1 MRL, 1 ATK, 1 engr regt

EQUIPMENT

MBT 3,895 (875 in store): 145 T-55, 2,250 T-64, 1,230 T-72, 270 T-80

RECCE some 600 BRDM-2

AIFV 3,048 (330 in store): 1,000 BMP-1, 450 BRM-1K, 1,460 BMP-2, 3 BMP-3, 60 BMD-1, 75 BMD-2

APC 1,770 (230 in store): 200 BTR-60, 1,080 BTR-70, 450 BTR-80, 40 BTR-D; plus 2,090 MT-LB, 4,700 'look-alikes'

TOTAL ARTY 3,680 (456 in store)

TOWED 1,115: **122mm**: 430 D-30; **152mm**: 215 D-20, 185 2A65, 285 2A36

SP 1,304: **122mm**: 640 2S1; **152mm**: 500 2S3, 24 2S5, 40 2S19, **203mm**: 100 2S7

COMBINED GUN/MOR 120mm: 62 2S9, 2 2B16

MRL 593: **122mm**: 340 BM-21, 20 9P138; **132mm**: 4 BM-13; **220mm**: 135 9P140; **300mm**: 94 9A52

MOR 604: **120mm**: 346 2S12, 257 PM-38; **160mm**: 1 M-160

SSM 72 *Scud* B, 50 FROG, 90 SS-21

ATGW AT-4 *Spigot*, AT-5 *Spandrel*, AT-6 *Spiral*

ATK GUNS 100mm: ε500 T-12/MT-12

AD GUNS 30mm: 70 2S6 SP; **57mm**: ε400 S-60

SAM 100 SA-4, 125 SA-8, 60 SA-11, ε150 SA-13

ATTACK HEL 247 Mi-24

SPT HEL 4 Mi-2, 31 Mi-6, 162 Mi-8, 11 Mi-26

SURV SNAR-10 (*Big Fred*), *Small Fred* (arty)

Navy† ε13,000

(incl nearly 2,500 Naval Aviation, 1,500 Naval Infantry, 2,000 conscripts)

On 31 May 1997, Russian President Yeltsin and Ukrainian President Kuchma signed an inter-governmental agreement on the status and terms of the Black Sea Fleet's deployment on the territory of Ukraine and parameters for the fleet's division. The Russian Fleet will lease bases in Sevastopol for the next 20 years. It is based at Sevastopol and Karantinnaya Bays and jointly with Ukrainian warships at Streletskaya Bay. The overall serviceability of the Fleet is very low

BASES Sevastopol, Donuzlav, Odessa, Kerch, Ochakov, Chernomorskoye (Balaklava, Nikolaev construction and repair yards)

SUBMARINES 1†

SSK 1 *Foxtrot* (Type 641) (non-op)

PRINCIPAL SURFACE COMBATANTS 4

CRUISERS 1†

CG 1 *Ukraina* (RF *Slava*) (in refit)

FRIGATES 3

FFG 2

2 *Mikolair* (RF *Krivak* I) 1 with 4 SS-N-14 *Silex* SSM/ASW, 2 SA-N-4 *Gecko* SAM, 4 × 76mm gun, 8 × 533mm TT

FF 1

1 *Sagaidachny* (RF *Krivak* III) 3 with 2 SA-N-4 *Gecko* SAM, 1 × 100mm gun, 8 × 533mm TT, 1 KA-27 hel

PATROL AND COASTAL COMBATANTS 10

2 *Grisha* II/V FS with 2 SA-N-4 *Gecko* SAM, 1 × 76mm gun, 4 × 533mm TT

2 *Pauk* 1 PFT with 4 SA-N-5 *Grail* SAM, 1 × 76mm gun, 4 × 406mm TT

5 *Matka* PHM with 2 SS-N-2C *Styx* SSM, 1 × 76mm gun

1 *Zhuk* PCI†

MINE COUNTERMEASURES 5

1 *Yevgenya* MHC, 2 *Sonya* MSC, 2 *Natya* MSC

AMPHIBIOUS 7

4 *Pomornik* ACV with 2 SA-N-5 capacity 30 tps and crew

1 *Ropucha* LST with 4 SA-N-5 SAM, 2 × 2 57mm gun, 92 mines; capacity 190 tps or 24 veh

1 *Alligator* LST with 2/3 SA-N-5 SAM capacity 300 tps and 20 tk
1 *Polnocny* LSM capacity 180 tps and 6 tk

SUPPORT AND MISCELLANEOUS 9

1 AO, 2 *Vytegrales* AK, 1 *Lama* msl spt, 1 Mod *Moma* AGI, 1 *Primore* AGI, 1 *Kashtan* buoytender, 1 *Elbrus* ASR; 1 AGOS

NAVAL AVIATION (2,500)

13 armed hel
ASW 11 Be-12, 2 Ka-27E
TPT 8 An-26, 1 An-24, 5 An-12, 5 Mi-6, 1 Il-18, 1 Tu-134
MISC HEL 28 Ka-25, 42 Mi-14 plus 14 Su-17 (non-op)

NAVAL INFANTRY (1,500)

2 inf bn

Air Force 96,000

911 cbt ac (incl 98 MiG-23, 31 Su-15 awaiting disposal), no attack hel
2 air corps, 1 hvy bbr div, 1 tpt div, 1 cbt trg centre, 1 trg institute
BBR 1 div HQ, 2 regt with 39 Tu-22M
FGA/BBR 2 div HQ, 5 regt (incl 1 trg) with 188 Su-24
FGA 2 regt with 73 Su-25
FTR 2 div, 8 regt with 3 Su-15, 51 MiG-23, 224 MiG-29 (206 operational, 2 trg, 16 in store), 66 Su-27
RECCE 4 regt with 1* Tu-22, 59* Su-17, 30* Su-24
CBT TRG 21* Tu-22M, 26* Su-24, 1* MiG-23
TPT 78 Il-76, 45 An-12/An-24/An-26/An-30/Tu-134, Il-78 (tkr/tpt)
TRG 6 regt with 337 L-39 (plus 293 in store), 1 regt with 16 Mi-8
SPT HEL 111 Mi-2, 23 Mi-6, 170 Mi-8
SAM 825: SA-2/-3/-5/-10/-12A/-300

Forces Abroad

UN AND PEACEKEEPING

CROATIA (UNMOP): 1 obs **DROC** (MONUC): 4 obs **LEBANON** (UNIFIL): 650 **YUGOSLAVIA** (KFOR): 240

Paramilitary

MVS (Ministry of Internal Affairs) 42,000, 4 regions, internal security tps, 85 ACV, 6 ac, 8 hel

NATIONAL GUARD 26,600 (to be disbanded)

4 div, 1 armd regt, 1 hel bde, 60 MBT, 500 ACV, 12 attack hel

BORDER GUARD 34,000

HQ and 3 regions, 200 ACV

MARITIME BORDER GUARD

The Maritime Border Guard is an independent subdivision of the State Commission for Border Guards, is not part of the Navy and is org with:
4 cutter, 2 river bde • 1 gunship, 1 MCM sqn • 1 aux ship gp • 1 trg div • 3 air sqn

PATROL AND COASTAL COMBATANTS 36

3 *Pauk* 1 with 4 SA-N-5 SAM, 1 76mm gun, 4 406mm TT
3 *Muravey* PHT with 1 76mm gun, 2 406mm TT
10 *Stenka* PFC with 4 30mm gun, 4 406mm TT
20 *Zhuk* PCI

AIRCRAFT

An-24, An-26, An-72, An-8, Ka-27

COAST GUARD 14,000

3 patrol boats, 1 water jet boat, 1 ACV, 1 landing ship, 1 OPV, 1 craft

CIVIL DEFENCE TROOPS (Ministry of Emergency Situations): some 9,500; 4 indep bde, 4 indep regt

Foreign Forces

Russia ε1,500 naval inf

Yugoslavia, Federal Republic of (Serbia–Montenegro) FRY

new dinar d		1998	1999	2000	2001
GDP	d	106bn	174bn		
	US$	17bn	13bn		
per capita	US$	5,300	4,300		
Growth	d	4.0	-20		
Inflation	d	15.0			
Debt	US$	13.7bn			
Def exp	d	ε17bn	ε19bn		
	US$	1.6bn	1.6bn		
Def bdgt	d	6.6bn	14.4bn	16.3bn	
	US$	599m	1.3bn	1.3bn	
US$1=d		10.9	11.5	12.1	

Population ε10,600,000

Serbia ε9,900,000 (Serb 66%, Albanian 17% (90% in Kosovo), Hungarian 4%, mainly in Vojvodina)
Montenegro ε700,000 (Montenegrin 62%, Serb 9%, Albanian 7%)
(ε2,032,000 Serbs were living in the other Yugoslav republics before the civil war)

Age	13–17	18–22	23–32
Men	415,000	425,000	837,000
Women	391,000	402,000	795,000

The data outlined below represents the situation prior to the implementation of *Operation Allied Force* on 24 March 1999. Details should therefore be treated with caution.

Total Armed Forces

ACTIVE some 97,700
(43,000 conscripts) *Terms of service* 12–15 months

RESERVES some 400,000

Non-NATO Europe
NATO and Non-NATO Europe

Army (JA) some 74,000

(incl 4,000 naval ground tps; 37,000 conscripts)
3 Army, 7 Corps (incl 1 capital def) • 3 div HQ • 6 tk bde • 1 gd bde (-), 1 SF bde • 2–4 mech bde • 1 AB bde • 8–12 mot inf bde (incl 1 protection) • 5 mixed arty bde • 7 AD bde • 1 SAM bde

RESERVES

27 mot inf, 42 inf, 6 mixed arty bde

EQUIPMENT

MBT 733 T-55, 239 M-84 (T-74; mod T-72), 63 T-72
RECCE 53 BRDM-2
AIFV 532 M-80
APC 137 M-60P, 58 BOV VP M-86
TOWED 105mm: 269 M-56, 18 M2A1; **122mm**: 108 M-38, 288 D-30; **130mm**: 270 M-46; **152mm**: 24 D-20, 50 M-84; **155mm**: 133 M-1
SP 122mm: 71 2S1
MRL 107mm; 122mm: BM-21; **128mm**: 98 M-63, 56 M-77, **262mm**: M-87 *Orkan*
MOR 82mm: 1,100; **120mm**: 6 M-38/-39, 123 M-52, 283 M-74, 802 M-75
SSM 4 FROG
ATGW 142 AT-3 *Sagger* incl SP (BOV-1, BRDM-1/2), AT-4 *Fagot*
RCL 57mm: 1,550; **82mm**: 1,500 M-60PB SP; **105mm**: 650 M-65
ATK GUNS 1,250 incl: **90mm**: M-36B2 (incl SP), M-3; **100mm**: 138 T-12, MT-12
AD GUNS 2,000: **20mm**: M-55/-75, BOV-3 SP triple; **30mm**: M-53, M-53/-59, BOV-30 SP; **57mm**: ZSU-57-2 SP
SAM 1,400: 60 SA-6, SA-7/-9/-13/-14/-16/-18

Navy 7,000

(incl 3,000 conscripts and 900 Marines)
BASES Kumbor, Tivat, Bar, Novi Sad (River Comd)
(Most former Yugoslav bases are now in Croatian hands)

SUBMARINES 5

SSK 2
1 *Sava* with 533mm TT
1 *Heroj* with 533mm TT
plus 3 *Una* SSI for SF ops (all non-op)

FRIGATES 4

FFG 4
2 *Kotor* with 4 SS-N-2C *Styx* SSM, 1 × 2 SA-N-4 *Gecko* SAM, 2 × 3 ASTT, 2 × 12 ASW RL
2 *Split* (Sov *Koni*) with 4 SS-N-2C *Styx* SSM, 1 × 2 SA-N-4 *Gecko* SAM, 2 × 12 ASW RL

PATROL AND COASTAL COMBATANTS 30

MISSILE CRAFT 8
5 *Rade Koncar* PFM with 2 SS-N-2B *Styx* SSM (some †)
3 *Mitar Acev* (Sov *Osa* I) PFM with 4 SS-N-2A *Styx* SSM
PATROL CRAFT 22†
PATROL, INSHORE 4 *Mirna* PCI
PATROL, RIVERINE about 18 < (some in reserve)

MINE WARFARE 10

MINE COUNTERMEASURES 10
2 *Vukov Klanac* MHC, 1 UK *Ham* MSI, 7 *Nestin* MSI

AMPHIBIOUS 1

1 *Silba* LCT/ML: capacity 6 tk or 300 tps, 1 × 4 SA-N-5 SAM, can lay 94 mines
plus craft:
8 Type 22 LCU
6 Type 21 LCU
4 Type 11 LCVP

SUPPORT AND MISCELLANEOUS 9

1 PO-91 *Lubin* tpt, 1 water carrier, 4 AT, 2 AK, 1 degaussing

MARINES (900)

2 mot inf 'bde' (2 regt each of 2 bn) • 1 lt inf bde (reserve) • 1 coast arty bde • 1 MP bn

Air Force 16,700

(incl 3,000 conscripts); 183 cbt ac, 33 armed hel
2 Corps (1 AD)
FGA 4 sqn with 30 *Orao* 2, 50 *Galeb*, 9 *Super Galeb* G-4
FTR 5 sqn with 20 MiG-21F/PF/M/bis, 10 MiG-21U, 5 MiG-29
RECCE 2 sqn with 17* *Orao*, 10* MiG-21R
ARMED HEL 30 Gazelle
ASW 1 hel sqn with 3* Ka-25
TPT 8 An-26, 4 CL-215 (SAR, fire-fighting), 2 *Falcon* 50 (VIP), 6 Yak-40
LIAISON ac 32 UTVA-66 **hel** 14 *Partizan*
TRG ac 16* *Super Galeb*, 16* *Orao*, 25 UTVA, 15 UTVA-75 **hel** 16 *Gazelle*
AAM AA-2 *Atoll*, AA-8 *Aphid*, AA-10 *Alamo*, AA-11 *Archer*
ASM AGM-65 *Maverick*, AS-7 *Kerry*
AD 8 SAM bn, 12 SA-3
15 regt AD arty

Paramilitary

MINISTRY OF INTERIOR TROOPS ε80,000
internal security; eqpt incl 150 AFV, 170 mor, 16 hel

UN and Peacekeeping

KFOR (Kosovo Peace Implementation Force): some 39,100 tps from 28 countries are deployed in Kosovo, a further 5,900 provide rear area spt in Albania, FYROM and Greece

MAJOR DEVELOPMENTS

The developments that most influenced Russia's military and security policy in late 1999 and in 2000 were:

- Russia's second military campaign in Chechnya, which tested the Russian armed forces' readiness, tactics and equipment;
- foreign- and security-policy reviews that recommended focusing on restructuring and reforming Russia's conventional armed forces so that they would be better equipped to deal with external and internal threats and with the decreasing importance of strategic nuclear forces in both organisational and resource-allocation terms;
- a steady rebuilding of Russia's security and military cooperation with the major Western powers, bilaterally and through NATO, after the breakdown of relations over NATO's 1999 military intervention in Yugoslavia. This effort included reviving the Russia–NATO Permanent Joint Council meetings and improving Russia–NATO cooperation within the Kosovo Force (KFOR);
- progress in the arms-control field made by newly elected President Vladimir Putin and the new Duma in 2000. For instance, the Strategic Arms Reduction Treaty (START) II and the Comprehensive Test Ban Treaty (CTBT), were ratified despite Russia's opposition to US plans for national missile defence.

Military Reform

After more than a decade of gradual decline and little public and political support, Russia's armed forces and security services are likely to benefit from more coordinated, strategically focused and economically viable military reform under Putin's regime. Unlike his predecessor, Putin clearly understands military policy issues and during his presidential campaign he repeatedly reaffirmed his political commitment to bolstering Russia's armed-forces readiness for dealing with domestic and external security threats. However, by mid-2000, the president's ability to undertake difficult and costly military reforms was being challenged by the continuing war in Chechnya, which is draining economic and military resources and losing public support, as well as by the demands of other key domestic reforms, particularly federal-government and economic restructuring. Military reform continues to lack not only economic resources but also the human talent capable of implementing radical modernisation.

Putin's appointment of more military and security-force officers to key regional and federal political positions calls into question his stated support for a civilian minister of defence. The open confrontation between Defence Minister Marshal Igor Sergeyev and Chief of the General Staff General Anatolyi Kvashnin over the strategic direction of military reform is likely to compel Putin to look for a new team to carry out his plans in this area. One such step was taken when Putin charged Russia's Security Council – headed by former KGB General Sergei Ivanov, Putin's longstanding colleague – with formulating the military-reform programme and overseeing its implementation.This decision removed responsibility for military reform from the General Staff (where it lay for most of Yeltsin's term) and transferred it to a political and multi-agency body directly answerable to the president.

The first attempt to impart political momentum to military reform was made at a special meeting of the Security Council on 11 August 2000. The president laid down two key principles at the meeting:

- that any military-reform policy must be backed by a clear and balanced economic assessment and that reforms should result in the more efficient use of funds in order eventually to decrease public spending on defence;
- the structure of the armed forces should correspond to the threats that Russia faces now and in the near future.

An important outcome of the Security Council meeting was the decision to give priority to improving the conventional forces, which have been significantly weakened by ill-planned reforms attempted in the 1990s. The need to enhance these forces' state of readiness and equipment has been demonstrated during the Chechnya campaign, where Russian forces continue to suffer from a shortage of well-trained ground forces as well as from out-dated weapons. Funding for land forces will be increased and priority will be given to increasing the number of permanent-readiness units. Another significant decision was to begin the move to a three-service structure, consisting of the Army, Air Force and Navy.

The Security Council decided not to replace intercontinental ballistic missile (ICBM) systems when they reach the end of their operational life and to allow the number of systems to decline. The plan is to reduce operational strategic nuclear warheads to 1,000–1,500 and 'review the status' of the Strategic Missile Troops (SMT) in 2006. The SMT might be merged with the Air Force, and it has already been decided that units operating satellite systems will be transferred to the Air Force by 2002. Production of the *Topol*-M mobile ICBM will continue, although at a reduced rate.

A third important decision was to eliminate the equal-funding principle for all services and to introduce a more rational system, under which the funding for each service would be determined by its tasks. The meeting set the long-term objective that, by 2015, half the defence budget would be allocated to operations and maintenance of the armed forces and half to research, development and procurement of weapon systems and equipment. The military tasks are drawn from the new national-security concept and associated military doctrines which, in essence, respond to the perception that local and regional conflicts are presenting a growing threat.

New Military Doctrines

The key documents that informed the August 2000 National Security Council decisions were:

- the National Security Concept published on 10 January 2000;
- the Naval Doctrine published on 4 March 2000, setting out the policy to 2010;
- the Military Doctrine of 21 April; and
- the Foreign Policy Concept of 28 June.

The military doctrine builds on the National Security Concept in recognising that the threat of direct major military aggression against Russia has declined, but that a number of external and internal security threats remain. These threats are seen to be increasing in certain areas, in particular the threat of local conflicts along Russia's southern borders. The doctrine still links external threats to NATO's enlargement and its readiness to undertake out-of-area operations. Also numbered among the external threats is foreign support for terrorist and guerrilla groups such as the Chechen insurgents.

Internal threats include terrorist and other violent actions perceived as undermining Russia's territorial integrity, as well as weapons proliferation and organised crime. The new doctrine outlines four priorities for military restructuring:

- establishing combined groups of the different types of armed forces of the Russian Federation (including those of the Ministry of Defence (MOD), Interior Ministry, Border Guard Service and other services);
- improving the mobilisation system and readiness of the MOD and other armed forces;

• improving staffing, equipment and training to enable conventional permanent-readiness forces to fulfil their missions;
• strengthening the armed forces' strategic-deterrence capabilities, including nuclear deterrence. The priority accorded to strategic nuclear forces over the others listed above was qualified at the August Security Council meeting.

The new military doctrine for the first time stipulates that the armed forces of the Ministry of Defence can be used not only to repel external aggression, but also in domestic operations to protect the state from 'anti-constitutional actions and illegal armed violence which threaten Russia's territorial integrity'. According to the doctrine, the armed forces will be used in five types of operations: strategic (large-scale and regional wars); armed conflicts (local wars and international armed conflicts); joint operations (with other services in internal armed conflicts); counter-terrorist operations and peacekeeping. Finally, the new doctrine gives priority to developing military cooperation within the Commonwealth of Independent States (CIS) and in particular within the framework of the Russia–Belorussia union. It also states that Russian service personnel who serve in joint military units with CIS states will be professionals (that is, on contract). The doctrine does not specify any plans to professionalise the main elements of the armed forces based in Russia.

Naval Doctrine

The new naval doctrine sets out the expanding role projected for the Navy within Russia's strategic nuclear forces. Once START II is implemented, the Navy will be responsible for 60% of Russia's total strategic capability. With regard to conventional naval forces, although primarily a land-based power, Russia has extensive maritime interests to protect and promote. These comprise many thousands of kilometres of coastline; maritime borders with several states; a vast economic-exclusion zone; considerable maritime trade and a large ocean-going merchant fleet. For these reasons, it is in Russia's strategic interests to have an effective ocean-going navy. The naval doctrine reflects these interests by stating that the naval priorities should be to:

- ensure the Russian Federation's guaranteed access to the world's oceans;
- prevent discriminatory actions against Russia or its allies by individual states or military or political alliances;
- prevent the domination of the sea by any states or alliances against Russian interests, especially in adjacent seas; and
- help settle international disputes over the use of the sea on terms advantageous to Russia.

While these have long been Russia's maritime interests, recent developments have highlighted the Navy's difficulty in fulfilling the doctrine's requirements. Like most major navies, Russia's has declined in size since the end of the Cold War. However, the Russian decline since 1990 has been particularly steep: the submarine and surface-combatant fleets have reduced by 80%. The Russian Navy now has about the same number of surface combatants as the UK's Royal Navy. However, its traditionally strong submarine fleet is significantly larger than the UK's. Even if the Navy receives a greater share of the budget than planned, it is unlikely that its capability will be expanded sufficiently by 2010 (the end of the time span covered by the new doctrine) to meet the objectives set out. Even the ability to maintain the current fleet in operational condition is in doubt. The Navy's biggest surface combatant, the aircraft carrier *Kuznetsov*, was supposed to deploy to the Mediterranean in August and September 2000, accompanied by a destroyer, a frigate and a support tanker. This has now been delayed until November 2000 because of funding difficulties. Whether it will go ahead at all has been called into question by the sinking of the *Oscar II*-class nuclear-powered submarine (SSN), the *Kursk*, in the Barents Sea in August. The submarine was taking part in an exercise by the Northern Fleet, as part of the latter's preparation for the deployment. Involving up to thirty ships and submarines, the Barents Sea exercise was the largest carried out by the Navy for several years. If more funds were forthcoming, the Navy's priorities after the ballistic-missile-carrying nuclear-fuelled submarines (SSBNs), which are separately funded, would probably be maintenance and upgrading of attack submarines, guided-missile destroyers and frigates and the *Kuznetsov* and support ships to enable global deployments.

The Second Chechen War

Many elements of the new military doctrine were tested during the second military campaign in Chechnya. This was termed an 'anti-terrorist operation' which allowed the government to use force without seeking approval from the *Duma* and to use MOD forces in support of the Civil Power and Interior Troops. It also enabled the government to play on international concerns about terrorism and international terrorist networks, particularly in relation to groups with Islamic connections. Branding it an anti-terrorist campaign enabled a clear link to be made between Chechnya and the terrorist bombings in Moscow, Volgodonsk and Buinakskand, thus gaining more public support for the military intervention. However, by any objective standards, the war in Chechnya is an internal armed conflict according to the accepted definitions in international law.

Chronology of the campaign The tactics applied by the Russian armed forces evolved throughout the campaign. There have been three distinct periods:

First stage: 2 August–30 September 1999: a counter-insurgency operation in Daghestan conducted by Russian forces with support from the local militia and self-defence forces;

Second stage: 1 October 1999–22 April 2000: Large-scale military intervention in Chechnya to gain control of territory, culminating in the storming of Grozny in December 1999–January 2000;

Third stage: From May 2000 onwards: Guerrilla warfare by small Chechen groups against Russian bases, checkpoints and convoys, as well as terrorist attacks in large towns.

In the first stage of the war, Russian MOD forces (106 Brigade), Interior Troops (102 Brigade), local and federal militia and self-defence forces conducted an operation to remove Chechen armed groups (with an estimated strength of 1,700–2,000) from the Botlikh region in south-western Daghestan. Mi-24 and Mi-8 attack helicopters were used for air and artillery attacks and the key mountain passes were mined extensively. Although there were shortcomings, Russian forces coordinated well between the commands of the MOD and Interior troops. Nevertheless, there were casualties from 'friendly fire' due to a lack of joint training for the Air Force and Interior Troops. During this stage of the campaign, Russian forces lost 118 servicemen, three helicopters and one Su-25 fighter, ground attack (FGA) aircraft.

During the war's second stage, troops engaged reached a peak of 100,000 in January, gradually reducing to about 80,000 by the end of April. The tactics included extensive use of artillery and air strikes to reduce the need for close combat. The need to minimise casualties and low training standards were the main reasons for this approach. As a result, every major offensive was characterised by Russian forces surrounding towns and villages and subjecting them to prolonged artillery barrages and air–strikes to try to force the guerrillas to abandon their positions. Many area weapons – such as fuel-air explosives and cluster bombs – were used against towns and villages, causing casualties among the civilian population. In purely military terms, this tactic proved successful in capturing major Chechen towns like Grozny, Argun and Gudermes. However, the Russian forces still incurred a significant number of casualties during this phase, with over 2,000 servicemen killed and about 5,800 wounded. The biggest losses occurred during the assault on Grozny, when 368 soldiers were killed and over 3,500 wounded. By May, the Russian forces had lost two Su-25 ground-attack fighters, two Su-24 maritime-reconnaissance aircraft and ten helicopters. As in the 1994–96 campaign, Chechen guerrillas used shoulder-launched *Strela*-3 and *Igla*-1 surface-to-air missiles (SAM). However, it was reported that many SAMs were not operational, due to lack of replacement batteries and because the Russian forces managed to cut off supply routes quickly.

The third phase of the campaign started on 18 April, when Deputy Chief of the General Staff and MOD spokesman on Chechnya, General Valery Manilov, announced the end of 'major military operations in Chechnya'. This meant that by late April, Russian forces claimed to control most of Chechnya, except for remote parts of the Vedensky and Nozhay–Yurtovsky regions, and the southern part of Argun gorge. However, bombings and clashes with small groups of Chechen continued throughout southern Chechnya, including large towns such as Grozny and Gudermes. Operations in these areas were mainly conducted by Special Forces (*Spetsnaz*) and police units of the Interior Ministry, while MOD forces continued to provide artillery and air support. Guerrilla attacks continue to claim the lives of Russian service personnel and slowly to undermine public support for the war. About 500 Russian forces were killed and over 2,000 wounded between mid-July and the end of August 2000.

However, the reduced tempo and changed nature of the operations allowed the MOD to start withdrawing its forces from Chechnya and they were down to about 48,000 by August 2000. In late April, many units were withdrawn in order to meet the south-flank limits of the Conventional Forces in Europe (CFE) Treaty. Forces withdrawn included four motorised infantry regiments, one artillery regiment, one airborne regiment, two tank battalions, one parachute battalion and one motorised brigade. A total of 167 tanks, 540 armoured vehicles and 134 artillery pieces and mortars were withdrawn by the end of June.

If the conflict continues to abate, the Russian forces stationed in Chechnya are to decrease to 25,000. The units withdrawn are likely to include the MOD's 42nd Motorised Rifle Division (about 15,000-strong) and the 16th brigade of the Interior Forces (8,000–10,000 strong). Construction of all military facilities is scheduled to be completed by 2001. To reinforce the military, a 2,500-strong police force has been formed from local civilians, with the aim of increasing it to 6,000.

According to the Russian MOD, the Russian armed forces (both MOD and Interior Troops) suffered 2,585 fatalities and 8,050 wounded (only 2,500 of those returned fit for military service). In comparison, reports show losses of 3,826 servicemen during the 1994–96 campaign. Losses among Chechen fighters are hard to estimate; although the MOD reported over 13,000 killed, this figure is likely to be inflated. The MOD assesses casualties among the civilian population at fewer than 1,000. While this is likely to be an underestimate, accurate data is unlikely to emerge until more refugees are able to return and civilian administration is more widely established. Civilian deaths are likely to be much fewer than during the 1994–95 war as, recalling the horrors of the earlier experience, many fled the province well before the major military operations began. The IISS estimates that civilian deaths in the 1999–2000 war are likely to total around 5,000.

Conduct of the Campaign It is clear that the Russian military learned many lessons from its humiliating defeat in 1994–96. That campaign revealed grave shortcomings in training standards and in the ability of different elements of the ground and air forces to coordinate joint operations. The intensive second phase of the current campaign demonstrated improvements in the following areas:

- *Joint operations*: There was better coordination between the different services: ground forces, the Air Force, interior forces (including the special forces of the Ministry of Internal Affairs), border guards, government communications troops, federal security service, emergency service, railway troops and police. During the campaign, the ground troops split into smaller operational groups at regimental and battalion level. The tactical groups benefited from artillery and helicopter support under their direct control and, most important, they underwent joint training before deployment. In the earlier campaign, to their cost in lives and materiel, many units were formed on an *ad hoc* basis and sent into battle immediately.

• *Reconnaissance*: More emphasis was placed on effective reconnaissance and intelligence-collection: from the air, using electronic intelligence (ELINT); by special forces; and by the internal security service. Air reconnaissance was conducted mainly by Su-24MR, Su-25, and MiG-25RB aircraft, while ELINT was gathered by An-30B and An-50 aircraft. In this campaign, Russian forces introduced upgraded *Pchela*-1T unmanned aerial vehicles (UAV) as a part of the *Story*-P UAV reconnaissance system. This system allowed Russian commanders to obtain real-time aerial-reconnaissance data on guerrilla positions. Better communications and reconnaissance allowed Russian forces to locate mobile groups of fighters and to restrict supplies of weapons and ammunition to the guerrillas from outside Chechnya.

• *More effective manpower*: Much better conditions of service for military personnel have been provided in this campaign compared with 1994–96 and public support and strong political leadership have meant morale has been much higher. The Russian armed forces have included more contract soldiers from permanent-readiness units and better-trained conscripts than previously. Many units have been given special training (including Special Forces training for mountain and urban warfare). Many troops had combat experience from the earlier war. However, there were still shortages of some specialist personnel, particularly airborne forces, forward air controllers and specialists in mountain operations, within both the MOD and the Interior Troops. As a result, the rotation time for these specialist troops during the current campaign has been 90 days for Interior Troops (as opposed to 45 days in the first war), and 60 days for MOD forces. One of the major lessons from the 1999–2000 conflict already acted upon is to increase the airborne forces from 40,000 to around 45,000. The turnover of conscripts has also been higher in the current campaign than in 1994–96, as their time in Chechnya has counted as double-time and they have also had longer periods of training before being sent on operations. Moreover, unlike during the first campaign, Russian servicemen have been given financial and social rewards for taking part in the operation, including higher pay. Soldiers have received 800–950 roubles (nearly $30) a day for each day spent on active operations. The average salary of an officer of the rank of Major is the equivalent of $400 per month.

• *Improved logistics*: A much better organised logistic system than in 1994–96 supported the recent campaign, although this remained one of the operation's weakest points. There were still shortages of munitions, medicine, food and fuel but they were less severe than before, helped by the MOD having prepared logistic units and facilities in the neighbouring regions of North Caucasus before the current campaign started.

• *New weapons*: In the course of the campaign, particularly towards the end of 1999, the armed forces started to receive new weapons and equipment, including night-vision equipment, artillery systems, light weapons, and reconnaissance and communications equipment. However, there have still been shortages, particularly of night-vision equipment for aircrew. New equipment was tested, such as the modernised Su-25T fighter, the KA-50 *Black Shark* helicopter and BMP-3 infantry fighting vehicle. However, Russian forces primarily used older equipment such as the Su-24 and Su-25 and the Mi-24 and Mi-8 helicopters. The newer aircraft, such as the KA-50 helicopter and modernised Su-30. MiG-29SMTs are ill-suited to missions in mountainous regions against targets such as small groups of guerrillas. The forces improved their ground-to-air communications by creating tactical aviation groups and training additional forward air controllers.

While military performance during phase two of the campaign was far better than in 1994–96, on the down side was the heavy and often indiscriminate use of artillery and air-strikes against urban and rural targets, causing unnecessary civilian casualties and high losses among Russian servicemen, despite attempts to minimise casualties. There has also been criticism of human-rights violations commited by the Russian military against the civilian, particularly male, population in Chechnya. According to the Russian military prosecutor's office, more than 467 servicemen are currently under investigation for their conduct. However, the most serious shortcoming of the current campaign is that it has relied solely on military means without a parallel attempt at a political process to prepare the ground for a lasting peace. As a result, since completion of the large-scale military operation, the forces have been embroiled in a draining partisan war that cannot be won by military means alone, particularly without substantial support from the local population.

Costs of the Chechnya campaign The Chechnya campaign cost R5bn in 1999, twice the amount allocated in the budget, according to the official figures. The first three months of 2000, when the majority of operations were carried out and force-deployment levels were at their highest, added a further R6bn. If converted to dollars using the purchasing power parity (PPP) rate of exchange, rather than the misleading official or market rate, the official R11bn cost to the end of March is equivalent to nearly $3bn. However, independent sources put the cost much higher than this, with estimates of monthly costs running at R4bn. It seems likely that the cost of the Russian campaign in Chechnya and Daghestan during the 12 months to August 2000 was at least R20bn or, measured at PPP rates, around $5bn. This compares with the estimated $3.8bn cost of the less intensive 1994–96 war.

DEFENCE SPENDING

The fortunes of Russia's armed forces necessarily hinge upon the performance of the Russian economy. The new national-security concept and draft military doctrine will be difficult to implement unless Putin manages to mobilise Russia's economic resources more effectively than the succession of Yeltsin governments. Putin has been fortunate in that the end of the Yeltsin era coincided with signs of an economic revival after a decade in the doldrums.

According to official figures, Russian gross domestic product (GDP) grew by 3.2% in real terms in 1999, thanks to the recovery in oil and other commodity prices at the same time as the heavy devaluation of the rouble in 1998 had improved the competitiveness of Russian exports. This performance was better than expected by the International Monetary Fund (IMF) and the Organisation for Economic Cooperation and Development (OECD), which had projected zero and 2% growth respectively.

After a decade of economic transition, even modest improvements in economic performance will generate substantial additional resources for the government, and make it possible, among other priorities, to take steps to check the steep decline in Russian military power. The half-hearted experiments in military reform of recent years may give way to a process of rebuilding and modernising Russia's armed forces.

Defence spending in 1999

The recovery in the Russian economy and associated rise in government revenues, particularly from oil, in 1999 allowed Russia's officially reported defence spending to rise, in nominal terms at least, for the first time since the formation of the Russian Federation. In contrast to the previous years of the decade, not only was the defence budget fully disbursed without any sequestration, but outlays, at R116bn, were 24% more than the R94bn in the original budget for the year. Spending on all law-enforcement agencies, including the Interior Troops active in Chechnya, also rose 10% above budget to R57bn. Official defence spending still declined in real terms compared to 1998, thanks to inflation of 37% according to official figures, but the decline was much less than the dramatic plummeting of previous years.

Table 16 Official Russian defence outlays and budgets, 1992–2000

(Redenominated roubles m)	Defence budget	Defence share of Federal budget (%)	Defence Outlay	Defence share of Federal outlay (%)	Defence outlay/ GDP (%)
1992	384	16.0	855	16.4	4.7
1993	3,116	16.6	7,210	20.7	4.4
1993 revised	8,327	n.a.	7,210	20.7	4.4
1994	40,626	20.9	28,028	16.4	4.6
1995	48,577	19.6	47,800	12.2	3.1
1995 revised	59,379	21.3	47,800	12.2	3.1
1996	80,185	18.4	63,900	14.2	3.0
1997	104,300	19.7	79,700	16.2	3.1
1997 revised	83,000	19.7	79,700	16.2	3.1
1998	81,765	16.4	56,700	12.7	2.1
1999	93,702	16.3	116,000	17.2	2.6
1999 revised	109,000	19.0	116,000	17.2	2.6
2000	143,000	n.a.	n.a.	n.a.	2.6

Note Military pensions (R11bn) were transferred from the Defence Budget to the Social Budget from 1998.

The defence budget for 2000

With economic conditions brighter and a need to raise military spending in real terms to finance the continued conflict in the Caucasus and the defence priorities set out by the Russian Security Council, the stage was set for a significant rise in the defence budget for 2000. At an estimated R143bn ($5.1bn), the allocation is more than 30% up on the revised budget for 1999. Procurement funding is set to increase to R62bn ($2.2bn), if spending on spare parts, repairs and maintenance, research and development as well as new equipment is included.

It now remains to be seen whether this defence budget, like others framed in election years, most notably in 1993, has been inflated by the need to win votes from the military and fails to materialise in a real rise in spending.

Table 17 Estimated Russian defence budget by function, 1999 and 2000

(redenominated roubles m)	1999	Share of total (%)	2000	Share of total (%)
Personnel	33,900	31.1	50,100	35.0
Operations and Maintenance	29,600	27.2	40,000	28.0
Procurement	23,800	21.8	27,300	19.1
Research and Development (R&D)	14,000	12.8	15,600	10.9
Infrastructure	3,500	3.2	4,000	2.8
Nuclear	1,900	1.7	3,000	2.1
MOD	500	0.5	1,000	0.7
Other	1800	1.7	2,000	1.4
Total official defence budget	**109,000**	**100**	**143,000**	**100**

Weapon programmes

The Russian inventory of major weapon systems in development, pre-production and production remains larger than in any other country apart from the US and is broadly comparable in technical capability to the combined resources of the major European NATO-member countries. The 2000 procurement-spending plan of R62bn excludes funding accruing to the government and industry from arms-export revenues, estimated at a net R30–40bn for 2000. Measured using the PPP exchange rate, the combined procurement budget for 2000 is the equivalent of over $25bn, compared to around $50bn in total for NATO European countries.

Series production of ballistic missiles is now confined to the new SS-27 (*Topol*-M2) intercontinental ballistic missile (ICBM), a further 10 of which were deployed in 1999, following the introduction of the first batch of 10 in 1998. Annual production may be less in 2000. The SS-NX-28 submarine-launched ballistic missile (SLBM) was cancelled in August 1998 after three test failures. Work was also suspended on the missile's intended platform, the *Dolgoruky* nuclear-powered ballistic-missile submarine (SSBN), until new design criteria for the missile and submarine are established. Refurbishment of the road-mobile SS-25 ICBM has continued, with the aim of prolonging its service life. The successor to the *Scud* short-range ballistic missile (SRBM), SS-X-26, should complete its development in 2000. Russia is strengthening its strategic aviation, beginning with the acquisition of 8 Tu-160 and 3 Tu-95 bombers from the Ukraine in 1999–2000 and the modernisation of the Tu-160 and Tu-95 fleets. In the past year, the strategic aviation force has devoted more resources to long-range exercises to enhance its operational capabilities, which were in decline in the 1990s. The MiG-29SMT upgrade programme, suspended in early 1999, restarted in 2000. The total order was for 180 aircraft, of which about 20 had been delivered. In 1999 the Air Force also ordered the Il-112V transport aircraft to replace its AN-26 fleet. Of the Air Force's total inventory, about 20% can be described as modern and only about half the total is fully operational. The equipment problems are compounded by the lack of resources to give aircrew sufficient flying hours. Apart from those engaged in Chechnya, a pilot receives an average of about 20 flying hours per year; for some branches of the Air Force this number might be even less.

In 1999, the Army was expecting to receive 30 T-90 main battle tanks (MBT) and the first consignment of new armoured vehicles based on the existing BTR-80. Development of a new MBT known as the T-95 commenced in 2000. Most of Russia's production of military equipment continues to be exported.

Defence Industry

The government's industrial strategy since 1993 has hinged on the twin policies of conversion and privatisation. Neither has been particularly successful. Using conversion to preserve Russia's defence industry was probably an expensive mistake. As a result, the industry remains a vast and unreformed collection of design plants and production facilities. Together they comprise around one-fifth of Russia's total industrial capacity and employ some 2m workers, of whom around 400,000 are engaged full-time in defence work. In 1999 there remained some 1,500 specialist defence companies, but the steadily declining defence budget means that two-thirds of these are effectively bankrupt. Only 670 firms received orders from the government in 1998, and this number reduced to around 500 in 1999 and 2000. Around 80 of these are reported to generate export revenue.

Russia's military doctrine outlines a set of priorities for the defence-industrial base. Proposals for measures to protect and sustain the industry include:

- preferential purchasing by the state from a selected group of enterprises;

- guaranteed fixed prices for equipment;
- retention of an independent research and development capability across the range of weapon systems and industrial subsidies where necessary to protect a threatened capability.

In 1999, arms-sales revenue, converted into roubles, was more than twice the combined defence-budget allocation to research and development and to equipment procurement.

Arms Exports

Given the fall in government contracts, arms exports provide one of the main sources of revenue for the Russian defence industry. Overall, Russian arms exports increased in 1999 to an estimated $3.5bn from $2.8bn in 1997. Sales are expected to grow to $4bn in 2000, despite the fiercely competitive nature of the global arms market. Weapons and military equipment account for 40% of Russia's total exports of machinery. The August 1998 rouble devaluation lowered the price of Russian equipment in international markets, which had a particular impact on demand from poorer countries, notably those at war in the Horn of Africa, Central Africa and Angola. About half of Russia's arms trade involves just two countries: China and India accounted for some $2bn of the trade in 1999. Iran also remains one of Russia's largest customers.

Table 18 **Russian arms deliveries, 1992–2000**

	Arms deliveries (US$bn)	Arms deliveries (Rbn)	Domestic procurement (Rbn)	Arms deliveries as % of domestic procurement	All Merchandise exports (US$bn)	Arms deliveries as % of merchandise exports
1992	2.5	0.5	0.2	197	42	5.9
1993	3.1	3.1	2.1	144	44	7.0
1994	2.7	5.9	8.4	70	67	4.0
1995	3.5	16.0	10.3	155	83	4.2
1996	3.5	17.9	13.2	136	91	3.8
1997	2.5	23.1	21.0	110	89	2.8
1998	2.8	27.2	17.0	159	74	3.8
1999	3.5	86.5	23.8	363	73	4.8
2000	4.0	114.4	27.3	419	n.a.	n.a.

Estimating Russian military expenditure

Estimating the real scale of Russian military spending is fraught with difficulty. Not only have the defence budgets of 1999 and 2000 been shrouded in secrecy, but also much that would normally appear in a defence budget appears elsewhere in the Russian federal budget. Press reports suggest that the 2000 defence budget accounts for 2.6% of GDP, but this excludes military pensions and funding for military reform. The MOD succeeded in having military pensions transferred from the defence budget to the social budget in 1998. Irrespective of budget headings, however, the government continues to classify military pensions as a defence cost and it seems likely that the costs of military reform are classified in the same way.

Taking into account military-related spending outside the MOD domain gives a significant boost to the total military spend. Industrial subsidies and science and technology allocations under the Ministry of the Economy budget, together with funding from regional and local governments and revenue from arms exports, further inflate the figure to around 5% of GDP. Estimates of Russia's GDP in 1998 range from $277bn at the average annual market exchange rate, through $600–700bn at middle-of-range purchasing power parity (PPP) estimates, rising to

$1,100bn at the high end of the PPP range. The latest IMF figures, published in October 1999, value Russia's GDP in 1998 at some $945bn at PPP, somewhat higher than the fund's previous estimates.

The key factor in these assessments is the real purchasing power of the rouble. The dollar–rouble exchange rate applies to the external sector only and gives a misleading impression. It is important to take account of what the rouble will buy, not in world markets but within Russia itself, and, where defence is concerned, what the rouble can buy from the country's still very substantial defence industry. Consumer prices of Russian goods and services in Moscow are 4–5 times lower than western equivalents, suggesting that wholesale and producer prices may be 6–7 times lower. They are probably lower still outside Russia's main cities.

Translated into dollars at the market exchange rate, Russia's official defence budget for 2000 amounts to $5bn – roughly equivalent to Singapore's annual defence spending. To attribute such a low value to the Russian military effort would be very inaccurate. *The Military Balance* estimates the real purchasing power of the rouble as five to the dollar, based on its own survey conducted in Moscow in early 2000. This relatively high value is approximately twice that assessed by the World Bank, which was used in the Stockholm International Peace Research Institute (SIPRI) assessment of 1998. More recently, the IMF has revised its PPP estimates for Russia, bringing them much closer to those of *The Military Balance.* While few would dispute that Russia's military effort has contracted enormously since the Cold War, the evidence suggests that real military spending accounted for around 5% of GDP or some $57bn in 1999. Only the US exceeds this level of expenditure in absolute terms. In relative terms, Russia's defence spending as a proportion of GDP is 2–3 times higher than that of the US and its major NATO allies.

Russia RF

rouble r		1998	1999	2000	2001
GDP[1]	r	2,696bn	4,545bn		
	US$	1,100bn	1,100bn		
per capita	US$	6,600	7,000		
Growth	%	-4.6	3.2		
Inflation	%	27.8	85.7		
Debt	US$	153bn	218bn		
Def exp[1]	US$	55bn	56bn		
Def bdgt[1]	r	82bn	112bn	143bn	
	US$	35bn	31bn	29bn	
FMA[2] (US)	US$	0.9m	0.9m	0.9m	0.9m
FMA (Ge)	US$	**1991–97**	5.0bn		
US$1=r[3]		12.0	24.7	28.4	

[1] PPP est
[2] Under the US Cooperative Threat Reduction programme, $2.8bn has been authorised by the US to support START implementation and demilitarisation in Russia, Ukraine, Belarus and Kazakstan. Russia's share is 60–65%.
[3] redenominated

Population 146,000,000
(Tatar 4%, Ukrainian 3%, Chuvash 1%, Bashkir 1%, Belarussian 1%, Moldovan 1%, other 8%)

Age	13–17	18–22	23–32
Men	5,918,000	5,618,000	10,102,000
Women	5,691,000	5,468,000	9,893,000

Total Armed Forces

ACTIVE 1,004,100
(incl about 200,000 MoD staff, centrally controlled units for EW, trg, rear services, not incl elsewhere; perhaps 330,000 conscripts, 145,000 women)
Terms of service 18–24 months. Women with medical and other special skills may volunteer

RESERVES some 20,000,000
some 2,400,000 with service within last 5 years; Reserve obligation to age 50

Strategic Deterrent Forces ε149,000

(incl 49,000 assigned from Air Force and Navy)

NAVY (ε13,000)
324 msl in 19 operational SSBN†
SSBN 19 (all based in Russian ports)
3 *Typhoon* with 20 SS-N-20 *Sturgeon* (60 msl)
7 *Delta* IV with 16 SS-N-23 *Skiff* (112 msl)
7 *Delta* III with 16 SS-N-18 *Stingray* (112 msl)
2 *Delta* I with 20 SS-N-8 *Sawfly* (40 msl)
(The following non-op SSBNs remain START-accountable, with a total of 180 msl)
3 *Typhoon* with 60 SS-N-20 *Sturgeon*
6 *Delta* III with 96 SS-N-18 *Stingray*
2 *Delta* I with 12 SS-N-8 *Sawfly* (24 msl)

In the 31 January START I declaration, Russia declared a total of 504 'deployed' SLBMs. The above figures represent holdings as of 31 January 2000.

STRATEGIC MISSILE FORCE TROOPS (ε100,000 incl 50,000 conscripts)
5 rocket armies equipped with silo and mobile missile launchers. 776 launchers with 3,540 nuclear warheads org in 19 div: launcher gp normally with 10 silos (6 for SS-18) and one control centre; 12 SS-24 rail each 3 launchers
ICBM 776
180 SS-18 *Satan* (RS-20) at 4 fields; mostly mod 4/5, 10 MIRV
160 SS-19 *Stiletto* (RS-18) at 4 fields; mostly mod 3, 6 MIRV
46 SS-24 *Scalpel* (RS-22) 10 MIRV; 10 silo, 36 rail
370 SS-25 *Sickle* (RS-12M); 360 mobile, single-warhead; 10 variant for silo launcher; 10 bases with some 40 units
20 SS-27 (*Topol*-M2)
ABM 100: 36 SH-11 (mod *Galosh*), 64 SH-08 *Gazelle*, S-400

WARNING SYSTEMS
ICBM/SLBM launch-detection capability, others include photo recce and ELINT
RADARS
OVER-THE-HORIZON-BACKSCATTER (OTH-B)
2 in the Ukraine, at Nikolaev and Mukachevo, covering US and polar areas. (While these facilities are functioning, they are not tied in with the Russian air-defence system because of outstanding legal difficulties with Ukraine.)
1 near Yeniseysk, covering China
LONG-RANGE EARLY-WARNING ABM-ASSOCIATED
7 long-range phased-array systems **Operational** Moscow, Olenegorsk (Kola), Gaballa (Azerbaijan), Pechora (Urals), Balkhash (Kazakstan), Mishelevka (Irkutsk). Azerbaijan and Kazakstan sites not functioning because of outstanding legal difficulties. **Under construction** Baranovichi (Belarus) – planned to become operational shortly
11 *Hen House*-series; range 6,000km, 6 locations covering approaches from the west and south-west, north-east and south-east and (partially) south. Engagement, guidance, battle management: 1 *Pill Box* phased-array at Pushkino (Moscow)

Army ε348,000

(incl ε190,000 conscripts)
7 Mil Districts (MD), 1 Op Strategic Gp
6 Army HQ, 3 Corps HQ
5 TD (3 tk, 1 motor rifle, 1 arty, 1 SAM regt; 1 armd recce bn; spt units)
21 MRD (incl trg) (3 motor rifle, 1 arty, 1 SAM regt; 1 indep tk, 1 ATK, 1 armd recce bn; spt units)

4 ABD (each 2/3 para, 1 arty regt) plus 1 AB trg centre (bde)
7 District trg centre (each = bde - 1 per MD)
7 MG/arty div
5 arty div (each up to 6 bde incl 1 MRL, 1 ATK)
18 indep arty bde (incl MRL)
12 indep bde (9 MR, 3 AB)
7 SF (*Spetsnaz*) bde
15 SSM bde (SS-21)
5 ATK bde, 3 ATK regt
19 SAM bde (incl 2 SA-4, 4 SA-11, 1 SA-12)
20 hel regt (9 attack, 6 aslt tpt, 5 trg)
Other Front and Army tps
engr, pontoon-bridge, pipe-line, signals, EW, CW def, tpt, supply bde/regt/bn

RESERVES (cadre formations, on mobilisation form)

2 TD, 15 MRD, 1 hy arty bde, 4 indep arty bde, 4 MR bde

EQUIPMENT

Figures in () were reported to CFE on 1 Jan 2000 and include those held by Naval Infantry and Coastal Defence units

MBT about 21,820 (5,275), 1,200 T-55 (20), 2,020 T-62 (91), 4,300 T-64A/-B (194), 9,700 T-72L/-M (1,909) 4,500 T-80/-U/UD/UM (3,058), 100 T-90 (2) (total incl ε8,900 in store - in Russia)
LT TK 150 PT-76 (3)
RECCE some 2,000 BRDM-2
TOTAL AIFV/APC ε25,975 (9,542)
AIFV 14,700 (6,308): 7,500 BMP-1 (1,498), 4,600 BMP-2 (3,109), 100 BMP-3 (103), some 1,800 BMD incl BMD-1 (704), BMD-2 (333), BMD-3 (26), 700 BRM-1K (503), BTR-80A (32) (total incl 900 in store)
APC 11,275 (3,234): 1,000 BTR-50, 4,900 BTR-60/-70/-80 incl BTR-60 (46), BTR-70 (842), BTR-80 (997), 575 BTR-D (524); 4,800 MT-LB (812), plus 'look-alikes' (total incl 1,150 in store)
TOTAL ARTY 20,746 (6,159), with ε6,213 in store
TOWED 10,065 (2,238) incl: **122mm**: 1,200 M-30 (13); 3,050 D-30 (1,054); **130mm**: 50 M-46 (1); **152mm**: 100 ML-20 (1); 700 M-1943 (D1); 1,075 D-20 (176), 1,100 2A36 (562), 750 2A65 (431); **203mm**: 40 B-4M; incl ε2,000 mainly obsolete types
SP 4,705 (2,358) incl: **122mm**: 1,725 2S1 (474); **152mm**: 1,600 2S3 (976), 700 2S5 (459), 550 2S19 (419); **203mm**: 130 2S7 (30)
COMBINED GUN/MOR 820+ (358): **120mm**: 790 2S9 SP (340), 2B16 (2), 30 2S23 (16)
MRL 2,606 (904) incl: **122mm**: 50 BM-13/-14/-16, 1,750 BM-21 (351), 25 9P138 (13); **220mm**: 675 (428) 9P140; **300mm**: 106 (106) 9A52
MOR 2,550 (301) incl: **120mm**: 920 2S12 (170), 900 PM-38 (103); **160mm**: 300 M-160; **240mm**: 430 2S4 SP (28)
SSM (nuclear-capable) ε200 SS-21 *Scarab* (*Tochka*), (all *Scud* and FROG in store)
ATGW AT-2 *Swatter*, AT-3 *Sagger*, AT-4 *Spigot*, AT-5 *Spandrel*, AT-6 *Spiral*, AT-7 *Saxhorn*, AT-9, AT-10
RL 64mm: RPG-18; **73mm**: RPG-7/-16/-22/-26; **105mm**: RPG-27/-29
RCL 73mm: SPG-9; **82mm**: B-10
ATK GUNS 57mm: ASU-57 SP; **76mm**; **85mm**: D-44/SD-44, ASU-85 SP; **100mm**: 526 T-12/-12A/M-55 towed
AD GUNS 23mm: ZU-23, ZSU-23-4 SP; **30mm**: 2S6; **37mm**; **57mm**: S-60, ZSU-57-2 SP; **85mm**: M-1939; **100mm**: KS-19; **130mm**: KS-30
SAM about 2,300
500 SA-4 A/B *Ganef* (twin) (Army/Front weapon - most in store)
400 SA-6 *Gainful* (triple) (div weapon)
400 SA-8 *Gecko* (2 triple) (div weapon)
200 SA-9 *Gaskin* (2 twin) (regt weapon)
250 SA-11 *Gadfly* (quad) (replacing SA-4/-6)
100 SA-12A/B (*Gladiator/Giant*)
350 SA-13 *Gopher* (2 twin) (replacing SA-9)
100 SA-15 (replacing SA-6/SA-8)
SA-19 (2S6 SP) (8 SAM, plus twin **30mm** gun)
SA-7, SA-14 being replaced by SA-16, SA-18 (man-portable)
HELICOPTERS ε2,108 (with 600 in store)
ATTACK ε900 Mi-24 (737), 8 Ka-50 *Hokum* (4)
RECCE 140 Mi-24
TPT some 1,060 incl 20 Mi-6, 990 Mi-8/-17 (some armed), 50 Mi-26 (hy)

Navy 171,500

(incl ε16,000 conscripts, ε13,000 Strategic Forces, ε35,000 Naval Aviation, 9,500 Coastal Defence Forces/Naval Infantry)

SUBMARINES 67
STRATEGIC 19 (see p. 120)
TACTICAL 43
SSGN 8 *Oscar* II with 24 SS-N-19 *Shipwreck* USGW (VLS); T-65 HWT
SSN 19
8 *Akula* with SS-N-21 *Sampson* SLCM, T-65 HWT
3 *Sierra* with SS-N-21 *Sampson* SLCM, T-65 HWT
1 *Yankee 'Notch'* with 20+ SS-N-21 *Sampson* SLCM
7 *Victor* III with SS-N-15 *Starfish* SSM, T-65 HWT
SSK 16
12 *Kilo*, 3 *Tango*, 1 *Foxtrot* (all with T-53 HWT)
OTHER ROLES 5
3 *Uniform* SSN, 1 *Yankee* SSN, 1 *X-Ray* SSK trials
RESERVE probably some *Foxtrot*, *Tango* and *Kilo*

PRINCIPAL SURFACE COMBATANTS 35

CARRIERS 1 *Kuznetsov* CV (67,500t) capacity 20 ac Su-33 and 15–17 ASW hel or 36 Su-33 with 12 SS-N-19 *Shipwreck* SSM, 4 × 6 SA-N-9 *Gauntlet* SAM, 8 *CADS-1* CIWS, 2 *RBU-12* mor
CRUISERS 7
CGN 2 *Kirov* with 20 SS-N-19 *Shipwreck* SSM, 12 SA-N-6 *Grumble* SAM, SA-N-4 *Gecko* SAM, 2 × 130mm

gun, 10 × 533mm ASTT, SS-N-15 *Starfish* SUGW, 3 Ka-25/-27 hel
CG 5
3 *Slava* with 8 × 2 SS-N-12 *Sandbox* SSM, 8 SA-N-6 *Grumble* SAM, 2 × 130mm gun, 8 × 533mm ASTT, 1 Ka-25/-27 hel
1 *Kara* with 2 × 2 SA-N-3 *Goblet* SAM, 2 SA-N-4 *Gecko* SAM, 10 × 533mm ASTT, 2 × 4 SS-N-14 *Silex* SUGW, 1 Ka-25 hel
1 *Kynda* with 8 SS-N-3B *Sepal* SSM, 2 SA-N-1 *Goa* SAM, 4 × 76mm gun, 6 × 533mm ASTT, 2 RBU 6000 mor
DESTROYERS 17
DDG 17
7 *Sovremennyy* with 2 × 4 SS-N-22 *Sunburn* SSM, 2 × 1 SA-N-7 *Gadfly* SAM, 2 × 2 130mm guns, 4 × 533mm TT, 1 Ka-25 (B) hel
1 mod *Kashin* with 8 SS-N-25 *Svezda* SSM, 2 × 2 SA-N-1 *Goa* SAM, 2 × 76mm gun, 5 × 533mm ASTT
1 *Kashin* with 2 × 2 SA-N-1 *Goa* SAM, 2 × 76mm gun, 5 × 533mm ASTT, 2 ASW RL
7 *Udaloy* with 8 SA-N-9 *Gauntlet* SAM, 2 × 100mm gun, 8 × 533mm ASTT, 2 × 4 SS-N-14 *Silex* SUGW, 2 Ka-27 hel
1 *Udaloy* II with 8 × 4 SS-N-22 *Sunburn* SSM, 8 SA-N-9 *Gauntlet* SAM, 8 SA-N-11 *Grisson* SAM, 2 CADS-N-1 CIWS, 2 × 100mm gun, 10 × 533mm ASTT
FRIGATES 10
FFG 10
2 *Krivak* II with 2 SA-N-4 *Gecko* SAM, 2 × 100mm gun, 8 × 533mm ASTT, 1 × 4 SS-N-14 *Silex* SUGW, 2 × 12 ASW RL
7 *Krivak* I (weapons as *Krivak* II, but with 2 twin 76mm guns)
1 *Neustrashimyy* with SA-N-9 *Gauntlet* SAM, 1 × 100mm gun, 6 × 533mm ASTT, 2 × 12 ASW RL
PATROL AND COASTAL COMBATANTS 108
CORVETTES 27
27 *Grisha* I, -III, -IV, -V, with SA-N-14 *Gecko* SAM, 4 × 533mm ASTT, 2 × 12 ASW RL
LIGHT FRIGATES 12
12 *Parchim* II (ASW) with 2 SA-N-5 *Grail* SAM, 1 × 76mm gun, 4 × 406mm ASTT, 2 × 12 ASW RL
MISSILE CRAFT 54
29 *Tarantul* PFM, 1 -I, 5 -II, both with 2 × 2 SS-N-2C *Styx* SSM; 22 -III with 2 × 2 SS-N-22 *Sunburn* SSM
20 *Nanuchka* PFM 4 -I, 17 -III and 1 -IV with 2 × 3 SS-N-9 *Siren* SSM
2 *Dergach* PHM with 8 SS-N-22 *Sunburn* SSM, 1 SAN-4 *Gecko* SAM, 1 × 76mm gun
3 *Matka* PHM with 2 × 1 SS-N-2C *Styx* SSM
TORPEDO CRAFT 8 *Turya* PHT with 4 × 533mm TT
1 *Mukha* PHT with 8 × 406mm TT
PATROL CRAFT 6
COASTAL 6 *Pauk* PFC with 4 ASTT, 2 ASW RL
MINE WARFARE about 72
MINE COUNTERMEASURES about 72
OFFSHORE 15
2 *Gorya* MCO
13 *Natya* I and -II MSO
COASTAL 27 *Sonya* MSC
INSHORE 30 MSI<
AMPHIBIOUS about 25
LPD 1 *Ivan Rogov* with 4–5 Ka-27 hel, capacity 520 tps, 20 tk
LST 23
19 *Ropucha*, capacity 225 tps, 9 tk
4 *Alligator*, capacity 300 tps, 20 tk
LSM 1 *Polnocny*, capacity 180 tps, 6 tk
Plus about 21 craft: about 6 *Ondatra* LCM; about 15 LCAC (incl 4 *Pomornik*, 3 *Aist*, 3 *Tsaplya*, 1 *Lebed*, 1 *Utenok*, 2 *Orlan* WIG and 1 *Utka* (wing-in-ground-experimental))
Plus about 80 smaller craft
SUPPORT AND MISCELLANEOUS about 436
UNDER WAY SUPPORT 28
1 *Berezina*, 5 *Chilikin*, 22 other AO
MAINTENANCE AND LOGISTIC about 271
some 15 AS, 38 AR, 20 AOT, 8 msl spt/resupply, 90 AT, 9 special liquid carriers, 8 AWT, 17 AK, 46 AT/ARS, 13 ARS, 7 AR/C
SPECIAL PURPOSES about 57
some 17 AGI (some armed), 1 msl range instrumentation, 7 trg, about 24 icebreakers (civil-manned), 4 AH, 4 specialist support vessels
SURVEY/RESEARCH about 80
some 19 naval, 61 civil AGOR

MERCHANT FLEET (aux/augmentation for sealift)
1,503 ocean-going vehicles over 1,000 tonnes; 275 AOT, 104 dry bulk; 24 AK, 8 ro-ro, 7pax; 1,085 other (breakbulk, partial AK, refrigerated AK, specialised AK and LASH)

NAVAL AVIATION (ε35,000)
some 244 cbt ac; 107 armed hel
Flying hours 40
HQ Naval Air Force
FLEET AIR FORCES 4
each org in air div, each with 2–3 regt of HQ elm and 2 sqn of 9–10 ac each; recce, ASW, tpt/utl org in indep regt or sqn
BBR some 45
5 regt with some 45 Tu-22M (AS-4 ASM)
FGA 52 Su-24, 10 Su-25, 52 Su-27
ASW ac 10 Tu-142, 26 Il-38, 4 Be-12 **hel** 3 Mi-14, 42 Ka-27
MR/EW ac 18 An-12 **hel** 8 Mi-8
CBT ASLT 12 Ka-29, 15 Mi-24
TPT ac 37 An-12, An-24, An-26
ASM AS-4 *Kitchen*, AS-7 *Kerry*, AS-10 *Karen*, AS-11 *Kilter*, AS-12 *Kegler*, AS-13 *Kingbolt*, AS-14 *Kedge*

COASTAL DEFENCE (12,500)
(incl Naval Infantry, Coastal Defence Troops)
NAVAL INFANTRY (Marines) (7,500)
1 inf 'div' (bde?) (2,500: 3 inf, 1 tk, 1 arty bn) (Pacific Fleet)
2 indep bde (4 inf, 1 tk, 1 arty, 1 MRL, 1 ATK bn), 1 indep regt, 3 indep bn
3 fleet SF bde (1 op, 2 cadre): 2–3 underwater, 1 para bn, spt elm
EQUIPMENT
MBT 130: T-55, T-72, T-80
RECCE 60 BRDM-2/*Sagger* ATGW
AIFV some BRM-1K
APC some 750: BTR-60/-70/-80, 250 MT-LB
TOTAL ARTY about 321
TOWED 122mm: D-30
SP 122mm: 96 2S1; **152mm**: 18 2S3
MRL 122mm: 96 9P138
COMBINED GUN/MOR 120mm: 100 2S9 SP, 2B16, 11 2S23 SP
ATGW 72 AT-3/-5
ATK GUNS 100mm: MT-12
AD GUNS 23mm: 60 ZSU-23-4 SP
SAM 250 SA-7, 20 SA-8, 50 SA-9/-13
COASTAL DEFENCE TROOPS (5,000)
(all units reserve status)
1 coastal defence div
1 coastal defence bde
1 arty regt
2 SAM regt
EQUIPMENT
MBT 350 T-64
AIFV 450 BMP
APC 280 BTR-60/-70/-80, 400 MT-LB
TOTAL ARTY 364 (152)
TOWED 280: **122mm**: 140 D-30; **152mm**: 40 D-20, 50 2A65, 50 2A36
SP 152mm: 48 2S5
MRL 122mm: 36 BM-21

NAVAL DEPLOYMENT

NORTHERN FLEET (Arctic and Atlantic)
(HQ Severomorsk)
BASES Kola peninsula, Severodvinsk
SUBMARINES 37
strategic 14 SSBN **tactical** 23 (4 SSGN, 14 SSN, 2 SSK, 3 SSN other roles)
PRINCIPAL SURFACE COMBATANTS 12
1 CV, 3 CG/CGN, 6 DDG, 2 FFG
OTHER SURFACE SHIPS about 26 patrol and coastal combatants, 18 MCM, 8 amph, some 130 spt and misc
NAVAL AVIATION
75 cbt ac; 30 armed hel
BBR 25 Tu-22M
FTR/FGA 7 Su-24/-25, 24 Su-27
ASW ac 11 Il-38 **hel** 25 Ka-27
MR/EW ac 2 An-12
CBT ASLT HEL 5 Ka-29
COMMS 6 Tu-142
TPT ac 25 An-12, An-24, An-26

BALTIC FLEET (HQ Kaliningrad)
BASES Kronstadt, Baltiysk
SUBMARINES 2 SSK
PRINCIPAL SURFACE COMBATANTS 6
2 DDG, 4 FFG
OTHER SURFACE SHIPS about 26 patrol and coastal combatants, 13 MCM, 5 amph, some 130 spt and misc
NAVAL AVIATION
68 cbt ac, 41 armed hel
FGA 2 regt: 20 Su-24, 28 Su-27
ASW hel 22* Ka-27
MR/EW ac 2 An-12, 25 Su-24
CBT ASLT HEL 4 Ka-29, 15 Mi-24

BLACK SEA FLEET (HQ Sevastopol)
The Russian Fleet is leasing bases in Sevastopol for the next 20 years; it is based at Sevastopol and Karantinnaya Bays, and, jointly with Ukrainian warships, at Streletskaya Bay. The Fleet's overall serviceability is low.
BASES Sevastopol, Temryuk, Novorossiysk
SUBMARINES 12 (only one op)
11 SSK, 1 SSK other roles
PRINCIPAL SURFACE COMBATANTS 7
3 CG/CGN, 2 DDG, 2 FFG
OTHER SURFACE SHIPS about 15 patrol and coastal combatants, 14 MCM, 5 amph, some 90 spt and misc
NAVAL AVIATION
30 cbt ac; 13 armed hel
BBR ac 22 Su-24 M
ASW ac 4* Be-12 **hel** 5* Ka-27
MR/EW ac 4 An-12 **hel** 8 Mi-8

CASPIAN SEA FLOTILLA
BASE Astrakhan (Russia)
The Caspian Sea Flotilla has been divided between Azerbaijan (about 25%), and Russia, Kazakstan and Turkmenistan, which are operating a joint flotilla under Russian command currently based at Astrakhan.
SURFACE COMBATANTS about 36
10 patrol and coastal combatants, 5 MCM, some 6 amph, about 15 spt

PACIFIC FLEET (HQ Vladivostok)
BASES Vladivostok, Petropavlovsk Kamchatskiy, Magadan, Sovetskaya Gavan, Fokino
SUBMARINES 16+
strategic 5 SSBN **tactical** 11 (5 SSGN, 5 SSN, 1 SSN other roles)
PRINCIPAL SURFACE COMBATANTS 10
1 CG/CGN, 7 DDG, 2 FFG
OTHER SURFACE SHIPS about 30 patrol and coastal combatants, 8 MCM, 4 amph, some 57 spt and misc

NAVAL AVIATION (Pacific Fleet Air Force)
(HQ Vladivostok) 71 cbt ac, 23 cbt hel
BBR 2 regt with 20 Tu-22M
ASW ac 10 Tu-142, 15 Il-38 **hel** 20 Ka-27; ashore 3 Mi-14
MR/EW ac some 10 An-12
CBT ASLT HEL 3 Ka-29
COMMS 16 Tu-142

Military Air Forces (VVS) ε184,600

The amalgamation of the Air Defence Troops (PVO) with the Air Force (VVS) under one Air Force Command is now complete. The Military Air Forces now comprise 2 Air Armies (Long Range Aviation (LRA) and Military Transport Aviation (VTA)), 7 Tactical/Air Defence formations and 70 air regts. Tactical/Air Defence roles includes air defence, interdiction, recce and tactical air support. LRA (6 div) and VTA (9 regt) are subordinated to central Air Force command. There is a Tactical/AD Army in each MD, except Volga and Ural MD, each of which has an Air Corps. These air formations are subordinated to commanders of MDs. A joint CIS Unified Air Defence System covers Russia, Armenia, Belarus, Kazakstan, Kyrgyzstan and Uzbekistan.
Flying hours Average annual flying time for LRA and Tactical/Air Defence is about 20 hours, and for VTA approximately 44 hours. Only aircrew operating over Chechnya fly 100+ hours

LONG-RANGE AVIATION COMMAND (37th Air Army)
2 div and 1 cbt conversion unit
BBR (START-accountable) 74 hy bbr; 68 Tu-95, 15 Tu-160 (Test ac: 7 Tu-95, 5 Tu-160)
158 Tu-22M/MR (92 in store)
TKR 20 Il-78
TRG 10 Tu-22M-2/3, 30 Tu-134

TACTICAL AVIATION
Flying hours 20
BBR/FGA some 575: 350 Su-24, 225 Su-25
FTR some 880: incl 260 MiG-29, 340 Su-27, 280 MiG-31
RECCE some 135: incl 15 MiG-25, 120 Su-24
AEW AND CONTROL 16 A-50/A-50U
ECM 60 Mi-8
TRG 1 centre for op conversion: some 90 ac incl 20 MiG-29, 35 Su-24, 15 Su-25
1 centre for instructor trg: 65 ac incl 10 MiG-25, 20 MiG-29, 15 Su-24, 10 Su-25, 10 Su-27
AAM AA-8 *Aphid*, AA-10 *Alamo*, AA-11 *Archer*
ASM AS-4 *Kitchen*, AS-7 *Kerry*, AS-10 *Karen*, AS-11 *Kilter*, AS-12 *Kegler*, AS-13 *Kingbolt*, AS-14 *Kedge*, As-15 *Kent*, AS-17 *Krypton*, AS-18 *Kazoo*
SAM 37 SAM regt, some 2,150 launchers in some 225 sites
50 SA-2 *Guideline* (being replaced by SA-10)
200 SA-5 *Gammon* (being replaced by SA-10)
some 1,900 SA-10/S-300. The first S-400 unit reportedly to be deployed near Moscow by the end of 2000.

MILITARY TRANSPORT AVIATION COMMAND (VTA) (61st Air Army)
2 div, each 4 regt, each 30 ac; 1 indep regt
EQUIPMENT
some 280 ac, incl Il-76M/MD, An-12, An-22, An-124
CIVILIAN FLEET 1,500 medium- and long-range passenger ac, incl some 350 An-12 and Il-76

AIR FORCE AVIATION TRAINING SCHOOLS
TRG 5 mil avn institutes subordinate to Air Force HQ: some 1,150 ac incl L-39, L-29, Tu-134, Mig-23, MiG-29, Su-22, Su-25, Su-27

OPERATIONAL COMBAT AIRCRAFT
based west of Urals (CFE totals as at 1 Jan 2000 for all air forces other than maritime)
ac 2,733: 188 Su-17 • 55 Su-22 • 450 Su-24 • 193 Su-25 • 295 Su-27 • 451 MiG-23 • 131 MiG-25 • 166 MiG-27 • 443 MiG-29 • 243 MiG-31 • 66 Tu-22M • 52 Tu-22. No armed hel
DECOMMISSIONED AIRCRAFT IN STORE over 1,000 ac incl Tu-22M, MiG-23, MiG-27, Su-17, Su-22

Deployment

Deployment of formations within the Atlantic to the Urals (ATTU) region is reported to be 2 TD, 8 MRD, perhaps 4 AB, 1 arty div, 9 indep arty, 3 MRL, 7 MR, 8 SSM, 12 SAm bde.
The manning state of Russian units is difficult to determine. The following assessment of units within the ATTU region is based on the latest available information. Above 75% – possibly 3 ABD, all MR bde and 1 AB bde; above 50% – possibly 1 TD, 6 MRD, 1 ABD, 1 arty bde. The remainder are assessed as 20–50%. Units outside the ATTU are likely to be at a lower level. All bde are maintained at or above 50%. TLE in each MD includes active and trg units and in store

KALININGRAD OPERATIONAL STRATEGIC GROUP
These forces are commanded by The Ground and Coastal Defence Forces of the Baltic Fleet.
GROUND 12,700: 2 MRD, 1 tk, 1 SSM bde, 1 SAM regt, 1 indep MRR (trg), 1 attack hel regt, 816 MBT, 850 ACV (plus 390 'look-a-likes'), 345 arty/MRL/mor, 18 *Scarab*, 20 attack hel
NAVAL INFANTRY (1,100)
1 regt (26 MBT, 220 ACV, 30 arty/MRL)(Kaliningrad)
COASTAL DEFENCE
2 arty regt (133 arty)
1 SSM regt: some 8 SS-C-1b *Sepal*
AD 1 regt: 28 Su-27 (Baltic Fleet)
SAM 50

RUSSIAN MILITARY DISTRICTS

LENINGRAD MD (HQ St Petersburg)

GROUND 38,100: 1 ABD; plus 2 indep MR bde, 2 arty bde/regt, 1 SSM, 1 SF, 4 SAM bde, 1 ATK, 1 MRL, 1 aslt tpt hel regt, 333 MBT, 500 ACV (plus 2,200 'look-alikes'), 940 arty/MRL/mor, 18 *Scarab*, 25 attack hel

NAVAL INFANTRY (1,300 – subordinate to Northern Fleet)

1 regt (74 MBT, 230 ACV, 45 arty)

COASTAL DEFENCE

1 Coastal Defence (360 MT-LB, 134 arty), 1 SAM regt

AIR 6th Air Army has 392 combat ac. It is divided into two PVO corps, 1 bbr div (66 Su-24), 1 recce regt (26 MiG-25, 20 Su-24), 1 ftr div (35 Su-27, 60 MiG-29), 1 hel ECM sqn (20 Mi-8)

AD 7 regt: 95 MiG-31, 90 Su-27

SAM 525

MOSCOW MD (HQ Moscow)

GROUND 95,900: 1 Army HQ, 1 Corps HQ, 2 TD, 2 MRD, 2 ABD, plus 1 arty div HQ, 5 arty bde/regt, 3 SSM, 1 indep MR, 1 SF, 4 SAM bde, 2 aslt tpt hel regt, 2,000 MBT, 2,200 ACV (plus 1,700 'look-alikes'), 1,800 arty/MRL/mor, 48 *Scarab*, 120 attack hel

AIR 2 PVO divs, one mixed div incl air aslt regt, one mil depot storing 43 MiG-25

550 cbt ac: MiG-25/-29/-31, Su-24/-25/-27: hel 2 ECM sqn with 46 Mi-8

SAM 850

VOLGA MD (HQ Samara) (to be merged with Ural MD)

GROUND 32,600: 1 MRD, 1 AB bde, 1 arty regt, 1 SSM bde, 1 SAM bde, 1 hel, 1 hel trg regt, 730 MBT, 1,000 ACV (plus 500 'look-alikes'), 650 arty/MRL/mor, 18 *Scarab*, 33 attack hel

AIR 1 regt attack/op trg, 48 MiG-29, 21 Su-25 **hel** Mi-8 communications

AD 3 regts ad/op trg: 118 MiG-23, 39 Su-27, 8 MiG-31

Air Force aviation schools (383 L-29, L-39, Mi-2), storage bases.

NORTH CAUCASUS MD (HQ Rostov-on-Don)

GROUND 82,500: 1 Army HQ, 1 Corps HQ, 2 MRD, 1 ABD, 3 MR, 3 AB, 1 SF, 1 arty bde, 1 SSM, 4 SAM bde, 1 ATK, 2 attack hel, 1 aslt tpt hel regt, 600 MBT, 1,900 ACV (plus 1,200 'look-alikes'), 750 arty/MRL/mor, 18 *Scarab*, 60 attack hel

NAVAL INFANTRY (ε1,400 - subordinate to Black Sea Fleet)

1 regt (59 ACV, 14 arty)

AIR 4th Air Army: 459 cbt ac, 1 bbr div (112 Su-24), 1 recce regt (50 Su-24), 1 air aslt div (98 Su-25, 35 Su-22), 1 ftr corps of 4 regt (105 MiG-29, 59 Su-27), 1 hel ECM sqn with 47 Mi-8, trg regt of tac aviation and Air Force aviation schools

SAM 125

URAL MD (HQ Yekaterinburg) (to be merged with Volga MD)

GROUND 1 TD, 1 MRD, 1 *Spetsnaz*, 2 arty bde/regt, 1 SSM bde, 1,300 MBT, 1,600 ACV, 900 arty/MRL/mor, 18 *Scarab*

AIR Air Force avn and trg schools

AD Ural and Volga assets cover Siberian and Far East MDs: MiG-23s, MiG-29s, Su-27s

SAM 600

SIBERIAN MD (HQ Novosibirsk)

GROUND 2 Corps HQ, 3 TD (1 trg), 2 MRD, 1 arty div, 2 MG/arty div, 3 MR bde, 10 arty bde/regt, 2 SSM, 2 SAM, 2 SF bde, 4 ATK, 1 attack hel, 4,468 MBT, 6,000 ACV, 4,300 arty/MRL/mor, 36 *Scud*/*Scarab*, 35 attack hel

AIR Covering Siberian Zone and Far Eastern MD:

FGA 315 Su-24/25

FTR 100 Su-27/MiG-29

RECCE 100 Su-24

AD See Volga/Ural MDs

FAR EASTERN MD (HQ Vladivostok) incl Pacific Fleet

GROUND 2 Army, 2 Corps HQ, 10 MRD (2 trg), plus 2 MG/arty div, 1 arty div, 9 arty bde/regt, 1 MR, 3 SSM, 5 SAM, 1 SF, 1 ATK bde, 2 attack hel, 2 aslt tpt hel regt, 3,900 MBT, 6,400 ACV, 3,000 arty/MRL/mor, 54 *Scarab*, 85 attack hel

NAVAL INFANTRY (2,500 - subordinate to Pacific Fleet)

1 div HQ, 3 inf, 1 tk and 1 arty bn

COASTAL DEFENCE

1 div

AIR See Siberian MD

AD See Volga/Ural MDs

Forces Abroad ε22,150

Declared str of forces deployed in Armenia and Georgia as at 1 Jan 2000 was 8,800. These forces are now subordinate to the North Caucasus MD. Total probably excludes locally enlisted personnel.

ARMENIA

GROUND 3,100; 1 mil base; 74 MBT, 17 APC, 148 ACV, 84 arty/MRL/mors

AD 1 sqn: 14 MiG-29, 1 SA-10 (S-300) bty

GEORGIA

GROUND 5,000; 3 mil bases (each = bde+); 140 T-72 MBT, 500 ACV, 173 arty incl **122mm** D-30, 2S1 SP; **152mm** 2S3; **122mm** BM-21 MRL; **120mm** mor, 5 attack hel. Plus 118 ACV and some arty deployed in Abkhazia

AD 60 SA-6

AIR 1 composite regt with some 35 **ac** An-12, An-26 **hel** Mi-8

MOLDOVA (Dniestr)

GROUND 2,600; 1 op gp; 117 MBT, 133 ACV, 128 arty/MRL/mor. These forces are now subordinate to the Moscow MD

TAJIKISTAN

GROUND ε8,200; 1 MRD, 190 MBT, 303 ACV, 180 arty/MRL/mor; plus 14,500 Frontier Forces (Russian officers, Tajik conscripts)

UKRAINE

NAVAL INFANTRY 1,500; 1 regt (212 ACV, 31 arty)

AFRICA 100

CUBA some 800 SIGINT and ε10 mil advisers

SYRIA 150

VIETNAM 700; naval facility and SIGINT station. Used by RF aircraft and surface ships on reduced basis

Peacekeeping

BOSNIA (SFOR II): 1,300; 2 AB bn
GEORGIA/ABKHAZIA 1,500; 1 AB regt
GEORGIA/SOUTH OSSETIA 500; 1 MR bn
MOLDOVA/TRANSDNIESTR 500; 1 MR bn
YUGOSLAVIA (KFOR): 3,600

UNITED NATIONS

BOSNIA (UNMIBH): 1 **CROATIA** (UNMOP): 1 obs **DROC** (MONUC): 5 obs **EAST TIMOR** (UNTAET): 2 obs **GEORGIA** (UNOMIG): 3 obs **IRAQ/KUWAIT** (UNIKOM): 1 obs **MIDDLE EAST** (UNTSO): 3 obs **SIERRA LEONE** (UNAMSIL): 24 incl 15 obs; 4 Mi-24 **WESTERN SAHARA** (MINURSO): 25 obs

Paramilitary ε423,000 active

FEDERAL BORDER GUARD SERVICE ε140,000

directly subordinate to the President; 10 regional directorates, 7 frontier gps

EQUIPMENT

1,000 ACV (incl BMP, BTR), 90 arty (incl 2S1, 2S9, 2S12)

ac some 70 Il-76, Tu-134, An-72, An-24, An-26, Yak-40, 16 SM-92 **hel** some 200+ Mi-8, Mi-24, Mi-26, Ku-27

PATROL AND COASTAL COMBATANTS about 237

PATROL, OFFSHORE 23

7 *Krivak*-III with 1 Ka-27 hel, 1 100mm gun, 12 *Grisha*-II, 4 *Grisha*-III

PATROL, COASTAL 35

20 *Pauk*, 15 *Svetlyak*

PATROL, INSHORE 95

65 *Stenka*, 10 *Muravey*, 20 *Zhuk*

RIVERINE MONITORS about 84

10 *Yaz*, 7 *Piyavka*, 7 *Vosh*, 60 *Shmel*

SUPPORT AND MISCELLANEOUS about 26

8 *Ivan Susanin* armed icebreakers, 18 *Sorum* armed AT/F

INTERIOR TROOPS 140,000

7 districts, some 11 'div' incl 5 indep special purpose div (ODON – 2 to 5 op regt), 29 indep bde incl 10 indep special designation bde (OBRON – 3 mech, 1 mor bn); 65 regt/bn incl special motorised units, avn

EQUIPMENT

incl 1,700 ACV (incl BMP-1/-2, BTR-80), 20 D-30, 45 PM-38, 4 Mi-24

FORCES FOR THE PROTECTION OF THE RUSSIAN FEDERATION 25,000

org incl elm of Ground Forces (1 mech inf bde, 1 AB regt)

FEDERAL SECURITY SERVICE ε4,000 armed incl Alfa, Beta and Zenit cdo units

FEDERAL PROTECTION SERVICE ε10,000 incl Presidential Guard regt

FEDERAL COMMUNICATIONS AND INFORMATION AGENCY ε54,000

RAILWAY TROOPS ε50,000 in 4 rly corps, 28 rly bde

MILITARY DEVELOPMENTS

With the underlying tensions of the Middle East and North Africa region far from resolved, this remains the world's leading arms market. The most significant regional military event in 2000 was the withdrawal in June of Israeli troops from southern Lebanon, but political progress in the peace process has been disappointing. Hopes for movement towards an Israel–Palestine settlement were raised by the accession of Prime Minister Ehud Barak's government in Israel in mid-1999, only to be dashed a year later by the failure of the Camp David talks. Even the Israeli–Syrian rapprochement faltered and was finally checked by the death of President Hafez al-Assad in June 2000. In the Gulf region, Iraq continued to defy the demands of the UN Security Council and by August 2000, nearly two years had passed without UN inspectors on the ground. In Iran, political reform is making only halting progress and fears of the country's advancing missile and weapons of mass destruction (WMD) programmes remain. In North Africa, the Algerian government remains locked in its deadly struggle with the *Groupe Islamique Armée* (GIA).

The Middle East

Peace Process Soon after coming to power in May 1999, Barak's government set out its 15-month plan to restart the Middle East peace process, not only with the Palestinians, but also with Syria. By August 2000, both bilateral efforts were more or less stalled. The Israeli–Syrian dialogue gained some momentum in late 1999, but this was lost following Assad's death on 10 June 2000 and is unlikely to be recovered until the new Syrian President, Assad's son Bashar, is confidently established. Despite the personal intervention of US President Bill Clinton at the Camp David talks in June 2000, the Israelis and Palestinians were unable to make progress on the vexed question of control of Jerusalem. Palestinian leader Yasser Arafat is under pressure from some Palestinian groups to make a unilateral declaration of statehood on 13 September 2000, but the counsel he has been receiving from the US, Russia and major European powers is to delay such a declaration until some agreement has been reached with Israel. At that point, the major powers might be in a better position to recognise and support a Palestinian state and the possibility of conflict incited by extremists on either side would be reduced. In a policy shift in August 2000, Barak indicated that Israel might be prepared to accept Palestinian statehood, but only as part of a final settlement.

South Lebanon Israel finally withdrew its forces from its 'security zone' in southern Lebanon in June 2000, precipitating the immediate collapse of the Israeli-backed South Lebanese Army (SLA). *Hizbollah* quickly filled the power vacuum in the area, but without major incident. By early August, a reinforced United Nations Interim Force in Lebanon (UNIFIL), now just over 5,500 strong, was deployed along the border. Clarity is needed on whether the UN or the Lebanese army and police are responsible for dealing with unrest at the border. However, so far there have only been minor disturbances such as stone throwing, and the Israeli Defence Force (IDF) has been restrained in its response. The days of repeated rocket attacks on northern Israel by *Hizbollah*, and Israeli air attacks throughout Lebanon seem to have ended, for the time being at least.

Missile Defence The Israeli Air Force has set up its first anti-ballistic missile battery about 30 miles south of Tel Aviv. The *Arrow* 2 system, which is being developed jointly with the US, is undergoing evaluation, but it is assessed to have an emergency-operational capability. When the system is fully operational – which may be by 2005 – two other batteries will be deployed, one in northern Israel and the other in the south. The areas covered by the three batteries will overlap where there are main population centres, providing these with the maximum defence. *Arrow* will initially be linked to Israel's lower-altitude *Patriot* air-defence system. The $1.6 billion *Arrow* 2 project is described as a 'theatre' missile-defence system, but as far as Israel is concerned, it is effectively a national defence system. Israel hopes to defray the development costs by selling *Arrow* to other countries, such as the UK, Turkey and Japan. The US will be comparing it with other systems under consideration for its own theatre-missile-defence requirement.

Israel's Other Military Capabilities Israel has commissioned all three of the *Dolphin*-class (German Type 800) diesel submarines to replace its three *Gal*-class vessels, which were over twenty years old. It has been reported that two of Israel's new *Dolphin* diesel submarines launched conventionally armed missiles in the Indian Ocean in May. These may have been the submarines' standard fit of *Sub-Harpoon* anti-surface ship missiles (ASSM). There have been reports, denied by the Israeli government, that the *Dolphin*-class submarines will be fitted with cruise missiles capable of being equipped with nuclear warheads. Israel's request for 50 *Tomahawk* land-attack missiles from the US was turned down. The upgrade of Israel's AH-64 combat helicopters to AH-64D *Longbow* standard, equipped with *Hellfire* missiles, continues, for use by the Army. A further 57 advanced medium-range air-to-air missiles (AMRAAM) have been ordered for the Air Force.

The Gulf

UNSCOM and UNMOVIC The United Nations Monitoring, Verification and Inspection Commission (UNMOVIC) was established on 17 December 1999 under UN Resolution 1284. It has taken over the responsibilities of the former United Nations Special Commission (UNSCOM) for ensuring Iraq's WMD programmes and prohibited missiles are eliminated. UNMOVIC has taken over UNSCOM's assets, liabilities and archives, and is mandated to establish and operate a reinforced monitoring and verification system, address unresolved disarmament issues and identify additional sites to be inspected under the new monitoring system. The Swedish former Director-General of the International Atomic Energy agency (IAEA), Dr Hans Blix, was appointed Chairman of the Commission on 1 March 2000 and its work programme and organisational plan were submitted to the Security Council and approved on 13 April 2000.

The new team of UNMOVIC inspectors began their training in July and the Baghdad monitoring centre (originally set up by UNSCOM) was scheduled to reopen and prepare for inspections by August. The IAEA is ready to resume inspections at any time. Meanwhile, Iraq has continued to declare that UNMOVIC inspectors are not welcome in the country. Baghdad has said that it will only consider admitting inspectors after international sanctions on Iraq are lifted and the US- and UK-enforced no-fly zones over Iraq are lifted. Under UN Resolution 1284, sanctions will only be lifted if Baghdad cooperates with UNMOVIC for 120 days, and will be automatically re-imposed if Iraq fails to show full cooperation. The Security Council will then review the progress of the inspection process every 120 days until it is satisfied that the disarmament issues have been resolved. Meanwhile, sanctions seem to be slowly eroding in some important areas. Iraq continues to export a significant amount of oil through Iranian territorial waters with the tacit acceptance of Tehran. Although Iran periodically stops tankers carrying Iraqi oil, this export could amount to as much as 20% of the bunker oil exported from the Gulf region, according to some estimates. In August 2000, also in contravention of sanctions, Iraq opened its international airport near Baghdad to flights other than those permitted to deliver humanitarian aid under the sanctions regime.

Iran Democratic reform in Iran is making only halting progress. There continue to be tensions within the clerical oligarchy over how to respond to the popular demands for political and economic reform. The reformist President Mohammed Khatami continues to try to improve relations abroad, in particular with the major European countries. Nevertheless, the hard-liners have a firm grip on the armed forces and the weapon programmes. Iran's military capabilities continue to advance, particularly in respect of missiles and naval forces. The Iranian Revolutionary Guard Command is reported to have established five units to operate the country's missile arsenal, and on 15 July 2000, Iran tested a *Shahab*-3 missile thought to have a 1,500km range. Iran also claims to have successfully fired an advanced version of the Chinese C-802 surface-to-surface missile (SSM) which are to be fitted to its *Houdong* fast patrol craft. Iran's three *Kilo*-class diesel submarines have had severe problems with their batteries, which were not adapted for the climatic conditions of the Gulf, but all three have been fitted with new and more effective batteries in the past year. It is not yet clear whether Iran is manufacturing these under licence or whether they are being obtained from abroad – if the latter, then Pakistan and China are among the most probable suppliers. Iran's naval capability has been further enhanced by the delivery of four Mi-8AMT (Mi-171) transport/attack helicopters from Russia. Up to 20 may be ordered if the initial batch are a success.

North Africa For Algeria, 2000 opened with a significantly reduced level of violence following the implementation of the government's 'Law on Civil Concord', which had been endorsed by a national referendum in September 1999. An amnesty for those associated with armed groups ended on 13 January 2000 and the cessation of armed action by the *Armée Islamique de Salut* (AIS) was confirmed. President Abdelaziz Bouteflika made strenuous efforts to court European countries for political support and financial aid with some success. He also made overtures to neighbouring Maghreb states, making the first visit to Tunisia by an Algerian head of state for six years in June 2000. The talks focused on boosting bilateral trade and reviving the Maghreb Arab Union. However, as the year progressed, the *Groupe Islamique Armée*, which had not responded to the amnesty, stepped up its military campaign. There are differences among the military leadership on how to deal with the renewed violence. Some want to continue taking a hard line with Islamic groups, others believe a more flexible approach should be taken to encourage the support of more moderate Islamic groups such as the *Front Islamique du Salut* (FIS), of which the AIS was the armed wing. The death toll continues to mount, with 2,000 killed as a direct result of the conflict in the year to August 2000. Since this phase of violence began in 1992 it has caused 82,000 deaths. There appears to be no early end in sight.

DEFENCE SPENDING

Regional military spending in 1999 was US$60bn, measured in constant 1999 dollars, virtually unchanged from 1998. Initial budgets for 1999 indicated that spending might fall by some 5%, but there was substantial supplementary spending during the year, probably stimulated by the continuing strength of oil prices. Official budgets for 2000 are similar to those originally set out for 1999, but there may well be another overspend as oil prices continued to rise to record levels during the year. Largely as a result of this, gross domestic product (GDP) grew by 4.7% across the region in 1999.

Israel

Israel's defence spending in 1999 was lower than in recent years as a percentage of GDP. Total expenditure of $8.8bn represented 8.9% of GDP compared with the average levels of 10–12% seen over the past five years. In 2000, the official budget is set to increase by around 3.9% to NS28.9bn ($7.0bn). The official budget understates real military spending, as it does not take into account expenditure under the US Foreign Military Assistance programmes (FMA). This is a substantial sum. In fiscal year 2001, US Foreign Military Financing (FMF) for Israel will rise from $1.92bn to

$1.98bn for procurement, but there will be a cut in the military-related economic assistance provided by the Economic Support Fund (ESF) from $950 million to $850m. It is intended that the ESF will be phased out by 2008, and there is therefore provision for Israel's FMF to increase to $2.4bn by 2008. Furthermore, in FY2001, Israel will be eligible to receive Excess Defence Articles (essentially a free grant of equipment) under section 516 of the US Foreign Assistance Act for defence maintenance, spare parts, support equipment and other requirements.

Jordan

FMF funding to Jordan peaked in FY2000 at $225m (including an additional $150m under the Wye Agreement) and it is projected to fall to $75m in FY2001. The biggest items of spending will be an upgrade of Jordan's air-defence systems and improvements in command, control and communications capabilities. Jordan will also receive $1.5m from the US for de-mining along its border with Syria.

Syria and Lebanon

Syria's defence budget, which covers personnel, operations and maintenance, but excludes procurement costs, rose from S£39.5bn in 1999 to S£42.3bn in 2000, although in dollar terms this translates to a $140m fall to $729m. Assad's visit to Moscow in July 1999 may well have helped further to defuse the dispute over Syria's debt to Russia hanging over from the Soviet era (estimated by Russia at $11bn) thus paving the way to renewed military dealings between the two countries. Damascus is believed to be seeking up to $2.5bn worth of equipment, including the S-300 surface-to-air missile (SAM), Mig-29 SMT fighters, T-80 main battle tanks (MBT) and Su-27 multi-role aircraft. Lebanon's defence budget of L£870bn (US$575m) in 2000 is virtually unchanged from 1999.

Gulf Co-operation Council States

UAE Defence spending by the United Arab Emirates has more than doubled since 1996, rising from $1.8bn to a budgeted figure of $3.9bn in 2000. The peak in procurement spending will probably occur in 2002, when $1.2bn will be needed to cover F-16 and *Mirage* deals. On 10 July 2000, the UAE finally ended months of speculation by signing a $6.4bn deal for the purchase of 80 F-16 Block 60 fighters from Lockheed Martin, which will be equipped with AMRAAM, high-speed anti-radiation missiles (HARM) and al-Hakim missiles. The F-16s, known as *Desert Falcons* are among the most sophisticated aircraft in operation. They will be delivered from May 2004 through to 2007. Another US company, Raytheon, will supply the AMRAAM, second-generation HARM and laser-guided bomb kits. Not only is the UAE Air Force getting newer and more sophisticated versions of the F-16s than those flown by the US Air Force, but Lockheed Martin has agreed to put up a $2bn performance bond. The bond safeguards the UAE from the cost of technological failures and from any adverse shifts in political tides of the type that thwarted the US F-16 deal with Pakistan. Neither this extraordinary financial arrangement nor the transfer of such advanced equipment would have been contemplated in the early 1990s. They show how far the US government will go to support its defence manufacturers. The UAE has also signed a $734m contract for the development and delivery of 50 Russian-made *Pantzyr*-S1 self-propelled air-defence systems; each is equipped with 12 SA-19 surface-to-air missiles (SAM) and 2 × 30mm cannon. This is one of the largest export orders ever won by a Russian defence company and demonstrates how buying countries want to diversify their sources of supply, thus sharpening the competition among defence manufacturers. The UAE continues to take deliveries of its $500m order for *Leclerc* tanks from France.

Kuwait Kuwait's defence budget for 2000 was set at D483m ($1.56bn). However, if spending on capital projects and procurement is taken into account the figure is boosted to D806m ($2.6bn), up from $2.3bn in 1999. The budget increase follows last year's fall in actual defence expenditure to $3.2bn from $3.6bn in 1998, as the delivery of new equipment from the rearmament programme was nearing completion.

This year's budget increase is mainly due to extra expenditure caused by increased deployments by allied forces. (Kuwait pays the cost of allied countries' forces deployed in its territory as part of its defence arrangements.) Kuwait is also making a contribution to the proposed $1.2bn US command, control and intelligence system due to be deployed with allied forces later this year. The only significant naval acquisition has been Kuwait's commissioning of the last four of the eight *Um Almaradim* (French *Combattante I*) missile patrol boats, which replace those lost in the Gulf War.

Oman Oman raised its defence budget for 2000 by 9%, from $1.6bn in 1999 to $1.75bn in 2000. If spending remains in line with budget, then Oman will have spent just over US$9bn on defence in the last five years, overshooting its 1996–2000 spending plan by around $500m. In the last five-year plan for 1991–95, defence spending was set at $7.4bn but spending over the period turned out to be $8.7bn.

Saudi Arabia Saudi Arabia's defence budget rose from R68.7bn ($18.3bn) in 1999 to R70bn ($18.7bn) in 2000. Expenditure in 1999 was 19% over budget. Saudi Arabia remains interested in buying diesel submarines, especially since Iran's vessels of this type have been made more effective. No formal negotiations on an order have begun. Saudi naval-procurement plans remain limited to the delivery of three modified *Lafayette*-class frigates, the first of which was launched in France in 2000 for expected delivery in 2002.

Qatar Surging oil prices are forecast to increase Qatar's state revenues by around 20% and it has correspondingly raised planned government spending by almost 9% for fiscal year 2000–01. The defence budget increased in line to QR5.1bn ($1.4bn).

Egypt

Egypt's defence budget is estimated to have increased from E£88.7bn ($2.5bn) in 1999 to E£9.6bn ($2.8bn) for 2000. In addition, Egypt is scheduled to receive US FMA for equipment purchases totalling $1.3bn and military related economic assistance of $227m. Deliveries of 21 F-16C/D's from the US (assembled in Turkey) continued through 1999 and 2000. Deliveries of 24 Block 40 F-16C/Ds and associated support equipment are due to start in early 2001 and run for 15 months. Egypt has ordered 80 K-8 basic trainer/light ground-attack jet aircraft from the Chinese Hong Aviation Industry Group, to replace its Czech L-29s. This is another example of diversification in the highly competitive arms trade where old customer loyalties count for less than they did.

Algeria and Morocco

Algeria once again increased its defence budget to deal with its continuing civil war, now entering its ninth year. The official budget rose from D121bn ($1.7bn) in 1999 to D138bn ($1.8bn) in 2000. However, spending was closer to D210bn ($3.1bn) in 1999 and is likely to exceed budget again in 2000. Defence spending in Morocco rose in 1999 to D17.5bn ($1.6bn) from D16bn ($1.6bn) in 1998.

Table 19 **Arms orders and deliveries, Middle East and North Africa, 1998–2000**

Country	Country supplier	Classification	Designation	Quantity	Order date	Delivery date	Comment
Algeria	Ukr	MBT	**T-72**	27	1997	1998	
	Ukr	AIFV	**BMP-2**	32	1997	1998	
	Ukr	cbt hel	**Mi-24**	14	1997	1998	
	Bel	FGA	**MiG-29**	36	1998	1999	Reportedly in exchange for 120 MiG-21s
	RF	ASSM	**Kh-35**	96	1998	1999	For FACs; in 2 batches of 48
	RF	FGA	**Su-24**	3	1999	2000	
	US	ESM	***Beech 1900***	6	2000		For SIGINT role
Bahrain	US	SAM	***Hawk***	8	1996	1997	Deliveries to 1998; 8 btys
	US	FGA	**F-16C/D**	10	1998	2000	With AMRAAM
	US	AAM	**AMRAAM**		1999		
Egypt	US	MBT	**M1A1**	555	1988	1993	Complete by end-1998
	US	hel	**AH-64**	36	1990	1994	Deliveries to 1999
	US	FF	***Perry***	4	1994	1996	Deliveries to 1998
	US	hel	**SH-2G**	10	1994	1997	Deliveries to 1999
	US	arty	**SP 122 SPG**	24	1996	1998	2nd order
	US	FGA	**F-16C/D**	21	1996	1999	2 delivered per month until 2000
	US	hel	**CH-47D**	4	1997	1999	Also updates for 6 CH-47Cs to D
	US	SAM	***Avenger***	50	1998	2001	
	US	ARV	**M88A2**	50	1998	2002	1st of 2 contracts; 2nd for parts
	dom	APC	***Al-Akhbar***		1998	2001	Development complete
	US	SAM	***Patriot***	384	1998	2001	384 missiles; 48 launchers
	RF	SAM	***Pechora***	50	1999	2003	Upgrade to *Pechora* 2
	US	LST	***Newport***	1	1999	2000	
	US	FGA	**F-16**	24	1999	2001	12 × 1 seater; 12 × 2 seater
	PRC	trg	**K-8**		1999	2001	
	US	AEW	**E-2C**	5	1999	2002	Upgrade
	US	MBT	**M1A1**	200	1999		Kits for local assembly
	US	SAM	**AMRAAM**		2000		
Iran	RF	MBT	**T-72**	100	1989	1998	Kits for local assembly
	RF	AIFV	**BMP-2**	200	1989	1998	Kits for local assembly
	dom	SSM	***Shahab* 2**		1994	1998	Dom produced *Scud*
	dom	SSM	***Shahab* 3**		1994	1999	Based on *No-dong* 1
	dom	MRBM	***Shahab* 4**		1994		Dev; reportedly based on RF SS-4
	dom	ICBM	***Shahab* 5**		1994		Dev; possibly based on *Taepo-dong*
	PRC	tpt	**Y-7**	14	1996	1998	Deliveries 1998–2006
	PRC	FGA	**F-7**	10	1996	1998	
	dom	hel	***Shahed-5***	20	1999		
Israel	dom	MBT	***Merkava* 3**		1983	1989	In production
	col	BMD	***Arrow***	2	1986	1999	Deployment to begin 1999; with US
	dom	PFM	***Saar* 4.5**	6	1990	1994	Deliveries to 1998
	dom	sat	***Ofek* 4**	1	1990	1999	Launch failed
	dom	MBT	***Merkava* 4**		1991	2001	In development

Country	Country supplier	Classification	Designation	Quantity	Order date	Delivery date	Comment
	dom	ATGW	**LAHAT**		1991	1999	Dev completed end-1999
	Ge	SSK	***Dolphin***	3	1991	1998	Final delivery 2000; funded by Ge
	col	BMD	***Nautilus***		1992	2000	Joint development with US
	US	MRL	**MLRS**	42	1994	1995	16 delivered 1997, completed 1998
	Fr	hel	**AS-565**	8	1994	1997	5 delivered 1997
	US	FGA	**F-15I**	25	1994	1998	4 in 1998, continue to 2000
	US	tpt hel	**S-70A**	15	1995	1998	1st 2 deliveries complete
	dom	UAV	***Silver Arrow***		1997		Prototype in 1998
	US	AAM	**AIM-120B**	64	1998	1999	
	US	cbt hel	**AH-64**	42	1998	2001	Upgrade to *Longbow* standard
	US	FGA	**F-16I**	50	1999	2003	With *Popeye* 2 and *Python* 4 AAM
	US	ASM	***Hellfire***	480	1999		
	US	cbt hel	**B200**	5	2000		
	US	AAM	**AMRAAM**	57	2000		
Jordan	US	FGA	**F-16A/B**	16	1995	1997	Deliveries complete April 1998
	US	hel	**UH-60L**	4	1995	1998	
	US	SP arty	**M-110**	18	1996	1998	
	UK	MBT	***Challenger* 1**	288	1999	1999	Ex-British Army
	UK	recce	***Scorpion***		1999	2001	Upgrade
	Tu	tpt	**CN-235**	2	1999	2001	One year lease
	Ukr	APC	**BTR-94**	50	1999	2000	Modified BTR-80
Kuwait	Fr	PFM	***Al Maradim***	8	1995	1998	Final 2 delivered 2000; with *Sea Skua*
	UK	ASSM	***Sea Skua***	60	1997	1998	
	US	ATGW	**TOW-2B**	728	1999		
	Aus	APC	**S-600**	22	1997	1998	
	US	cbt hel	**AH-64**	16	1997	2000	*Longbow* radar not fitted
	US	SP arty	**M-109A6**	48	1998		Includes spt veh; order frozen late 1998
Morocco	Fr	FF	***Floreal***	2	1998	2001	
Oman	Fr	APC	**VBL**	51	1995	1997	Deliveries to 1998
	UK	ftr	***Jaguar***	15	1997	1999	Upgrade
	UK	MBT	***Challenger* 2**	18	1997	1999	
	UK	radar	**S743D**		1999	2002	
	CH	trg	**PC-9**	12	1999	2000	
Qatar	Fr	FGA	***Mirage* 2000-5**	12	1994	1997	3 delivered 1997, 8 1998
	UK	APC	***Piranha* 2**	36	1995	1997	2 delivered 1997, 26 1998
	UK	trg	***Hawk* 100**	15	1996	1999	
	Fr	MBT	**AMX-30**	10	1997	1998	Military aid from Fr
Saudi Arabia							
	Ca	LAV	**LAV-25**	1,117	1990	1992	Delivery completed 1999
	UK	FGA	***Tornado* IDS**	48	1993	1996	Deliveries completed 1998
	Fr	FFG	**F-3000**	3	1994	2001	1st in 2001, 2nd 2003, 3rd 2005
	US	Construction	***Jizan***	1	1996	1999	Military city and port

Country	Country supplier	Classification	Designation	Quantity	Order date	Delivery date	Comment
	Fr	hel	**AS-532**	12	1996	1998	4 delivered 1998
	US	AWACS	**E-3**	5	1997	2000	Upgrade
	It	SAR hel	**AB-412TP**	40	1998	2000	
Syria	RF	ATGW	**AT-14**	1,000	1997	1998	Missiles
	RF	SAM	**S-300**		1997		Unconfirmed
	RF	FGA	**Su-27**			2000	4 delivered
	RF	FGA	**MiG-29**			2000	Order unnanounced
Tunisia	US	hel	**HH-3**	4	1996	1998	
United Arab Emirates							
	UK	ALCM	***Al-Hakim***	416	1992	1998	All delivered 1998
	Fr	MBT	***Leclerc***	390	1993	1994	Also 46 ARVs; deliveries to 2000
	Fr	hel	**AS-565**	6	1995	1998	For *Kortenaer* frigates
	Nl	FFG	***Kortenaer***	2	1996	1997	Ex-Nl; 2nd delivery 1998
	Fr	hel	**AS-332**	5	1996	1998	Upgrade
	Ge	trg	**G-115 TA**	12	1996	1997	Option for further 12
	RF	tpt	**Il-76**	4	1997	1998	On lease
	Tu	APC	**M-113**	136	1997	1999	
	Indo	tpt	**CN-235**	7	1997		
	US	cbt hel	**AH-64A**	10	1997	1999	
	Fr	hel	***Gazelle***	5	1997	1999	Option for further 5
	Fr	FGA	***Mirage* 2000-09**	30	1997	2000	*Mirage* 2000-9
	Fr	FGA	***Mirage* 2000**	33	1997	2000	Upgrade to 2000-9 standard
	Fr	ALCM	***Black Shahine***		1998	2000	For *Mirage* 2000-9
	UK	trg	***Hawk*-200**	18	1998	2001	Following delivery of 26 1992–6
	Indo	MPA	**CN-235**	4	1998		
	UK	PFC	***Protector***	2	1998	1999	
	Fr	trg	***Alpha Jet***		1999		
	Fr	trg	**AS 350B**	14	1999		
	US	FGA	**F-16**	80	2000	2002	
	Ca	tpt	**Bell 407**	1	1999	2000	
	Fr	trg	**AS-350B**	14	1999	2000	
Yemen	Fr	APC	**AML**	5	1996	1998	
	Fr	PCI	***Vigilante***	6	1996	1997	Commissioning delayed
	Cz	trg	**L-39C**	12	1999	1999	Deliveries began late 1999
	RF	FGA	**Su-27**	14	1999	2001	

Algeria Ag

dinar D		1998	1999	2000	2001
GDP	D	2.8tr	3.2tr		
	US$	47.3bn	46.8bn		
per capita	US$	6,600	6,800		
Growth	%	2.4	2.8		
Inflation	%	2.9	3.1		
Debt	US$	32bn	28.3bn		
Def exp	D	180bn	210bn		
	US$	3.1bn	3.1bn		
Def bdgt	D	112bn	121bn	138bn	
	US$	1.9bn	1.8bn	1.8bn	
FMA (US)	US$	0.1m	0.1m	0.1m	0.1m
US$1=D		58.7	68.0	74.6	

Population			30,200,000
Age	13–17	18–22	23–32
Men	1,955,000	1,787,000	2,871,000
Women	1,818,000	1,668,000	2,697,000

Total Armed Forces

ACTIVE ε124,000

(incl ε75,000 conscripts)

Terms of service **Army** 18 months (6 months basic, 12 months civil projects)

RESERVES

Army some 150,000, to age 50

Army 107,000

(incl ε75,000 conscripts)

6 Mil Regions; re-org into div structure on hold

2 armd div (each 3 tk, 1 mech regt) • 2 mech div (each 3 mech, 1 tk regt) • 5 AB regt • 1 indep armd bde • 4 indep mot/mech inf bde, 25 indep inf, 6 arty, 6 ATK, 8 AD, 6 AAA bn

EQUIPMENT

MBT 1,006: 324 T-54/-55, 332 T-62, 350 T-72

RECCE 75 BRDM-2

AIFV 700 BMP-1, 257 BMP-2

APC 530 BTR-50/-60, 150 OT-64, some BTR-80 (reported)

TOWED ARTY 122mm: 28 D-74, 100 M-1931/37, 60 M-30 (M-1938), 198 D-30; **130mm**: 10 M-46; **152mm**: 20 ML-20 (M-1937)

SP ARTY 185: **122mm**: 150 2S1; **152mm**: 35 2S3

MRL 122mm: 48 BM-21; **140mm**: 48 BM-14-16; **240mm**: 30 BM-24

MOR 82mm: 150 M-37; **120mm**: 120 M-1943; **160mm**: 60 M-1943

ATGW AT-2 *Swatter*, AT-3 *Sagger*, AT-4 *Spigot*, AT-5 *Spandrel*

RCL 82mm: 120 B-10; **107mm**: 58 B-11

ATK GUNS 57mm: 156 ZIS-2; **85mm**: 80 D-44; **100mm**: 12 T-12, 50 SU-100 SP

AD GUNS 14.5mm: 80 ZPU-2/-4; **20mm**: 100; **23mm**: 100 ZU-23 towed, 210 ZSU-23-4 SP; **37mm**: 150 M-1939; **57mm**: 75 S-60; **85mm**: 20 KS-12; **100mm**: 150 KS-19; **130mm**: 10 KS-30

SAM SA-7/-8/-9

Navy ε7,000

(incl ε500 Coast Guard)

BASES Mers el Kebir, Algiers, Annaba, Jijel

SUBMARINES 2

SSK 2 Sov *Kilo* with 533mm TT

FRIGATES 3

FF 3 *Mourad Rais* (Sov *Koni*) with SA-N-4 *Gecko* SAM, 4 × 76mm gun, 2 × 12 ASW RL

PATROL AND COASTAL COMBATANTS 17

CORVETTES 5

3 *Rais Hamidou* (Sov *Nanuchka* II) FSG with 4 SS-N-2C *Styx* SSM, SA-N-4 *Gecko* SAM

2 *Djebel Chinoise* FS with 3 × 76mm gun

MISSILE CRAFT 9 *Osa* with 4 SS-N-2 *Styx* SSM (plus 2 non-op)

PATROL CRAFT 3

COASTAL 3 *El Yadekh* PCC

AMPHIBIOUS 3

2 *Kalaat beni Hammad* LST: capacity 240 tps, 10 tk, hel deck

1 *Polnocny* LSM: capacity 180 tps, 6 tk

SUPPORT AND MISCELLANEOUS 3

1 div spt, 1 *Poluchat* TRV, 1 *El Idrissi* AGHS

COAST GUARD (ε500)

Some 7 PRC *Chui-E* PCC, about 6 *El Yadekh* PCC, 16 PCI<, 1 spt, plus boats

Air Force 10,000

214 cbt ac, 65 armed hel

Flying hours ε160

FGA 3 sqn

1 with 13 Su-24MK, 2 with 40 MiG-23BN

FTR 5 sqn

1 with 10 MiG-25, 1 with 30 MiG-23B/E, 3 with 70 MiG-21MF/bis (12+ MiG-29C/UB possibly serving with 2 sqn)

RECCE 1 sqn with 4* MiG-25R, 1 sqn with 6* MiG-21

MR 2 sqn with 15 *Super King Air* B-200T

TPT 2 sqn with 10 C-130H, 6 C-130H-30, 3 Il-76MD, 6 Il-76TD

VIP 2 *Falcon* 900, 3 *Gulfstream* III, 2 F-27

HELICOPTERS

ATTACK 35 Mi-24, 1 with 30 Mi-8/-17

TPT 2 Mi-4, 5 Mi-6, 46 Mi-8/17, 10 AS 355

TRG 6 T-34C, 30 ZLIN-142, 3* MiG-21U, 5* MiG-23U, 3* MiG-25U, 30* L-39 hel: 20 Mi-2

UAV *Seeker*

AAM AA-2, AA-6, AA-7

AD GUNS 3 bde+: 725 **85mm, 100mm, 130mm**
SAM 3 regt with 100 SA-3, SA-6, SA-8

Forces Abroad

UN AND PEACEKEEPING
DROC (MONUC): 13 obs

Paramilitary ε181,200

GENDARMERIE 60,000 (Ministry of Defence)
6 regions; 44 Panhard AML-60/M-3, BRDM-2 recce, 200 *Fahd* APC **hel** Mi-2
NATIONAL SECURITY FORCES 20,000 (Directorate of National Security)
small arms
REPUBLICAN GUARD 1,200
AML-60, M-3 recce
LEGITIMATE DEFENCE GROUPS ε100,000
self-defence militia, communal guards

Opposition

GROUPE ISLAMIQUE ARMÉE (GIA) small groups each ε50–100; total less than 1,500
SALAFIST GROUP FOR CALL AND COMBAT small groups; total less than 500

Bahrain Brn

dinar D		1998	1999	2000	2001
GDP	D	2.0bn	2.2bn		
	US$	5.3bn	5.7bn		
per capita	US$	8,600	9,000		
Growth	%	-2.8	5.0		
Inflation	%	-0.2	1.0		
Debt	US$	2.0bn			
Def exp	D	151m	166m		
	US$	402m	441m		
Def bdgt[a]	D	112m	115m		
	US$	298m	306m		
FMA (US)	US$	0.3m	0.2m	0.2m	0.2m
US$1=D		0.38	0.38	0.38	

[a] Excl procurement

Population			**639,000**
(Nationals 63%, Asian 13%, other Arab 10%, Iranian 8%, European 1%)			
Age	13–17	18–22	23–32
Men	33,000	26,000	40,000
Women	32,000	25,000	40,000

Total Armed Forces

ACTIVE 11,000

Army 8,500

1 armd bde (-) (2 tk, 1 recce bn) • 1 inf bde (2 mech, 1 mot inf bn) • 1 arty 'bde' (1 hy, 2 med, 1 lt, 1 MRL bty) • 1 SF, 1 *Amiri* gd bn • 1 AD bn (2 SAM, 1 AD gun bty)

EQUIPMENT
MBT 106 M-60A3
RECCE 22 AML-90, 8 *Saladin*, 8 *Ferret*, 8 Shorland
AIFV 25 YPR-765 (with **25mm**)
APC some 10 AT-105 *Saxon*, 110 Panhard M-3, 220 M-113A2
TOWED ARTY 105mm: 8 lt; **155mm**: 28 M-198
SP ARTY 203mm: 62 M-110
MRL 227mm: 9 MLRS
MOR 81mm: 18; **120mm**: 12
ATGW 15 TOW
RCL 106mm: 30 M-40A1; **120mm**: 6 MOBAT
AD GUNS 35mm: 15 Oerlikon; **40mm**: 12 L/70
SAM 40+ RBS-70, 18 *Stinger*, 7 *Crotale*, 8 I HAWK

Navy 1,000

BASE Mina Sulman
PRINCIPAL SURFACE COMBATANTS 1
FRIGATES
FFG 1 *Sabah* (US OH *Perry*) with 4 *Harpoon* SSM, 1 *Standard* SM-1MR SAM, 1 × 76mm gun, 2 × 3 ASTT
PATROL AND COASTAL COMBATANTS 10
CORVETTES 2 *Al Manama* (Ge Lürssen 62m) FSG with 2 × 2 MM-40 *Exocet* SSM, 1 × 76mm gun, hel deck
MISSILE CRAFT 4 *Ahmad el Fateh* (Ge Lürssen 45m) PFM with 2 × 2 MM-40 *Exocet* SSM, 1 × 76mm gun
PATROL CRAFT 4
COASTAL/INSHORE 4
2 *Al Riffa* (Ge Lürssen 38m) PFC
2 *Swift* FPB-20 PCI<
SUPPORT AND MISCELLANEOUS 5
4 *Ajeera* LCU-type spt
1 *Tiger* ACV, **hel** 2 B-105

Air Force 1,500

34 cbt ac, 40 armed hel
FGA 1 sqn with 8 F-5E, 4 F-5F
FTR 2 sqn with 18 F-16C, 4 F-16D
TPT 2 *Gulfstream* (1 -II, 1 -III; VIP), 1 Boeing 727
HEL 1 sqn with 12 AB-212 (10 armed), 3 sqn with 24 AH-1E*, 6 TAH-1P*, 1 VIP unit with 3 Bo-105, 1 UH-60L (VIP), 1 S-70A (VIP)
MISSILES
ASM AS-12, AGM-65D/G *Maverick*
AAM AIM-9P *Sidewinder*, AIM-7F *Sparrow*
ATGW BGM-71 TOW

Paramilitary ε10,150

POLICE 9,000 (Ministry of Interior)

2 Hughes 500, 2 Bell 412, 1 BO-105 hel
NATIONAL GUARD ε900
3 bn
COAST GUARD ε250 (Ministry of Interior)
1 PCI, some 20 PCI<, 2 spt/landing craft, 1 hovercraft

Foreign Forces

US Air Force periodic detachments of ftr and support ac **Navy** (HQ CENTCOM and 5th Fleet): 680 **Marine** 220
UK RAF 40 (*Southern Watch*), 2 VC-10 tkr

Egypt Et

pound E£		1998	1999	2000	2001
GDP	E£	283bn	302bn		
	US$	83bn	89bn		
per capita	US$	4,600	5,000		
Growth	%	5.6	6.0		
Inflation	%	3.5	3.1		
Debt	US$	32bn	29bn		
Def exp	E£	ε9.6bn	ε10.1bn		
	US$	2.8bn	3.0bn		
Def bdgt	E£	8.2bn	8.7bn		
	US$	2.4bn	2.5bn		
FMA[a] (US)	US$	2.1bn	2.1bn	2.0bn	2.0bn
US$1=E£		3.39	3.39	3.44	

[a] UNTSO **1997** US$27m **1998** US$27m

Population			62,100,000
Age	13–17	18–22	23–32
Men	3,634,000	3,218,000	5,067,000
Women	3,437,000	3,036,000	4,768,000

Total Armed Forces

ACTIVE 448,500
(incl some 322,000+ conscripts)
Terms of service 18 months–3 years (selective)

RESERVES 254,000
Army 150,000 **Navy** 14,000 **Air Force** 20,000 **AD** 70,000

Army 320,000

(perhaps 250,000+ conscripts)
4 Mil Districts, 2 Army HQ • 4 armd div (each with 2 armd, 1 mech, 1 arty bde) • 8 mech inf div (each with 2 mech, 1 armd, 1 arty bde) • 1 Republican Guard armd bde • 4 indep armd bde • 4 indep mech bde • 1 air-mobile bde • 2 indep inf bde • 1 para bde • 6 cdo gp • 15 indep arty bde • 2 SSM bde (1 with FROG-7, 1 with *Scud*-B)
EQUIPMENT[a]
MBT 895 T-54/-55, 260 *Ramses* II (mod T-54/55), 550 T-62, 1,700 M-60 (400 M-60A1, 1,300 M-60A3), 555 M1A1 *Abrams*
RECCE 300 BRDM-2, 112 *Commando Scout*
AIFV 220 BMP-1 (in store), 220 BMR-600P, 310 YPR-765 (with **25mm**)
APC 600 *Walid*, 165 *Fahd*/-30, 1,075 BTR-50/OT-62 (most in store), 2,320 M-113A2 (incl variants), 70 YPR-765
TOWED ARTY 122mm: 36 M-1931/37, 359 M-1938, 156 D-30M; **130mm**: 420 M-46
SP ARTY 122mm: 76 SP 122, **155mm**: 175 M-109A2
MRL 122mm: 96 BM-11, 60 BM-21/*as-Saqr*-10/-18/-36
MOR 82mm: 540 (some 50 SP); **120mm**: 1,800 M-1938; **160mm**: 60 M-160
SSM 9 FROG-7, *Saqr*-80 (trials), 9 *Scud*-B
ATGW 1,400 AT-3 *Sagger* (incl BRDM-2); 220 *Milan*; 200 *Swingfire*; 530 TOW (incl I-TOW, TOW-2A (with 52 on M-901, 210 on YPR-765 SP))
RCL 107mm: B-11
AD GUNS 14.5mm: 200 ZPU-4; **23mm**: 280 ZU-23-2, 118 ZSU-23-4 SP, 36 *Sinai*; **37mm**: 200 M-1939; **57mm**: some S-60, 40 ZSU-57-2 SP
SAM 600+ SA-7/'*Ayn as-Saqr*, 20 SA-9, 26 M-54 SP *Chaparral*, *Stinger*, *Avenger*
SURV AN/TPQ-37 (arty/mor), RASIT (veh, arty), *Cymbeline* (mor)
UAV R4E-50 *Skyeye*

[a] Most Sov eqpt now in store, incl MBT and some cbt ac

Navy ε18,500

(incl ε2,000 Coast Guards and ε12,000 conscripts)
BASES Mediterranean Alexandria (HQ), Port Said, Mersa Matruh, Safaqa, Port Tewfig **Red Sea** Hurghada (HQ)
SUBMARINES 4
SSK 4 *Romeo* with *Harpoon* SSM and 533mm TT
PRINCIPAL SURFACE COMBATANTS 11
DESTROYERS 1 DD *El Fateh* (UK 'Z') (trg) with 4 × 114mm guns, 5 × 533mm TT
FRIGATES 10
FFG 10
4 *Mubarak* (ex-US *OH Perry*) with 4 *Harpoon* SSM, *Standard* SM-1-MR SAM, 1 × 76mm gun, 2 hel
2 *El Suez* (Sp *Descubierta*) with 2 × 4 *Harpoon* SSM, 1 × 76mm gun, 2 × 3 ASTT, 1 × 2 ASW RL
2 *Al Zaffir* (PRC *Jianghu* I) with 2 CSS-N-2 (*HY* 2) SSM, 2 ASW RL
2 *Damyat* (US *Knox*) with 8 *Harpoon* SSM, 1 × 127mm gun, 4 × 324mm TT
PATROL AND COASTAL COMBATANTS 40
MISSILE CRAFT 25
6 *Ramadan* with 4 *Otomat* SSM
5 Sov *Osa* I with 4 SS-N-2A *Styx* SSM (1 may be non-op)
6 *6th October* with 2 *Otomat* SSM
2 Sov *Komar* with 2 SSN-2A *Styx*

6 PRC *Hegu* (*Komar*-type) with 2 SSN-2A *Styx* SSM

PATROL CRAFT 15

4 PRC *Hainan* PFC with 6 × 324mm TT, 4 ASW RL (plus 4 in reserve)

6 Sov *Shershen* PFC; 2 with 4 × 533mm TT and BM-21 (8-tube) 122mm MRL; 4 with SA-N-5 and 1 BM-24 (12-tube) 240mm MRL

5 PRC *Shanghai* II PFC

MINE COUNTERMEASURES 13

6 *Assiout* (Sov T-43 class) MSO (op status doubtful)

4 *Aswan* (Sov *Yurka*) MSC

3 *Swiftship* MHI

plus 1 route survey boat

AMPHIBIOUS 3

3 Sov *Polnocny* LSM, capacity 100 tps, 5 tk

plus craft: 9 *Vydra* LCU

SUPPORT AND MISCELLANEOUS 20

7 AOT (small), 5 trg, 6 AT, 1 diving spt, 1 *Tariq* (ex-UK FF) trg

NAVAL AVIATION

24 armed Air Force **hel** 5 *Sea King* Mk 47, 9 SA-342, 10 SH-2G *Super Sea-Sprite* with *Mk 46* LWT

COASTAL DEFENCE (Army tps, Navy control)

GUNS 130mm: SM-4-1

SSM *Otomat*

Air Force 30,000

(incl 10,000 conscripts); 580 cbt ac, 129 armed hel

FGA 7 sqn

2 with 41 *Alpha Jet*, 2 with 44 PRC J-6, 2 with 28 F-4E, 1 with 20 *Mirage* 5E2

FTR 22 sqn

2 with 25 F-16A/10 F-16B, 6 with 40 MiG-21, 7 with 135 F-16C/29 F-16D, 3 with 53 *Mirage* 5D/E, 3 with 53 PRC J-7, 1 with 18 *Mirage* 2000C

RECCE 2 sqn with 6* *Mirage* 5SDR, 14* MiG-21R

EW ac 2 C-130H (ELINT), 4 Beech 1900 (ELINT) **hel** 4 *Commando* 2E (ECM)

AEW 5 E-2C

MR 2 Beech 1900C surv ac

TPT 19 C-130H, 5 DHC-5D, 1 *Super King Air*, 3 *Gulfstream* III, 1 *Gulfstream* IV, 3 *Falcon* 20

HELICOPTERS

ASW 9* SA-342L, 5* *Sea King* 47, 10* SH-2G (with Navy)

ATTACK 4 sqn with 69 SA-342K (44 with HOT, 25 with 20mm gun), 36 AH-64A

TAC TPT hy 15 CH-47C, 14 CH-47D **med** 66 Mi-8, 25 *Commando* (3 VIP), 2 S-70 (VIP) **lt** 12 Mi-4, 17 UH-12E (trg), 2 UH-60A, 2 UH-60L (VIP), 3 AS-61

TRG incl 4 DHC-5, 54 EMB-312, 36 *Gumhuria*, 16* JJ-6, 40 L-29, 48 L-39, 30* L-59E, first of 80 K-8 being delivered to replace L-29, 10* MiG-21U, 5* *Mirage* 5SDD, 3* *Mirage* 2000B

UAV 29 Teledyne-Ryan 324 *Scarab*

MISSILES

ASM AGM-65 *Maverick*, AGM-84 *Harpoon*, *Exocet* AM-39, AS-12, AS-30, AS-30L HOT, AGM-119 *Hellfire*

ARM *Armat*

AAM AA-2 *Atoll*, AIM-7E/F/M *Sparrow*, AIM-9F/L/P *Sidewinder*, MATRA R-530, MATRA R-550 *Magic*

Air Defence Command 80,000

(incl 50,000 conscripts)

4 div: regional bde, 100 AD arty bn, 40 SA-2, 53 SA-3, 14 SA-6 bn, 12 bty I HAWK, 12 bty *Chaparral*, 14 bty *Crotale*

EQUIPMENT

AD GUNS some 2,000: **20mm, 23mm, 37mm, 57mm, 85mm, 100mm**

SAM some 282 SA-2, 212 SA-3, 78 I HAWK, 36 *Crotale*

AD SYSTEMS some 18 *Amoun* (*Skyguard*/RIM-7F *Sparrow*, some 36 twin **35mm** guns, some 36 quad SAM); *Sinai*-23 short-range AD (Dassault 6SD-20S radar, **23mm** guns, *'Ayn as-Saqr* SAM)

Forces Abroad

Advisers in Oman, Saudi Arabia, Zaire

UN AND PEACEKEEPING

DROC (MONUC): 12 obs **EAST TIMOR** (UNTAET): 83 incl 10 obs **GEORGIA** (UNOMIG): 3 obs **SIERRA LEONE** (UNAMSIL): 10 obs **WESTERN SAHARA** (MINURSO): 19 obs

Paramilitary 230,000 active

CENTRAL SECURITY FORCES 150,000 (Ministry of Interior)

110 *Hotspur Hussar*, *Walid* APC

NATIONAL GUARD 60,000

8 bde (each of 3 bn; cadre status); lt wpns only

BORDER GUARD FORCES 20,000

19 Border Guard Regt; lt wpns only

COAST GUARD ε2,000 (incl in Naval entry)

PATROL, INSHORE 40

20 *Timsah* PCI, 9 *Swiftships*, 5 *Nisr*†, 6 *Crestitalia* PFI<, plus some 60 boats

Foreign Forces

PEACEKEEPING

MFO Sinai: some 1,896 from **Aus, Ca, Co, Fji, Fr, Hu, It, No, NZ, Ury, US**

Iran Ir

rial r		1998	1999	2000	2001
GDP[a]	r	327tr	401tr		
	US$	89bn	109bn		
per capita	US$	6,300	6,600		
Growth	%	2.5	5.5		
Inflation	%	22	22		
Debt	US$	20.6bn	12.1bn		
Def exp[a]	r	10.1tr	10.0tr		
	US$	5.8bn	5.7bn		
Def bdgt	r	10.1tr	10.0tr	13.1tr	
	US$	5.8bn	5.7bn	7.5bn	
US$1=r[b]		1,753	1,753	1,753	

[a] Excl defence industry funding
[b] Market rate **2000** US$1=r9,400

Population			**72,664,000**
(Persian 51%, Azeri 24%, Gilaki/Mazandarani 8%, Kurdish 7%, Arab 3%, Lur 2%, Baloch 2%, Turkman 2%)			
Age	13–17	18–22	23–32
Men	4,587,000	3,827,000	5,771,000
Women	4,395,000	3,695,000	5,445,000

Total Armed Forces

ACTIVE 513,000
(perhaps 220,000 conscripts)
Terms of service 21 months

RESERVES
Army 350,000, ex-service volunteers

Army 325,000

(perhaps 220,000 conscripts)
4 Corps HQ • 4 armd div (each 3 armd, 1 mech bde, 4–5 arty bn) • 6 inf div (each 4 inf bde, 4-5 arty bn) • 1 cdo div, 1 SF div • 1 AB bde • some indep armd, inf, cdo bde • 5 arty gps • Army avn

EQUIPMENT† (overall totals incl those held by Revolutionary Guard Corps Ground Forces)

- **MBT** some 1,135 incl: 500 T-54/-55 and PRC Type-59, some 75 T-62, 120 T-72, 140 *Chieftain* Mk 3/5, 150 M-47/-48, 150 M-60A1, some *Zulfiqar*
- **LT TK** 80 *Scorpion*
- **RECCE** 35 EE-9 *Cascavel*
- **AIFV** 300 BMP-1, 140 BMP-2
- **APC** 300 BTR-50/-60, 250 M-113, ε40 *Boragh*
- **TOWED** 1,950: **105mm**: 130 M-101A1; **122mm**: 400 D-30, 100 PRC Type-54; **130mm**: 1,100 M-46/Type-59; **152mm**: 30 D-20; **155mm**: 20 WAC-21, 70 M-114; 80 GHN-45; **203mm**: 20 M-115
- **SP** 290: **122mm**: 60 2S1; **155mm**: 160 M-109; **170mm**: 10 M-1978; **175mm**: 30 M-107; **203mm**: 30 M-110
- **MRL** 664+: **107mm**: 500 PRC Type-63; *Haseb*, *Fadjr* 1; **122mm**: 50 *Hadid/Arash/Noor*, 100 BM-21, 5 BM-11; **240mm**: 9 M-1985, *Fadjr* 3
- **MOR** 6,500 incl: **60mm; 81mm; 82mm; 107mm**: 4.2in M-30; **120mm**
- **SSM** ε10 *Scud*-B/-C (300 msl), ε25 CSS-8 (150 msl), *Fadjr* 5, *Oghab*, *Shahin* 1/-2, *Nazeat*, some *Shehab* 3
- **ATGW** TOW, AT-3 *Sagger* (some SP), some AT-5 *Spandrel*
- **RL 73mm**: RPG-7
- **RCL 75mm**: M-20; **82mm**: B-10; **106mm**: M-40; **107mm**: B-11
- **AD GUNS** 1,700: **14.5mm**: ZPU-2/-4; **23mm**: ZU-23 towed, ZSU-23-4 SP; **35mm; 37mm**: M-1939, PRC Type-55; **57mm**: ZSU-57-2 SP
- **SAM** SA-7
- **UAV** *Mohajer* II/III/IV
- **AC** incl 50 Cessna (150, 180, 185, 310), 19 F-27, 8 *Falcon* 20
- **HEL** 100 AH-1J **attack**; 40 CH-47C **hy tpt**; 130 Bell 214A, 35 AB-214C; 70 AB-205A; 130 AB-206; 12 AB-212; 30 Bell 204; 5 Hughes 300C; 9 RH-53D; 17 SH-53D, 10 SA-319; 45 UH-1H

Revolutionary Guard Corps (*Pasdaran Inqilab*) some 125,000

GROUND FORCES some 100,000
grouped into perhaps 16–20 div incl 2 armd, 5 mech, 10 inf, 1 SF and 15–20 indep bde, incl inf, armd, para, SF, 5 arty gp (incl SSM), engr, AD and border defence units, serve indep or with Army; eqpt incl 470 tk, 620 APC/ACV, 360 arty, 40 RL and 140 AD guns, all incl in army inventory; controls *Basij* (see *Paramilitary*) when mob

NAVAL FORCES some 20,000
BASES Al Farsiyah, Halul (oil platform), Sirri, Abu Musa, Larak
some 40 Swe Boghammar Marin boats armed with ATGW, RCL, machine guns; 10 *Hudong* with C-802 SSM; controls coast-defence elm incl arty and CSSC-3 (*HY* 2) *Seersucker* SSM bty. Under joint command with Navy

MARINES some 5,000 1 bde

AIR FORCES
Few details are known of this organisation, which is commanded by a Brig Gen

Navy 18,000

(incl Naval Air and 2,600 Marines)
BASES Bandar Abbas (HQ), Bushehr, Kharg, Bandar-e-Anzelli, Bandar-e-Khomeini, Chah Bahar

SUBMARINES 5
SSK 3 Sov *Kilo* with 6 × 533mm TT (TEST 71/96 HWT/LWT)
SSI 2

PRINCIPAL SURFACE COMBATANTS 3
FRIGATES
FFG 3 *Alvand* (UK Vosper Mk 5) with 2 × 2 C-802

SSM, 1 × 114mm gun, 1 × 3 *Limbo* ASW RL

PATROL AND COASTAL COMBATANTS 63

CORVETTES 2 *Bayandor* FS (US PF-103) with 2 × 76mm gun

MISSILE CRAFT 20

10 *Kaman* (Fr *Combattante* II) PFM; 5 of which have 2 or 4 C-802 SSM

10 *Houdong* PFM with 4 C-802 SSM (manned by IRGC)

PATROL, COASTAL 3

3 *Parvin* PCC

PATROL, INSHORE 38

3 *Zafar* PCI<, some 35 PFI<, plus some 14 hovercraft< (not all op), 200+ small craft

MINE WARFARE 7

MINE LAYERS

2 *Hejaz* LST

MINE COUNTERMEASURES 5

1 *Shahrokh* MSC (in Caspian Sea as trg ship)

2 *292* MSC

2 *Riazi* (US *Cape*) MSI

AMPHIBIOUS 9

4 *Hengam* LST, capacity 225 tps, 9 tk, 1 hel

3 *Iran Hormuz 24* (ROK) LSM, capacity 140 tps, 9 tk

2 *Fouque* LSL

Plus craft: 3 LCT, 6 ACV

SUPPORT AND MISCELLANEOUS 30

1 *Kharg* AO with 2 hel, 2 *Bandar Abbas* AO with 1 hel; 4 AWT, 7 *Delvar* spt, 13 *Hendijan* spt; 1 AT, 2 trg craft

NAVAL AIR (2,000)

5 cbt ac, 29 armed hel

MR 5 P-3F, 5 Do-228

ASW 1 hel sqn with ε14 SH-3D, 6 AB-212 ASW

MCM 1 hel sqn with 9 RH-53D

TPT 1 sqn with 4 *Commander*, 4 F-27, 1 *Falcon* 20 hel, 5 AB 205a, 4 Mi-171, AB-206

MARINES (2,600) 2 bde

Air Force ε45,000

(incl 15,000 Air Defence); some 291 cbt ac (serviceability probably about 60% for US ac types and about 80% for Chinese/Russian ac); no armed hel

FGA 9 sqn

4 with some 66 F-4D/E, 4 with some 60 F-5E/F, 1 with 24 Su-24MK (including former Irq ac), 7 Su-25K (former Irq ac)

FTR 7 sqn

2 with 25 F-14, 1 with 24 F-7M, 2 with 30 MiG-29A/UB (incl former Irq ac)

(Some F-7 operated by Pasdaran air arm)

MR 5* C-130H-MP

AEW 1 Il-76 (former Irq ac)

RECCE 1 sqn (det) with some 15* RF-4E

TKR/TPT 1 sqn with 3 Boeing 707, 1 Boeing 747

TPT 5 sqn with 6 Boeing 747F, 1 Boeing 727, 18 C-130E/H, 3 *Commander* 690, 15 F-27, 4 *Falcon* 20, 1 *Jetstar*, 10 PC-6B, 2 Y-7, some Il-76 (former Irq ac), 9 Y-12(II)

HEL 2 AB-206A, 39 Bell 214C, Shabaviz 2061 and 2-75 (indigenous versions in production), 5 CH-47

TRG incl 26 Beech F-33A/C, 10 EMB-312, 40 PC-7, 7 T-33, 15* FT-7, 20* F-5B, 8 TB-21, 4 TB-200

MISSILES

ASM AGM-65A *Maverick*, AS-10, AS-11, AS-14, C-801

AAM AIM-7 *Sparrow*, AIM-9 *Sidewinder*, AIM-54 *Phoenix*, probably AA-8, AA-10, AA-11 for MiG-29, PL-2A, PL-7

SAM 16 bn with 100 I HAWK, 5 sqn with 30 *Rapier*, 15 *Tigercat*, 45 HQ-2J (PRC version of SA-2), 10 SA-5, FM-80 (PRC version of *Crotale*), SA-7, *Stinger*

Forces Abroad

LEBANON ε150 Revolutionary Guard

SUDAN mil advisers

Paramilitary 40,000 active

BASIJ ('Popular Mobilisation Army') (R) ε200,000

peacetime volunteers, mostly youths; str up to 1,000,000 during periods of offensive ops. Small arms only; not currently embodied for mil ops

LAW-ENFORCEMENT FORCES (Ministry of Interior) ε40,000

incl border-guard elm **ac** Cessna 185/310 lt **hel** ε24 AB-205/-206; about 90 patrol inshore, 40 harbour craft

Opposition

NATIONAL LIBERATION ARMY (NLA) some 15,000

Iraq-based; org in bde, armed with captured eqpt. Perhaps 160+ T-54/-55 tanks, BMP-1 AIFV, D-30 **122mm** arty, BM-21 **122mm** MRL, Mi-8 hel

KURDISH DEMOCRATIC PARTY OF IRAN (KDP-Iran) ε1,200–1,800

KURDISH COMMUNIST PARTY OF IRAN (KOMALA-Iran) based in Iraq ε200

Iraq — Irq

dinar D		1998	1999	2000	2001
GDP	US$	ε19bn			
Growth	%	ε12			
Inflation	%	ε45			
Debt	US$	ε23bn			
Def exp	US$	ε1.3bn	ε1.4bn		
US$1=D		0.31	0.31	0.31	
Population					**22,300,000**

(Arab 75–80% (of which Shi'a Muslim 55%, Sunni Muslim 45%), Kurdish 20–25%)

Age	13–17	18–22	23–32
Men	1,498,000	1,281,000	1,894,000
Women	1,433,000	1,229,000	1,833,000

Total Armed Forces

ACTIVE ε429,000

Terms of service 18–24 months

RESERVES ε650,000

Army ε375,000

(incl ε100,000 recalled Reserves)

7 corps HQ • 3 armd div, 3 mech div[a] • 11 inf div[a] • 6 Republican Guard Force div (3 armd, 1 mech, 2 inf) • 4 Special Republican Guard bde • 7 cdo bde • 2 SF bde

EQUIPMENT[b]

MBT perhaps 2,200, incl 1,500 T-55/-62 and PRC Type-59, 700 T-72

RECCE ε1,000: BRDM-2, AML-60/-90, EE-9 *Cascavel*, EE-3 *Jararaca*

AIFV perhaps 1,000 BMP-1/-2

APC perhaps 2,400, incl BTR-50/-60/-152, OT-62/-64, MTLB, YW-701, M-113A1/A2, Panhard M-3, EE-11 *Urutu*

TOWED ARTY perhaps 1,900, incl **105mm**: incl M-56 pack; **122mm**: D-74, D-30, M-1938; **130mm**: incl M-46, Type 59-1; **155mm**: some G-5, GHN-45, M-114

SP ARTY 150, incl **122mm**: 2S1; **152mm**: 2S3; **155mm**: M-109A1/A2, AUF-1 (GCT)

MRL perhaps 500, incl **107mm; 122mm**: BM-21; **127mm**: ASTROS II; **132mm**: BM-13/-16; **262mm**: *Ababeel*

MOR 81mm; 120mm; 160mm: M-1943; **240mm**

SSM up to 6 *Scud* launchers (ε27 msl) reported

ATGW AT-3 *Sagger* (incl BRDM-2), AT-4 *Spigot* reported, SS-11, *Milan*, HOT (incl 100 VC-TH)

RCL 73mm: SPG-9; **82mm**: B-10; **107mm**

ATK GUNS 85mm; 100mm towed

HELICOPTERS ε500 (120 armed)

ATTACK ε120 Bo-105 with AS-11/HOT, Mi-24, SA-316 with AS-12, SA-321 (some with *Exocet*), SA-342

TPT ε350 **hy** Mi-6 **med** AS-61, Bell 214 ST, Mi-4, Mi-8/-17, SA-330 **lt** AB-212, BK-117 (SAR), Hughes 300C, Hughes 500D, Hughes 530F

SURV RASIT (veh, arty), *Cymbeline* (mor)

[a] All divisions less Republican Guard at reported 50% cbt effectiveness

[b] 50% of all eqpt lacks spares

Navy ε2,000

BASES Basra (limited facilities), Az Zubayr, Umm Qasr (currently closed for navy; commercials only)

PATROL AND COASTAL COMBATANTS 6

MISSILE CRAFT 1 Sov *Osa* I PFM with 4 SS-N-2A *Styx* SSM

PATROL, INSHORE 5†

1 Sov *Bogomol* PFI<, 3 PFI<, 1 PCI< (all non-op) plus 80 boats

MINE COUNTERMEASURES 4

2 Sov *Yevgenya*, 2 *Nestin* MSI

SUPPORT AND MISCELLANEOUS 3

1 *Damen* AG, 1 *Aka* (Yug *Spasilac*-class) AR, 1 yacht with hel deck

Air Force ε35,000

Serviceability of fixed-wg ac about 55%, serviceability of hel poor

Flying hours snr pilots 90–120, jnr pilots as low as 20

BBR ε6, incl H-6D, Tu-22

FGA ε130, incl MiG-23BN, *Mirage* F1EQ5, Su-7, Su-20, 9 Su-24 MK, Su-25

FTR ε180 incl F-7, MiG-21, MiG-23, MiG-25, *Mirage* F-1EQ, MiG-29

RECCE incl MiG-25

TKR incl 2 Il-76

TPT incl An-2, An-12, An-24, An-26, Il-76

TRG incl AS-202, EMB-312, some 50 L-39, *Mirage* F-1BQ, 25 PC-7, 15 PC-9

MISSILES

ASM AM-39, AS-4, AS-5, AS-11, AS-9, AS-12, AS-30L, C-601

AAM AA-2/-6/-7/-8/-10, R-530, R-550

Air Defence Command ε17,000

AD GUNS ε6,000: **23mm**: ZSU-23-4 SP; **37mm**: M-1939 and twin; **57mm**: incl ZSU-57-2 SP; **85mm; 100mm; 130mm**

SAM 575 launchers SA-2/-3/-6/-7/-8/-9/-13/-14/-16, *Roland*, *Aspide*

Paramilitary 45–50,000

SECURITY TROOPS ε15,000

BORDER GUARDS ε20,000

lt wpns and mor only

SADDAM'S *FEDAYEEN* ε10–15,000

Opposition

KURDISH DEMOCRATIC PARTY (KDP) ε15,000

(plus 25,000 tribesmen); small arms, some Iranian lt arty, MRL, mor, SAM-7

PATRIOTIC UNION OF KURDISTAN (PUK) ε10,000

(plus 22,000 tribesmen); 450 mor (**60mm, 82mm, 120mm**); **106mm** RCL; some 200 **14.5mm** AA guns; SA-7 SAM

SOCIALIST PARTY OF KURDISTAN ε500

SUPREME COUNCIL FOR ISLAMIC RESISTANCE IN IRAQ (SCIRI)

ε4,000; ε1 'bde'; Iran-based; Iraqi dissidents, ex-prisoners of war

Foreign Forces

UN (UNIKOM): some 908 tps and 203 mil obs from 30 countries

Israel II

new sheqalim NS		1998	1999	2000	2001
GDP	NS	376bn	410bn		
	US$	99bn	99bn		
per capita	US$	18,200	18,700		
Growth	%	2.0	2.2		
Inflation	%	7.0	7.0		
Debt	US$	55bn	56bn		
Def exp	NS	ε34.8bn	ε36.4bn		
	US$	9.2bn	8.9bn		
Def bdgt	NS	25.1bn	27.6bn	28.9bn	
	US$	6.6bn	6.7bn	7.0bn	
FMA[a] (US)	US$	3bn	3bn	3bn	3bn
US$1=NS		3.80	4.12	4.12	

[a] UNDOF **1997** US$38m **1998** US$35m

Population[b]			6,200,000
(Jewish 82%, Arab 14%, Christian 3%, Druze 2%, Circassian ε3,000)			
Age	13–17	18–22	23–32
Men	281,000	270,000	526,000
Women	265,000	258,000	518,000

[b] Incl ε180,000 Jewish settlers in Gaza and the West Bank, ε217,000 in East Jerusalem and ε15,000 in Golan

Total Armed Forces

ACTIVE 172,500

(107,500 conscripts)

Terms of service **officers** 48 months **men** 36 months **women** 21 months (Jews and Druze only; Christians, Circassians and Muslims may volunteer). Annual trg as cbt reservists to age 41 (some specialists to age 54) for men, 24 (or marriage) for women

RESERVES 425,000

Army 400,000 **Navy** 5,000 **Air Force** 20,000. Reserve service can be followed by voluntary service in Civil Guard or Civil Defence

Strategic Forces

Israel is widely believed to have a nuclear capability with up to 100 warheads. Delivery means could include ac, *Jericho* 1 SSM (range up to 500km), *Jericho* 2 (range ε1,500–2,000km)

Army 130,000

(85,000 conscripts, male and female); some 530,000 on mob

3 territorial, 1 home front comd • 3 corps HQ • 3 armd div (2 armd, 1 arty bde, plus 1 armd, 1 mech inf bde on mob) • 2 div HQ (op control of anti-*intifada* units) • 3 regional inf div HQ (border def) • 4 mech inf bde (incl 1 para trained) • 3 arty bn with MLRS

RESERVES

8 armd div (2 or 3 armd, 1 affiliated mech inf, 1 arty bde) • 1 air-mobile/mech inf div (3 bde manned by para trained reservists) • 10 regional inf bde (each with own border sector)

EQUIPMENT

MBT 3,900: 800 *Centurion*, 300 M-48A5, 300 M-60/A1, 600 M-60A3, 400 *Magach* 7, 200 Ti-67 (T-54/-55), 100 T-62, 1,200 *Merkava* I/II/III

RECCE about 400, incl RAMTA RBY, BRDM-2, ε8 *Fuchs*

APC 5,500 M-113A1/A2, ε200 *Nagmashot* (*Centurion*), ε200 *Achzarit*, *Puma*, BTR-50P, 4,000 M-2/-3 half-track (most in store)

TOWED ARTY 520: **105mm**: 70 M-101; **122mm**: 100 D-30; **130mm**: 100 M-46; **155mm**: 50 Soltam M-68/-71, 50 M-839P/-845P, 50 M-114A1, 100 Soltam M-46

SP ARTY 855: **155mm**: 150 L-33, 530 M-109A1/A2; **175mm**: 140 M-107; **203mm**: 35 M-110

MRL 198: **122mm**: 50 BM-21; **160mm**: 50 LAR-160; **227mm**: 48 MLRS; **240mm**: 30 BM-24; **290mm**: 20 LAR-290.

MOR 60mm: ε5,000; **81mm**: 700; **120mm**: 530; **160mm**: 240 (some SP)

SSM 20 *Lance* (in store), some *Jericho* 1/2

ATGW 300 TOW-2A/-B (incl *Ramta* (M-113) SP), 1,000 *Dragon*, AT-3 *Sagger*, 25 *Mapats*, *Gill/Spike*

RL 82mm: B-300

RCL 106mm: 250 M-40A1

AD GUNS 20mm: 850: incl TCM-20, M-167 *Vulcan*, 35 M-163 *Vulcan*/M-48 *Chaparral* gun/msl, *Machbet Vulcan*/*Stinger* gun/msl SP system; **23mm**: 150 ZU-23 and 60 ZSU-23-4 SP; **37mm**: M-39; **40mm**: 150 L-70

SAM 250 *Stinger*, 1,000 *Redeye*, 48 *Chaparral*

SURV EL/M-2140 (veh), AN/TPQ-37 (arty), AN/PPS-15 (arty)

Navy ε6,500

(incl 2,500 conscripts), 11,500 on mob

BASES Haifa, Ashdod, Eilat

SUBMARINES 2

SSK 2 *Dolphin* (Ge Type 212 variant) with *Sub-Harpoon* USGW, 4 × 650mm ASTT, 6 × 533mm ASTT

PATROL AND COASTAL COMBATANTS 47

CORVETTES 3 *Eilat* (*Sa'ar* 5) FSG with 8 *Harpoon* SSM, 8 *Gabriel* II SSM, 2 *Barak* VLS SAM (2 × 32 mls), 1 × 76mm gun, 6 × 324mm ASTT, 1 SA-366G hel

MISSILE CRAFT 12

2 *Aliya* PFM with 4 *Harpoon* SSM, 4 *Gabriel* SSM, 1 SA-366G *Dauphin* hel

6 *Hetz* (*Sa'ar* 4.5) PFM with 8 *Harpoon* SSM, 6 *Gabriel* SSM, 6 *Barak* VLS SAM, 1 × 76mm gun

4 *Reshef* (*Sa'ar* 4) PFM with 8 *Harpoon* SSM, 6 *Gabriel* SSM, 1 × 76mm gun

PATROL, INSHORE 32

11 *Super Dvora* PFI<, some with 2 × 324mm TT, 3 *Nashal* PCI, 15 *Dabur* PFI< with 2 × 324mm TT, 3 Type-1012 *Bobcat* catamaran PCC

AMPHIBIOUS craft only

1 *Ashdod* LCT, 1 US type LCM

NAVAL COMMANDOS ε300

Air Force 36,000

(20,000 conscripts, mainly in AD), 57,000 on mob; 446 cbt ac (plus perhaps 250 stored including significant number of *Kfir* C7), 133 armed hel

Flying hours regulars: 180; reserves: 80

FGA/FTR 12 sqn

2 with 50 F-4E-2000, 20 F-4E

2 with 73 F-15 (38 -A, 8 -B, 16 -C, 11 -D)

1 with 25 F-15I

7 with 237 F-16 (92 -A, 17 -B, 79 -C, 49 -D)

FGA 1 sqn with 25 A-4N

RECCE 10* RF-4E

AEW 6 Boeing 707 with *Phalcon* system

EW 3 Boeing 707 (ELINT/ECM), 6 RC-12D, 3 IAI-200, 15 Do-28, 10 *King Air* 2000

MR 3 IAI-1124 *Seascan*

TKR 3 KC-130H

TPT 1 wg incl 5 Boeing 707 (3 tpt/tkr), 12 C-47, 22 C-130H

LIAISON 2 *Islander*, 20 Cessna U-206, 10 *Queen Air* 80

TRG 77 CM-170 *Tzukit*, 28 *Super Cub*, 9* TA-4H, 17* TA-4J, 4 *Queen Air* 80

HELICOPTERS

ATTACK 21 AU-1G, 36 AH-1F, 30 Hughes 500MD, 42 AH-64A

ASW 4* AS-565A, 2 × SA-366G

TPT 38 CH-53D, 10 UH-60; 15 S-70A *Blackhawk*, 54 Bell 212, 43 Bell 206

UAV *Scout*, *Pioneer*, *Searcher*, *Firebee*, *Samson*, *Delilah*, *Hunter Silver Arrow*

MISSILES

ASM AGM-45 *Shrike*, AGM-62A *Walleye*, AGM-65 *Maverick*, AGM-78D *Standard*, AGM-114 *Hellfire*, TOW, *Popeye* I + II, (GBU-31 JDAM undergoing IAF op/integration tests)

AAM AIM-7 *Sparrow*, AIM-9 *Sidewinder*, AIM-120B AMRAAM, R-530, *Shafrir*, *Python* III, *Python* IV

SAM 17 bty with MIM-23 I HAWK, 3 bty *Patriot*, 1 bty Arrow 2, 8 bty *Chapparal*, *Stinger*

Forces Abroad

TURKEY occasional det of Air Force F-16 ac to Akinci air base

Paramilitary ε8,050

BORDER POLICE ε8,000

some *Walid* 1, 600 BTR-152 APC

COAST GUARD ε50

1 US PBR, 3 other patrol craft

Foreign Forces

UN (UNTSO): 143 mil obs from 22 countries

Jordan HKJ

dinar D		1998	1999	2000	2001
GDP	D	5.2bn	5.5bn		
	US$	7.1bn	7.7bn		
per capita	US$	4,500	4,500		
Growth	%	2.2	1.5		
Inflation	%	4.5	0.6		
Debt	US$	7.7bn	8.2bn		
Def exp	D	390m	403m		
	US$	548m	569m		
Def bdgt	D	318m	347m		
	US$	447m	488m		
FMA[a] (US)	US$	52m	47m	77m	78m
US$1=D		0.71	0.71	0.71	

[a] Excl US military debt waiver **1997** US$15m **1998** US$12m

Population	**5,173,000** (Palestinian ε50–60%)		
Age	13–17	18–22	23–32
Men	274,000	245,000	447,000
Women	267,000	238,000	433,000

Total Armed Forces

ACTIVE ε103,880

RESERVES 35,000 (all services)

Army 30,000 (obligation to age 40)

Army 90,000

2 armd div (each 2 tk, 1 mech inf, 1 arty, 1 AD bde)
2 mech inf div (each 2 mech inf, 1 tk, 1 arty, 1 AD bde)
1 indep Royal Guard bde
1 SF bde (2 SF, 2 AB, 1 arty bn)
1 fd arty bde (4 bn)
Southern Mil Area (3 inf, 1 recce bn)

EQUIPMENT

MBT 1,246: 78 M-47/-48A5 (in store), 310 M-60A1/A3, 274 *Khalid/Chieftain*, 296 *Tariq* (*Centurion*), 288

Challenger 1 (*Al Hussein* (being delivered))
LT TKS 19 *Scorpion*
AIFV some 32 BMP-2
APC 1,400 M-113, 50 BTR-94 (BTR-80)
TOWED ARTY 132: **105mm**: 54 M-102; **155mm**: 38 M-114 towed, 18 M-59/M-1; **203mm**: 22 M-115 towed (in store)
SP ARTY 412: **105mm**: 35 M-52; **155mm**: 23 M-44, 234 M-109A1/A2; **203mm**: 120 M-110
MOR 81mm: 450 (incl 130 SP); **107mm**: 50 M-30; **120mm**: 300 Brandt
ATGW 330 TOW (incl 70 M-901 ITV), 310 *Dragon*
RL 94mm: 2,500 LAW-80; **112mm**: 2,300 APILAS
AD GUNS 416: **20mm**: 100 M-163 *Vulcan* SP; **23mm**: 52 ZSU-23-4 SP; **40mm**: 264 M-42 SP
SAM SA-7B2, 52 SA-8, 92 SA-13, 300 SA-14, 240 SA-16, 260 *Redeye*
SURV AN-TPQ-36/-37 (arty, mor)

Navy ε480

BASE Aqaba
PATROL AND COASTAL COMBATANTS 6
PATROL CRAFT, INSHORE 6
3 *Al Hussein* (Vosper 30m) PFI
3 *Al Hashim* (Rotork) PCI<; plus 3 (ex US Coastguard) cutters

Air Force 13,400

(incl 3,400 AD); 106 cbt ac, 16 armed hel
Flying hours 180
FGA/RECCE 4 sqn
3 with 50 F-5E/F
1 with 15 *Mirage* F-1EJ
FTR 2 sqn
1 with 25 *Mirage* F-1 CJ/BJ
1 with 16 F-16A/B (12 -A, 4 -B)
TPT 1 sqn with 8 C-130 (3 -B, 5 -H), 4 C-212A
VIP 1 royal flt with **ac** 2 *Gulfstream* IV, 1 L-1011 **hel** 3 S-70
HELICOPTERS 3 sqn
ATTACK 2 with 16 AH-1F (with TOW ASM)
TPT 1 with 9 AS-332M, 36 UH-1H, 3 Bo-105 (operated on behalf of police)
TRG 3 sqn with ac: 16 *Bulldog*, 15 C-101, hel: 7 Hughes 500D
AD 2 bde: 14 bty with 80 I HAWK
MISSILES
ASM TOW, AGM-65D *Maverick*
AAM AIM-9 *Sidewinder*, MATRA R-530, MATRA R-550 *Magic*

Forces Abroad

UN AND PEACEKEEPING
CROATIA (UNMOP): 1 obs **DROC** (MONUC): 2 obs **EAST TIMOR** (UNTAET): 720 incl 5 obs **GEORGIA** (UNOMIG): 6 obs **SIERRA LEONE** (UNAMSIL): 1,831 incl 5 obs **YUGOSLAVIA** (KFOR): 61

Paramilitary ε10,000 active

PUBLIC SECURITY DIRECTORATE (Ministry of Interior) ε10,000
(incl Police Public Sy bde); some *Scorpion* lt tk, 25 EE-11 *Urutu*, 30 *Saracen* APC
CIVIL MILITIA 'PEOPLE'S ARMY' (R) ε35,000
(to be 5,000) **men** 16–65 **women** 16–45

Kuwait Kwt

dinar D		1998	1999	2000	2001
GDP	D	7.7bn	9.0bn		
	US$	25.1bn	29.5bn		
per capita	US$	13,100	14,600		
Growth	D	-14.0	13.0		
Inflation	D	0.1	3.0		
Debt	US$	9.0bn			
Def exp[a]	D	1.1bn	1.0bn		
	US$	3.5bn	3.2bn		
Def bdgt	D	711m	700m	800m	
	US$	2.3bn	2.3bn	2.6bn	
US$1=D		0.31	0.31	0.31	

[a] UNIKOM **1997** US$50m **1998** US$52m

Population 2,274,000
(Nationals 35%, other Arab 35%, South Asian 9%, Iranian 4%, other 17%)

Age	13–17	18–22	23–32
Men	120,000	103,000	147,000
Women	89,000	77,000	111,000

Total Armed Forces

ACTIVE 15,300
(some conscripts)
Terms of service voluntary, conscripts 2 years

RESERVES 23,700
obligation to age 40; 1 month annual trg

Land Force 11,000

(incl 1,600 foreign personnel)
3 armd bde • 2 mech inf bde • 1 recce (mech) bde • 1 force arty bde • 1 force engr bde

ARMY
1 reserve bde • 1 Amiri gd bde • 1 cdo bn
EQUIPMENT
MBT 150 M-84 (ε50% in store), 218 M-1A2, 17 *Chieftain* (in store)

AIFV 46 BMP-2, 55 BMP-3, 254 *Desert Warrior*
APC 60 M-113, 40 M-577, 40 *Fahd* (in store)
SP ARTY 155mm: 23 M-109A2, 18 GCT (in store), 18 F-3
MRL 300mm: 27 *Smerch* 9A52
MOR 81mm: 44; **107mm**: 6 M-30; **120mm**: ε12
ATGW 118 TOW/TOW II (incl 8 M-901 ITV; 66 HMMWV)

Navy ε1,800

(incl 400 Coast Guard)
BASE Ras al Qalaya
PATROL AND COASTAL COMBATANTS 10
MISSILE CRAFT 8
6 *Um Almaradim* PFM (Fr P-37 BRL) with 4 *Sea Skua* SSM, 1 × 6 Sadral SAM
1 *Istiqlal* (Ge Lürssen FPB-57) PFM with 2 × 2 MM-40 *Exocet* SSM
1 *Al Sanbouk* (Ge Lürssen TNC-45) PFM with 2 × 2 MM-40 *Exocet* SSM
PATROL CRAFT 2
2 *Al Shaheed* PCC
plus about 30 boats
SUPPORT AND MISCELLANEOUS 6
2 LCM, 4 spt

Air Force ε2,500

82 cbt ac, 20 armed hel
Flying hours 210
FTR/FGA 40 F/A-18 (-C 32, -D 8)
FTR 14 *Mirage* F1-CK/BK
CCT 1 sqn with 12 *Hawk* 64, 16 Shorts *Tucano*
TPT ac 3 L-100-30, 1 DC-9 **hel** 4 AS-332 (tpt/SAR/attack), 8 SA-330
TRG/ATK hel 16 SA-342 (with HOT)

AIR DEFENCE
4 *Hawk* Phase III bty with 24 launchers
6 bty *Amoun* (each bty, 1 *Skyguard* radar, 2 *Aspide* launchers, 2 twin **35mm** Oerlikon), 48 *Starburst*

Paramilitary 5,000 active

NATIONAL GUARD 5,000
3 gd, 1 armd car, 1 SF, 1 mil police bn; 20 VBL recce, 70 *Pandur* APC (incl variants)
COAST GUARD
4 *Inttisar* (Aust 31.5m) PFC, 3 LCU
Plus some 30 armed boats

Foreign Forces

UN (UNIKOM): some 908 tps and 203 obs from 30 countries
UK Air Force (Southern Watch): 12 Tornado-GR1/1A
US 5,190 **Army** 3,000; prepositioned eqpt for 1 armd bde (2 tk, 1 mech, 1 arty bn) **Air Force** 2,100 (Southern Watch); Force structure varies with aircraft detachments **Navy** 10 **USMC** 80

Lebanon RL

pound LP		1998	1999	2000	2001
GDP	LP	24.5tr	25.2tr		
	US$	16.2bn	16.7bn		
per capita	US$	4,900	4,900		
Growth	%	3.0	-1.0		
Inflation	%	2.0	1.0		
Debt	US$	6.2bn	7.5bn		
Def exp	LP	871bn			
	US$	575m			
Def bdgt	LP	901bn	846bn		
	US$	594m	560m		
FMA[a] (US)	US$	0.6m	0.6m	0.6m	0.6m
US$1=LP		1,515	1,512	1,512	

[a] UNIFIL **1997** US$121m **1998** US$143m

Population			**4,345,000**
(Christian 30%, Druze 6%, Armenian 4%, excl ε300,000 Syrian nationals and ε500,000 Palestinian refugees)			
Age	13–17	18–22	23–32
Men	213,000	195,000	391,000
Women	217,000	200,000	402,000

Total Armed Forces

ACTIVE 63,570 (incl 22,600 conscripts)
Terms of Service 1 year

Army 60,670 (incl conscripts)

5 regional comd
11 mech inf bde (-) • 1 Presidential Guard bde, 1 MP bde, 1 cdo/Ranger, 5 SF regt • 1 air aslt regt • 1 mne cdo regt • 2 arty regt
EQUIPMENT
MBT 115 M-48A1/A5, 212 T-54/-55
LT TK 36 AMX-13
RECCE 67 AML, 22 *Saladin*
APC 1,164 M-113A1/A2, 81 VAB-VCI, 81 AMX-VCI, 12 Panhard M3/VTT
TOWED ARTY 105mm: 13 M-101A1; **122mm**: 36 M-1938, 26 D-30; **130mm**: 11 M-46; **155mm**: 12 Model 50, 18 M-114A1, 35 M-198
MRL 122mm: 23 BM-21
MOR 81mm: 158; **82mm**: 111; **120mm**: 108
ATGW ENTAC, *Milan*, 20 BGM-71A TOW
RL 85mm: RPG-7; **89mm**: M-65
RCL 106mm: M-40A1
AD GUNS 20mm; 23mm: ZU-23; **40mm**: 10 M-42A1

Navy 1,200

BASES Jounieh, Beirut

PATROL AND COASTAL COMBATANTS 7

PATROL CRAFT, INSHORE 7

5 UK *Attacker* PCI<, 2 UK *Tracker* PCI<, plus 27 armed boats

AMPHIBIOUS 2

2 *Sour* (Fr *Edic*) LST, capacity 96 tps

Air Force 1,700

All ac grounded and in store

EQUIPMENT

FGA 6 *Hunter* F9, 10 *Mirage* EL3 reportedly sold to Pak

HEL 16 UH-1H, 1 SA-318, 3 SA-316, 5 Bell-212, 3 SA-330, 2 SA-342

TRG 5 CM-170, 3 *Bulldog*

Paramilitary ε13,000 active

INTERNAL SECURITY FORCE ε13,000 (Ministry of Interior)

(incl Regional and Beirut *Gendarmerie* coy plus Judicial Police); 30 *Chaimite* APC

CUSTOMS

2 *Tracker* PCI<, 5 *Aztec* PCI<

Opposition

MILITIAS

Most militias, except *Hizbollah*, have been substantially disbanded and hy wpn handed over to the National Army.

HIZBOLLAH ('Party of God'; Shi'a, fundamentalist, pro-Iranian): ε3–500 (-) active; about 3,000 in spt

EQUIPMENT arty, MRL, RL, RCL, ATGW (AT-3 *Sagger*, AT-4 *Spigot*), AA guns, SAM

Foreign Forces

UN (UNIFIL): 5,619; 7 inf bn, 1 each from **Fji**, **Gha**, **Ind**, **Irl**, **N**, **SF**, **Ukr**, plus spt units from **Fr**, **It**, **Pl**, **Swe**

IRAN ε150 Revolutionary Guard

SYRIA 22,000 **Beirut** elm 1 mech inf bde, 5 SF regt **Metn** elm 1 mech inf bde **Bekaa** 1 mech inf div HQ, elm 2 mech inf, elm 1 armd bde **Tripoli** 1 SF regt **Batrum** 1 SF Regt **Kpar Fallus** elm 3 SF regt

Libya — LAR

dinar D		1998	1999	2000	2001
GDP	US$	ε26bn	ε35bn		
per capita	US$	5,600	5,700		
Growth	%	ε-1.8	ε5.4		
Inflation	%	ε24	ε6.0		
Debt	US$	ε3.8bn	ε3.8bn		
Def exp	D	ε560m			
	US$	1.5bn			
Def bdgt	D	ε495m	ε580m	ε600m	
	US$	1.3bn	1.3bn	1.2bn	
US$1=D[a]		0.38	0.45	0.5	

[a] Market rate **2000** US$1=D2–3

Population			6,401,000
Age	13–17	18–22	23–32
Men	375,000	310,000	475,000
Women	360,000	300,000	456,000

Total Armed Forces

ACTIVE 76,000

(incl ε40,000 conscripts)

Terms of service selective conscription, 1–2 years

RESERVES some 40,000

People's Militia

Army 45,000

(ε25,000 conscripts)

11 Border Def and 4 Sy Zones • 5 elite bde (regime sy force) • 10 tk bn • 22 arty bn • 21 inf bn • 8 AD arty bn • 8 mech inf bn • 15 para/cdo bn • 5 SSM bde

EQUIPMENT

MBT 560 T-55, 280 T-62, 145 T-72 (plus some 1,040 T-54/-55, 70 T-62, 115 T-72 in store†)

RECCE 250 BRDM-2, 380 EE-9 *Cascavel*

AIFV 1,000 BMP-1

APC 750 BTR-50/-60, 100 OT-62/-64, 40 M-113, 100 EE-11 *Urutu*, some BMD

TOWED ARTY some 720: **105mm**: some 60 M-101; **122mm**: 270 D-30, 60 D-74; **130mm**: 330 M-46; **152mm**: M-1937

SP ARTY: 450: **122mm**: 130 2S1; **152mm**: 60 2S3, 80 DANA; **155mm**: 160 *Palmaria*, 20 M-109

MRL 107mm: Type 63; **122mm**: 350 BM-21/RM-70, 300 BM-11

MOR 82mm; 120mm: M-43; **160mm**: M-160

SSM launchers: 40 FROG-7, 80 *Scud*-B

ATGW 3,000: *Milan*, AT-3 *Sagger* (incl BRDM SP), AT-4 *Spigot*

RCL 84mm: *Carl Gustav*; **106mm**: 220 M-40A1

AD GUNS 600: **23mm**: ZU-23, ZSU-23-4 SP; **30mm**: M-53/59 SP

SAM SA-7/-9/-13, 24 quad *Crotale*

SURV RASIT (veh, arty)

Navy 8,000

(incl Coast Guard)

BASES Major: Tripoli, Benghazi, Tobruk, Khums
Minor: Derna, Zuwurah, Misonhah

SUBMARINES 2†

SSK 2 *Al Badr* † (Sov *Foxtrot*) with 533mm and 406mm TT (plus 2 non-op)

FRIGATES 2

FFG 2 *Al Hani* (Sov *Koni*) with 4 SS-N-2C *Styx* SSM, 4 ASTT, 2 ASW RL

PATROL AND COASTAL COMBATANTS 16

CORVETTES 3

3 *Ean al Gazala* (Sov *Nanuchka* II) FSG with 2 × 2 SS-N-2C *Styx* SSM

MISSILE CRAFT 13

7 *Sharaba* (Fr *Combattante* II) PFM with 4 *Otomat* SSM, 1 × 76mm gun (plus 2 non-op)

6 *Al Katum* (Sov *Osa* II) PFM with 4 SS-N-2C *Styx* SSM (plus 6 non-op)

MINE COUNTERMEASURES 6

6 *Ras al Gelais* (Sov *Natya*) MSO (plus 2 non-op)

(*El Temsah* and about 5 other ro-ro tpt have mine-laying capability)

AMPHIBIOUS 4

2 *Ibn Ouf* LST, capacity 240 tps, 11 tk, 1 SA-316B hel

2 Sov *Polnocny* LSM, capacity 180 tps, 6 tk (plus 1 non-op)

Plus craft: 3 LCT

SUPPORT AND MISCELLANEOUS 9

1 *El Temsah* tpt, about 5 other ro-ro tpt, 1 *Zeltin* log spt; 1 ARS, 1 diving spt

COASTAL DEFENCE

1 SSC-3 *Styx* bty

NAVAL AVIATION

32 armed hel

HEL 2 sqn

1 with 25 Mi-14 PL (ASW), 1 with 7 SA-321 (Air Force assets)

Air Force 23,000

(incl Air Defence Command; ε15,000 conscripts) 426 cbt ac, 52 armed hel (many ac in store, number n.k.) **Flying hours** 85

BBR 1 sqn with 6 Tu-22

FGA 13 sqn

12 with 40 MiG-23BN, 15 MiG-23U, 30 *Mirage* 5D/DE, 14 *Mirage* 5DD, 14 *Mirage* F-1AD, 12 Su-24 MK, 45 Su-20/-22

1 with 30 J-1 *Jastreb*

FTR 9 sqn with 50 MiG-21, 75 MiG-23, 60 MiG-25, 3 -25U, 15 *Mirage* F-1ED, 6 -BD

RECCE 2 sqn with 4* *Mirage* 5DR, 7* MiG-25R

TPT 9 sqn with 15 An-26, 12 Lockheed (7 C-130H, 2 L-100-20, 3 L-100-30), 16 G-222, 20 Il-76, 15 L-410

ATTACK HEL 40 Mi-25, 12 Mi-35

TPT HEL hy 18 CH-47C **med** 34 Mi-8/17 **lt** 30 Mi-2, 11 SA-316, 5 AB-206

TRG ac 80 *Galeb* G-2 **hel** 20 Mi-2 **other ac** incl 1 Tu-22, 150 L-39ZO, 20 SF-260WL

MISSILES

ASM AT-2 *Swatter* ATGW (hel-borne), AS-7, AS-9, AS-11

AAM AA-2 *Atoll*, AA-6 *Acrid*, AA-7 *Apex*, AA-8 *Aphid*, R-530, R-550 *Magic*

AIR DEFENCE COMMAND

Senezh AD comd and control system

4 bde with SA-5A: each 2 bn of 6 launchers, some 4 AD arty gun bn; radar coy

5 Regions: 5–6 bde each 18 SA-2; 2–3 bde each 12 twin SA-3; ε3 bde each 20–24 SA-6/-8

Forces Abroad

UN AND PEACEKEEPING

DROC (MONUC): 4 obs

Paramilitary

CUSTOMS/COAST GUARD (Naval control)

a few patrol craft incl in naval totals, plus armed boats

Mauritania RIM

Mauritanian ougiya OM

		1998	1999	2000	2001
GDP	OM	188bn	203bn		
	US$	1.2bn	1.0bn		
per capita	US$	1,800	1,900		
Growth	%	3.9	4.1		
Inflation	%	7.1	4.0		
Debt	US$	2.5bn			
Def exp	OM	4.7bn			
	US$	26m			
Def bdgt	OM	ε4.7bn	ε5.4bn	ε5.7bn	
	US$	25.5m	26m	23.6m	
FMA (Fr)	US$	1.0m	1.2m		
US$1=OM		184	207	242	

Population			**2,609,000**
Age	13–17	18–22	23–32
Men	145,000	119,000	189,000
Women	142,000	114,000	184,000

Total Armed Forces

ACTIVE ε15,650

Terms of service conscription 24 months authorised

Army 15,000

6 Mil Regions • 7 mot inf bn • 8 inf bn • 1 para/cdo bn • 1 Presidential sy bn • 2 Camel Corps bn • 3 arty bn • 4 AD arty bty • 1 engr coy • 1 armd recce sqn

EQUIPMENT

MBT 35 T-54/-55
RECCE 60 AML (20 -60, 40 -90), 40 *Saladin*, 5 *Saracen*
TOWED ARTY 105mm: 35 M-101A1/HM-2; **122mm**: 20 D-30, 20 D-74
MOR 81mm: 70; **120mm**: 30
ATGW *Milan*
RCL 75mm: M-20; **106mm**: M-40A1
AD GUNS 23mm: 20 ZU-23-2; **37mm**: 15 M-1939; **57mm**: S-60; **100mm**: 12 KS-19
SAM SA-7

Navy ε500

BASES Nouadhibou, Nouakchott
PATROL CRAFT 7
OFFSHORE 2
1 *Aboubekr Ben Amer* (Fr OPV 54) PCO
1 *N'Madi* (UK *Jura*) PCO (fishery protection)
COASTAL 1
1 *El Nasr* (Fr *Patra*) PCC
INSHORE 4
4 *Mandovi* PCI<

Air Force 150

7 cbt ac, no armed hel
CCT 5 BN-2 *Defender*, 2 FTB-337 *Milirole*
MR 2 *Cheyenne* II
TPT 2 Cessna F-337, 1 DHC-5D, 1 *Gulfstream* II, 2 Y-12 (II)

Paramilitary ε5,000 active

GENDARMERIE (Ministry of Interior) ε3,000
6 regional coy
NATIONAL GUARD (Ministry of Interior) 2,000
plus 1,000 auxiliaries
CUSTOMS
1 *Dah Ould Bah* (Fr *Amgram* 14)

Morocco Mor

dirham D		1998	1999	2000	2001
GDP	D	347bn	352bn		
	US$	36bn	35bn		
per capita	US$	3,800	3,900		
Growth	%	6.7	0.6		
Inflation	%	2.7	0.7		
Debt	US$	23bn	18bn		
Def exp	D	16bn	17.5bn		
	US$	1.7bn	1.8bn		
Def bdgt	D	16.0bn	17.3bn		
	US$	1.6bn	1.7bn		
FMA[a] (US)	US$	0.9m	2.9m	3.2m	3.0m
US$1=D		9.60	9.94	10.5	

[a] MINURSO **1997** US$29m **1998** US$23m

Population			30,355,000
Age	13–17	18–22	23–32
Men	1,750,000	1,583,000	2,698,000
Women	1,693,000	1,531,000	2,570,000

Total Armed Forces

ACTIVE 198,500
(incl ε100,000 conscripts)
Terms of service conscription 18 months authorised; most enlisted personnel are volunteers

RESERVES
Army 150,000; obligation to age 50

Army 175,000

(ε100,000 conscripts)
2 Comd (Northern Zone, Southern Zone) • 3 mech inf bde • 1 lt sy bde • 2 para bde • 8 mech inf regt • Indep units
11 armd bn • 2 cav bn • 39 inf bn • 1 mtn inf bn • 2 para bn • 3 mot (camel corps) bn • 9 arty bn • 7 engr bn • 1 AD gp • 7 cdo units

ROYAL GUARD 1,500
1 bn, 1 cav sqn

EQUIPMENT

MBT 224 M-48A5, 420 M-60 (300 -A1, 120 -A3)
LT TK 100 SK-105 *Kuerassier*
RECCE 16 EBR-75, 80 AMX-10RC, 190 AML-90, 38 AML-60-7, 20 M-113
AIFV 60 *Ratel* (30 -20, 30 -90), 45 VAB-VCI, 10 AMX-10P
APC 420 M-113, 320 VAB-VTT, some 45 OT-62/-64 may be op
TOWED ARTY 105mm: 35 L-118, 20 M-101, 36 M-1950; **130mm**: 18 M-46; **155mm**: 20 M-114, 35 FH-70, 26 M-198
SP ARTY 105mm: 5 Mk 61; **155mm**: 126 F-3, 44 M-109, 20 M-44; **203mm**: 60 M-110
MRL 122mm: 39 BM-21
MOR 81mm: 850; **120mm**: 600 (incl 20 VAB SP)
ATGW 440 *Dragon*, 80 *Milan*, 150 TOW (incl 80 on M-901), 50 AT-3 *Sagger*
RL 89mm: 150 3.5in M-20
RCL 106mm: 350 M-40A1
ATK GUNS 90mm: 28 M-56; **100mm**: 8 SU-100 SP
AD GUNS 14.5mm: 200 ZPU-2, 20 ZPU-4; **20mm**: 40 M-167, 60 M-163 *Vulcan* SP; **23mm**: 90 ZU-23-2; **100mm**: 15 KS-19 towed
SAM 37 M-54 SP *Chaparral*, 70 SA-7
SURV RASIT (veh, arty)
UAV R4E-50 *Skyeye*

Navy 10,000

(incl 1,500 Marines)

BASES Casablanca, Agadir, Al Hoceima, Dakhla, Tangier

FRIGATES 1 *Lt Col. Errhamani* (Sp *Descubierta*) FFG with *Aspide* SAM, 1 × 76mm gun, 2 × 3 ASTT (Mk 46 LWT), 1 × 2 375mm AS mor (fitted for 4 MM-38 *Exocet* SSM)

PATROL AND COASTAL COMBATANTS 27

MISSILE CRAFT 4 *Cdt El Khattabi* (Sp *Lazaga* 58m) PFM with 4 MM-38 *Exocet* SSM, 1 × 76mm gun

PATROL CRAFT 23

COASTAL 17

2 *Okba* (Fr PR-72) PCC with 1 × 76mm gun

6 *LV Rabhi* (Sp 58m B-200D) PCC

4 *El Hahiq* (Dk *Osprey* 55) PCC (incl 2 with customs)

5 *Rais Bargach* (navy marine for fisheries dept)

INSHORE 6 *El Wacil* (Fr P-32) PFI< (incl 4 with customs)

AMPHIBIOUS 4

3 *Ben Aicha* (Fr *Champlain* BATRAL) LSM, capacity 140 tps, 7 tk

1 *Sidi Mohammed Ben Abdallah* (US Newport) LST, capacity 400 troops

Plus craft: 1 *Edic*-type LCT

SUPPORT AND MISCELLANEOUS 4

2 log spt, 1 tpt, 1 AGOR (US lease)

MARINES (1,500)

2 naval inf bn

Air Force 13,500

89 cbt ac, 24 armed hel

Flying hours F-5 and *Mirage*: over 100

FGA 10 F-5A, 3 F-5B, 16 F-5E, 4 F-5F, 14 *Mirage* F-1EH

FTR 1 sqn with 15 *Mirage* F-1CH

RECCE 2 C-130H (with side-looking radar), 4* 0V-10

EW 2 C-130 (ELINT), 1 *Falcon* 20 (ELINT)

TKR 1 Boeing 707, 2 KC-130H (tpt/tkr)

TPT 11 C-130H, 7 CN-235, 3 Do-28, 3 *Falcon 20*, 1 *Falcon* 50 (VIP), 2 *Gulfstream* II (VIP), 5 *King Air* 100, 3 *King Air* 200

HELICOPTERS

ATTACK 24 SA-342 (12 with HOT, 12 with cannon)

TPT hy 7 CH-47 **med** 27 SA-330, 27 AB-205A **lt** 20 AB-206, 3 AB-212

TRG 10 AS-202, 2 CAP-10, 4 CAP-230, 12 T-34C, 23* *Alpha Jet*

LIAISON 2 *King Air 200*, 2 UH-60 *Blackhawk*

AAM AIM-9B/D/J *Sidewinder*, R-530, R-550 *Magic*

ASM AGM-65B *Maverick* (for F-5E), HOT

Forces Abroad

UN AND PEACEKEEPING

BOSNIA (SFOR II): ε800; 1 mot inf bn **DROC** (MONUC): 4 obs **YUGOSLAVIA** (KFOR): 279

Paramilitary 42,000 active

GENDARMERIE ROYALE 12,000

1 bde, 4 mobile gp, 1 para sqn, air sqn, coast guard unit

EQPT 18 boats **ac** 2 *Rallye* **hel** 3 SA-315, 3 SA-316, 2 SA-318, 6 *Gazelle*, 6 SA-330, 2 SA-360

FORCE AUXILIAIRE 30,000

incl 5,000 Mobile Intervention Corps

CUSTOMS/COAST GUARD

4 *Erraid* PCI, 32 boats, 3 SAR craft

Opposition

POLISARIO ε3–6,000

Mil wing of Sahrawi People's Liberation Army, org in bn

EQPT 100 T-55, T-62 tk; 50+ BMP-1, 20–30 EE-9 *Cascavel* MICV; 25 D-30/M-30 **122mm** how; 15 BM-21 **122mm** MRL; 20 **120mm** mor; AT-3 *Sagger* ATGW; 50 ZSU-23-2, ZSU-23-4 **23mm** SP AA guns; SA-6/-7/-8/-9 SAM (Captured Moroccan eqpt incl AML-90, *Eland* armd recce, *Ratel* 20, Panhard APC, Steyr SK-105 *Kuerassier* lt tks)

Foreign Forces

UN (MINURSO): some 27 tps, 204 mil obs in Western Sahara from 25 countries

Oman O

rial R		1998	1999	2000	2001
GDP	R	5.5bn	5.8bn		
	US$	14.2bn	15.0bn		
per capita	US$	8,300	8,600		
Growth	%	-7.0	5.5		
Inflation	%	-5.7	0.3		
Debt	US$	3.9bn	3.9bn		
Def exp	R	675m	627m		
	US$	1.8bn	1.6bn		
Def bdgt[a]	R	655m	613m	673m	
	US$	1.7bn	1.6bn	1.75bn	
FMA[b] (US)	US$	0.2m	0.2m	0.3m	0.25m
US$1=R		0.38	0.38	0.38	

[a] Five-year plan 1996–2000 allocates R3.3bn (US$8.6bn) for defence

[b] Excl εUS$100m over 1990–99 from US Access Agreement renewed in 1990

Population		**2,296,000** (expatriates 27%)	
Age	13–17	18–22	23–32
Men	131,000	106,000	154,000
Women	127,000	103,000	144,000

Total Armed Forces

ACTIVE 43,500

(incl Royal Household tps, and some 3,700 foreign personnel)

Army 25,000

(regt are bn size)
1 armd, 2 inf bde HQ • 2 armd regt (3 tk sqn) • 1 armd recce regt (3 sqn) • 4 arty (2 fd, 1 med (2 bty), 1 AD (2 bty)) regt • 1 inf recce regt (3 recce coy), 2 indep recce coy • 1 fd engr regt (3 sqn) • 1 AB regt • Musandam Security Force (indep rifle coy)

EQUIPMENT

MBT 6 M-60A1, 73 M-60A3, 38 *Challenger* 2
LT TK 37 *Scorpion*
RECCE 41 VBL
APC 6 *Spartan*, 13 *Sultan*, 4 *Stormer*, 160 *Piranha*
TOWED ARTY 91: **105mm**: 42 ROF lt; **122mm**: 30 D-30; **130mm**: 12 M-46, 12 Type 59-1
SP ARTY 155mm: 24 G-6
MOR 81mm: 69; **107mm**: 20 4.2in M-30
ATGW 18 TOW, 50 *Milan* (incl 2 VCAC)
AD GUNS 23mm: 4 ZU-23-2; **35mm**: 10 GDF-005 with *Skyguard*; **40mm**: 12 Bofors L/60
SAM *Blowpipe*, 28 *Javelin*, 34 SA-7

Navy 4,200

BASES Seeb (HQ), Wudam (main base), Salalah, Ghanam Island, Alwi

PATROL AND COASTAL COMBATANTS 13

CORVETTES 2 *Qahir Al Amwaj* FSG with 8 MM-40 *Exocet* SSM, 8 *Crotale* SAM, 1 76mm gun, 6 × 324mm TT, hel deck)
MISSILE CRAFT 4 *Dhofar* PFM, 1 with 2 × 3 MM-40 *Exocet* SSM, 3 with 2 × 4 MM-40 *Exocet* SSM
PATROL CRAFT, COASTAL/INSHORE 7
3 *Al Bushra* (Fr P-400) PCC with 1 × 76m gun, 4 × 406mm TT
4 *Seeb* (Vosper 25m) PCI<

AMPHIBIOUS 1

1 *Nasr el Bahr* LST†, capacity 240 tps, 7 tk, hel deck
Plus craft: 3 LCM, 1 LCU

SUPPORT AND MISCELLANEOUS 4

1 *Al Sultana* AK, 1 *Al Mabrukah* trg with hel deck (also used in offshore patrol role), 1 supply, 1 AGHS

Air Force 4,100

40 cbt ac, no armed hel
FGA 2 sqn, each with 8 *Jaguar* S(O) Mk 1, 4 T-2 (being progressively upgraded to (S01) GR-3 standard)
FGA/RECCE 12 *Hawk* 203
CCT 1 sqn with 12* PC-9, 4* *Hawk* 103
TPT 3 sqn
1 with 3 BAC-111
2 with 15 *Skyvan* 3M (7 radar-equipped, for MR), 3 C-130H
HEL 2 med tpt sqn with 19 AB-205, 3 AB-206, 3 AB-212, 5 AB-214
TRG 4 AS-202-18, 7 MFI-17B *Mushshak*
AD 2 sqn with 28 *Rapier* SAM, *Martello* radar
AAM AIM-9M *Sidewinder*

Royal Household 6,500

(incl HQ staff) 2 SF regt (1,000)
Royal Guard bde (5,000) 9 VBC-90 lt tk, 14 VAB-VCI APC, 9 VAB-VDAA, *Javelin* SAM
Royal Yacht Squadron (based Muscat) (150) 1 Royal Yacht *Al Said*, 3,800t with hel deck, 1 *Fulk Al Salamah* tps and veh tpt with up to 2 AS-332C *Puma* hel, 1 *Zinat Al Bihaar Dhow*
Royal Flight (250) **ac** 2 Boeing-747 SP, 1 DC-8-73CF, 2 *Gulfstream* IV **hel** 3 AS-330, 2 AS-332C, 1 AS-332L

Paramilitary 4,400 active

TRIBAL HOME GUARD (*Firqat*) 4,000
org in teams of ε100
POLICE COAST GUARD 400
3 CG 29 PCI, plus 14 craft
POLICE AIR WING
ac 1 Do-228, 2 CN 235M, 1 BN-2T Islander **hel** 3 Bell 205A, 6 Bell 214ST

Foreign Forces

US 690 **Air Force** 630 **Navy** 60

Palestinian Autonomous Areas of Gaza and Jericho GzJ

		1998	1999	2000	2001
GDP	US$	ε3.5bn			
per capita	US$	1,600			
Growth	%	ε2.0			
Inflation	%	ε7.0			
Debt	US$	ε500m	ε479m		
Sy bdgt	US$	ε300m	ε500m		
FMA (US)	US$	85m	100m	100m	

Population **ε3,000,000**

(*West Bank and Gaza excluding East Jerusalem* ε2,900,000 (Israeli ε180,000 excl East Jerusalem) *Gaza* ε1,200,000 (Israeli ε6,100) *West Bank excl East Jerusalem* ε1,700,000 (Israeli ε174,000) *East Jerusalem* Israeli ε217,000, Palestinian ε86–200,000)

Age	13–17	18–22	23–32
Men	163,000	140,000	233,000
Women	158,000	134,000	222,000

Total Armed Forces

ACTIVE Nil

Paramilitary ε35,000

PUBLIC SECURITY 6,000 Gaza, 8,000 West Bank

CIVIL POLICE 4,000 Gaza, 6,000 West Bank

PREVENTIVE SECURITY 1,200 Gaza, 1,800 West Bank

GENERAL INTELLIGENCE 3,000

MILITARY INTELLIGENCE 500

PRESIDENTIAL SECURITY 1,000

Others include **Coastal Police, Civil Defence, Air Force, Customs and Excise Police Force, University Security Service**

EQPT incl small arms, 45 APC **ac** 1 Lockheed *Jet Star* **hel** 2 Mi-8, 2 Mi-17

PALESTINIAN GROUPS

All significant Palestinian factions are listed irrespective of where they are based. Est number of active 'fighters' are given; these could perhaps be doubled to give an all-told figure. In 1991, the Lebanon Armed Forces (LAF), backed by Syria, entered refugee camps in southern Lebanon to disarm many Palestinian groups of their heavier weapons, such as tk, arty and APCs. The LAF conducted further disarming operations against *Fatah* Revolutionary Council (FRC) refugee camps in spring 1994.

PLO (Palestine Liberation Organisation) **Leader** Yasser Arafat

FATAH Political wing of the PLO

PNLA (Palestine National Liberation Army) ε2,000 Effectively mil wing of the PLO **Based** Ag, Et, RL, LAR, HKJ, Irq, Sdn, Ye. Units closely monitored by host nations' armed forces.

PLF (Palestine Liberation Front) ε300–400 **Leader** Al Abas; **Based** Irq

DFLP (Democratic Front for the Liberation of Palestine) ε100 **Leader** Hawatmah; **Based** Syr, RL, elsewhere **Abd Rabbu faction** ε150–200 **Based** HKJ

PFLP (Popular Front for the Liberation of Palestine) ε100 **Leader** n.k.; **Based** Syr, RL, Occupied Territories

PSF (Popular Struggle Front) ε50 **Leader** Samir Ghansha; **Based** Syr

ARAB LIBERATION FRONT ε300 **Based** RL, Irq

GROUPS OPPOSED TO THE PLO

***FATAH* DISSIDENTS** (Abu Musa gp) ε1,000 **Based** Syr, RL

FRC (*Fatah* Revolutionary Council, Abu Nidal Group) ε300 **Based** RL, Syr, Irq, elsewhere

PFLP (GC) (Popular Front for the Liberation of Palestine (General Command)) ε300 **Leader** Jibril

PFLP (SC) (Popular Front for the Liberation of Palestine – Special Command) str n.k. **Based** RL, Irq, Syr

SAIQA ε300 **Leader** al-Khadi; **Based** Syr

HAMAS ε500 **Based** Occupied Territories

PIJ (Palestine Islamic *Jihad*) ε500 all factions **Based** Occupied Territories

PALESTINE LIBERATION FRONT Abd al-Fatah Ghanim faction **Based** Syr

PLA (Palestine Liberation Army) ε2,000 **Based** Syr

Qatar Q

rial R		1998	1999	2000	2001
GDP	R	32bn	35bn		
	US$	8.7bn	9.7bn		
per capita	US$	14,300	15,300		
Growth	%	-7.4	6.5		
Inflation	%	2.6	2.0		
Debt	US$	10.0bn	12.2bn		
Def exp	R	ε4.9bn	ε5.1bn		
	US$	1.3bn	1.4bn		
Def bdgt	R	ε4.5bn	ε4.7bn		
	US$	1.2bn	1.3bn		
US$1=R		3.64	3.64	3.64	

Population 691,000

(nationals 25%, expatriates 75%, of which Indian 18%, Iranian 10%, Pakistani 18%)

Age	13–17	18–22	23–32
Men	25,000	21,000	35,000
Women	28,000	23,000	33,000

Total Armed Forces

ACTIVE ε12,330

Army 8,500

1 Royal Guard regt • 1 tk bn • 4 mech inf bn • 1 fd arty regt • 1 mor bn • 1 SF 'bn' (coy)

EQUIPMENT

MBT 44 AMX-30

RECCE 16 VBL, 12 AMX-10RC, 8 V-150

AIFV 40 AMX-10P

LAV 36 *Piranha* II

APC 160 VAB, 12 AMX-VCI

TOWED ARTY 155mm: 12 G5

SP ARTY 155mm: 28 F-3

MRL 4 ASTROS II

MOR 81mm: 24 L16 (some SP); **120mm**: 15 Brandt

ATGW 100 *Milan*, HOT (incl 24 VAB SP)

RCL 84mm: *Carl Gustav*

Navy ε1,730

(incl Marine Police)

BASE Doha

PATROL AND COASTAL COMBATANTS 7

MISSILE CRAFT 7

3 *Damsah* (Fr *Combattante* III) PFM with 2 × 4 MM-40

Exocet SSM

4 *Barzan* (UK *Vita*) PFM with 8 *Exocet* SSM, 6 *Mistral* SAM, 1 × 76mm gun

Plus some 40 small craft operated by Marine Police

COASTAL DEFENCE

4 × 3 *quad* MM-40 *Exocet* SSM bty

Air Force 2,100

18 cbt ac, 19 armed hel

FGA/FTR 2 sqn

1 with 6 *Alpha* jets

1 with 12 *Mirage* 2000-5 (9 EDA, 3 DDA)

TPT 1 sqn with 2 Boeing 707, 1 Boeing 727, 2 *Falcon* 900, 1 *Airbus* A340

ATTACK HEL 11 SA-342L (with HOT), 8 *Commando* Mk 3 (*Exocet*)

TPT 4 *Commando* (3 Mk 2A tpt, 1 Mk 2C VIP)

MISSILES

ASM *Exocet* AM-39, *HOT*, *Apache*

AAM MATRA R550 *Magic*, MATRA *Mica*

SAM 9 *Roland* 2, 24 *Mistral*, 12 *Stinger*, 20 SA-7 *Grail*

Foreign Forces

US Army 30; prepositioned eqpt for 1 armd bde (forming)

Saudi Arabia Sau

rial R		1998	1999	2000	2001
GDP	R	482bn	529bn		
	US$	128bn	141bn		
per capita	US$	8,500	9,400		
Growth	%	-10.8	9.0		
Inflation	%	-0.2	-1.2		
Debt	US$	25bn	26bn		
Def exp	R	78bn	81bn		
	US$	20.9bn	21.8bn		
Def bdgt	R	78bn	69bn	70bn	
	US$	20.9bn	18.4bn	18.7bn	
US$1=R		3.75	3.75	3.75	

Population 21,661,000

(nationals 73%, of which Bedouin up to 10%, Shi'a 6%, expatriates 27%, of which Asians 20%, Arabs 6%, Africans 1% and Europeans <1%)

Age	13–17	18–22	23–32
Men	1,348,000	1,133,000	1,670,000
Women	1,207,000	1,010,000	1,435,000

Total Armed Forces

ACTIVE ε126,500

(plus 75,000 active National Guard)

Army 75,000

3 armd bde (each 3 tk, 1 mech, 1 fd arty, 1 recce, 1 AD, 1 ATK bn) • 5 mech bde (each 3 mech, 1 tk, 1 fd arty, 1 AD, 1 spt bn) • 1 AB bde (2 AB bn, 3 SF coy) • 1 Royal Guard regt (3 bn) • 8 arty bn • 1 army avn comd

EQUIPMENT

MBT 315 M-1A2 *Abrams* (ε200 in store), 290 AMX-30 (50% in store), 450 M60A3

RECCE 300 AML-60/-90

AIFV 570+ AMX-10P, 400 M-2 *Bradley*

APC 1,750 M-113 A1/A2/A3 (incl variants), 150 Panhard M-3

TOWED ARTY 105mm: 100 M-101/-102; **155mm**: 50 FH-70 (in store), 90 M-198, M-114; **203mm**: 8 M-115 (in store)

SP ARTY 155mm: 110 M-109A1B/A2, 90 GCT

MRL 60 ASTROS II

MOR 400, incl: **107mm**: 4.2in M-30; **120mm**: 110 Brandt

SSM some 10 PRC CSS-2 (40 msl)

ATGW TOW-2 (incl 200 VCC-1 SP), 1,000 M-47 *Dragon*, HOT (incl 100 AMX-10P SP)

RCL 84mm: 300 *Carl Gustav*; **90mm**: 100 M-67; **106mm**: 50 M-40A1

HEL 12 AH-64, 12 S-70A-1, 22 UH-60 (tpt, 4 medevac), 6 SA-365N (medevac), 15 Bell 406CS

SAM *Crotale*, *Stinger*, 500 *Redeye* **SURV** AN/TPQ-36/-37 (arty, mor)

Navy 15,500

(incl 3,000 Marines)

BASES Riyadh (HQ Naval Forces) **Western Fleet** Jiddah (HQ), Yanbu **Eastern Fleet** Al-Jubayl (HQ), Ad-Dammam, Ras al Mishab, Ras al Ghar, Jubail

FRIGATES 4

FFG 4

4 *Madina* (Fr F-2000) with 8 *Otomat* 2 SSM, 8 *Crotale* SAM, 1 × 100mm gun, 4 × 533mm ASTT, 1 SA 365F hel

CORVETTES 4

4 *Badr* (US *Tacoma*) FSG with 2 × 4 *Harpoon* SSM, 1 × 76mm gun, 2 × 3 ASTT (Mk 46 LWT)

PATROL AND COASTAL COMBATANTS 26

MISSILE CRAFT 9 *Al Siddiq* (US 58m) PFM with 2 × 2 *Harpoon* SSM, 1 × 76mm gun

PATROL CRAFT 17 US *Halter Marine* PCI< (some with Coast Guard) plus 40 craft

MINE COUNTERMEASURES 7

3 *Al Jawf* (UK *Sandown*) MHO

4 *Addriyah* (US *MSC-322*) MCC†

AMPHIBIOUS (craft only)

4 LCU, 4 LCM

SUPPORT AND MISCELLANEOUS 7

2 *Boraida* (mod Fr *Durance*) AO with 1 or 2 hel, 3 AT/F, 1 ARS, 1 Royal Yacht with hel deck

NAVAL AVIATION

21 armed hel
19 AS-565 (4 SAR, 15 with AS-15TT ASM), 12 AS-332B/F (6 tpt, 6 with AM-39 *Exocet*)

MARINES (3,000)

1 inf regt (2 bn) with 140 BMR-600P

Air Force 20,000

417 cbt ac
FGA 7 sqn
3 with 56 F-5E, 21 F-5F
3 with 76 *Tornado* IDS
FTR 1 sqn with 24 *Tornado* ADV, 5 sqn with 70 F-15C/24 F-15D, 3 sqn with 72 F-15S
RECCE 1 sqn with 10* RF-5E (plus 10 *Tornado* GR.1A recce configuration on FGA sqn)
AEW 1 sqn with 5 E-3A
TKR 8 KE-3A, 8 KC-130H (tkr/tpt)
OCU 2 sqn with 14* F-5B
TPT 3 sqn with 36 C-130 (7 -E, 29 -H), 3 L-100-30HS (hospital ac)
HEL 2 sqn with 22 AB-205, 13 AB-206A, 17 AB-212, 40 AB-41EP (SAR), 12 AS-532A2 (CSAR)
TRG 3 sqn with 30* *Hawk* Mk 65, 2 sqn with 20* *Hawk* Mk 65A, 2 sqn with 50 PC-9, 1 sqn with 1 *Jetstream* 31, 1 sqn with 13 Cessna 172
ROYAL FLT ac 1 Boeing-747SP, 1 Boeing-737-200, 4 BAe 125–800, 2 C-140, 4 CN-235, 2 *Gulfstream* III, 2 *Learjet* 35, 6 VC-130H, 1 Cessna 310 **hel** 3 AS-61, AB-212, 1 S-70

MISSILES

ASM AGM-65 *Maverick*, AS-15, AS-30, *Sea Eagle*, *Shrike* AGM-45, ALARM
AAM AIM-9J/L/P *Sidewinder*, AIM-7F *Sparrow*, *Skyflash*

Air Defence Forces 16,000

33 SAM bty
16 with 128 I HAWK
17 with 68 *Shahine* fire units and AMX-30SA 30mm SP AA guns
73 *Shahine/Crotale* fire units as static defence

EQUIPMENT

AD GUNS 20mm: 92 M-163 *Vulcan*; **30mm**: 50 AMX-30SA; **35mm**: 128; **40mm**: 150 L/70 (in store)
SAM 141 *Shahine*, 128 MIM-23B I HAWK, 40 *Crotale*

National Guard 100,000

(75,000 active, 25,000 tribal levies)
3 mech inf bde, each 4 all arms bn
5 inf bde
1 ceremonial cav sqn

EQUIPMENT

LAV 1,117 LAV (incl 384 LAV-25, 182 LAV-CP, 130 LAV-AG, 111 LAV-AT, 73 LAV-M, 47 LAV plus 190 spt vehs)
APC 290 V-150 *Commando* (plus 810 in store), 440 *Piranha*
TOWED ARTY 105mm: 40 M-102; **155mm**: 30 M-198
MOR 81mm; 120mm: incl 73 on LAV-M
RCL 106mm: M-40A1
ATGW TOW incl 111 on LAV

Paramilitary 15,500+ active

FRONTIER FORCE 10,500

COAST GUARD 4,500

EQPT 4 *Al Jouf* PFI, about 30 PCI<, 16 hovercraft, 1 trg, 1 Royal Yacht (5,000t) with 1 Bell 206B hel, about 350 armed boats

GENERAL CIVIL DEFENCE ADMINISTRATION UNITS

10 KV-107 **hel**

SPECIAL SECURITY FORCE 500

UR-416 APC

Foreign Forces

PENINSULAR SHIELD FORCE ε7,000

1 inf bde (elm from all GCC states)
FRANCE (Southern Watch): 170; 5 *Mirage* 2000C,3 F-1CR, 3 C 135FR
UK (Southern Watch): ε200; 6 *Tornado* GR-1A
US 5,720 **Army** 650 incl 1 *Patriot* SAM, 1 sigs unit and those on short-term duty (6 months) **Air Force** (Southern Watch) 4,800; units on rotational det, numbers vary (incl: F-15, F-16, F-117, C-130, KC-135, U-2, E-3) **Navy** 20 **USMC** 250

Syria — Syr

pound S£		1998	1999	2000	2001
GDP	S£	755bn	802bn		
	US$	37bn	39bn		
per capita	US$	6,100	6,100		
Growth	%	-3.3	2.0		
Inflation	%	1.0	2.5		
Debt	US$	21bn	16bn		
Def exp	S£	44bn	45bn		
	US$	1.7bn	1.9bn		
Def bdgt	S£	39bn	39bn	42bn	
	US$	1.7bn	1.7bn	1.8bn	
US$1=S£[a]	11.2	11.2	11.2		

[a] Market rate **2000** US$1=S£58

Population			**16,934,000**
Age	13–17	18–22	23–32
Men	1,042,000	853,000	1,242,000
Women	1,003,000	828,000	1,210,000

Total Armed Forces

ACTIVE ε316,000

Terms of service conscription, 30 months

RESERVES (to age 45) 396,000

Army 300,000 **Navy** 4,000 **Air Force** 92,000

Army ε215,000

(incl conscripts)

3 corps HQ • 7 armd div (each 3 armd, 1 mech bde, 1 arty regt) • 3 mech div (-) (each 2 armd, 2 mech bde, 1 arty regt) • 1 Republican Guard div (3 armd, 1 mech bde, 1 arty regt) • 1 SF div (3 SF regt) • 4 indep inf bde • 1 Border Guard bde • 2 indep arty bde • 2 indep ATK bde • 1 indep tk regt • 10 indep SF regt • 3 SSM bde (each of 3 bn): 1 with FROG, 1 with *Scud*-B/-C, 1 with SS-21 • 1 coastal def SSM bde with SS-C-1B *Sepal* and SS-C-3 *Styx*

RESERVES

1 armd div HQ, 4 armd bde, 2 armd regt

31 inf, 3 arty regt

EQUIPMENT

MBT 4,850 (incl some 1,200 in static positions and in store): 2,150 T-55/MV, 1,000 T-62M/K, 1,700 T-72/-72M

RECCE 850 BRDM-2, 85 BRDM-2 Rkh

AIFV 2,250 BMP-1, 100 BMP-2, BMP-3

APC 1,500 BTR-50/-60/-70/-152

TOWED ARTY some 1,600, incl: **122mm**: 100 M-1931/-37 (in store), 150 M-1938, 500 D-30; **130mm**: 800 M-46; **152mm**: 20 D-20, 50 M-1937; **180mm**: 10 S23

SP ARTY 122mm: 400 2S1; **152mm**: 50 2S3

MRL 107mm: 200 Type-63; **122mm**: 280 BM-21

MOR 82mm: 200; **120mm**: 350 M-1943; **160mm**: 100 M-160; **240mm**: ε8 M-240

SSM launchers: 18 FROG-7, some 18 SS-21, 26 *Scud*-B/-C; 4 SS-C-1B *Sepal*, 6 SS-C-3 *Styx* coastal

ATGW 3,500 AT-3 *Sagger* (incl 2,500 SP), 150 AT-4 *Spigot*, 200 AT-5 *Spandrel*, AT-7 *Saxhorn*, 2,000 AT-10, AT-14 *Kornet* and 200 *Milan*

AD GUNS 2,060: **23mm**: 650 ZU-23-2 towed, 400 ZSU-23-4 SP; **37mm**: 300 M-1939; **57mm**: 675 S-60, 10 ZSU-57-2 SP; **100mm**: 25 KS-19

SAM 4,000 SA-7, 20 SA-9, 35 SA-13

Navy ε6,000

BASES Latakia, Tartus, Minet el-Baida

FRIGATES 2

FF 2 Sov *Petya* II with 5 × 533mm TT, 4 ASW RL

PATROL AND COASTAL COMBATANTS 18

MISSILE CRAFT 10

10 Sov *Osa* I and II PFM with 4 SS-N-2 *Styx* SSM

PATROL CRAFT, INSHORE 8

8 Sov *Zhuk* PFI<

MINE COUNTERMEASURES 5

1 Sov T-43 MSO, 1 *Sonya* MSC, 3 *Yevgenya* MSI

AMPHIBIOUS 3

3 *Polnocny* LSM, capacity 100 tps, 5 tk

SUPPORT AND MISCELLANEOUS 4

1 spt, 1 trg, 1 div spt, 1 AGOR

NAVAL AVIATION

24 armed hel

ASW 20 Mi-14, 4 Ka-28 (Air Force manpower)

Air Force 40,000

589 cbt ac; 87 armed hel (some may be in store)

Flying hours 30

FGA 9/10 sqn

5 with 90 Su-22, 2 with 44 MiG-23 BN, 2 with 20 Su-24, 1 possibly forming with Su-27

FTR 17 sqn

8 with 170 MiG-21, 5 with 90 MiG-23, 2 with 30 MiG-25, 2 with 20 MiG-29

RECCE 6* MiG-25R, 8* MiG-21H/J

TPT ac 5 An-26, 2 *Falcon* 20, 4 Il-76, 7 Yak-40, 1 *Falcon* 900, 6 Tu-134 **hel** 10 Mi-2, 100 Mi-8/-17

ATTACK HEL 48 Mi-25, 39 SA-342L

TRG incl 80* L-39, 20 MBB-223, 20* MiG-21U, 6* MiG-23UM, 5* MiG-25U, 6 *Mashshak*

MISSILES

ASM AT-2 *Swatter*, AS-7 *Kerry*, AS-12, HOT

AAM AA-2 *Atoll*, AA-6 *Acrid*, AA-7 *Apex*, AA-8 *Aphid*, AA-10 *Alamo*

Air Defence Command ε55,000

25 AD bde (some 150 SAM bty)

Some 600 SA-2/-3, 200 SA-6 and 4,000 AD arty

2 SAM regt (each 2 bn of 2 bty) with some 48 SA-5, 60 SA-8, S-300 on order

Forces Abroad

LEBANON 22,000; 1 mech div HQ, elm 1 armd, 4 mech inf bde, elm 10 SF, 2 arty regt

Paramilitary ε108,000

GENDARMERIE 8,000 (Ministry of Interior)

WORKERS' MILITIA (PEOPLE'S ARMY) (*Ba'ath* Party) ε100,000

Foreign Forces

UN (UNDOF): 1,035; contingents from **A** 367 **Ca** 186 **J** 30 **Pl** 358 **Slvk** 93

RUSSIA ε150 advisers, mainly AD

Tunisia Tn

dinar D		1998	1999	2000	2001
GDP	D	23bn	25bn		
	US$	20bn	21bn		
per capita	US$	6,400	6,800		
Growth	%	5.0	6.2		
Inflation	%	3.1	2.7		
Debt	US$	11bn	12.5bn		
Def exp	D	405m			
	US$	355m			
Def bdgt	D	398m	421m	461m	
	US$	340m	351m	365m	
FMA (US)	US$	0.8m	0.9m	2.9m	3.2m
US$1=D		1.17	1.20	1.36	

Population			9,465,000
Age	13–17	18–22	23–32
Men	524,000	496,000	856,000
Women	502,000	476,000	830,000

Total Armed Forces

ACTIVE ε35,000

(incl ε23,400 conscripts)

Terms of service 12 months selective

Army 27,000

(incl 22,000 conscripts)

3 mech bde (each with 1 armd, 2 mech inf, 1 arty, 1 AD regt) • 1 Sahara bde • 1 SF bde • 1 engr regt

EQUIPMENT

MBT 54 M-60A3, 30 M-60A1
LT TK 55 SK-105 *Kuerassier*
RECCE 24 *Saladin*, 35 AML-90
APC 140 M-113A1/-A2, 18 EE-11 *Urutu*, 110 Fiat F-6614
TOWED ARTY 105mm: 48 M-101A1/A2; **155mm**: 18 M-114A1, 57 M-198
MOR 81mm: 95; **107mm**: 66 4.2in; **120mm**: 18 Brandt
ATGW 100 TOW (incl 35 M-901 ITV), 500 *Milan*
RL 89mm: 300 LRAC-89, 300 3.5in M-20
RCL 57mm: 140 M-18; **106mm**: 70 M-40A1
AD GUNS 20mm: 100 M-55; **37mm**: 15 Type-55/-65
SAM 48 RBS-70, 25 M-48 *Chaparral*
SURV RASIT (veh, arty)

Navy ε4,500

(incl ε700 conscripts)

BASES Bizerte, Sfax, Kelibia

PATROL AND COASTAL COMBATANTS 19

MISSILE CRAFT 6

3 *La Galite* (Fr *Combattante* III) PFM with 8 MM-40 *Exocet* SSM, 1 × 76mm gun
3 *Bizerte* (Fr *P-48*) PFM with 8 SS-12M SSM

PATROL, COASTAL/INSHORE 13

3 *Utique* (mod PRC *Shanghai* II) PCC, some 10 PCI<

SUPPORT AND MISCELLANEOUS 2

1 *Salambo* (US *Conrad*) survey/trg, 1 AGS

Air Force 3,500

(incl 700 conscripts); 44 cbt ac, 7 armed hel

FGA 15 F-5E/F
CCT 3 MB-326K, 2 MB-326L
TPT 5 C-130B, 2 C-130H, 1 *Falcon* 20, 3 LET-410
LIAISON 2 S-208M
TRG 18 SF-260 (6 -C, 12* -W), 5 MB-326B, 12* L-59
ARMED HEL 5 SA-341 (attack) 2 HH-3 (ASW)
TPT HEL 1 wg with 15 AB-205, 6 AS-350B, 1 AS-365, 6 SA-313, 3 SA-316, 2 UH-1H, 2 UH-1N
AAM AIM-9J *Sidewinder*

Forces Abroad

UN AND PEACEKEEPING

DROC (MONUC): 6 obs

Paramilitary 12,000

NATIONAL GUARD 12,000 (Ministry of Interior)

incl Coastal Patrol with 5 (ex-GDR) *Kondor* I-class PCC, 5 (ex-GDR) *Bremse*-class PCI<, 4 *Gabes* PCI<, plus some 10 other PCI< **ac** 5 P-6B **hel** 8 SA-318/SA-319

United Arab Emirates UAE

dirham D		1998	1999	2000	2001
GDP	D	170bn	190bn		
	US$	46bn	52bn		
per capita	US$	15,400	16,700		
Growth	%	-5.6	10.0		
Inflation	%	3.1	2.4		
Debt	US$	13.0bn	15.5bn		
Def exp	D	ε11.0bn	ε11.3bn		
	US$	3.0bn	3.2bn		
Def bdgt[a]	D	ε13.7bn	ε14.0bn	ε14.5bn	
	US$	3.7bn	3.8bn	3.9bn	
US$1=D		3.67	3.67	3.67	

[a] Including extra-budgetary funding for procurement

Population			2,700,000
(nationals 24%, expatriates 76%, of which Indian 30%, Pakistani 20%, other Arab 12%, other Asian 10%, UK 2%, other European 1%)			
Age	13–17	18–22	23–32
Men	86,000	84,000	143,000
Women	85,000	80,000	110,000

Total Armed Forces

The Union Defence Force and the armed forces of the UAE (Abu Dhabi, Dubai, Ras Al Khaimah and Sharjah) were formally merged in 1976 and centred on Abu Dhabi. Dubai still maintains its independence, as do other emirates to a smaller degree.

ACTIVE ε65,000 (perhaps 30% expatriates)

Army 59,000

(incl **Dubai** 15,000) **MoD** Dubai **GHQ** Abu Dhabi
INTEGRATED 1 Royal Guard 'bde' • 2 armd bde • 3 mech inf bde • 2 inf bde • 1 arty bde (3 regt)
NOT INTEGRATED 2 inf bde (Dubai)

EQUIPMENT

MBT 45 AMX-30, 36 OF-40 Mk 2 (*Lion*), ε250 *Leclerc*
LT TK 76 *Scorpion*
RECCE 49 AML-90, 20 *Saladin* (in store)
AIFV 18 AMX-10P, 415 BMP-3
APC 80 VCR (incl variants), 370 Panhard M-3, 120 EE-11 *Urutu*, ε50 AAPC
TOWED ARTY 105mm: 73 ROF lt; **130mm**: 20 PRC Type-59-1
SP ARTY 155mm: 18 Mk F-3, 72 G-6, 87 M-109A3
MRL 70mm: 18 LAU-97; **122mm**: 48 FIROS-25 (ε24 op)
MOR 81mm: 114 L16; **120mm**: 21 Brandt
SSM 6 *Scud*-B (Dubai only)
ATGW 230 *Milan*, *Vigilant*, 25 TOW, 50 HOT (20 SP)
RCL 84mm: 250 *Carl Gustav*; **106mm**: 12 M-40
AD GUNS 20mm: 42 M-3VDA SP; **30mm**: 20 GCF-BM2
SAM 20+ *Blowpipe*, *Mistral*

Navy ε2,000

BASE Abu Dhabi
NAVAL FACILITIES Dalma, Mina Zayed, Ajman **Dubai** Mina Rashid, Mina Jabal, Al Fujairah **Ras al Khaimah** Mina Sakr **Sharjah** Mina Khalid, Khor Fakkan

FRIGATES

FFG 2 *Abu Dhabi* (NL *Kortenaer*) with 8 *Harpoon* SSM, 8 *Sea Sparrow* SAM, 1 × 76mm gun, 4 × 324mm TT, 2 AS565 hel

PATROL AND COASTAL COMBATANTS 16

CORVETTES 2 *Muray Jip* FSG (Ge Lürssen 62m) with 2 × 2 MM-40 *Exocet* SSM, 1 SA-316 hel
MISSILE CRAFT 8
6 *Ban Yas* (Ge Lürssen TNC-45) PFM with 2 × 2 MM-40 *Exocet* SSM, 1 × 76mm gun
2 *Mubarraz* (Ge Lürssen 45m) PFM with 2 × 2 MM-40 *Exocet* SSM, 1 × 76mm gun
PATROL, COASTAL 6
6 *Ardhana* (UK Vosper 33m) PCC

AMPHIBIOUS (craft only)
3 *Al Feyi* LCT, 2 other LCT

SUPPORT AND MISCELLANEOUS 2
1 div spt, 1 AT

NAVAL AVIATION
4 SA-316 *Alouette* hel, 6 AS 585 *Panther* hel

Air Force 4,000

(incl Police Air Wing) 101 cbt ac, 49 armed hel
Flying hours 110
FGA 3 sqn
1 with 9 *Mirage* 2000E
1 with 17 *Hawk* 102
1 with 17 *Hawk* Mk 63/63A/63C (FGA/trg)
FTR 1 sqn with 22 *Mirage* 2000 EAD
CCT 1 sqn with 8 MB-326 (2 -KD, 6 -LD), 5 MB-339A
OCU 5* *Hawk* Mk 61, 4* MB-339A, 6* *Mirage* 2000 DAD
RECCE 8* *Mirage* 2000 RAD
TPT incl 1 BN-2, 4 C-130H, 2 L-100-30, 4 C-212, 7 CN-235M-100, 4 Il-76 (on lease)

HELICOPTERS
ATTACK 5 AS-332F (anti-ship, 3 with *Exocet* AM-39), 10 SA-342K (with HOT), 7 SA-316/-319 (with AS-11/-12), 20 AH-64A, 7 AS-565 *Panther*
TPT 2 AS-332 (VIP), 1 AS-350, 27 Bell (8 -205, 9 -206, 5 -206L, 4 -214, 1 -407), 10 SA-330, 2 *King Air* 350 (VIP)
SAR 3 Bo-105, 3 *Agusta* -109 K2
TRG 30 PC-7, 5 SF-260 (4 -TP, 1 -W), 12 Grob G-115TA

MISSILES
ASM HOT, AS-11/-12, AS-15 *Exocet* AM-39, *Hellfire*, Hydra-70, PGM1, PGM2
AAM R-550 *Magic*, AIM 9L

AIR DEFENCE
1 AD bde (3 bn)
5 bty I HAWK
12 *Rapier*, 9 *Crotale*, 13 RBS-70, 100 *Mistral* SAM

Forces Abroad

UN AND PEACEKEEPING
YUGOSLAVIA (KFOR): 1,250; 3 AIFV coy, 1 MBT sqn, 1 arty bty, 1 ATK hel flt

Paramilitary

COAST GUARD (Ministry of Interior)
some 40 PCI<, plus boats

Foreign Forces

US Air Force 390

Yemen, Republic of — Ye

rial R		1998	1999	2000	2001
GDP	R	844bn	988bn		
	US$	6.0bn	6.7bn		
per capita	US$	1,400	1,500		
Growth	%	2.7	3.3		
Inflation	%	11.1	7.0		
Debt	US$	4.8bn			
Def exp	R	53bn	63bn		
	US$	396m	429m		
Def bdgt	R	53bn	55bn		
	US$	395m	374m		
FMA (US)	US$	0.05m	0.1m	0.1m	0.1m
US$1=R		129	148	159	

Population 17,766,000 (North 79% South 21%)

Age	13–17	18–22	23–32
Men	974,000	788,000	1,293,000
Women	948,000	758,000	1,180,000

Total Armed Forces

ACTIVE 66,300
(incl conscripts)
Terms of service conscription, 3 years

RESERVES perhaps 40,000
Army

Army 61,000

(incl conscripts)
9 armd bde • 1 SF bde • 18 inf bde • 7 mech bde • 2 AB/cdo bde • 3 SSM bde • 5 arty bde • 1 central guard force • 3 AD arty bn • 4 AD bn (1 with SA-2 SAM)

EQUIPMENT

MBT 990: 150 T-34, 500 T-54/-55, 250 T-62, 60 M-60A1, 30 T-72
RECCE 100 AML-90, 100 BRDM-2
AIFV 200 BMP-1/-2
APC 60 M-113, 380 BTR-40/-60/-152
TOWED ARTY some 412: **100mm**: 20 M-1944; **105mm**: 35 M-101A1; **122mm**: 30 M-1931/37, 100 M-1938, 130 D-30; **130mm**: 75 M-46; **152mm**: 10 D-20; **155mm**: 12 M-114
ASLT GUNS 100mm: 30 SU-100
COASTAL ARTY 130mm: 36 SM-4-1
MRL 122mm: 185 BM-21; **140mm**: BM-14
MOR ε600 incl **81mm**: 200; **82mm**; **107mm**: 12; **120mm**: 100; **160mm**: ε100
SSM 12 FROG-7, 12 SS-21, 6 *Scud*-B
ATGW 12 TOW, 24 *Dragon*, 35 AT-3 *Sagger*
RL 66mm: M72 LAW
RCL 75mm: M-20; **82mm**: B-10; **107mm**: B-11
ATK GUNS 85mm: D-44; **100mm**
AD GUNS 20mm: 40 M-167, 20 M-163 *Vulcan* SP; **23mm**: 100 ZSU-23-4; **37mm**: 150 M-1939; **57mm**: 120 S-60; **85mm**: 40 KS-12
SAM SA-7/-9/-13/-14

Navy 1,800

BASES Aden, Hodeida
FACILITIES Al Mukalla, Perim Island, Socotra (these have naval support equipment)

PATROL AND COASTAL COMBATANTS 12

MISSILE CRAFT 4
3 *Huangfen* with C-801 SSM (only 4 C-801 between the 3 craft)
1 *Tarantul* 1 PFM with 4 SS-N-2C *Styx* SSM (plus 1 non-op)
plus 6 boats

PATROL, INSHORE 8
3 *Sana'a* (US *Broadsword* 32m) (1 non-op) PFI, 5 Sov *Zhuk* PFI< (3 non-op)

MINE COUNTERMEASURES 6
1 Sov *Natya* MSO
5 Sov *Yevgenya* MHC

AMPHIBIOUS 1
1 *Ropucha* LST, capacity 190tps/10 tks
plus craft: 2 Sov *Ondatra* LCM

AUXILIARIES 2
2 *Toplivo* AOT

Air Force 3,500

49 cbt ac (plus some 40 in store), 8 attack hel
FGA 10 F-5E, 17 Su-20/-22
FTR 11 MiG-21, 5 MiG-29
TPT 2 An-12, 6 An-26, 3 C-130H, 4 IL-14, 3 IL-76
HEL 2 AB-212, 14 Mi-8, 1 AB-47, 8 Mi-35 (attack)
TRG 2* F-5B,4* MiG-21U, 14 YAK-11, 12 L-39C

AIR DEFENCE 2,000
SAM some SA-2, SA-3, SA-6
AAM AA-2 *Atoll*, AIM-9 *Sidewinder*

Paramilitary 70,000

MINISTRY OF THE INTERIOR FORCES 50,000
TRIBAL LEVIES at least 20,000
COAST GUARD
(slowly being established)
5 Fr *Interceptor* PCI<

MILITARY DEVELOPMENTS

Regional Trends

Central and South Asian countries continue to use more government resources on military expenditure than any region other than the Middle East. The pattern of regional tensions and conflicts are little changed. Relations between India and Pakistan remain tense and terrorism continues in Kashmir. The interminable war in Sri Lanka continues to drain the country's human and material capital. In Afghanistan, the *Taleban* struggles to eliminate the remaining opposition in the north. In Central Asia, government forces, Islamic fighters and drug gangs clash in Tajikistan, Uzbekistan and, increasingly, Kyrgyzstan.

India and Pakistan

In 2000, there has been no positive movement towards improved relations between India and Pakistan. While there were few major incidents across the Line of Control in Kashmir, terrorism by Islamic groups in Indian-held Kashmir continued unabated, despite a brief cease-fire in July–August 2000. The IISS estimates that 1,000 people were killed by terrorist acts in Kashmir over the year to August 2000, bringing the total since 1989 to 23,000. On 24 July, the leader of the armed Islamic group *Hizbul Mujahidin*, Abdul Majid Dar, announced a unilateral cease-fire, following the Indian government's release of several prominent separatist leaders and statements from senior Indian ministers that they were ready to open a dialogue with the militant groups. On 29 July, soon after the cease-fire announcement, India suspended military operations against the separatists. The *Hizbul Mujahidin* began talks with government representatives in Srinagar on 3 August. Ninety people were killed in a surge of violence perpetrated by guerrilla groups opposed to the dialogue. Despite the violence the talks made a promising start, but they stalled because New Delhi refused the *Hizbul's* demand to include Pakistani representatives. The talks ended on 4 August and the Indian armed forces resumed military operations against the insurgents.

The nuclear capabilities of India and Pakistan were little changed during 2000. India was far from acquiring the capabilities needed to meet the demands of the ambitious draft nuclear doctrine, published by the government's Strategic Policy Advisory Board in 1999. New Delhi has not formally endorsed the doctrine and, while there has been an increase in defence-budget plans, only modest steps are being taken towards improving nuclear-delivery capabilities by aircraft and missile. The *Agni*-2 missile has not been tested since April 1999. There have been tests of the land- and sea-launched 150–250 kilometre range *Prithvi* missile, but these are not thought to be nuclear-capable. The land, sea and air delivery capabilities set out in the draft doctrine would require substantially more spending than currently envisaged. It would probably cost in the order of $500 million a year over the next ten years to develop the warheads, missile capabilities and command-and-control systems laid out in the document.

Pakistan's missile capabilities have continued to advance. The 2,400km-range *Shaheen* 2 is ready for flight-testing. The longer-range version of the *Hatf* 1 surface-to-surface missile (SSM) tested successfully over its 100km range. The new design permits a greater payload, improved accuracy and a greater flexibility in warheads. A total of 30 600km-range *Hatf* 3 (based on the Chinese M-11) are reported to be in service. There are also thought to be 12 1,500km-range *Ghauri* 1 missiles operational. A 2,500km-range *Ghauri* 2, which would be capable of striking anywhere in India, has undergone static-engine testing. These high-priority programmes go some way towards counterbalancing India's superiority in conventional forces, which budget plans for the next five years will increase further. Increased demands have been placed on these forces; in particular, ensuring that Pakistani-supported guerrillas do not repeat the 1999 incursion into the Kargil area of Indian-held Kashmir. India maintains a greater military presence in that region than before. It has set up a new Army corps, XIV Corps, based in Leh and Nimu, to be responsible for the northern border areas. XV Corps remains headquartered in Srinagar, focusing on counter-terrorism operations in Kashmir. Internal security problems place continuing demands on military resources. In the Assam region, for example, security forces are engaged in a campaign against separatist groups such as the National Democratic Front of Bodoland (NDFB) and the United Liberation Front of Assam (ULFA).

Central Asia

In August 1999, hostages were seized in the Batken region of Kyrgyzstan, by an Uzbek terrorist group led by Juma Namangoni of the Islamic Movement of Uzbekistan (IMU). This, and subsequent events, have led to increased resources devoted to border defence and countering Islamic militancy. Among the hostages seized by Namangoni's group were the deputy commander of the Kyrgyz Interior Troops and four geologists from Japan. Namangoni made a number of demands, including that President Islam Abdughanievich Karimov of Uzbekistan should release 50,000 prisoners, mostly Muslims, held on terrorism charges. At the same time, he and his estimated 400 supporters claimed that they intended to launch an Islamic crusade against Uzbekistan. In an already insecure region, suffering the depredations of criminal gangs involved in the drug trade, these events further exposed the weakness of the area's security forces. The international nature of the incident also excited the interest of major powers both within and beyond the region. China, France, India, Russia, Turkey and the US, which are all sensitive to perceived Islamic threats, have supported countervailing action. For example, in April 2000, the US announced that it had earmarked $10m to provide training and equipment for Uzbek counter-terrorism and anti-drug units on the Afghan border. The US has offered similar packages to Kazakstan and Kyrgyzstan. In May 2000, China agreed an estimated 11m yuan ($1.3m) aid programme to help equip Kazakstan's armed forces, as well as a similar arrangement for Tajikistan to the value of 5m yuan ($0.6m) in July. Also in July, French Defence Minister Alain Richard signed a military-aid agreement that included the establishment of a joint commission on 'military-technical co-operation and defence technology'. In the same month, the Chief of the Turkish General Staff, General Huseyin Kivrikolgu, agreed to an aid package involving military-technical cooperation reportedly worth $1m. Russia has stepped up its programme of assistance and exercises through the Commonwealth of Independent States (CIS) network. In March and April 2000, Russia ran *Exercise Southern Shield*, involving the forces of Kazakstan, Kyrgyzstan and Tajikistan in counter-terrorist operations. Even Uzbek forces took part, although only on their own territory. This was an unusual step, as Uzbekistan normally stands aside from CIS activities and is normally particularly sensitive about Russian military activities in the region. The rising Islamic militancy in 2000, particularly the Batken incident, has made such activities more acceptable. However, the object of the foreign donors – to strengthen the region's armed forces – is unlikely to promote stability. Indeed, there was a fresh surge of violence in the Batken area during August 2000, in which ten Kyrgyz soldiers and 30 IMU rebels were reported killed. Uzbek security forces also caught members of the IMU infiltrating the border into Uzbekistan. There continue to be tensions among all the regional states, particularly between Uzbekistan, Kyrgyzstan and Tajikistan, exacerbated by the problem that state borders bear little relation to the geographic dispersion of different ethnic groups and clans. Also, the flow of drugs and associated criminal gangs from Afghanistan through the Fergana valley is unlikely to abate in the near future. The drought in Afghanistan during 2000 will significantly reduce the opium crop; however, this will not reduce drug-gangs' activities or the accompanying violence, but simply raise the price of the drugs.

In Afghanistan, the *Taleban* have increased their pressure on the Northern Group of forces led by Ahmad Shah Masood with a vigorous summer 2000 offensive. Their campaign focused on Taloqan, an important Northern Group base, and further north towards Eshkamesh. Even if the *Taleban* capture Taloqan, it is questionable whether they can hold it until winter sets in. They have not succeeded in capturing the base before and have not been able to hold territory captured in the area in previous years. The object of their military offensive is clearly to put a stranglehold on the supply routes to Masood's forces in the Panjshir valley and from the Tajikistan border. Nevertheless, Masood continues to receive support from Iran, Russia and Uzbekistan, and there seems to be no end in sight for this conflict. Over the year to August 2000, 10,000 people were killed as a direct result of conflict in Afghanistan, bringing the total since 1992 to 76,000. Despite US pressure, Pakistan has been unable to exert any real influence on the *Taleban* regime to moderate its excesses or to deliver up the Saudi dissident Usama bin Laden to help bring an end to the international terrorist activities of his group.

Sri Lanka

In Sri Lanka, the 17-year civil war has claimed 66,000 lives. The Liberation Tigers of Tamil Eelam (LTTE) launched a major offensive on the Jaffna peninsula in April 2000, but this lost momentum and government forces inflicted substantial casualties on the rebels. Air power was an important factor in blunting the LTTE attacks, both in the form of bombing raids and in the use of aircraft to send supplies to the beleaguered government forces trapped on the peninsula. In addition to their attacks on military bases, mainly in the north, the LTTE continue their terrorist campaign, carrying it to the capital Colombo. One of the more dramatic attacks in Colombo in 2000 was the killing of Industry Minister C.V. Gooneratne and 20 others by a suicide bomber during June celebrations honouring the country's war heroes. President Chandrika Kumaratunga's government put a devolution plan before parliament in August that contained a new constitution granting the provinces considerable autonomy and effectively turning the country into a federation. Kumaratunga hoped this could lead to peace talks with the LTTE; however, the plan was decisively voted down by the opposition United National Party. While attempting a political solution, the government has also strengthened the armed forces. In 2000, the Air Force took delivery of eight *Kfir* combat aircraft from Israel as well as delivery, at short notice, of four MiG-27 fighter, ground-attack (FGA) aircraft from Ukraine. The MiG-27s were soon in action against rebel forces.

DEFENCE SPENDING

Regional defence spending increased in 1999 by 3.1% in real terms to $21.7bn (measured in constant 1999 US dollars). Economic performance in the area remained strong, with gross domestic product (GDP) higher by over 5% in real terms, driven mainly by India's steady growth. India accounted for most of the regional defence-spending increase with a 10.2% rise to $13.9bn, measured in constant 1999 US dollars. This was well over the budget of $12.4bn. The 1999 defence budgets of Pakistan and Sri Lanka fell by 13% and 18% respectively in real terms. Budget allocations have increased in terms of national currency, but since these two countries import nearly all their major equipment, the depreciation of their currencies has hit them hard. The defence budgets of Central Asian countries remain difficult to access, although spending is known to be increasing, boosted by foreign aid.

India

India's defence budget for 2000 rose by nearly 30% to Rs709bn ($15.9bn) in nominal terms or 20% in real terms over the previous year. The increase – the biggest ever – will be partly financed by an increase in income tax, for the second year running. The latest defence budget amounts to 2.8% of GDP compared with 2.4% in 1998.

The Army will receive Rs349bn ($7.8bn), which is Rs30bn more than in 1999. It plans to acquire unmanned aerial vehicles, battlefield radar, improved artillery and up to 310 T-90 main battle tanks (MBT) from Russia. The additional costs of the Army's deployment in Kashmir following the Kargil border conflict, which is estimated at Rs100m per day, will be met by an extra allocation of Rs17.3bn.

The Indian Air Force will receive Rs143bn to help fund 66 advanced jet trainers, ten more *Mirage* 2000D fighters and the continued upgrade of its MiG-21 fighters. The trainers are urgently needed to curb the increasing number of flying accidents. However no decision on which aircraft to buy had been made by mid-2000. Such characteristic delay bedevils the Indian procurement system, which the Chief of the Army Staff, General V. P. Malik, has described as 'tedious, time consuming procedures' that hold up acquisitions even when parliament has allocated the funds. The main contenders remain the British *Hawk*, the French *Alphajet* and the Russian MiG-AT. Another regular cause of accidents is that the ageing MiG-21 fleet is desperately in need of the upgrade programme now underway. The 60 Jaguars are also being upgraded. A significant advance in capability was marked by the delivery in 2000 of the last of 40 Russian Su-30MK FGA aircraft.

The Indian Navy receives an increase of Rs10bn in the 2000 budget, bringing its allocation to Rs81bn ($1.8bn). The bulk of the extra funds are to develop naval aviation capabilities. India continues to negotiate with Russia about the transfer of the 45,000-tonne carrier *Admiral Gorshkov*. A Memorandum of Understanding between the two countries was signed in December 1999 and it is believed that the ship is currently being refitted in St Petersburg at India's expense. It is also believed that India wants to acquire about 20 MiG-29Ks from Russia for the carrier and forgo upgrading the *Sea Harrier* aircraft, at a cost of $200m, in order buy the MiGs. However, doubts remain about India's ability to finance the running of the carrier. Moreover, if the plan to have two carriers by 2010 is to be fulfilled, the *Viraat*, currently in refit, will have to be replaced within the decade. This is a financial burden that the Navy is unlikely to be able to bear. In other naval aviation developments, India is in negotiation with Russia to upgrade its 13 maritime-reconnaissance aircraft (eight Tu-142 and five Il-138). Linked to this deal is a negotiation to lease at least four Tu-22M3s for four years from Russia. If this arrangement goes ahead, it is not clear whether these aircraft would be operated in a maritime role or for wider tasks.

Further enhancements to the Indian Navy's surface combatants are based on Russian designs but are mostly built in India. The first *Brahmaputra*-class guided-missile frigate was commissioned in early 2000; two more are to follow. However, the class is without its main weapon system, the *Trishul* surface-to-air missile, which has not yet started trials. The third of the *Delhi*-class guided-missile destroyers will be commissioned in late 2000; it is hoped to build another three. In May 2000, the first of the *Kashmir*-class (*Krivak* III design) guided-missile frigates was launched in St Petersburg and it should be delivered to India in early 2002. Two more will be delivered by late 2003. Construction of an improved *Kashmir*-class frigate will start in India in late 2000 for first delivery in 2007. Two more of the *Kashmir*-class are on order. They are general-purpose frigates but will have a strong anti-submarine capability. The tenth and last *Kilo*-class diesel submarine was commissioned in mid-2000 and is armed with *Klub* anti-surface-ship missiles. It has been reported that the *Kilos* are not as effective as expected due to problems with their batteries.

Table 20 Indian defence budget by service/department, 1995–2000

(1998 US$m)	1995	%	1996	%	1997	%	1998	%	1999	%	2000	%
Army	4,673	*53.0*	4,630	*53.4*	5,663	*57.2*	5,218	*52.2*	5,816	*48.5*	7,074	*46.1*
Air Force	2,274	*25.8*	2,221	*25.6*	2,468	*24.9*	2,271	*22.7*	2,329	*19.4*	3,126	*20.4*
Navy	1,246	*14.1*	1,175	*13.5*	1,168	*11.8*	1,448	*14.5*	1,538	*12.8*	1,776	*11.6*
R&D	454	*5.1*	429	*4.9*	365	*3.7*	431	*4.3*	632	*5.3*	670	*4.4*
DP&S,other	165	*1.9*	221	*2.6*	237	*2.4*	618	*6.2*	1,673	*14.0*	2,705	*17.6*
Total	8,812	*100*	8,676	*100*	9,901	*100*	9,986	*100*	11,988	*100*	15,351	*100*
% Change		*9.0*		*-1.6*		*14.1*		*0.9*		*20.0*		*28.0*

Table 21 Indian defence and military-related spending by function, 1998–2000

(US$m)	1998 outturn	1999 outturn	2000 budget
Personnel, Operations & Maintenance			
MoD	84	75	81
Defence Pensions	1,762	2,560	2,702
Army	5,351	5,719	6,005
Navy	761	835	910
Air Force	1,336	1,430	1,778
Defence ordnance factories	N.A.	1,173	1,288
Recoveries & receipts	-1,846	-1,298	-1,337
Sub-Total	7,448	10,494	11,427
R&D, Procurement and Construction			
Tri-Service Defence R&D	138	151	186
Army	667	1,446	1,867
Navy	740	781	938
Air Force	882	971	1,475
Other	111	52	81
Sub-Total	2,538	3,401	4,547
Total Defence Budget	9,986	13,895	15,974
Other military-related funding			
Paramilitary forces	891	918	953
Department of Atomic Energy	586	363	461
Department of Space	366	342	382
Intelligence Bureau	57	71	74
Total	1,900	1,694	1,870

Although not published in the defence budget, there is increased funding in 2000 for the atomic energy and space programmes, both featuring military-specific projects. Together the two divisions are budgeted to receive $843m in 2000, up from $705m in 1999.

Pakistan

Pakistan's official defence budget rose from Rs142 ($2.9bn) in 1999 to Rs170bn ($3.2bn) in 2000, but as usual no detailed breakdown is available. *The Military Balance* estimates that spending in 1999 was above budget (and official outlay figures) at $3.5bn. The figure would probably have been higher still but for a $134m reduction in spending to divert funds to public works in rural areas.

Pakistan took delivery of a further eight upgraded *Mirage* 3 and *Mirage* 5 combat aircraft from France. A joint programme with China for the development and production of the FC-1 combat aircraft continues, with a planned in-service date of 2005. In the meantime, it is reported that Pakistan took delivery of a part-order for 50 F-7MG FGA aircraft from China in 2000.

In late 1999, the Pakistan Navy commissioned its first *Khalid*-class (French *Agosta B*) diesel submarine. Two more are being built under licence in Karachi, to be ready in 2002. They will replace the ageing *Hangor*-class boats first commissioned in 1969. It is still uncertain whether air-independent propulsion will be fitted; even without, the new vessels will greatly enhance Pakistan's submarine capabilities. If indigenous construction is successful, Pakistan may export them, with Saudi Arabia and Qatar as possible buyers. Funding, however, remains difficult for the Pakistani Navy; it cannot yet afford to replace the *Atlantique* maritime-reconnaissance aircraft shot down by India in 1999.

Sri Lanka

The war between government forces and the LTTE resulted in 1999 defence spending of Rs57.2bn ($807m), according to official figures, which was approximately Rs17.2bn ($242m) over budget. The official budget for 2000 has been set at Rs45bn ($699m). Given the increased tempo and scale of military operations, this budget too will almost certainly be overspent.

Bangladesh

Concerned by the military build-up of its neighbours, Bangladesh has decided to upgrade its ageing fleet of combat aircraft. In 2000 eight air-defence MiG-29s were delivered from Russia in a contract reputedly worth $115m. They will all be based at Dhaka and will replace obsolescent MiG-21s and Chinese copies of the MiG-19. The Navy is also expected to take delivery of a South Korean *Ulsan* frigate.

Table 22 Arms orders and deliveries, Central and South Asia, 1998–2000

Country	Country supplier	Classification	Designation	Quantity	Order date	Delivery date	Comment
Bangladesh							
	SF	PCO	***Madhumati***	1	1995	1998	
	PRC	FGA	**F-7**	24	1996	1997	Deliveries to 1999
	RF	radar	**IL-117 3-D**	2	1996	1999	Requirement for 3 more
	RF	hel	**Mi-17**	4	1997	1999	Following delivery of 12 1992–96
	PRC	trg	**FT-7B**	4	1997	1999	
	US	tpt	**C-130B**	4	1997	1999	
	RF	FGA	**MiG-29B**	8	1999	1999	Order placed 1999 after delay
	ROK	FF	***Ulsan***	1	1998	2002	
	Cz	trg	**L-39ZA**	4	1999	2000	Following delivery of 8 in 1995
India	dom	SSN	**ATV**	1	1982	2007	
	dom	ICBM	***Surya***		1983		Development
	dom	SLBM	***Dhanush***		1983	2003	Failed test firing April 2000
	dom	SLCM	***Sagarika***		1983	2003	300km range. May be ballistic
	dom	MRBM	***Agni* 1**		1983	1998	
	dom	MRBM	***Agni* 2**		1983	2000	Tested April 1999

Country	Country supplier	Classification	Designation	Quantity	Order date	Delivery date	Comment
	dom	MRBM	***Agni* 3**		1983		Dev. Range 3,500km
	dom	SSM	***Prithvi* 150**	75	1983	1995	Low-volume prod continues
	dom	SSM	***Prithvi***		1983	1999	Naval variant. Deployed Jan 1999
	dom	SSM	***Prithvi* 350**		1983	1998	Land and naval variants in dev
	dom	SAM	***Akash***		1983	1999	Development. High-altitude SAM
	dom	SAM	***Trishul***		1983	1999	In development
	dom	ATGW	***Nag***		1983	1999	Ready for production mid-1999
	dom	AAM	***Astra***		1999	2002	Dev. 1st test planned July 1999
	dom	FGA	***LCA***	7	1983	2005	
	RF	SSK	***Kilo***	10	1983	2000	Last of 10 delivered in 2000
	dom	FFG	***Brahmaputra***	3	1989	2000	1st delivered in 2000
	dom	hel	**ALH**	12	1984	2000	Delivery may slip to 2001
	dom	ELINT	**HS-748**		1990		Development
	dom	FSG	***Kora***	2	1990	1998	2nd delivered in 1999
	dom	UAV	***Nishant***	14	1991	1999	Dev. 3 prototypes built. 14 pre-production units on order
	dom	DD	***Delhi***	3	1986	1997	1st in 1997, 2nd 1998, 3rd 2000
	dom	LST	***Magyar***	2	1991	1997	1 more under construction
	RF	AD	**2S6**	24	1994	1996	12 units in 1996, 12 1998–99
	dom	FSG	***Kora***	2	1994	2000	
	dom	sat	***Ocean sat***	1	1995	1999	Remote sensing
	dom	AGHS	***Sandhayak***	2	1995	1999	Following delivery of 6 1981–93
	RF	TKR AC	**IL-78**	6	1996	1998	First 2 delivered early 1998
	RF	ASSM	**SS-N-25**	16	1996	1997	Deliveries continue
	RF	FGA	**Su-30MK**	40	1996	1997	Delivery ended in 2000
	Il	PFC	***Super Dvora* MK3**	6	1996	1998	First delivery 1998. Il designation T-81
	RF	FF	***Krivak* 3**	3	1997	2002	1 for delivery by 2002, 2 by 2003
	RF	hel	**KA-31**	3	1997	2002	
	Ge	SS	**Type 209**	2	1997	2003	To be built in Ind
	US	MPA	**P-3C**	3	1997		Delayed due to sanctions
	UK	FGA	***Harrier* TMk4**	2	1997	1999	2 ex-RN ac for delivery 1999
	RSA	APC	***Casspir***	90	1998	1999	
	RF	SLCM	**SS-NX-27**		1998	2004	For *Krivak* 3 frigate. First export
	UK	FGA	***Jaguar***	18	1998	2001	Upgrade for up to 60
	RF	FGA	**MiG-21**	125	1999	2003	Upgrade. Fr and Il avionics
	dom	MBT	***Arjun***	124	1999	2001	
	Fr	FGA	***Mirage* 2000**	10	1999	2002	Approved but not contracted
	dom	trg	**HJT-36**	200	1999	2004	
	Pl	trg	**TS-11**	12	1999	2000	Option on 8 more
	dom	CV	***Viraat***	1	1999	2001	Upgrade
	RF	CV	***Admiral Gorshkov***	1	1999	2003	MoU signed
	Slvk	ARV	**T-72 VT**	42	1999	2001	
	Pl	ARV	**WZT-3**	43	1999	2001	
	Il	arty	**M-46**	35	1999	2000	Il upgrade

Country	Country supplier	Classification	Designation	Quantity	Order date	Delivery date	Comment
	dom	AAM	***Astra***		1999		Live Firing due 2001
	dom	MPA	**Do-228**	7	1999		Deliveries completed by 2003
	Il	arty	**M-46**	35	1999	2000	Requirement for further 500
	RF	hel	**Mi-17iB**	40	2000	2001	
	RF	MBT	**T-90**	310	2000		186 to be built in Ind
	Il	UAV	***Searcher* 2**	20	2000		In addition to 8 delivered in 1999
Kazakstan	RF	FGA	**Su-27**	16	1997	1999	4 delivered early 1999, 10 in 1997
	RF	SAM	**S-300**		1997	2000	
Pakistan	dom	sat	***Badar* 2**				Development
	dom	sat	***Badar* 1**				Multi-purpose sat. In operation
	US	APC	**M113**	775	1989	1990	Licensed prod; deliveries to 1999
	dom	MBT	***Al-Khalid***		1991	1998	In acceptance trials
	Fr	MHC	***Munsif***	3	1992	1992	Second delivered 1996. Third 1998
	PRC	FGA	**FC-1**		1993	2005	With PRC, req for up to 150
	dom	MRBM	***Ghauri* 1**		1993	1998	Range 1,500km. Aka *Hatf* 5
	dom	MRBM	***Ghauri* 2**		1993	1999	Dev. Aka *Hatf* 6
	dom	MRBM	***Ghauri* 3**		1993		Dev. Based on *Taepo-dong* 2
	dom	SSM	***Hatf* 3**		1994	1999	In-service. Based on M-11
	dom	SSM	***Shaheen* 1**		1994	1999	Prod 1999. Based on M-9. Aka *Hatf* 4
	Fr	SSK	***Khalid***	3	1994	1999	1st in 1999, 2nd 2001, 3rd 2002
	Fr	FGA	***Mirage* III**	40	1996	1998	Upgrade. 8 delivered by 1999
	Ukr	MBT	**T-80UD**	320	1996	1996	Final 105 delivered in 1999
	dom	PFM	**Mod. *Larkana***	1	1996	1997	Commissioned 14 August 1997
	PRC	PFM	***Shujat* 2**	1	1997	1999	
	PRC	FGA	**F-7MG**	50	1999	2001	Unconfirmed
Sri Lanka	Il	UAV	***Super Scout***				
	Ukr	cbt hel	**Mi-24**	2	1995	1996	1 delivered 1998
	UK	ACV	**M10**		1995	1999	Hovercraft
	RF	cbt hel	**Mi-35**	2	1997	1999	May be 4. 5 delivered previously
	US	tpt	**C-130**	3	1997	1999	
	Ukr	cbt hel	**Mi-24**	2	1998	1999	
	PRC	arty	**152mm**	36	1999	2000	
	UK	tpt	**C-130**	2	1999	1999	
	Il	FGA	***Kfir***	8	2000	2000	
	Ukr	FGA	**MiG-27**	4	2000	2000	
	US	tpt hel	**Bell-41ZEP**	2	2000	2000	

Afghanistan Afg

afghani Afs		1998	1999	2000	2001
GDP	US$	ε1.7bn			
per capita	US$	ε700			
Growth	%	ε6			
Inflation	%	ε14			
Debt	US$	ε6.2bn			
Def exp	US$	ε250m			
US$1=Afs[a]		3,000	3,000	3,000	

[a] Market rate **2000** εUS$1 = Afs4,700

Population[b]			**ε24,000,000**
(Pashtun 38%, Tajik 25%, Hazara 19%, Uzbek 12%, Aimaq 4%, Baluchi 0.5%)			
Age	13–17	18–22	23–32
Men	1,451,000	1,178,000	2,014,000
Women	1,395,000	1,119,000	1,894,000

[b] Includes ε1,500,000 refugees in Pakistan, ε1,000,000 in Iran, ε150,000 in Russia and ε50,000 in Kyrgyzstan

Total Armed Forces

There are no state-constituted armed forces. The *Taleban* now controls 85–90% of Afghanistan. It continues to mount mil ops against an alliance of Ahmad Shah Massoud, deposed President Burhanuddin Rabbani and the National Islamic Movement (NIM) of General Abdul Rashid Dostum.

EQUIPMENT

It is impossible to show the division of ground force equipment among the different factions. The list below represents weapons known to be in the country in April 1992. Individual weapons quantities are unknown.

MBT ε1,000: T-54/-55, T-62
LT TK PT-76
RECCE BRDM-1/-2
AIFV BMP-1/-2
APC ε1,000: BTR-40/-60/-70/-80/-152
TOWED ARTY 76mm: M-1938, M-1942; **85mm**: D-48; **100mm**: M-1944; **122mm**: M-30, D-30; **130mm**: M-46; **152mm**: D-1, D-20, M-1937 (ML-20)
MRL ε125: **122mm**: BM-21; **140mm**: BM-14; **220mm**: 9P140 *Uragan*
MOR 82mm: M-37; **107mm; 120mm**: M-43
SSM ε20–30: *Scud*, FROG-7
ATGW AT-1 *Snapper*, AT-3 *Sagger*
RCL 73mm: SPG-9; **82mm**: B-10
AD GUNS: 14.5mm; 23mm: ZU-23, ZSU-23-4 SP; **37mm**: M-1939; **57mm**: S-60; **85mm**: KS-12; **100mm**: KS-19
SAM SA-7/-13

Air Force

Only the former government–NIM alliance and *Taleban* have aircraft. These groups have a quantity of Su-17/22 and MiG-21s and both have some Mi-8/17. The inventory shows ac in service in April 1992. Since then, an unknown number of fixed-wing ac and hel have either been shot down or destroyed on the ground. It is believed that the *Taleban* have about 20 MiG-21 and Su-22, and 5 L-39, all being used in the FGA role. The NIM have about 30 Su-17/22, 30 MiG-21 and 10 L-39. The number of helicopters on each side is unknown.

FGA 30 MiG-23, 80 Su-7/-17/-22
FTR 80 MiG-21F
ARMED HEL 25 Mi-8, 35 Mi-17, 20 Mi-25
TPT ac 2 Il-18D; 50 An-2, An-12, An-26, An-32 **hel** 12 Mi-4
TRG 25 L-39*, 18 MiG-21*

AIR DEFENCE

SAM 115 SA-2, 110 SA-3, *Stinger*, SAM-7, SAM-14, **37mm**, **85mm** and **100mm** guns
AD guns some 200–300

Opposition Groups

In the midst of a civil war, this section lists armed groups operating in the country.

TALEBAN str n.k. **Leaders** Mullah Mohamed Omar, Mullah Mohamed Rabbani **Area** now control 85–90% of Afghanistan **Ethnic group** Pashtun. Formed originally from religious students in Madrassahs (mostly Pashtun)

Northern Alliance

The Northern Alliance represents the armed grouping of the 'United Islamic Front for the Salvation of Afghanistan', comprising:

ISLAMIC SOCIETY (*Jamia't-i-Isla'mi*) str n.k. **Leaders** Ahmad Shah Massoud and deposed President Burhanuddin Rabbani **Area** north of Kabul and Panshir Valley **Ethnic groups** Turkoman, Uzbek, Tajik
NATIONAL ISLAMIC MOVEMENT (NIM)[a] (*Jumbesh-i-Milli Islami*) str n.k. **Leader** General Abdul Rashid Dostum. Formed in March 1992, mainly from troops of former Afghan Army Northern Comd. Predominantly Uzbek, Tajik, Turkoman, Ismaeli and Hazara Shi'a.
ISLAMIC UNITY PARTY (*Hizbi Wahdat-Khalili*) **Leader** Abdul Karim Khalili

Other Groups

ISLAMIC PARTY (*Hezbi Islami-Gulbuddin*) Gulbuddin *Hekmatyar* faction
ISLAMIC PARTY (*Hizbi Islami-Khalis*) Yunis Khalis faction
ISLAMIC UNION FOR THE LIBERATION OF AFGHANISTAN (*Ittihad-i-Islami Barai Azadi Afghanistan*) **Leader** Abdul Rasul Sayyaf
ISLAMIC REVOLUTIONARY MOVEMENT (*Harakat-Inqilab-i-Islami*) **Leader** Mohammed Nabi Mohammadi

AFGHANISTAN NATIONAL LIBERATION FRONT (*Jabha-i-Najat-i-Milli Afghanistan*) **Leader** Sibghatullad Mojaddedi
NATIONAL ISLAMIC FRONT (*Mahaz-i-Milli-Islami*) **Leader** Sayed Aha Gailani
ISLAMIC UNITY PARTY (*Hizbi Wahdat-Akbari* faction) **Leader** Mohammed Akbar Akbari
ISLAMIC MOVEMENT (*Harakat-i-Islami*) **Leader** Mohammed Asif Mohseni
These smaller groups occasionally support the *Taleban* as well as at times supporting the Northern Alliance

HEZBI-WAHDAT (Unity Party) Shi'a umbrella party of which the main groups are:
Sazman-e-Nasr str n.k. **Ethnic group** Hazara
Shura-Itifaq-Islami str n.k. **Area Ethnic group** Hazara
Haraka't-e-Islami str n.k. **Ethnic group** Pashtun, Tajik, Uzbek
These Shi'a groups have at times been allied with the Northern Alliance, at others were attacked by them. The Hazara group enjoy support from Iran.

[a] Form the Supreme Coordination Council

Bangladesh Bng

taka Tk		1998	1999	2000	2001
GDP	Tk	2.0tr	2.2tr		
	Tk	42bn	44bn		
per capita	US$	1,700	1,700		
Growth	%	5.6	4.4		
Inflation	%	8.3	6.3		
Debt	US$	14.0bn	15.1bn		
Def exp	Tk	29bn			
	US$	619m			
Def bdgt	Tk		30bn		
	US$		612m		
FMA (US)	US$	0.3m	0.4m	0.4m	0.4m
US$1=taka		46.9	49.0	51.0	

Population		**132,417,000** (Hindu 12%)	
Age	13–17	18–22	23–32
Men	8,038,000	7,564,000	11,988,000
Women	7,701,000	7,104,000	11,351,000

Total Armed Forces

ACTIVE 137,000

Army 120,000

7 inf div HQ • 17 inf bde (some 26 bn) • 1 armd bde (2 armd regt) • 2 armd regt • 1 arty div (6 arty regt) • 1 engr bde • 1 AD bde

EQUIPMENT†
MBT 100 PRC Type-59/-69, 100 T-54/-55
LT TK some 40 PRC Type-62
APC 60 BTR-70, 20 BTR-80, some MT-LB, ε50 YW531
TOWED ARTY 105mm: 30 Model 56 pack, 50 M-101; **122mm**: 20 PRC Type-54; **130mm**: 40+ PRC Type-59
MRL 122mm: reported
MOR 81mm; 82mm: PRC Type-53; **120mm**: 50 PRC Type-53
RCL 106mm: 30 M-40A1
ATK GUNS 57mm: 18 6-pdr; **76mm**: 50 PRC Type-54
AD GUNS 37mm: 16 PRC Type-55; **57mm**: PRC Type-59
SAM some HN-5A

Navy† 10,500

BASES Chittagong (HQ), Dhaka, Khulna, Kaptai
FRIGATES 4
FFG 1 *Osman* (PRC *Jianghu* I) with 2 × 2 CSS-N-2 *Hai Ying* 2 SSM, 2 × 2 100mm gun, 2 × 5 ASW mor
FF 3
1 *Umar Farooq* (UK *Salisbury*) with 1 × 2 115mm gun, 1 × 3 *Squid* ASW mor
2 *Abu Bakr* (UK *Leopard*) with 2 × 2 115mm guns
PATROL AND COASTAL COMBATANTS 33
MISSILE CRAFT 10
5 *Durdarsha* (PRC *Huangfeng*) PFM with 4 *HY* 2 SSM
5 *Durbar* (PRC *Hegu*) PFM< with 2 *SY*-1 SSM
TORPEDO CRAFT 4
4 PRC *Huchuan* PHT< with 2 × 533mm TT
PATROL, OFFSHORE 2
1 *Madhumati* (J *Sea Dragon*) PCO with 1 × 76mm gun
1 *Durjoy* (PRC *Hainan*) PCO with 4 × 5 ASW RL
PATROL, COASTAL 8
2 *Meghna* fishery protection
2 *Karnaphuli* PCC
4 *Shahead Daulat* PFC
PATROL, INSHORE 4
1 *Bishkali* PCI<, 1 *Bakarat* PCI<, 2 *Akshay* PCI<
PATROL, RIVERINE 5 *Pabna*< PCR
MINE COUNTERMEASURES 4
3 *Shapla* (UK *River*) MSI, 1 *Sagar* MSO
AMPHIBIOUS craft only
7 LCU, 4 LCM, 3 LCVP
SUPPORT AND MISCELLANEOUS 8
1 coastal AOT, 1 AR, 1 AT/F, 1 AT, 2 *Yuch'in* AGHS, 1 *Shaibal* AGOR (UK *River*) (MCM capable), 1 *Shaheed Ruhul Amin* (trg)

Air Force† 6,500

83 cbt ac, no armed hel **Flying hours** 100–120
FGA/FTR 4 sqn with 8 MiG-29, 18 A-5C *Fantan*, 16 F-

6, 23 F-7M/FT-7B *Airguard*, 1 OCU with 10 FT-6, 8 L-39ZA
TPT 3 An-32
HEL 3 sqn with 11 Bell 212, 1 Mi-8, 15 Mi-17
TRG 20 PT-6, 12 T-37B, 8 CM-170, 2 Bell 206L
AAM AA-2 *Atoll*

Forces Abroad

UN AND PEACEKEEPING

CROATIA (UNMOP): 1 obs **DROC** (MONUC): 16 obs **EAST TIMOR** (UNTAET): 570 incl 25 obs **GEORGIA** (UNOMIG): 7 obs **IRAQ/KUWAIT** (UNIKOM): 815 incl 5 obs **SIERRA LEONE** (UNAMSIL): 792 incl 12 obs **WESTERN SAHARA** (MINURSO): 6 obs

Paramilitary 55,200

BANGLADESH RIFLES 30,000
border guard; 41 bn
ARMED POLICE 5,000
rapid action force (forming)
ANSARS (Security Guards) 20,000+
A further 180,000 unembodied

COAST GUARD 200
(HQ Chittagong and Khulma)
1 *Bishkhali* PCI
(force in its infancy and expected to expand)

India — Ind

rupee Rs		1998	1999	2000	2001
GDP	Rs	17.0tr	18.9tr		
	US$	412bn	440bn		
per capita	US$	1,700	1,800		
Growth	%	6.7	5.9		
Inflation	%	13.2	4.7		
Debt	US$	94bn	99bn		
Def exp[a]	Rs	580bn	610bn		
	US$	14.1bn	14.2bn		
Def bdgt	Rs	412bn	533bn	709bn	
	US$	10.0bn	12.4bn	15.9bn	
FMA[b] (US)	US$	0.2m	0.5m	0.5m	0.5m
FMA (Aus)	US$	0.2m	0.2m		
US$1=Rs		41.3	43.0	44.4	

[a] Incl exp on paramil org
[b] UNMOGIP **1997** US$7m **1998** US$8m

Population			**1,016,242,000**
(Hindu 80%, Muslim 14%, Christian 2%, Sikh 2%)			
Age	13–17	18–22	23–32
Men	53,812,000	49,257,000	87,033,000
Women	50,432,000	45,713,000	79,562,000

Total Armed Forces

ACTIVE 1,303,000

RESERVES 535,000
Army 300,000 (first-line reserves within 5 years' full-time service, a further 500,000 have commitment until age 50) **Territorial Army** (volunteers) 40,000 **Air Force** 140,000 **Navy** 55,000

Army 1,100,000

HQ: 5 Regional Comd, 4 Fd Army, 12 Corps
3 armd div (each 2–3 armed, 1 SP arty (2 SP fd, 1 med regt) bde) • 4 RAPID div (each 2 inf, 1 mech bde) • 18 inf div (each 2–5 inf, 1 arty bde; some have armd regt) • 9 mtn div (each 3–4 bde, 1 or more arty regt) • 1 arty div (3 bde) • 15 indep bde: 7 armd, 5 inf, 2 mtn, 1 AB/cdo • 1 SSM regt (*Prithvi*) • 4 AD bde (plus 14 cadre) • 3 engr bde
These formations comprise
59 tk regt (bn) • 355 inf bn (incl 25 mech, 8 AB, 3 cdo) • 190 arty regt (bn) reported: incl 1 SSM, 2 MRL, 50 med (11 SP), 69 fd (3 SP), 39 mtn, 29 AD arty regt; perhaps 2 SAM gp (3–5 bty each) plus 15 SAM regt • 22 hel sqn: incl 5 ATK

RESERVES
Territorial Army 25 inf bn, plus 29 'departmental' units
EQUIPMENT
MBT ε3,414 (ε1,100 in store): some 700 T-55 (450 op), ε1,500 T-72/M1, 1,200 *Vijayanta*, ε14 *Arjun*
LT TK ε90 PT-76
RECCE ε100 BRDM-2
AIFV 350+ BMP-1, 1,000 BMP-2 (*Sarath*)
APC 157 OT-62/-64 (in store), some *Casspir*
TOWED ARTY 4,175 (perhaps 600 in store) incl: **75mm**: 900 75/24 mtn, 215 FRY M-48; **105mm**: some 1,300 IFG Mk I/II, 50 M-56; **122mm**: some 550 D-30; **130mm**: 750 M-46; **155mm**: 410 FH-77B
SP ARTY 105mm: 80 *Abbot* (ε30 in store); **130mm**: 100 mod M-46 (ε70 in store); **152mm**: some 2S19
MRL 122mm: ε100 incl BM-21, LRAR; **214mm**: *Pinacha* (being deployed)
MOR 81mm: L16A1, E1; **120mm**: 500 Brandt AM-50, E1; **160mm**: 500 M-1943
SSM *Prithvi* (3–5 launchers)
ATGW *Milan*, AT-3 *Sagger*, AT-4 *Spigot* (some SP), AT-5 *Spandrel* (some SP)
RCL 84mm: *Carl Gustav*; **106mm**: 1,000+ M-40A1
AD GUNS some 2,400: **20mm**: Oerlikon (reported); **23mm**: 300 ZU 23-2, 100 ZSU-23-4 SP; **30mm**: 24 2S6 SP; **40mm**: 1,200 L40/60, 800 L40/70
SAM 180 SA-6, 620 SA-7, 50 SA-8B, 400 SA-9, 45 SA-3, SA-13, 500 SA-16
SURV MUFAR, *Green Archer* (mor)
UAV *Searcher*, *Nishant*
HEL 120 *Chetak*, 40 *Cheetah*

LC 2 LCVP

DEPLOYMENT

North 3 Corps with 8 inf, 2 mtn div **West** 3 Corps with 1 armd, 5 inf div, 3 RAPID **Central** 1 Corps with 1 armd, 1 inf, 1 RAPID **East** 3 Corps with 1 inf, 7 mtn div **South** 2 Corps with 1 armd, 3 inf div

Navy 53,000

(incl 5,000 Naval Aviation and 1,000 Marines, ε2,000 women)

PRINCIPAL COMMAND Western, Southern, Eastern (incl Far Eastern sub command)

SUB-COMMAND Submarine, Naval Air

BASES Mumbai (Bombay) (HQ Western Comd), Goa (HQ Naval Air), Karwar (under construction), Kochi (Cochin) (HQ Southern Comd), Vishakhapatnam (HQ Eastern), Calcutta, Madras, Port Blair (Andaman Is) (HQ Far Eastern Comd), Arakonam (Naval Air)

FLEETS Western base Bombay **Eastern base** Visakhapatnam

SUBMARINES 16

SSK 16

10 *Sindhughosh* (Sov *Kilo*) with 533mm TT

4 *Shishumar* (Ge T-209/1500) with 533mm TT

2 *Kursura* (Sov *Foxtrot*)† with 533mm TT (plus 3 in reserve)

PRINCIPAL SURFACE COMBATANTS 26

CARRIERS 1 *Viraat* (UK *Hermes*) (29,000t) CVV

Air group typically **ac** 6 *Sea Harrier* ftr/attack **hel** 6 *Sea King* ASW/ASUW (*Sea Eagle* ASM) (in refit until April 2001)

DESTROYERS 8

DDG 8

5 *Rajput* (Sov *Kashin*) with 4 SS-N-2C *Styx* SSM, 2 × 2 SA-N-1 *Goa* SAM, 2 × 76mm gun, 5 × 533mm ASTT, 2 ASW RL, 1 Ka-25 or 28 hel (1 in refit)

3 *Delhi* with 16 SS-N-25 *Switchblade* SSM, 2 × SA-N-7 *Gadfly* SAM, 1 × 100mm gun, 5 × 533mm ASTT, 2 hel

FRIGATES 12

FFG 4

1 *Brahmaputra* with 8 × SS-N-25 *Switchblade* SSM, 20 SA-N-4 *Gecko* SAM, 1 × 76mm gun, 2 × 3 324mm ASTT, 1 hel

3 *Godavari* with SS-N-2D *Styx* SSM, 1 × 2 SA-N-4 *Gecko* SAM, 2 × 3 324mm ASTT, 1 *Sea King* hel

FF 8

4 *Nilgiri* (UK *Leander*) with 2 × 114mm guns, 2 × 3 ASTT, 1 × 3 *Limbo* ASW mor, 1 *Chetak* hel (2 with 1 *Sea King*)

1 *Krishna* (UK *Leander*) (trg role)

3 *Arnala* (Sov *Petya*) with 4 × 76mm gun, 3 × 533mm ASTT, 4 ASW RL

CORVETES 5

4 *Khukri* FSG with 2 or 4 SS-N-2C *Styx* SSM, 1 × 76mm gun, hel deck

1 mod *Khukri* FSG with 8 × SS-N-25 *Switchblade* SSM, SA-N-5 *Grail* SAM, 1 × 76mm gun

PATROL AND COASTAL COMBATANTS 38

CORVETTES 14

1 *Vijay Durg* (Sov *Nanuchka* II) FSG with 4 SS-N-2C *Styx* SSM, SA-N-4 *Gecko* SAM (plus 1 non-op)

3 *Veer* (Sov *Tarantul*) FSG with 4 *Styx* SSM, SA-N-5 *Grail* SAM, 1 × 76mm gun (plus 2 non-op)

6 *Vibhuti* (similar to *Tarantul*), armament as *Veer*

4 *Abhay* (Sov *Pauk* II) FS with SA-N-5 *Grail* SAM, 1 × 76mm gun, 4 × 533mm ASTT, 2 ASW mor

MISSILE CRAFT 6 *Vidyut* (Sov *Osa* II) with 4 *Styx* SSM†

PATROL, OFFSHORE 7 *Sukanya* PCO

PATROL, INSHORE 11

7 SDB Mk 3

4 *Super Dvora* PCI<

MINE WARFARE 17

MINELAYERS 0

none, but *Kamorta* FF and *Pondicherry* MSO have minelaying capability

MINE COUNTERMEASURES 17

11 *Pondicherry* (Sov *Natya*) MSO, 6 *Mahé* (Sov *Yevgenya*) MSI<

AMPHIBIOUS 9

2 *Magar* LST, capacity 500 tps, 18 tk, 1 hel

7 *Ghorpad* (Sov *Polnocny* C) LSM, capacity 140 tps, 6 tk

Plus craft: 10 *Vasco da Gama* LCU

SUPPORT AND MISCELLANEOUS 26

1 *Adiyta* (mod *Deepak*) AO, 1 *Deepak* AO, 1 *Jyoti* AO, 4 small AOT; 1 YDT, 1 *Tir* trg, 2 AT/F, 3 TRV, 1 AH; 6 *Sandhayak* AGHS, 4 *Makar* AGHS, 1 *Sagardhwani* AGOR

NAVAL AVIATION (5,000)

37 cbt ac, 72 armed hel **Flying hours** some 180

ATTACK 2 sqn with 23 *Sea Harrier* FRS Mk-51, 1 T-60 trg* plus 2 T-4 (on order)

ASW 6 hel sqn with 24 *Chetak*, 7 Ka-25, 14 Ka-28, 25 *Sea King* Mk 42A/B

MR 3 sqn with 5 Il-38, 8 Tu-142M *Bear* F, 19 Do-228, 18 BN-2 *Defender*

COMMS 1 sqn with **ac** 10 Do-228 **hel** 3 *Chetak*

SAR 1 hel sqn with 6 *Sea King* Mk 42C

TRG 2 sqn with **ac** 6 HJT-16, 8 HPT-32 **hel** 2 *Chetak**, 4 Hughes 300

MISSILES

AAM R-550 *Magic* I and II

ASM *Sea Eagle, Sea Skua*

MARINES (1,200)

1 regt (3 gp)

Air Force 150,000

774 cbt ac, 34 armed hel **Flying hours** 150

Five regional air commands: **Central** (Allahabad),

Central and South Asia

Western (New Delhi), **Eastern** (Shillong), **Southern** (Tiruvettipuram), **South-Western** (Gandhinagar); 2 spt cmds: trg and maint

FGA 18 sqn
- 1 with 10 Su-30K, 3 with 53 MiG-23 BN/UM, 4 with 88 *Jaguar* S(I), 6 with 147 MiG-27, 4 with 69 MiG-21 MF/PFMA

FTR 20 sqn
- 4 with 66 MiG-21 FL/U, 10 with 169 MiG-21 bis/U, 1 with 26 MiG-23 MF/UM, 3 with 64 MiG-29, 2 with 35 *Mirage* 2000H/TH (believed to have secondary GA capability), 8 Su-30MK

ECM 4 *Canberra* B(I) 58 (ECM/target towing, plus 2 *Canberra* TT-18 target towing)
ELINT 2 Boeing 707, 2 Boeing 737
AEW 4 HS-748
TANKER 6 IL-78
MARITIME ATTACK 6 *Jaguar* S(I) with *Sea Eagle*
ATTACK HEL 3 sqn with 32 Mi-25
RECCE 2 sqn
- 1 with 8 *Canberra* (6 PR-57, 2 PR-67)
- 1 with 6* MiG-25R, 2* MiG-25U

MR/SURVEY 2 *Gulfstream* IV SRA, 2 *Learjet* 29
TRANSPORT
- **ac** 12 sqn
  - 6 with 105 An-32 *Sutlej*, 2 with 45 Do-228, 2 with 28 BAe-748, 2 with 25 Il-76 *Gajraj*
- **hel** 11 sqn with 73 Mi-8, 50 Mi-17, 10 Mi-26 (hy tpt)

VIP 1 HQ sqn with 2 Boeing 737-200, 7 BAe-748, 6 Mi-8
TRG ac 28 BAe-748 (trg/tpt), 120 *Kiran* I, 56 *Kiran* II, 88 HPT-32, 38 *Hunter* (20 F-56, 18 T-66), 14* *Jaguar* B(1), 9* MiG-29UB, 44 TS-11 *Iskara* **hel** 20 *Chetak*, 2 Mi-24, 2* Mi-35

MISSILES
- **ASM** AS-7 *Kerry*, AS-11B (ATGW), AS-12, AS-30, *Sea Eagle*, AM 39 *Exocet*, AS-17 *Krypton*
- **AAM** AA-7 *Apex*, AA-8 *Aphid*, AA-10 *Alamo*, AA-11 *Archer*, R-550 *Magic*, *Super* 530D
- **SAM** 38 sqn with 280 *Divina* V75SM/VK (SA-2), *Pechora* (SA-3), SA-5, SA-10

Forces Abroad

UN AND PEACEKEEPING

DROC (MONUC): 12 obs **IRAQ/KUWAIT** (UNIKOM): 6 obs **LEBANON** (UNIFIL): 618 **SIERRA LEONE** (UNAMSIL): 3,161 incl 14 obs

Paramilitary 1,069,000 active

NATIONAL SECURITY GUARDS 7,400
(Cabinet Secretariat)
Anti-terrorism contingency deployment force, comprising elements of the armed forces, CRPF and Border Security Force

SPECIAL PROTECTION GROUP 3,000
Protection of VVIP

SPECIAL FRONTIER FORCE 9,000
(Cabinet Secretariat)
mainly ethnic Tibetans

RASHTRIYA RIFLES 36,000 (Ministry of Defence)
36 bn in 12 Sector HQ

DEFENCE SECURITY CORPS 31,000
provides security at Defence Ministry sites

INDO-TIBETAN BORDER POLICE 30,000 (Ministry of Home Affairs)
28 bn, Tibetan border security

ASSAM RIFLES 52,000 (Ministry of Home Affairs)
7 HQ, 31 bn, security within north-eastern states, mainly Army-officered; better trained than BSF

RAILWAY PROTECTION FORCES 70,000

CENTRAL INDUSTRIAL SECURITY FORCE 88,600 (Ministry of Home Affairs)[a]
guards public-sector locations

CENTRAL RESERVE POLICE FORCE (CRPF) 160,000 (Ministry of Home Affairs)
130–135 bn incl 10 rapid action, 2 *Mahila* (women); internal security duties, only lightly armed, deployable throughout the country

BORDER SECURITY FORCE (BSF) 174,000 (Ministry of Home Affairs)
some 150 bn, small arms, some lt arty, tpt/liaison air spt

HOME GUARD (R) 472,000
authorised, actual str 416,000 in all states except Arunachal Pradesh and Kerala; men on lists, no trg

STATE ARMED POLICE 400,000
For duty primarily in home state only, but can be moved to other states, incl 24 bn India Reserve Police (commando-trained)

CIVIL DEFENCE 394,000 (R)
in 135 towns in 32 states

COAST GUARD over 8,000
- **PATROL CRAFT** 36
  - 3 *Samar* PCO, 9 *Vikram* PCO, 21 *Jija Bai*, 3 SDB-2 plus 16 boats
- **AVIATION**
  - 3 sqn with **ac** 14 Do-228, **hel** 15 *Chetak*

[a] Lightly armed security guards only

Opposition ε2,000+

HIZBUL MUJAHIDEEN: str n.k. Operates in Indian Kashmir
HARKAT-UL-MUJAHIDEEN: str n.k. Operates from Pakistan Kashmir
LASHKAR-E-TOIBA: str n.k. Operates from Pakistan Kashmir
TEHRIK-E-JIHAD: str n.k. Operates from Pakistan Kashmir
AL-BADR: str n.k. Operates in Indian Kashmir

Foreign Forces

UN (UNMOGIP): 46 mil obs from 8 countries

Kazakstan Kaz

tenge t		1998	1999	2000	2001
GDP	t	1.8tr	1.9tr		
	US$	22bn	14.5bn		
per capita	US$	3,600	3,800		
Growth	%	-2.5	1.7		
Inflation	%	7.0	8.2		
Debt	US$	5.7bn	7.9bn		
Def exp[a]	t	39bn	65bn		
	US$	498m	504m		
Def bdgt	t	19.9bn	15.1bn	16.5bn	
	US$	259m	117m	115m	
FMA[b] (US)	US$	0.6m	0.6m	0.6m	0.6m
US$1=t		78.3	128.9	142.5	

[a] Incl exp on paramilitary forces

[b] Excl US Cooperative Threat Reduction Programme funds for nuclear dismantlement and demilitarisation. Bdgt **1993–99** εUS$300m. Programme continues through 2000.

Population 15,000,000

(Kazak 51%, Russian 32%, Ukrainian 5%, German 2%, Tatar 2%, Uzbek 2%)

Age	13–17	18–22	23–32
Men	905,000	813,000	1,359,000
Women	883,000	800,000	1,333,000

Total Armed Forces

ACTIVE 64,000

Terms of service 31 months

Army 45,000

1 Mil District (3 more to form)
2 Army Corps (third to form)
1 with 1 mech div, 3 MR bde, 1 arty bde
1 with 1 mech div, 1 MR bde, 1 arty bde, 1 trg centre
1 air aslt, 1 SSM, 1 arty bde

EQUIPMENT

MBT 650 T-72, 280 T-62
RECCE 140 BRDM
ACV 508 BMP-1/-2, 65 BRM AIFV, 84 BTR-70/-80, 686 MT-LB APC (plus some 1,000 in store)
TOWED ARTY 505: **122mm**: 161 D-30; **152mm**: 74 D-20, 90 2A65, 180 2A36
SP ARTY 163: **122mm**: 74 2S1; **152mm**: 89 2S3
COMBINED GUN/MOR 120mm: 26 2S9
MRL 147: **122mm**: 57 BM-21; **220mm**: 90 9P140 *Uragan*
MOR 145: **120mm**: 2B11, M-120
SSM 12 SS-21
ATK GUNS 100mm: 68 T-12/MT-12

In 1991, the former Soviet Union transferred some 2,680 T-64/-72s, 2,428 ACVs and 6,900 arty to storage bases in Kazakstan. This eqpt is under Kazak control, but has deteriorated considerably. An eqpt destruction programme is about to begin.

Air Force 19,000

(incl Air Defence)
1 Air Force div, 131 cbt ac **Flying hours** 100
FTR 1 regt with 40 MiG-29
FGA 3 regt
1 with 14 Su-25
1 with 25 Su-24
1 with 14 Su-27
RECCE 1 regt with 12 Su-24*
TRG 12 L-39, 4 Yak-18
HEL numerous Mi-8, Mi-29
STORAGE some 75 MiG-27/MiG-23/MiG-23UB/MiG-25/MiG-29/SU-27
AIR DEFENCE
FTR 1 regt with 43 MiG-31, 16 MiG-25
SAM 100 SA-2, SA-3, 27 SA-4, SA-5, 20 SA-6, S-300
MISSILES
ASM AS-7 *Kerry*, AS-9 *Kyle*, S-10 *Karen*, AS-11 *Killer*
AAM AA-6 *Acrid*, AA-7 *Apex*, AA *Aphid*

Forces Abroad

TAJIKISTAN 300: 1 border gd bn

Paramilitary 34,500

STATE BORDER PROTECTION FORCES ε12,000 (Ministry of Defence) incl
MARITIME BORDER GUARD (3,000)
BASE Aktau
PATROL AND COASTAL COMBATANTS 10
5 *Guardian* PCI, 1 *Dauntless* PCI, 4 *Almaty* PCI, plus 2 boats
INTERNAL SECURITY TROOPS ε20,000 (Ministry of Interior)
PRESIDENTIAL GUARD 2,000
GOVERNMENT GUARD 500

Kyrgyzstan Kgz

som s		1998	1999	2000	2001
GDP	s	34.0bn	43.5bn		
	US$	1.8bn	1.1bn		
per capita	US$	2,100	2,100		
Growth	%	2.0	3.6		
Inflation	%	10	39		
Debt	US$	1,700m			
Def exp[a]	s	1,350m	1,972m		
	US$	65m	51m		
Def bdgt	s	573m	950m	1.4bn	
	US$	28m	24m	29m	
FMA (US)	US$	0.3m	0.3m	0.4m	0.4m
US$1=s		20.8	39.0	48.0	

[a] Incl exp on paramilitary forces

Population **4,852,000**

(Kyrgyz 56%, Russian 17%, Uzbek 13%, Ukrainian 3%)

Age	13–17	18–22	23–32
Men	285,000	241,000	365,000
Women	280,000	238,000	361,000

Total Armed Forces

ACTIVE 9,000

Terms of service 18 months

RESERVES 57,000

Army 6,600

1 MRD

2 indep MR bde (mtn), 1 AD bde, 1 AAA regt, 3 SF bn

EQUIPMENT

- **MBT** 190 T-72
- **RECCE** 30 BRDM-2
- **AIFV** 260 BMP-1, 100 BMP-2
- **APC** 24 BTR-70. 10 BTR-80
- **TOWED ARTY** 161: **100mm**: 18 M-1944 (BS-3); **122mm**: 72 D-30, 35 M-30; **152mm**: 16 D-1
- **SP ARTY 122mm**: 18 2S1
- **COMBINED GUN/MOR 120mm**: 24 2S9
- **MRL 122mm**: 15 BM-21
- **MOR 120mm**: 6 2S12, 48 M-120
- **ATGW** 26 AT-3 *Sagger*
- **ATK GUNS 100mm**: 18 T-12/MT-12
- **AD GUNS 23mm**: 16 ZSU-23-4SP; **57mm**: 24 S-60
- **SAM** SA-7

Air Force 2,400

ac and hel assets inherited from Sov Air Force trg school; Kgz failed to maintain pilot trg for foreign students.

AC 24 L-39, 50 MiG-21, 2 An-12, 2 An-26

HEL 11 Mi-24, 19 Mi-8

AIR DEFENCE

- **SAM** SA-2, SA-3, 12 SA-4

Forces Abroad

UN AND PEACEKEEPING

SIERRA LEONE (UNAMSIL): 2 obs

Paramilitary ε5,000

BORDER GUARDS ε5,000 (Kyrgyz conscripts, Russian officers)

Nepal N

rupee NR		1998	1999	2000	2001
GDP	NR	296bn	335bn		
	US$	4.4bn	4.9bn		
per capita	US$	1,500	1,500		
Growth	%	3.9	3.3		
Inflation	%	10.0	8.1		
Debt	US$	2.7bn	2.7bn		
Def exp	NR	1.3bn	1.8bn		
	US$	65m	51m		
Def bdgt	NR		1.0bn	1.4bn	
	US$		24m	29m	
FMA (US)	US$	0.2m	0.2m	0.2m	0.2m
US$1=NR		66.0	67.4	70.2	

Population **22,600,000**

(Hindu 90%, Buddhist 5%, Muslim 3%)

Age	13–17	18–22	23–32
Men	1,481,000	1,228,000	1,841,000
Women	1,400,000	1,143,000	1,679,000

Total Armed Forces

ACTIVE 46,000 (to be 50,000)

Army 46,000

1 Royal Guard bde (incl 1 MP bn) • 7 inf bde (16 inf bn) • 44 indep inf coy • 1 SF bde (incl 1 AB bn, 2 indep SF coy, 1 cav sqn (*Ferret*)) • 1 arty bde (1 arty, 1 AD regt) • 1 engr bde (4 bn)

EQUIPMENT

- **RECCE** 40 *Ferret*
- **TOWED ARTY**† **75mm**: 6 pack; **94mm**: 5 3.7in mtn (trg); **105mm**: 14 pack (ε6 op)
- **MOR 81mm**; **120mm**: 70 M-43 (ε12 op)
- **AD GUNS 14.5mm**: 30 PRC Type 56; **37mm**: PRC **40mm**: 2 L/60

AIR WING (215)

no cbt ac, or armed hel

TPT ac 1 BAe-748, 2 *Skyvan* **hel** 2 SA-316B *Chetak*, 1 SA-316B, 1 AS-332L (*Puma*), 2 AS-332L-1 (*Super Puma*), 1 Bell 206, 2 Bell 206L, 2 AS-350 (*Ecureuil*)

Forces Abroad

UN AND PEACEKEEPING

CROATIA (UNMOP): 1 obs **CYPRUS** (UNFICYP): 1 **DROC** (MONUC): 9 obs **EAST TIMOR** (UNTAET): 164 incl 5 obs **LEBANON** (UNIFIL) 712: 1 inf bn **SIERRA LEONE** (UNAMSIL): 6 obs

Paramilitary 40,000

POLICE FORCE 40,000

Opposition

COMMUNIST PARTY OF NEPAL (United Marxist and Leninist): armed wing ε1–1,500

Foreign Forces

UK **Army** 90 (Gurkha trg org)

Pakistan Pak

rupee Rs		1998	1999	2000	2001
GDP	Rs	2.8tr	3.0tr		
	US$	60.8bn	61.6bn		
per capita	US$	2,400	2,500		
Growth	%	4	3.1		
Inflation	%	6.2	4.1		
Debt	US$	32bn	34.5bn		
Def exp	Rs	180bn	173bn		
	US$	4.0bn	3.5bn		
Def bdgt	Rs	145bn	142bn	170bn	
	US$	3.2bn	2.9bn	3.3bn	
FMA[a] (US)	US$	1.5m	2.9m	0.4m	–
FMA (Aus)	US$	0.02m	0.02m		
US$1=Rs		45.0	49.1	52.0	

[a] UNMOGIP **1997** US$7m **1998** US$8m

Population	**148,012,000** (less than 3% Hindu)		
Age	13–17	18–22	23–32
Men	8,755,000	7,501,000	12,112,000
Women	8,337,000	6,815,000	10,735,000

Total Armed Forces

ACTIVE 612,000

RESERVES 513,000

Army ε500,000; obligation to age 45 (men) or 50 (officers); active liability for 8 years after service **Navy** 5,000 **Air Force** 8,000

Army 550,000

9 Corps HQ • 2 armd div • 9 Corps arty bde • 19 inf div • 7 engr bde • 1 area comd (div) • 3 armd recce regt • 7 indep armd bde • 1 SF gp (3 bn) • 9 indep inf bde • 1 AD comd (3 AD gp: 8 bde)

AVN 17 sqn

7 ac, 8 hel, 1 VIP, 1 obs flt

EQUIPMENT

MBT 2,285+: 15 M-47, 250 M-48A5, 50 T-54/-55, 1,200 PRC Type-59, 250 PRC Type-69, 200+ PRC Type-85, 320 T-80UD

APC 1,000+ M-113

TOWED ARTY 1,467: **85mm**: 200 PRC Type-56; **105mm**: 300 M-101, 50 M-56 pack; **122mm**: 200 PRC Type-60, 250 PRC Type-54; **130mm**: 227 PRC Type-59-1; **155mm**: 30 M-59, 60 M-114, 124 M-198; **203mm**: 26 M-115

SP ARTY 105mm: 50 M-7; **155mm**: 150 M-109A2; **203mm**: 40 M-110A2

MRL 122mm: 45 *Azar* (PRC Type-83)

MOR 81mm: 500; **120mm**: 225 AM-50, M-61

SSM 80 *Hatf* 1, 30 *Hatf* 3 (PRC M-11), *Shaheen* 1, 12 *Ghauri*

ATGW 800 incl: *Cobra*, 200 TOW (incl 24 on M-901 SP), *Green Arrow* (PRC *Red Arrow*)

RL 89mm: M-20 3.5in

RCL 75mm: Type-52; **106mm**: M-40A1

AD GUNS 2,000+ incl: **14.5mm; 35mm**: 200 GDF-002; **37mm**: PRC Type-55/-65; **40mm**: M1, 100 L/60; **57mm**: PRC Type-59

SAM 350 *Stinger*, *Redeye*, RBS-70, 500 *Anza* Mk-1/-2

SURV RASIT (veh, arty), AN/TPQ-36 (arty, mor)

AIRCRAFT

SURVEY 1 *Commander* 840

LIAISON 1 Cessna 421, 2 *Commander* 690, 80 *Mashshaq*, 1 F-27, 2 Y-12 (II)

OBS 40 O-1E, 50 *Mashshaq*

HELICOPTERS

ATTACK 20 AH-1F (TOW)

TPT 12 Bell 47G, 7 -205, 10 -206B, 16 Mi-8, 6 IAR/SA-315B, 23 IAR/SA-316, 35 SA-330, 5 UH-1H

Navy 22,000

(incl Naval Air, ε1,200 Marines and ε2,000 Maritime Security Agency (see *Paramilitary*))

BASE Karachi (Fleet HQ) (2 bases being built at Gwadar and Ormara)

SUBMARINES 10

SSK 7

1 *Khalid* (Fr *Agosta* 90B) with 533mm TT, *Exocet* SM39 USGW

2 *Hashmat* (Fr *Agosta*) with 533mm TT (F-17 HWT), *Harpoon* USGW

4 *Hangor* (Fr *Daphné*) with 533mm TT (L-5 HWT), *Harpoon* USGW

SSI 3 MG110 (SF delivery)

PRINCIPAL SURFACE COMBATANTS 8

FRIGATES 8

FFG 6 *Tariq* (UK *Amazon*) with 4 × *Harpoon* SSM (in 3

Central and South Asia

of class), 1 × *LY-60N* SAM (in 3 of class), 1 × 114mm gun, 6 × 324mm ASTT, 1 *Lynx* HAS-3

FF 2 *Shamsher* (UK *Leander*) with 2 × 114mm guns, 1 × 3 ASW mor, 1 SA-319B hel

PATROL AND COASTAL COMBATANTS 9

MISSILE CRAFT 5

4 *Sabqat* (PRC *Huangfeng*) PFM with 4 *HY* 2 SSM

1 × *Jalalat* II with 4 C-802 SSM

PATROL, COASTAL 1 *Larkana* PCC

PATROL, INSHORE 3

2 *Quetta* (PRC *Shanghai*) PFI

1 *Rajshahi* PCI

MINE COUNTERMEASURES 3

3 *Munsif* (Fr *Eridan*) MHC

SUPPORT AND MISCELLANEOUS 9

1 *Fuqing* AO, 1 *Moawin* AO, 2 *Gwadar* AOT, 1 *Attack* AOT; 3 AT; 1 *Behr Paima* AGHS

NAVAL AIR

5 cbt ac (all operated by Air Force), 9 armed hel

ASW/MR 1 sqn with 3 *Atlantic* plus 2 in store, 2 P-3C (operated by Air Force)

ASW/SAR 2 hel sqn with 6 *Sea King* Mk 45 (ASW), 3 *Lynx* HAS Mk-3 (ASW)

COMMS 5 Fokker F-27 **ac** (Air Force) **hel** 4 SA-319B

ASM *Exocet* AM-39

MARINES (ε1,200)

1 cdo/SF gp

Air Force 40,000

353 cbt ac, no armed hel **Flying hours** some 210

3 regional cmds: **Northern** (Peshawar) **Central** (Sargodha) **Southern** (Faisal). The Composite Air Tpt Wg, Combat Cdrs School and PAF Academy are Direct Reporting Units.

FGA 6 sqn

1 with 16 *Mirage* (13 IIIEP (some with AM-39 ASM), 3 IIIDP (trg))

3 (1 OCU) with 52 *Mirage* 5 (40 -5PA/PA2, 10 5PA3 (ASuW), 2 5DPA/DPA2)

2 with 42 Q-5 (A-5III *Fantan*), some FT-6

FTR 12 sqn

3 (1 OCU) with 40 F-6/FT-6 (J-6/JJ-6), 2 (1 OCU) with 32 F-16 (22 -A, 10 -B), 6 (1 OCU) with 77 F-7P/FT-7 (J-7), 1 with 43 *Mirage* IIIO/7-OD

RECCE 1 sqn with 11* *Mirage* IIIRP

ELINT/ECM 2 *Falcon* DA-20

SAR 1 hel sqn with 15 SA-319

TPT ac 12 C-130 (11 B/E, 1 L-100), 2 Boeing 707, 1 Boeing 737, 1 *Falcon* 20, 2 F-27-200 (1 with Navy), 1 Beech *Super King Air* 200, 2 Y-12 (II), **hel** 15 SA 316/319, 4 Cessna 172, 1 Cessna 560 *Citation*, 1 *Piper* PA-34 *Seneca*, 4 MFI-17B *Mashshaq*

TRG 30 FT-5, 15 FT-6, 13 FT-7, 40* MFI-17B *Mashshaq*, 30 T-37B/C, 12 K-8

AD 7 SAM bty

6 each with 24 *Crotale*, 1 with 6 CSA-1 (SA-2)

MISSILES

ASM AM-39 *Exocet*, AGM-65 *Maverick*, AS 30, AGM-84 *Harpoon*

AAM AIM-7 *Sparrow*, AIM-9L/P *Sidewinder*, R-530 *Magic*

ARM AGM-88 *Harm*

Forces Abroad

UN AND PEACEKEEPING

CROATIA (UNMOP): 1 obs **DROC** (MONUC): 29 obs **EAST TIMOR** (UNTAET): 804 incl 30 obs **GEORGIA** (UNOMIG): 7 obs **IRAQ/KUWAIT** (UNIKOM): 6 obs **SIERRA LEONE** (UNAMSIL): 10 obs **WESTERN SAHARA** (MINURSO): 6 obs

Paramilitary ε288,000 active

NATIONAL GUARD 185,000

incl *Janbaz* Force, *Mujahid* Force, National Cadet Corps, Women Guards

FRONTIER CORPS up to 65,000 reported (Ministry of Interior)

11 regt (40 bn), 1 indep armd car sqn; 45 UR-416 APC

PAKISTAN RANGERS ε25,000–30,000 (Ministry of Interior)

NORTHERN LIGHT INFANTRY ε12,000; 3 bn

MARITIME SECURITY AGENCY ε1,000

1 *Alamgir* (US *Gearing* DD) (no ASROC or TT), 4 *Barkat* PCO, 2 (PRC *Shanghai*) PFI<

COAST GUARD

some 23 craft

Foreign Forces

UN (UNMOGIP): 46 mil obs from 8 countries

Sri Lanka Ska

rupee Rs		1998	1999	2000	2001
GDP	Rs	1,029bn	1,113bn		
	US$	16bn	15.7bn		
per capita	US$	4,100	4,200		
Growth	%	5.6	4.2		
Inflation	%	9.4	4.7		
Debt	US$	8.5bn	8.9bn		
Def exp	Rs	63bn	57bn		
	US$	975m	807m		
Def bdgt	Rs	47bn	45bn	52bn	
	US$	733m	635m	700m	
FMA (US)	US$	0.2m	0.2m	0.2m	0.2m
US$1=Rs		64.6	70.9	74.9	

Population			**19,035,000**
(Sinhalese 74%, Tamil 18%, Moor 7%; Buddhist 69%, Hindu 15%, Christian 8%, Muslim 8%)			
Age	13–17	18–22	23–32
Men	927,000	916,000	1,589,000
Women	890,000	880,000	1,555,000

Total Armed Forces

ACTIVE some 110–115,000
(incl recalled reservists)

RESERVES 4,200
Army 1,100 **Navy** 1,100 **Air Force** 2,000
Obligation 7 years, post regular service

Army ε90–95,000

(incl 42,000 recalled reservists; ε1,000 women)
10 div • 3 mech inf bde • 1 air mobile bde • 23 inf bde • 1 indep SF bde • 1 cdo bde • 1 armd regt • 3 armd recce regt (bn) • 4 fd arty (1 reserve) • 4 fd engr regt (1 reserve)

EQUIPMENT
MBT ε25 T-55 (perhaps 18 op)
RECCE 26 *Saladin*, 15 *Ferret*, 12 Daimler *Dingo*
AIFV 16 BMP (12 -1, 4 -2) (trg)
APC 35 PRC Type-85, 10 BTR-152, 31 *Buffel*, 30 *Unicorn*, 10 Shorland, 6 *Hotspur*, 30 *Saracen*, some BTR-80A (reported)
TOWED ARTY 76mm: 12 FRY M-48; **85mm**: 12 PRC Type-56; **88mm**: 12 25-pdr; **122mm**: some; **130mm**: 12 PRC Type-59-1; **152mm**: 33 PRC Type-66
MRL 122mm: 16 RM-70
MOR 81mm: 276; **82mm**: 100+; **107mm**: 12; **120mm**: 36 M-43
RCL 105mm: 15 M-65; **106mm**: 34 M-40
AD GUNS 40mm: 24 L-40; **94mm**: 3 3.7in
SURV 2 AN/TPQ-36 (arty)
UAV 1 *Seeker*

Navy 10,000

(incl 1,100 recalled reservists)
BASES Colombo (HQ), Trincomalee (main base), Karainagar, Tangalle, Kalpitiya, Galle, Welisara
PATROL AND COASTAL COMBATANTS 39
PATROL, OFFSHORE 2
1 *Jayesagara* PCO
1 *Parakrambahu* PCO
PATROL, COASTAL 5
2 *Rana* PCC
3 *Sooraya* PCC
PATROL, INSHORE 32
3 *Dvora* PFI<
8 *Super Dvora* PFI<
3 ROC *Killer* PFI<
10 *Colombo* PFI<
6 *Trinity Marine* PFI<
2 *Shaldag* PFI<
plus some 36 boats
AMPHIBIOUS 1
1 *Wuhu* LSM
plus 7 craft: 2 LCM, 2 LCU, 1 ACV, 2 fast personnel carrier

Air Force 10,000

26 cbt ac, 19 armed hel **Flying hours** 420
FGA 4 F-7M, 1 FT-7, 2 FT-5, 12 *Kfir* (8 -C2, 4 -C7), 1 *Kfir*-TC2, 4 MiG-27
ARM AC 8 SF-260TP, 2 FMA IA58A *Pucara*
ATTACK HEL 11 Bell 212, 6 Mi-24V, 2 Mi-35
TPT 1 sqn with **ac** 3 BAe 748, 2 C-130C, 1 Cessna 421C, 1 *Super King Air*, 1 Y-8, 9 Y-12 (II), 4 An-24, 7 An-32B, 1 Cessna 150 **hel** 3 Bell 412 (VIP)
HEL 9 Bell 206, 3 Mi-17 (plus 6 in store)
TRG incl 4 DHC-1, 4 SF-260 W, 3 Bell 206
RESERVES Air Force Regt, 3 sqn; Airfield Construction, 1 sqn
UAV 5 *Superhawk*

Paramilitary ε88,600

POLICE FORCE (Ministry of Defence) 60,600
incl 30,400 reserves, 1,000 women and Special Task Force: 3,000-strong anti-guerrilla unit
NATIONAL GUARD ε15,000
HOME GUARD 13,000

Opposition

LIBERATION TIGERS OF TAMIL EELAM (LTTE) ε6,000
1 Robinson R-44 *Astro* light helicopter plus 2 light aircraft for reconnaisance and liaison
Leader Velupillai Prabhakaran

Tajikistan Tjk

rouble Tr		1998	1999	2000	2001
GDP[a]	Tr	1,025bn	1,256bn		
	US$	1.2bn	1.2bn		
per capita	US$	900	1,000		
Growth	%	5.3	3.7		
Inflation	%	41	23		
Debt	US$	1,069m			
Def exp[a]	US$	ε100m	ε95m		
Def bdgt[a]	US$		18m		
US$1=Tr		850	1,035	1,436	

[a] UNMOT **1997** US$8m **1998** US$8m

Population			6,105,000
(Tajik 67%, Uzbek 25%, Russian 2%, Tatar 2%)			
Age	13–17	18–22	23–32
Men	422,000	334,000	485,000
Women	409,000	326,000	474,000

Total Armed Forces

ACTIVE some 6,000

Terms of service 24 months

A number of potential officers are being trained at the Higher Army Officers and Engineers College, Dushanbe. It is planned to form an Air Force sqn and to acquire Su-25 from Belarus, 5 Mi-24 and 10 Mi-8 have been procured.

Army some 6,000

2 MR bde (incl 1 trg), 1 mtn bde
1 SF bde, 1 SF det (εbn+)
1 SAM regt

EQUIPMENT

MBT 35 T-72, 3 T-62
AIFV 11 BMP-1, 24 BMP-2
APC 2 BTR-60, 7 BTR-70, 32 BTR-80
TOWED ARTY 122mm: 13 D-30
MRL 122mm: 3 BM-21
MOR 122mm: 9
SAM 20 SA-2/-3
HEL 15 attack, 40 utility

Paramilitary ε1,200

BORDER GUARDS ε1,200 (Ministry of Interior)

Opposition

ISLAMIC MOVEMENT OF TAJIKISTAN some 5,000

Signed peace accord with government on 27 June 1997. Integration with govt forces slowly proceeding

Foreign Forces

UN (UNMOT): 18 mil obs from 10 countries
RUSSIA Frontier Forces ε14,500 (Tajik conscripts, Russian officers) **Army** 8,200; 1 MRD

EQUIPMENT

MBT 190 T-72
AIFV/APC 313 BMP-2, BRM-1K, BTR-80
SP ARTY 122mm: 66 2S1; **152mm**: 54 2S3
MRL 122mm: 12 BM-21; **220mm**: 12 9P140
MOR 120mm: 36 PM-38

AIR DEFENCE

SAM 20 SA-8

KAZAKSTAN ε300: 1 border gd bn

Turkmenistan Tkm

manat		1998	1999	2000	2001
GDP	US$	ε2.9bn	ε3.3bn		
per capita	US$	2,000	2,200		
Growth	%	4.5	18.5		
Inflation	%	16.8	27		
Debt	US$	1,665m			
Def exp	US$	ε93m	ε109m		
Def bdgt	US$	98m	108m	157m	
FMA (US)	US$	0.3m	0.3m	0.3m	0.3m
US$1=manat		4,700	5,350	5,350	

Population			5,000,000
(Turkmen 77%, Uzbek 9%, Russian 7%, Kazak 2%)			
Age	13–17	18–22	23–32
Men	267,000	222,000	354,000
Women	260,000	219,000	349,000

Total Armed Forces

ACTIVE 17,500

Terms of service 24 months

Army 14,500

5 Mil Districts • 4 MRD (1 trg) • 1 arty bde • 1 MRL regt • 1 ATK regt •1 engr bde • 2 SAM bde • 1 indep air aslt bn

EQUIPMENT

MBT 690 T-72
RECCE 170 BRDM/BRDM-2
AIFV 930 BMP-1/-2, 12 BRM
APC 840 BTR (-60/-70/-80)
TOWED ARTY 122mm: 197 D-30; **152mm**: 17 D-1, 72 D-20
SP ARTY 122mm: 40 2S1
COMBINED GUN/MOR 120mm: 17 2S9
MRL 122mm: 56 BM-21, 9 9P138
MOR 82mm: 31; **120mm**: 66 PM-38
ATGW 100 AT-3 *Sagger*, AT-4 *Spigot*, AT-5 *Spandrel*, AT-6 *Spiral*
ATK GUNS 100mm: 72 T-12/MT-12
AD GUNS 23mm: 48 ZSU-23-4 SP; **57mm**: 22 S-60
SAM 40 SA-8, 13 SA-13

Navy none

Has announced intention to form a Navy/Coast Guard. Caspian Sea Flotilla (see **Russia**) is operating as a joint RF, Kaz and Tkm flotilla under RF comd based at Astrakhan.

Air Force 3,000

(incl Air Defence)
243 cbt ac (plus 172 in store)
FGA/FTR 1 composite regt with 22 MiG-29, 2 MiG-

29U, 65 Su-17, some Su-25

FTR 2 regt with 120 MiG-23, 10 MiG-23U, 24 MiG-25 (both regt non-operational)

TPT/GENERAL PURPOSE 1 composite sqn with 3 An-12, 1 An-26, 10 Mi-24, 8 Mi-8

TRG 1 unit with 3 Su-7B, 2 L-39

AIR DEFENCE

SAM 50 SA-2/-3/-5

IN STORE 172 MiG-23

Uzbekistan Uz

som s		1998	1999	2000	2001
GDP	s	1,344bn	1,942bn		
	US$	14.8bn	15.9bn		
per capita	US$	2,700	2,800		
Growth	%	4.4	4.4		
Inflation	%	23	20		
Debt	US$	3.1bn			
Def exp[a]	US$	657m	615m		
Def bdgt	US$	343m	285m		
FMA (US)	US$	0.4m	0.5m	0.5m	0.5m
US$1=s[b]		91.4	122	133	

[a] Incl exp on paramilitary forces

[b] Market rate **2000** US$1=εs775

Population **24,100,000**

(Uzbek 73%, Russian 6%, Tajik 5%, Kazak 4%, Karakalpak 2%, Tatar 2%, Korean <1%, Ukrainian <1%)

Age	13–17	18–22	23–32
Men	1,508,000	1,259,000	1,926,000
Women	1,476,000	1,242,000	1,921,000

Total Armed Forces

ACTIVE some 59,100

(incl MoD staff and centrally controlled units)

Terms of service conscription, 18 months

Army 50,000

4 Mil Districts, 2 op comd, 1 Tashkent comd

1 tk, 11 MR, 1 lt mtn, 1 AB, 3 air aslt, 5 engr bde

1 National Guard bde

EQUIPMENT

MBT 190 T-62, 100 T-64, 60 T-72

RECCE 13 BRDM-2

AIFV 160 BMP-2, 120 BMD-1, 9 BMD-2, 6 BRM

APC 25 BTR-70, 24 BTR-60, 210 BTR-80, 120 BTR-D

TOWED ARTY 122mm: 70 D-30; **152mm**: 140 2A36

SP ARTY 122mm: 18 2S1; **152mm**: 17 2S3, 2S5 (reported); **203mm**: 48 2S7

COMBINED GUN/MOR 120mm: 54 2S9

MRL 122mm: 36 BM-21, 24 9P138; **220mm**: 48 9P140

MOR 120mm: 18 PM-120, 19 2S12, 5 2B11

ATK GUNS 100mm: 36 T-12/MT-12

(In 1991 the former Soviet Union transferred some 2,000 tanks (T-64), 1,200 ACV and 750 arty to storage bases in Uzbekistan. This eqpt is under Uzbek control, but has deteriorated considerably.)

Air Force some 9,100

7 regts

135 cbt ac, 42 attack hel

BBR/FGA 1 regt with 20 Su-25/Su-25BM, 26 Su-17MZ/Su-17UMZ, 1 regt with 23 Su-24, 11 Su-24MP (recce)

FTR 1 regt with 30 MiG-29/MiG-29UB, 1 regt with 25 Su-27/Su-27UB

TPT/ELINT 1 regt with 26 An-12/An-12PP, 13 An-26/An-26RKR

TPT 1 Tu-134, 1 An-24

TRG 14 L-39

HELICOPTERS

1 regt with 42 Mi-24 (attack), 29 Mi-8 (aslt/tpt), 1 Mi-26 (tpt)

1 regt with 26 Mi-6 (tpt), 2 Mi-6AYa (cmd post), 29 Mi-8 (aslt/tpt)

MISSILES

AAM AA-8, AA-10, AA-11

ASM AS-7, AS-9, AS-10, AS-11, AS-12

SAM 45 SA-2/-3/-5

Paramilitary ε18–20,000

INTERNAL SECURITY TROOPS (Ministry of Interior) ε17–19,000

NATIONAL GUARD (Ministry of Defence) 1,000

1 bde

Opposition

ISLAMIC MOVEMENT OF UZBEKISTAN

Leader Tahir Yoldosh **Based** near Kunduz, Afghanistan; sometimes supported by Juma Numangoni, warlord, based in Tajikistan or Afghanistan

East Asia and Australasia

MILITARY DEVELOPMENTS

Regional Trends

The East Asia and Australasia region remains the second-largest regional arms market after the Middle East and North Africa. Military spending overall has recovered along with the Asian economy and in 1999 it was up on the previous year by just over 6% to $135 billion. As long as fundamental insecurities persist, this rising trend will continue. Indonesia, the second most populous country in East Asia, remains beset by conflict, although some order has been restored to East Timor, which is now under UN administration. After years of relative calm, long-standing Islamist and separatist insurgencies have resurfaced in the Philippines. In north-east Asia, the military situation is essentially unchanged, yet there have been positive political developments, in particular the summit meeting between the leaders of the two Koreas in June 2000, which seemed finally to end North Korea's diplomatic isolation. China, despite tensions over Taiwan and US plans for a national missile defence (NMD), remains absorbed with its internal political and economic development. Military confrontation over Taiwan seems a remote possibility, provided the key actors remain preoccupied by their other priorities. Outside the arena of major-power confrontation, ethnic violence erupted in the South Pacific region in Fiji and the Solomon Islands, arising from long-standing grievances that remain unresolved.

North-east Asia

The Summit Without doubt the region's most important political event was the summit meeting in Pyongyang on 13–15 June 2000, between President Kim Jong Il of North Korea and President Kim Dae Jung of South Korea. By August, some progress appears to have been made on the aims set out by the two leaders in the 15 June Joint Declaration.

Groups of senior officials from the North and South had met by the end of July. The two Koreas are establishing working groups to deal with the economic, political and cultural issues and endeavouring to install military hotlines and to reconnect suspended railway lines. At this stage, however, no officials dealing directly with military affairs are taking part in the talks. Both sides have stopped propaganda broadcasts directed at each other and the first North–South family reunions took place on 15 August. The summit's reverberations have gone beyond the elements in the Joint Declaration. On 19 June, the US announced its intention to lift the embargo on trade between the two countries, which has been in force for 50 years. Imports and exports of most goods will be allowed except for military items and technology with potential military uses. For the time being, the US has indicated that it will continue to block loans by international financial institutions to North Korea, because it has appeared on the US list of states that sponsor terrorism since the 1987 bombing of a Korean Air passenger plane by North Korean agents. Talks began on 9 August 2000 in Pyongyang between US and North Korean officials on how the country can meet the conditions for removal from the list. For impoverished North Korea, being no longer listed opens up the prospect of large-scale loans from such institutions as the World Bank and the Asian Development Bank.

In the past five years, international help has brought North Korea back from the brink of mass starvation. There are concerns that the aid funds are subject to insufficient controls and could be used to maintain and develop military capabilities. According to South Korean government sources, trade between the two Koreas totalled $333 million in 1999; most of this 'trade' was humanitarian aid to the North. Aid will increase as more countries open diplomatic relations with

Pyongyang. Greater exposure to outside influences could compel Kim Jong Il to spend more on economic rehabilitation and less on defence. Such a shift, however, could strain relations between the leader and the military on which he depends for support. This could explain Kim's insistence that North Korea would continue to trade in missile technology with countries such as Iran and Pakistan. Some estimates indicate that these exports bring in at least $500m a year, a sum that almost certainly goes to the military. The technology exports and advancing North Korean missile programme form a powerful lever for bargaining with the US and Japan. However, for North Korea to benefit from the engagement process, it must take more substantial steps on non-proliferation than the curious offer, conveyed to the Tokyo meeting of the G-8 group of leading industrial nations in July 2000. The proposal, via Russian President Vladimir Putin, was that Pyongyang would stop its missile programme in return for satellite-launch technology. At this stage Kim Jong Il seems to be concentrating his efforts on attracting foreign trade, aid and investment, which do not yet require what the leadership perceives to be destabilising transparency or demilitarisation.

China Beijing's main preoccupation in 2000 has been dealing with domestic economic reform, in particular preparing for membership of the World Trade Organisation (WTO) and its impact on China's political development. Another domestic preoccupation has been how to respond to the perceived threat to the leadership posed by the *Falun Gong* quasi-religious movement. These concerns have probably contributed to the muted response Beijing has made to other events, such as the outcome of the 2000 presidential election in Taiwan and US plans for NMD. Neither of these last developments has yet substantively changed China's plans for a steady improvement in its military organisation and capability.

The election of President Chen Shui-bian in Taiwan in March 2000, given his pro-independence background, shocked the Beijing leadership, but its response was muted. Chen has so far been careful to avoid public statements that risk raising tensions. There has been no significant change in the balance of forces in 2000 beyond a modest increase in the short-range missile capabilities along the coast opposite Taiwan in line with plans laid in the late 1990s. The US has played a role in reducing military tensions by turning down a Taiwanese request to buy *Aegis* missile cruisers, which are capable of being fitted with a theatre missile defence (TMD) system. However, this has not prevented a certain amount of megaphone diplomacy by China, condemning US NMD plans. On this front, China and Russia have made common cause. In a joint statement on 18 July 2000, the Russian and Chinese presidents denounced the US NMD plans and vowed to strengthen a strategic partnership between the two countries. China is the biggest customer for Russian arms exports.

Compared to other nuclear powers, China's strategic military capabilities remain limited. It has approximately 25 land-based intercontinental ballistic missiles (ICBM) operational at any one time; its submarine-launched capability is a single nuclear-fuelled ballistic-missile submarine (SSBN), still not fully in service. The longer-range power-projection capabilities of the Army and Air Force remain limited, although increasing in some respects. The Army is continuing to cut its regular and reserve forces, while improving the mobility and training of field formations. This continues the slow process begun in the early 1990s with the move away from territorial defence, involving large forces stationed throughout the country, to a more mobile, smaller force able to respond rapidly to internal and external threats. The process of reorganising six of the 21 Group Armies has continued, with conversion from a division-based structure to a more flexible brigade-based structure. This reform is scheduled for completion before the end of 2001. The introduction of more up-to-date equipment continues, including a significant number of Type-88C tanks, followed by a smaller number of Type-98. Although China's power-projection capability is limited at present, its military activities beyond its territory, however modest, can excite tensions, given regional concerns about China's potential. For example, in July 2000, the Japanese government issued a report citing a sharp increase in Chinese naval vessels seen in and around Japan's territorial waters: 28 sightings in 1999 compared with two in 1998. Reports of these activities so outraged the Japanese public that Tokyo went as far as to consider cancelling plans for $105m-worth of special-development assistance to China. To the south-west, China's activities in Myanmar have caused India concern for some years. China has equipped Myanmar's armed forces with most of their major weapons. It has also helped to strengthen the country's river, rail and road infrastructures and continues to man the electronic listening posts that they have established on the Indian Ocean coastline. At this stage, however, these developments are not likely to lead to a military confrontation between India and China, but more to a vying for influence over Myanmar's military rulers.

South-east Asia and Pacific

Indonesia Indonesia's internal-security situation has continued to deteriorate, despite hopes aroused by the election of President Abdurrahman Wahid in October 1999. The government's failure to stem the tide of violence is one of the principal reasons why, in August 2000, Wahid declared he was handing the daily running of his government over to his vice-president, Megawati Sukarnoputri. He could still have a major influence on policy, but will be less in public view. Indonesia's strife results from a complex mixture of separatist movements and sectarian violence, principally between Muslims and Christians. The armed forces and police in many instances have supported one side against the other and exacerbated the problem rather than bringing security and asserting central-government authority. Aceh witnessed the worst fighting of this kind in late 1999 and early 2000, after the government refused to countenance a referendum on the future status of the province. The government and the Free Aceh movement ultimately signed a cease-fire on 12 May, which was implemented in June. By the middle of 2000, the separatist movements appeared to have been at least temporarily contained; however, there is sporadic violence that could worsen if no substantive political progress is made. Wahid conducted negotiations with other separatist movements in Irian Jaya and Papua in June and July 2000. The situation has been relatively calm in these areas during 2000. By far the worst violence has been in the Moluccan Islands, particularly around the capital, Ambon, where it is believed that more than 3,000 people have been killed in fighting between Christians and Muslims. The security forces seem unable to restore calm, and the outlook is bleak.

The Indonesian Army has been the loser in the political changes because of its association with the Suharto regime, involvement in corruption and appalling human-rights abuses during its effort to combat the violence over the past two years. The Army's personnel numbers are likely to be cut substantially. The Special Forces (KOPASSUS) will almost certainly be cut back sharply, primarily because of their involvement in some of the worst human-rights violations. The Strategic Reserve (KOSTRAD), at present some 30,000 strong, is thought likely to emerge as the main element of a more professional army; it already contains more high quality specialist troops, such as its airborne brigade. The restructuring of the armed forces is focusing on the Navy and its Marines, and the Air Force. Wahid has created a cabinet-level post to oversee maritime affairs and identified upgrading the Navy as his most important defence priority – partly because only about 20–25% of its inventory are operational. The government is making plans to increase naval personnel from 47,000 to 67,000 during the next five years and to buy maritime helicopters and corvettes. Furthermore, in late 1999, a naval officer was appointed commander of the Indonesian armed forces; previously it was always an army appointment. There are plans afoot to expand the Marines from 12,000 to some 20,000 in the next five years, with the immediate aim of creating a third brigade. This restructuring in favour of maritime and air forces is long overdue for a country that is an archipelago of 13,000 islands. However, equipment funding for the restructured forces, particularly for land forces, is likely to be extremely limited for at least the next five years because of the poor state of the economy.

East Timor On 26 February 2000, the UN Transitional Administration in East Timor (UNTAET) took over the military operations of the Australian-led International Force in East Timor (INTERFET). The force had been providing security for those engaged in humanitarian assistance and the UN administration in the territory following the withdrawal of Indonesian forces. Since February the 9,000-strong UN force has faced direct attacks from militias operating from West Timor. Two UN soldiers have been killed and several others wounded. Calls on the Indonesian government by the UN Security Council in August 2000 to clamp down on the militia had no immediate results.

Philippines After four years of relative peace, separatist violence has escalated, forcing the Army to undertake major operations against the Moro Islamic Liberation Front (MILF). The security forces have also had to contend with the *Abu Sayyaf* group who seized more than 30 local and foreign hostages in March 2000. These military commitments have delayed the planned restructuring of the Army, including proposals to reduce its eight infantry divisions to three and to reorganise them as rapid-deployment forces. Eight independent brigades are to be created to form the nucleus of eight reserve divisions. Other proposed changes include reorganising the Armoured Brigade with heavier armoured fighting vehicles (AFVs) and the creation of new artillery, air-defence and aviation battalions. The formation of a Special Forces Command has already taken place and includes Special Forces and Scout Ranger regiments.

Piracy Incidents of piracy continue to rise in South-east Asia, although the general incidence is low in relation to the high volume of shipping in the area. In March and April, Japan canvassed a proposal for its coastguard to contribute to a regional operation to counter piracy. Should the proposal be implemented, Japanese vessels would be deployed more than 1,000 miles away from home waters. Most members of the Association of South-east Asian Nations (ASEAN) were in favour of the proposal; China, however, was vehemently opposed. The Japanese government is also to fund a feasibility study into building a canal across Thailand, which would obviate the need for ships to transit the Straits of Malacca. Indonesia made its own tentative contribution to tackling piracy off its shores in early 2000, announcing plans to create a special centre to track shipping in the Straits of Malacca. The centre would depend on reporting by merchant vessels and upon a rapid-reaction fleet of fast ships and aircraft.

South Pacific Longstanding ethnic rivalries surfaced in Fiji and the Solomon Islands during 2000. On 19 May, supporters of nationalist rebel leader George Speight stormed Fiji's parliament, demanding that ethnic Indians be stripped of political power. They took hostage Prime Minister Mahendra Chaudhry and members of his cabinet. Ten days after the coup attempt, Fiji's military assumed power and declared martial law. It took until 9 July for Fiji's military to make a deal with Speight. Many of the rebel leader's demands aimed at preserving indigenous Fijian political dominance and stripping ethnic Indians of a political role were met in a provisional agreement. Four days later, on 13 July, the deposed prime minister and 17 other MP's were released from captivity. On 27 July, the military arrested Speight on charges of arms offences. A caretaker government, led by Prime Minister Laisenia Qarase, and containing no Speight supporters, was sworn in on 28 July.

On 5 June 2000, armed rebels seized control of the Soloman Islands' capital, Honiara, held Prime Minister Bartholomew Ulufa'ulu hostage and demanded his resignation. The Malaitan Eagles (ME) militia has been fighting for the rights of people originally from the island of Malaita, who now live on neighbouring Guadalcanal Island. By 11 June, however, the conflict spread to another island where a third rebel group seized control of the western town of Gizo. These militiamen were reportedly linked to separatists on neighbouring Bougainville Island. On 30 June, by a narrow majority, parliament elected the former opposition leader, Manasseh Sogavare, as Prime Minister. An agreement to hold talks aimed at ending the conflict was reached on 12 July. Overall about 60 people were killed in the fighting and around 20,000 migrants from the island of Malaita were expelled from Guadalcanal. Outside military intervention in either the Solomon Islands or Fiji was never seriously considered, other than to rescue foreign nationals if necessary.

DEFENCE SPENDING

Regional Trends

Regional military spending in 1999 increased in real terms by 6.2% to $135bn measured in constant 1999 US dollars. Following increased budgetary allocations and with the major regional currencies stabilised, the trend is set to continue in 2000. Published budgets for 2000 indicate a further 4% increase in military spending. Asian economies rebounded in 1999 from the crisis which hit the region in mid-1997. Regional gross domestic product (GDP) which fell by over 7% in 1998, increased by 10% in 1999 when measured in US dollars at average exchange rates. The only economy to decline during the year was that of Japan, where GDP fell by 1.4%. Indonesia's economy was static, while all the other countries in the region saw a strong recovery, particularly Thailand where GDP rose by 4% and Malaysia where it was up 5.4%.

Japan The Japanese defence budget saw its first increase in yen terms for three years. The Japanese cabinet approved a budget for 2000 of ¥4,935bn, a rise of 0.3% over the 1999 budget of ¥4,920bn. However, due to the strength of the yen this translates to a nominal rise in dollar terms to $45.6bn compared to $43.2bn. Rising personnel and maintenance costs predominantly account for the ¥15bn increase in the budget. Provision for equipment procurement has fallen by ¥34.4bn, mainly due to the Japanese Defence Agency's shift in emphasis towards spending more on training, intelligence capabilities and readiness. Under the joint TMD programme with the US, the budget includes ¥2bn ($17m) to fund research into the project.

Table 23 **Japan defence budget by function and selected other budgets, 1993–2000**

(1999 US$bn)	1993	1994	1995	1996	1997	1998	1999	2000
Personnel	17.4	19.5	22.0	19.1	17.6	16.6	19.0	20.2
Supplies	24.3	26.3	28.2	25.5	23.3	21.0	24.2	24.9
of which								
Procurement	9.7	9.8	9.2	8.4	7.7	7.2	8.2	8.4
R&D	1.1	1.2	1.5	1.4	1.3	1.0	1.2	1.1
Maintenance	6.8	7.8	8.8	8.0	7.4	6.9	7.8	8.2
Infrastructure	6.7	7.5	8.6	7.6	6.8	6.1	6.9	6.5
Total defence budget	**41.7**	**45.8**	**50.2**	**44.5**	**40.8**	**37.6**	**43.2**	**45.2**
SACO				0.1	0.1	0.1	0.1	
Coast Guard	1.4	1.5	1.7	1.5	1.4	1.3	1.5	1.6
Veterans	13.9	15.1	16.1	13.4	11.6	10.3	11.5	11.7
Space	1.5	1.7	2.0	2.2	2.5	2.7	3.4	n.a.

Table 24 **Japan: selected major weapons procurement, 1993–2000**

(1999 US$bn)	1993	1994	1995	1996	1997	1998	1999	2000
Japanese Defense Agency	37.1	40.5	44.2	39.3	36.1	33.4	38.1	39.8
of which								
Ground Self-Defense Force	15.0	16.7	19.0	16.5	15.0	14.2	16.1	17.0
Japan Coast Guard	9.8	10.9	11.2	10.4	9.4	8.6	10.0	10.1
Air Self-Defense Force	10.6	11.1	11.7	10.3	9.7	8.8	9.8	10.5
Other	1.7	1.9	2.2	2.1	2.1	1.8	2.2	2.3
Defense Facility Administration Agency	4.7	5.3	6.0	5.3	4.8	4.3	5.1	5.4
Total defence budget	41.7	45.8	50.2	44.5	40.8	37.6	43.2	45.2

Partly in response to concerns about the increasing presence in Japan's territorial waters of Chinese warships, and the North Korean incursion into South Korean waters in 1999, Tokyo plans to set up a patrol-helicopter unit in order to improve maritime surveillance. Japan is also forming its first naval special-operations unit, the 60-strong Special Guard Force, in March 2001. Japan's substantial surface fleet continues to be updated with the commissioning in 2000 of the fifth and sixth of nine *Murasame* destroyers. Four improved *Murasame* have been ordered and should be in service in 2003–04. An amphibious capability is slowly being developed as delivery of the second and third *Osumi* landing platform, dock (LPD) and landing platform helicopter (LPH) should be made in 2002–03, which will give a small helicopter-carrying and troop-lift capability.

China China's official defence budget increased from Y105bn ($12.6bn) in 1999 to Y120bn ($14.5bn) in 2000. Chinese defence spending remains non-transparent, and official accounts substantially understate real military expenditure, estimated at more than $40bn in 1999, which was three times the official figure.

Under US pressure, Israel cancelled the sale of the *Phalcon* airborne early-warning (AEW) system to China. This advanced system was to be fitted to the A-50 (IL-76) AEW aircraft under a contract worth $250m. To improve its AEW capability, China could upgrade the UK-supplied *Searchwater* system, designed specifically for maritime surveillance. The major improvement for the Air Force is the acquisition from Russia of 50 Su-30s. This aircraft is due for delivery in 2002, with a substantial proportion being assembled in China. The order complements the deal to acquire around 200 Russian Su-27 air-superiority fighters, the first deliveries of which were made in 1999. When these two orders are completed, China's ability to dominate the airspace around Taiwan will be greatly improved.

China's naval capabilities are set to advance over the coming years. The surface fleet has been enhanced by the commissioning of the first of two Russian *Sovremennyy* destroyers, a more advanced vessel than any in China's inventory. The second destroyer will be delivered at the end of 2000. Negotiations are in progress for another two of these destroyers, although no contract has yet been signed. There have been reports that China plans to develop an aircraft carrier for launch in 2005. However, even if true, it would take at least two decades before the Navy could fully operate a carrier-based aviation capability. At present, China seems more concerned about reinforcing its strategic and tactical submarine fleet. Work has started on developing a new SSBN (the Type 094); reportedly six of these craft are to be built. If the plan goes ahead, each would be fitted with the *Julang*-2 submarine-launched ballistic missile (SLBM), which is still in development. The Type 093 nuclear-powered submarine (SSN), being built in China with Russian help, is still not operational. If these plans are fulfilled, then China's capability as an ocean-going navy will be considerably enhanced, which will inevitably raise anxieties among her immediate neighbours, particularly, Taiwan.

Taiwan The Taiwan defence budget increased significantly from NT$357bn ($10.9bn) to NT$395bn ($12.8bn). It is proposed to allocate an additional $50bn to the budget over the next 10 years in order to dispense with the previous system of supplementary budgets to finance one-off procurement projects. Taiwan's main plans are to improve missile defence, including radar, command-and-control equipment and six more *Patriot* Advanced Capability (PAC)-2 systems and possibly PAC-3 missiles. The US has turned down Taiwan's request to purchase DDG-51 *Aegis* destroyers, which for the US Navy will be fitted with the Navy Theater-Wide missile-defence system.

South Korea The improved political climate on the Korean peninsula has yet to affect South Korean military spending. The country's budget for 2000 was set at won15.4tr ($13bn). The economic recovery and the improvement of the won against the US dollar are likely to mean that the planned spending can be sustained. For the Army, there will be more Type-88 tanks and new K-9 155mm self-propelled guns and multiple-launch rocket systems (MLRS), including army tactical missile systems (ATACMS). South Korea has almost completed its naval expansion by commissioning the eighth of nine *Chang Boo* (German Type 209) diesel submarines (SSKs) and the third and last *Okapi* destroyer. However, there is some doubt about the follow-on KDX-2 guided-missile destroyer programme for three ships; it may be replaced with a locally built *Aegis* escort. For the Air Force, the fulfilment of the current F-16C/D order by 2003 will complete its fleet of 120 aircraft. Plans remain in place to acquire, with US support, a 500km-range surface-to-surface missile (SSM) that will reach all parts of North Korea.

ASEAN Singapore's budget for 2000 is up from S$7.3bn ($4.2bn) in 1999 to S$7.4bn ($4.4bn). Its largest defence order, in dollar terms, was for six modified *Lafayette* frigates with area surface-to-air missiles (FFG) from France. The new infantry-fighting vehicle (IFV) BIONIX has entered service in two versions, designated IFV-25 and IFV-40/50, and currently equips at least one battalion. This is the first such vehicle to be developed in South-east Asia and the world's first to enter service with hydro-pneumatic suspension. BIONIX will augment upgraded M-113 armoured personnel carriers (APCs) and provide a significant increase in armour, mobility and firepower.

Singapore is the only ASEAN country with plans for substantial naval expansion with new equipment, with the six general-purpose frigates being ordered from France for delivery between 2005–09 being a major component. The first will be built in France, the last five under licence in Singapore. The frigates will also be the first ships to carry helicopters in Singapore's naval history. Singapore's amphibious capability has been updated with the commissioning, in early 2000, of two indigenously built *Endurance*-class tank landing ships (LST) which replace her ageing US *County* class LSTs; two more are to be ordered. The first of four Swedish *Sjoormen* SSKs was commissioned in July 2000. A second is due to be delivered in 2001 and the other two will remain in Sweden for crew training. These submarines are to be used for developing an operational capability before ordering new submarines for service from around 2010.

The Malaysian defence budget for 2000 fell from RM6.9bn ($1.8bn) to RM6.0bn ($1.6bn), excluding procurement. Two *Lekiu*-class frigates from UK were delivered. With Malaysia's economy rejuvenated, defence spending is to be maintained and possibly increased. Procurement programmes held over from 1998 could be resurrected to include main battle tanks and armoured personnel carriers. This equipment, however, is still subject to government approval and funding. Further development of the Rapid Deployment Force and army aviation is also considered likely. The Malaysian navy would like to expand but still lacks the necessary financial support. It is seeking funding for the initial purchase of one or two SSKs under the '8th Malaysia Plan', covering the period 2001–05, and has mooted the idea of buying two *Agosta* SSKs from France but no governmental approval has yet been forthcoming. Malaysia would also like to obtain at least three more surface ships and one LPD capable of transporting a battalion plus equipment along with two LSTs; but again lacks money to do so. The only naval order to be made in recent years – for six patrol vessels from Germany – has encountered problems. The first two were supposed to be built in Germany with the following four built in Malaysia. Malaysia will probably be unable to build the last four itself and they will probably all have to be built in Germany, making the programme more expensive and possibly placing it in jeopardy.

In Thailand the budget for 2000 is unchanged from 1999 at b77bn (down in dollar terms from $2.1bn to $2.0bn), but it is set to show a 14% jump in 2001 to b88bn ($2.3bn). An order was placed for eight F-16s in October 1999. This replaces an earlier cancelled order for eight F/A-18C/Ds.

Australasia

The Australian defence budget for 1999 was set at A$11.2bn ($7.4bn), with A$6.5bn ($4.2bn) allocated for maintaining and acquiring equipment capabilities, including command, control and intelligence functions., Australia instigated a major defence review in 2000. The Australian Defence Forces (ADF) have fallen in strength from 70,000 to around 50,000 in the past decade. Major capabilities will need replacing over the next ten years, as they become obsolete. Australia's leadership of INTERFET and major combat contribution in East Timor in 1999 will strongly influence the outcome of the review. The ADF still provides a substantial contribution to UNTAET, a commitment that could last for some time. The government has produced a public discussion paper 'Defence Review 2000 – Our Future Defence Force'. For land forces, the paper presents a number of possible choices in respect of size, equipment and maintenance for regional-security operations, including a possible requirement for an enhanced amphibious capability. This capability could include a multi-role support ship for Australia's naval forces that would also have a marine-helicopter landing capability. Hard decisions will have to be taken over air and naval forces. Where its air forces are concerned, the ADF faces a period, between 2007 and 2015 when the F/A-18 *Hornet* fighters, the P-3C maritime-patrol aircraft and the C-130H transport aircraft all become obsolete or can no longer be maintained cost-effectively. The ADF plans to keep the F-111 long-range strike and reconnaissance aircraft operational until at least 2015. Australia is slowly upgrading its Navy's surface fleet with the commissioning of two of eight ANZAC frigates in 2000. Eleven *Kaman* SH-2G (A) *Super Sea Sprite* helicopters are being acquired to operate from the ANZAC frigates, for anti-surface warfare (ASW) and anti-surface-unit warfare (ASUW) operations (reviving a capability lost 20 years ago). These were due to be delivered in late 2000, but will probably not arrive until 2001. Other improvements to the surface fleet, such as the purchase of US *Kidd*-class destroyers and a new fleet of patrol boats to replace the *Freemantle* class, have either been cancelled or delayed. Two *Collins* diesel submarines are being upgraded to operate more effectively and should be fully operational by the end of 2000. This means that five of the six should be in service by the end of 2000.

Table 25 **Arms orders and deliveries, East Asia and Australasia, 1998–2000**

Country	Country supplier	Classification	Designation	Quantity	Order date	Delivery date	Comment
Australia	dom	SSK	***Collins***	6	1987	1996	Swe license. Deliveries to 2000
	dom	FGA	**F-111**	71	1990	1999	Upgrade of F/RF-111C
	Ca	LACV	**ASLAV**	276	1992	1996	2nd batch of 150 for delivery 2001
	dom	MHC	***Huon***	6	1994	1999	Last delivery 2002
	dom	FGA	**F-111**	36	1995	1996	Upgrade continuing

Country	Country supplier	Classification	Designation	Quantity	Order date	Delivery date	Comment
	US	MPA	**P-3C**	17	1996	1999	Upgrade to AP-3C
	US	tpt	**C-130J**	12	1996	1999	Deliveries to 2000; 2-yr slippage
	US	hel	**SH-2G**	11	1997	2000	Deliveries to 2002
	UK	trg	***Hawk*-100**	33	1997	1999	Final delivery 2006
	US	hel	**CH-47D**	2	1997	1999	
	UK	FGA	**F/A-18**	71	1998	2005	Upgrade
	dom	FFG	***Adelaide***	6	1999	2001	Upgrade to 2006
	dom	LACV	***Bushmaster***	370	1999	2001	
	US	AEW	**B-737**	5	1999	2004	
	No	ASSM	***Penguin***		1999	2003	
	US	AMRAAM	**AIM-120**		2000		
Brunei	UK	FSG	**FSG**	3	1995	2000	Scaled-down version of *Leiku* FF
	UK	trg	***Hawk* 100/20**	10	1996	1999	
	Indo	MPA	**CN-235**	3	1996	1999	Req for up to 12
	Fr	ASSM	***Exocet***	59	1997	1999	
	UK	FAC	***Waspada***	3	1997	1998	Upgrade
	Fr	SAM	***Mistral***	16	1998	1999	Launchers
Cambodia	Il	trg	**L-39**	8	1994	1996	2nd-hand; 2 delivered by Jan 1998
China	dom	ICBM	**DF-41**		1985	2005	Dev; range 12,000km
	dom	ICBM	**DF-31**		1985	2005	Dev; range 8,000km; tested 1999
	dom	SLBM	**JL-2**		1985	2008	Dev; range 8,000km
	dom	SSGN	**Type 093**	1	1985	2002	Similar to RF *Victor* 3; launch expected 2000
	dom	SSBN	**Type 094**	4	1985	2008	Dev programme
	dom	ASSM	**C701**			1999	Dev completed
	dom	bbr	**H-6**			1998	Still in production
	Fr	hel	**AS-365**	50	1986	1989	Local production continues
	dom	SRBM	**DF-11**	100	1988	1996	Prod continuing
	dom	SRBM	**DF-15**	300	1988	1996	Prod continuing
	dom	FGA	**FC-1**		1990	2005	With Pak (150 units); 1st flight 2000
	col	hel	**EC-120**		1990		In dev with Fr and Sgp
	RF	SAM	**S-300**	30	1990	1992	Continued in 1998
	dom	FGA	**F-8IIM**		1993	1996	Modernisation completed 1999
	dom	FGA	**F-10**		1993		Dev continues
	RF	SS	***Kilo***	4	1993	1995	Deliveries to 1999
	dom	SS	***Song***	2	1994	2002	
	dom	SS	***Ming***	6	1994	1997	All 6 units under construction
	RF	SAM	**SA-15**	35	1995	1997	Deliveries to 2000
	dom	AGI	***Shiyan* 970**	1	1995	1999	Sea trials in 1999
	RF	FGA	**SU-27**	200	1996	1998	15 units for prod 1998–2000
	dom	ATGW	***Red Arrow* 8E**		1996	1998	Modernised *Red Arrow* ATGW
	dom	DDG	***Luhai***	2	1996	1999	
	RF	DDG	***Sovremennyy***	2	1996	1999	1st in 1999, 2nd in 2000
	dom	SLCM	**C-801(mod)**		1997		Dev (aka YJ-82)
	col	ASM	**KR-1**		1997		In dev with RF; Kh-31P variant

Country	Country supplier	Classification	Designation	Quantity	Order date	Delivery date	Comment
	UK	MPA	***Jetstream***	2	1997	1998	For Hong Kong Government
	Il	AEW	**Il-76**	4	1997		
	RF	hel	**Ka-28**	12	1998	2000	For DDG operation
	RF	SAM	**FT-2000**		1998		
	RF	tkr ac	**Il-78**	4	1998		
	RF	SSM	**SSN-24**	24	1998	2002	For *Sovremenny*
	dom	FFG	***Jiangwei* II**	8	1998	1998	6 delivered
	RF	FGA	**SU-30**	50	1999	2002	
	dom	IRBM	**DF-21X**		1999		Modernised DF-15
	RF	FGA	**Su-27UB**	20	2000		
Indonesia	UK	lt tk	***Scorpion***	50	1995	1997	39 delivered 1998 incl 9 Stormers
	Slvk	APC	**BVP-2**	9	1996	1998	Incl 2 BVP-2K, orig from Ukr
	UK	FGA	***Hawk* 209**	16	1996	1999	12 were to be delivered in 1999
	Nl	tpt	**F-28**	8	1996	1998	
	dom	MPA	**CN-235MP**	3	1996	1999	
	US	hel	**NB-412**	1	1996	1998	Licence-produced
	Ge	hel	**BO-105**	3	1996	1998	40 delivered between 1980–93
Japan	US	AEW	**B-767**	4	1991	1998	
	dom	DD	***Murasame***	9	1991	1994	7 Delivered by 2000
	dom	SSK	***Oyasio* Class**	8	1993	2000	3 Delivered by 2000
	dom	AAM	**XAAM-5**		1994	2001	Development
	dom	LST	***Oosumi* Class**	3	1994	1997	1 delivered by 2000
	dom	BMD	**TMD**		1997		Dev; Joint with US from late 1998
	dom	recce	**sat**	4	1998	2002	Dev Prog. 2 optical, 2 radar
	dom	mor	**L16**	42	1999	2000	
	dom	mor	**120mm**	27	1999	2000	
	dom	SP arty	**Type 99 155 mm**	4	1999	2000	Replaces Type-75; prod in 1999
	col	arty	**FH70**		1999	2000	40 req under 1996–2000 MTDP
	dom	MRL	**MLRS**	9	1999	2000	45 req under 1996–2000 MTDP
	dom	AAA	**Type-87**	1	1999	2000	1 delivered 1998
	dom	MBT	**Type-90**	17	1999	2000	90 req under 1996–2000 MTDP
	dom	AIFV	**Type-89**	2	1999	2000	2 delivered 1998
	dom	APC	**Type-96**	28	1999	2000	157 req under 1996–2000 MTDP
	dom	APC	**Type-82**	1	1999	2000	1 delivered 1998
	dom	recce	**Type-87**	1	1999	2000	1 delivered 1998
	dom	hel	**AH-1S**		1999	2000	3 req under 1996–2000 MTDP
	dom	hel	**OH-1**	3	1999	2000	
	dom	hel	**UH-60JA**	3	1999	2000	
	dom	hel	**CH-47JA**	2	1999	2000	9 req under 1996–2000 MTDP
	dom	recce	**LR-2**	1	1999	2000	Cost $24m
	dom	SAM	***Hawk***		1999	2000	
	dom	ASSM	**Type-88**	4	1999	2000	24 req under 1996–2000 MTDP
	dom	DD	***Murasame***	4	1999	2001	Improved *Murasame*
	dom	MCMV	***Sugashima***	2	1999	2000	
	dom	FAC		2	1999	2000	

Country	Country supplier	Classification	Designation	Quantity	Order date	Delivery date	Comment
	dom	AK		1	1999	2000	
	dom	hel	**SH-60J**	9	1999	2000	37 req under 1996–2000 MTDP
	dom	FGA	**F-2**	8	1999	2000	45 req under 1996–2000 MTDP
	dom	hel	**CH-47J**	2	1999	2000	4 req under 1996–2000 MTDP
	dom	SAR	**U-125A**	2	1999	2000	Cost $76m
	dom	hel	**UH-60J**	2	1999	2000	Cost $59m
	dom	trg	**T-4**	10	1999	2000	54 req under 1996–2000 MTDP
	dom	trg	**T-400**		1999	2000	
	dom	tpt	**U-4**		1999	2000	
	dom	tpt	**C-X**		2000		Replacement for C-1A
	dom	MPA	**MPA-X**		2000		Replacement for P3
North Korea							
	dom	MRBM	***Taepo-dong* 1**				Tested October 1998
	dom	MRBM	***Taepo-dong* 2**				Testing programme delayed
	RF	hel	**Mi-17**	5	1998	1998	
	Kaz	FGA	**MiG-21**	30	1999	1999	Also spare parts for existing fleet
	RF	FGA	**MiG-21**	10	1999	2000	
South Korea							
	dom	APC	**KIFV**	2,000	1981	1985	Produced in 1998, incl exports
	Ge	SSK	**Type 209**	9	1987	1993	8 delivered by 2000
	US	hel	**UH-60P**	138	1988	1990	Deliveries to 1999
	US	FGA	**F-16C/D**	120	1992	1995	Licence; deliveries to 1999
	dom	AIFV	**M-113**	175	1993	1994	Deliveries to 1998
	dom	sat	**KITSAT-3**		1995	1999	
	RF	APC	**BTR-80**	20	1995	1996	Deliveries to 1998
	RF	AIFV	**BMP-3**	23	1995	1996	Deliveries to 1999
	RF	MBT	**T-80**	33	1995	1996	Deliveries to 1999
	US	sigint	***Hawker* 800**	10	1996	1999	
	US	AAM	**AMRAAM**	190	1996	1998	
	US	AAM	***Sidewinder***	284	1996	1998	
	Il	AAM	***Popeye***	100	1996	1999	
	dom	DDG	***Okpo***	3	1996	1998	3 delivered by end of 1999
	US	MRL	**MLRS**	29	1997	1999	10 delivered in 1998
	Il	UAV	***Harpy***	100	1997	1999	
	dom	trg	**KTX-2**	94	1997	2005	Dev
	Fr	utl	**F-406**	5	1997	1999	
	dom	SAM	***Pegasus***		1997	1999	Dev
	Fr	SAM	***Mistral***	1,294	1997	1998	Missiles
	Il	UAV	***Searcher***	3	1997	1998	
	RF	SAM	**SA-16**		1997	1999	
	RF	ATGW	**AT-7**		1997	1999	
	UK	hel	***Lynx***	13	1997	1999	
	Indo	tpt	**CN-235**	8	1997	1999	Delivery delayed
	US	AEW	**B-767**	4	1998		Delivery delayed

Country	Country supplier	Classification	Designation	Quantity	Order date	Delivery date	Comment
	dom	SAM	**M-SAM**		1998	2008	Development
	Ge	hel	**BO-105**	12	1998	1999	
	US	AAV	**AAV7A1**	57	1998	2001	Licence
	dom	SP arty	**XK9**	68	1998	1999	Being delivered
	RF	tpt	**Be-200**	1	1998	2000	
	dom	SAM	**P-SAM**		1998	2003	Dev
	dom	SSM	***Hyonmu***		1999		Dev; 300km and 500km variants
	US	FGA	**F-16C/D**	20	1999	2003	Follow on after orders for 120
Malaysia	UK	FF	***Lekiu*-class**	2	1992	1999	2 delivered in 1999
	It	FSG	***Assad***	4	1995	1997	Originally for Irq. Deliveries 1997–99
	It	ASSM	***Otomat***	4	1996	1998	
	Fr	ASSM	***Exocet***	4	1996	1998	
	Indo	tpt	**CN-235**	6	1995	1999	
	US	hel	**S-70A**	2	1996	1998	
	Ge	OPV	***Meko* A 100**	6	1997	2002	Lic built. Req for 27 over 20 years
	RF	FGA	**Mig-29**	18	1997	1999	Upgrade
	It	trg	**MB-339**	2	1998	1999	
	RF	hel	**Mi-17**	10	1998	1999	
	UK	hel	***Super Lynx***	6	1999	2001	
Myanmar	PRC	FGA	**F-7**	21	1996	1998	Following deliveries of 36 1991–96
	PRC	trg	**K-8**	4	1998	2000	
New Zealand							
	A	FF	***Anzac***	2	1989	1997	With Aus; 2nd delivered 1999
	US	ASW	**P3-K**	6	1995	1998	Upgrade; 1 delivered
	US	trg	**CT-4E**	13	1997	1998	11 delivered; lease programme
	US	hel	**SH-2G**	5	1997	2000	
	US	tpt	**C-130J**	5	1999		Lease of 5 to 7; delayed
Papua New Guinea							
	Indo	hel	**BO-105**	1	1998	1999	
Philippines	US	tpt	**C-130B**	2	1995	1998	
	US	trg	**T-41**	5	1997	1998	
	ROC	FGA	**F-5A**	5	1997	1998	Ex-ROK
	ROC	FGA	**F-5E**	40	1999		
Singapore	dom	AIFV	**IFV**	500	1991	1999	Two batches: 300 then 200
	dom	OPV	***Fearless***	12	1993	1996	Deliveries to 1999
	US	FGA	**F-16C/D**	42	1995	1998	First order for 18, follow-on for 24
	Swe	SSK	***Sjoormen***	4	1995	1997	One delivered
	dom	LST	***Endurance***	4	1997	1999	Deliveries to 2000
	RF	SAM	**SA-18**		1997	1998	
	US	tkr ac	**KC-135**	4	1997	2000	
	US	hel	**CH-47D**	8	1997	2000	
	US	cbt hel	**AH-64D**	8	2000	2003	
	Fr	FFG	***Lafayette***	6	2000	2005	1st to be built in Fr; final in 2009

Country	Country supplier	Classification	Designation	Quantity	Order date	Delivery date	Comment
Taiwan	dom	FGA	**IDF**	130	1982	1994	Deliveries completed 1999
	US	FF	***Knox***	8	1989	1993	Final delivery in 1998
	Fr	FF	***Lafayette***	6	1992	1996	Deliveries to 1998
	US	FGA	**F-16A/B**	150	1992	1997	60 delivered in 1997
	Fr	FGA	***Mirage* 2000**	60	1992	1997	Deliveries completed in 1999
	US	SAM	***Patriot***	6	1993	1997	Completed 1998. PAC-3 standard
	US	tpt	**C-130**	12	1993	1995	Deliveries continue
	US	SAR hel	**S-70C**	4	1994	1998	
	US	arty	**M-109A5**	28	1995	1998	
	US	SAM	***Avenger***	70	1996	1998	
	US	MPA	**P-3**		1996		With *Harpoon* SAM
	US	hel	**TH-67**	30	1996	1998	
	Sgp	recce	**RF-5E**		1996	1998	Some F-5E entered service as RF-5E
	US	SAM	***Stinger***	1,600	1996	1998	
	dom	trg	**AT-3**	40	1997		Order resheduled
	US	ASW hel	**S-70C**	11	1997	2000	
	US	hel	**OH-58D**	13	1998	2001	Following deliveries of 26 1994–95
	US	ASSM	***Harpoon***	58	1998		
	US	hel	**CH-47SD**	9	1999	2002	Following deliveries of 7 1993–97
	US	radar	***Pave Paws***		1999	2002	
	US	LSD	***Anchorage***	1	1999	2000	USS *Pensacola* to replace 2 LSDs
	dom	FF	***Chengkung***	1	1999	2003	Based on US *Perry*
	US	AEW	**E-2T**	4	1999	2002	Following delivery of 4 in 1995
	US	hel	**CH-47SD**	9	2000		3 plus long lead time for further 6
Thailand	A	arty	**155mm**	36	1995	1996	18 delivered 1996, 18 1998
	US	APC	**M113**	82	1995	1997	63 delivered 1997
	Indo	tpt	**CN-235**	2	1996		Delayed
	It	MHC	***Lat Ya***	2	1996	1998	Deliveries to December 1999
	dom	corvette		3	1996	2000	2 delivered by 2000
	US	FF		1	1996	1998	
	It	MCMV	***Gaeta***	2	1996	1998	Deliveries finished
	Il	UAV	***Searcher***	4	1997		
	Fr	APC	**VAB NG**		1997		Selected to replace 300 M-113.
	Fr	sat			1997		Recce sat order delayed late 1997
	US	hel	**SH-2F**	10	1999	2002	
	Ge	FGA	***Alpha Jet***		1999		To replace OV-10
	US	FGA	**F-16 A**	18	2000	2002	
Vietnam	RF	FGA	**Su-27**	6	1995	1997	Deliveries to 1998
	RF	corvette	***Taruntul* 2**	2	1997	1999	Follow delivery of 2 *Taruntul* 1995
	DPRK	SSM	***Scud***		1999	1999	Probably *Scud*-Cs; quantity n.k.

Australia Aus

Australian dollar A$		1998	1999	2000	2001
GDP	A$	578bn	610bn		
	US$	363bn	399bn		
per capita	US$	22,000	22,900		
Growth	%	4.9	3.0		
Inflation	%	0.8	1.5		
Publ Debt	%	33.4	31.3		
Def exp	A$	12.0bn	11.9bn		
	US$	8.1bn	7.8bn		
Def bdgt	A$	10.4bn	11.1bn	12.2bn	12.2bn
	US$	7.0bn	7.2bn	7.1bn	7.1bn
US$1=A$		1.59	1.53	1.72	

Population			19,292,000
(Asian 4%, Aborigines <1%)			
Age	13–17	18–22	23–32
Men	692,000	683,000	1,503,000
Women	654,000	649,000	1,456,000

Total Armed Forces

ACTIVE 50,600
(incl 8,500 women)

RESERVES 20,200
GENERAL RESERVE
Army 17,450 **Navy** 950 **Air Force** 1,800

Army 24,150

integrated = formation/unit comprising active and reserve personnel
(incl 2,600 women)
1 Land HQ, 1 Joint Force HQ, 1 Task Force HQ (integrated), 1 bde HQ
1 armd regt (integrated), 2 recce regt (1 integrated), 1 SF (SAS) regt, 6 inf bn (2 integrated), 1 cdo bn (integrated), 2 indep APC sqn (1 integrated), 1 med arty regt, 2 fd arty regt (1 integrated), 1 AD regt (integrated), 3 cbt engr regt (1 integrated), 2 avn regt

RESERVES
GENERAL RESERVE
1 div HQ, 7 bde HQ, 1 cdo, 2 recce, 1 APC, 1 med arty, 3 fd arty, 3 cbt engr, 2 engr construction regt, 13 inf bn; 1 indep fd arty bty; 1 recce, 3 fd engr sqn; 3 regional force surveillance units

EQUIPMENT
MBT 71 *Leopard* 1A3 (excl variants)
LAV 111 ASLAV-25
APC 463 M-113 (excl variants, 364 being upgraded, 119 in store)
TOWED ARTY 105mm: 246 M2A2/L5, 104 *Hamel*; **155mm**: 35 M-198
MOR 81mm: 296
RCL 84mm: 577 *Carl Gustav*; **106mm**: 74 M-40A1
SAM 19 *Rapier*, 17 RBS-70
AC 4 *King Air* 200, 2 DHC-6 (all on lease)
HEL 35 S-70 A-9, 40 Bell 206 B-1 *Kiowa* (to be upgraded), 25 UH-1H (armed), 17 AS-350B, 6 CH-47D
MARINES 15 LCM
SURV 14 RASIT (veh, arty), AN-TPQ-36 (arty, mor)

Navy 12,500

(incl 990 Fleet Air Arm; 1,970 women)
Maritime Comd, Support Comd, Training Comd
BASES Sydney, (Maritime Comd HQ) Stirling, Cairns, Darwin

SUBMARINES 3
3 *Collins* SSK with *sub-Harpoon* USGW and Mk 48 HWT

PRINCIPAL SURFACE COMBATANTS 9
DESTROYERS DDG 1 *Perth* (US *Adams*) with 1 SM-1 MR SAM/*Harpoon* SSM launcher, 2 × 127mm guns, 2 × 3 ASTT (Mk 32 LWT)
FRIGATES 8
FFG 6
6 *Adelaide* (US *Perry*), with SM-1 MR SAM, *Harpoon* SSM, 1 × 76mm gun, 2 × 3 ASTT (Mk 32 LWT), 2 S 70B *Sea Hawk* hel
FF 2
2 *Anzac* with *Sea Sparrow* VLS SAM, 1 × 127mm gun, 6 × 324mm ASTT (Mk 32 LWT), 1 S 70B *Sea Hawk* hel

PATROL AND COASTAL COMBATANTS 15
PATROL, OFFSHORE 15 *Freemantle* PCO

MINE COUNTERMEASURES 4
2 *Rushcutter* MHI, 2 *Huon* MHC, plus 2 *Bandicoot* MSA, 2 *Kooraaga* MSA, 1 *Brolga* MSA

AMPHIBIOUS 4
1 *Jervis Bay* catamaran (leased until mid-2001)
1 *Tobruk* LST, capacity 500 tps, 2 LCM, 2 LCVP
2 *Kanimbla* (US *Newport*) LPA, capacity 450 tps, 2 LCM, hel 4 Army *Blackhawk* or 3 *Sea King*, no beach-landing capability
plus 5 *Balikpapan* LCH

SUPPORT AND MISCELLANEOUS 13
1 *Success* AO, 1 *Westralia* AO; 1 sail trg, 5 AT, 3 TRV; 2 *Leuwin* AGHS plus 4 craft

FLEET AIR ARM (990)
no cbt ac, 16 armed hel
ASW 1 hel sqn with 16 S-70B-2 *Sea Hawk*
UTL/SAR 1 sqn with 6 AS-350B, 3 Bell 206B and 2 BAe-748 ac (EW trg), 1 hel sqn with 7 *Sea King* Mk 50/50A

Air Force 13,950

(incl 2,700 women); 148 cbt ac incl MR, no armed hel
3 Cmds - Air, Trg, Log

Flying hours F-111, 200; F/A-18, 175
STK/RECCE GP 2 sqn with 35 F-111 (13 F-111C, 4 F-111A (C), 14 F-111G, 4 RF-111C), 2 EP-3C
TAC/FTR GP 3 sqn (plus 1 OCU) with 71 F/A-18 (55 -A, 16 -B)
TAC TRG 2 sqn with 25 MB-326H (to be replaced by 33 *Hawk* 127 lead-in fighter trainers, 12 of which have been delivered)
FAC 1 flt with 3 PC-9A
MP GP 2 sqn with 17* P-3C, 3 TAP-3B
AIRLIFT GP
TPT/TKR 7 sqn
2 with 24 C-130 (12 -H, 12 -J)
1 with 5 Boeing 707 (4 tkr)
2 with 14 DHC-4 (*Caribou*)
1 VIP with 5 *Falcon* 900
1 with 10 HS-748 (8 for navigation trg, 2 for VIP tpt), 2 Beech-200 *Super King Air*, 1 Beech 1900-D
TRG 59 PC-9
AD *Jindalee* OTH radar: Radar 1 at Longreach (N Queensland), Radar 2 at Laverton (W Australia), third development site at Alice Springs, 3 control and reporting units (1 mobile)
MISSILES
ASM AGM-84A, AGM-142
AAM AIM-7 *Sparrow*, AIM-9M *Sidewinder*, ASRAAM

Forces Abroad

Advisers in **Fiji, Indonesia, Solomon Islands, Thailand, Vanuatu, Tonga, Western Samoa, Kiribati**
MALAYSIA Army: ε115; 1 inf coy (on 3-month rotational tours) **Air Force**: 33; det with 2 P-3C **ac**
PAPUA NEW GUINEA: 38; trg unit

UN AND PEACEKEEPING

EAST TIMOR (UNTAET): 1,620 incl 19 obs and 4 SA-70A hel **EGYPT** (MFO): 26 obs **MIDDLE EAST** (UNTSO): 11 obs **PAPUA NEW GUINEA**: 150 (Bougainville Peace Monitoring Group)

Paramilitary

AUSTRALIAN CUSTOMS SERVICE

ac 3 DHC-8, 3 *Reims* F406, 6 BN-2B-20, 1 *Strike Aerocommander* 500 **hel** 1 Bell 206L-4; about 6 boats

Foreign Forces

US Air Force 260; **Navy** 40; joint facilities at NW Cape, Pine Gap and Nurrungar
NEW ZEALAND Air Force 47; 6 A-4K/TA-4K, (trg for Australian Navy); 9 navigation trg
SINGAPORE 230; Flying Training School with 27 S-211 **ac**

Brunei — Bru

Brunei dollar B$		1998	1999	2000	2001
GDP	B$	9.5bn	10.2bn		
	US$	5.7bn	6.0bn		
per capita	US$	7,700	7,800		
Growth	%	3.0	1.8		
Inflation	%	5.0	3.2		
Debt	US$	1.5bn			
Def exp	B$	632m	684m		
	US$	378m	402m		
Def bdgt	B$	596m	ε620m	ε484m	
	US$	357m	365m	281m	
US$1=B$		1.6	1.7	1.7	

Population 332,000
(Muslim 71%; Malay 67%, Chinese 16%, non-Malay indigenous 6%)

Age	13–17	18–22	23–32
Men	16,000	14,000	28,000
Women	15,000	15,000	26,000

Total Armed Forces

ACTIVE 5,000
(incl 600 women)

RESERVES 700
Army 700

Army 3,900

(incl 250 women)
3 inf bn • 1 spt bn with 1 armd recce, 1 engr sqn

EQUIPMENT

LT TK 16 *Scorpion*
APC 26 VAB, 2 *Sultan*, 24 AT-104 (in store)
MOR 81mm: 24
RL *Armbrust* (reported)

RESERVES

1 bn

Navy 700

BASE Muara
PATROL AND COASTAL COMBATANTS 6
MISSILE CRAFT 3 *Waspada* PFM with 2 MM-38 *Exocet* SSM
PATROL, INSHORE 3 *Perwira* PFI†
PATROL, RIVERINE boats
AMPHIBIOUS craft only
4 LCU; 1 SF sqn plus boats

Air Force 400

no cbt ac, 5 armed hel

HEL 2 sqn
1 with 10 Bell 212, 1 Bell 214 (SAR), 4 S-70A, 1 S-70C (VIP)
1 with 5 Bo-105 armed hel (**81mm** rockets)
TRG 2 sqn
1 with 2 SF-260W, 6 PC-7, 2 Bell 206B
1 with 1 CN-235M
AIR DEFENCE 1 sqn with 12 *Rapier* (incl *Blindfire*), 16 *Mistral*

Paramilitary ε3,750

GURKHA RESERVE UNIT ε2,000+
2 bn
ROYAL BRUNEI POLICE 1,750
7 PCI<

Foreign Forces

UK Army some 1,050; 1 Gurkha inf bn, 1 hel flt, trg school
SINGAPORE 500; trg school incl hel det (5 UH-1)

Cambodia Cam

riel r		1998	1999	2000	2001
GDP	r	ε11.2tr	ε13.2tr		
	US$	3.0bn	3.5bn		
per capita	US$	700	725		
Growth	%	0.0	4.0		
Inflation	%	14.7	4.1		
Debt	US$	2.2bn	2.5bn		
Def exp	r	ε570bn	ε670bn		
	US$	152m	176m		
Def bdgt	r	ε281bn	ε330bn	ε460bn	
	US$	75m	87m	120m	
FMA (US)	US$		1.5m	2.6m	2.7m
FMA (Aus)	US$	0.2m	0.1m		
FMA (PRC)	US$	3.0m			
US$1=r		3,744	3,807	3,836	

Population			10,879,000
(Khmer 90%, Vietnamese 5%, Chinese 1%)			
Age	13–17	18–22	23–32
Men	621,000	490,000	882,000
Women	608,000	482,000	867,000

Total Armed Forces

ACTIVE ε140,000 (to reduce)
(incl Provincial Forces, perhaps only 19,000 cbt capable)
Terms of service conscription authorised but not implemented since 1993

Army ε90,000

6 Mil Regions (incl 1 special zone for capital) • 22 inf div[a] • 3 indep inf bde • 1 protection bde (4 bn) • 9 indep inf regt • 3 armd bn • 1 AB/SF regt • 4 engr regt (3 fd, 1 construction) • some indep recce, arty, AD bn

EQUIPMENT
MBT 100+ T-54/-55, 50 PRC Type-59
LT TK PRC Type 62, 20 PRC Type 64
RECCE BRDM-2
APC 160 BTR-60/-152, M-113, 30 OT-64 (SKOT)
TOWED ARTY some 400: **76mm**: M-1942; **122mm**: M-1938, D-30; **130mm**: Type 59
MRL 107mm: Type-63; **122mm**: 8 BM-21; **132mm**: BM-13-16; **140mm**: 20 BM-14-16
MOR 82mm: M-37; **120mm**: M-43; **160mm**: M-160
RCL 82mm: B-10; **107mm**: B-11
AD GUNS 14.5mm: ZPU 1/-2/-4; **37mm**: M-1939; **57mm**: S-60
SAM SA-7

[a] Inf div established str 3,500, actual str some 1,500

Navy ε3,000

(incl 1,500 Naval Infantry)
BASES Ream (maritime), Prek Ta Ten (river)
PATROL AND COASTAL COMBATANTS 4
PATROL, COASTAL 2
2 Sov *Stenka* PFC
RIVERINE 2
2 *Kaoh Chhlam* PCR

NAVAL INFANTRY (1,500)
7 inf, 1 arty bn

Air Force 2,000

24 cbt act; no armed hel
FTR 1 sqn with 19† MiG-21 (14 -bis, 5 -UM) (up to 9 to be upgraded by IAI: 2 returned but status unclear)
TPT 2 sqn with 1 An-26, 2 Y-12, 1 BN-2, 1 Cessna 401, 1 Cessna 421, 1 Socata Tobago
HEL 1 sqn with 14 Mi-8/Mi-17 (incl 1 VIP Mi-8P), 1 AS-355
TRG 5* L-39, 5 *Tecnam* P-92

Provincial Forces some 45,000

Reports of at least 1 inf regt per province, with varying numbers of inf bn with lt wpn

Paramilitary

POLICE 67,000 (incl *gendarmerie*)

Opposition

***FUNCINPEC*/KHMER PEOPLE'S LIBERATION FRONT (KPNLF)** ε1–2,000
alliance between Prince Ranariddh's party and the KPNLF

China, People's Republic of PRC

yuan Y		1998	1999	2000	2001
GDP[ab]	Y	7.9tr	8.2tr		
	US$	703bn	732bn		
per capita	US$	3,700	4,000		
Growth	%	7.8	7.1		
Inflation	%	-0.8	-1.3		
Debt	US$	154bn	154bn		
Def exp[a]	US$	ε37.5bn	ε39.5bn		
Def bdgt[c]	Y	90.9bn	104.7bn	120.5bn	
	US$	11.0bn	12.6bn	14.5bn	
US$1=Y		8.28	8.28	8.28	

[a] PPP est incl extra-budgetary mil exp
[b] Excl Hong Kong: GDP **1998** HK$1,289bn (US$166bn)
[c] Def bdgt shows official figures at market rates

Population 1,255,000,000
(Tibetan, Uighur and other non-Han 8% *Xinjiang* Muslim ε60% of which Uighur ε44% *Tibet* Chinese ε60%, Tibetan ε40%)

Age	13–17	18–22	23–32
Men	51,995,000	48,103,000	120,380,000
Women	49,241,000	44,960,000	112,976,000

Total Armed Forces

ACTIVE some ε2,470,000 (being reduced)
(incl perhaps 1,000,000 conscripts, some 136,000 women; about 130,000 MOD staff, centrally-controlled units not included elsewhere)
Terms of service selective conscription; all services 2 years

RESERVES some 500–600,000 all services
militia reserves being formed on a province-wide basis

Strategic Missile Forces

OFFENSIVE (100,000)+
org in 6 bases (army level) plus 1 testing base, with bde/regt incl 1 msl testing and trg regt; org varies by msl type but may incl up to 5 base spt regt per msl base
ICBM 20+
20+ DF-5A (CSS-4)
IRBM 100+
20+ DF-4 (CSS-3)
30+ DF-3A (CSS-2)
50+ DF-21 (CSS-5). At least 3 bde deployed
SLBM 1 *Xia* SSBN with 12 CSS-N-3 (JL-1)
SRBM about 20 DF-15 launchers with 200+ missiles (CSS-6/M-9) (range 600km). 1 bde deployed
40 DF-11 (CSS-7/M-11) (range 120–300+km). 2 bde deployed

DEFENSIVE
Tracking stations Xinjiang (covers Central Asia) and Shanxi (northern border)
Phased-array radar complex ballistic-missile early-warning

Army ε1,700,000

(perhaps 800,000 conscripts) (reductions continue)
7 Mil Regions, 27 Provinicial Mil Districts, 4 Garrison Comd
21 Integrated Group Armies (3 to disband) GA: from 40–89,000, equivalent to Western corps, org varies, normally with 2–3 inf div/bde, 1 tk, 1 arty, 1 AAA bde or 2–3 inf, 1 tk div/bde, 1 arty, 1 AAA bde, cbt readiness category varies with 10 GA at Category A and 11 at Category B (reorg to bde structure in progress)
Summary of cbt units
Group Army 44 inf div (incl 7 mech inf) 3 with national level rapid-reaction role and at least 9 with regional rapid-reaction role ready to mobilise in 24–48 hours; 10 tk div, 12 tk bde, 13 inf bde, 5 arty div, 20 arty bde, 7 hel regt
Independent 5 inf div, 1 tk, 2 inf bde, 1 arty div, 3 arty bde, 4 AAA bde
Local Forces (Garrison, Border, Coastal) 12 inf div, 1 mtn bde, 4 inf bde, 87 inf regt/bn
AB (manned by Air Force) ε35,000: 1 corps of 3 div
Support Troops incl 50 engr, 50 sigs regt

EQUIPMENT
MBT some ε7,060: incl ε5,500 Type-59-I/-II, ε150 Type-69-I, 500 Type-79, 500 Type-88B, 400 Type-88C, 10+ Type-98
LT TK ε700 incl Type-63, Type-62/62I
AIFV/APC 4,800 incl 1,800 Type-63A/I/II, some Type-77 (BTR-50PK), 1,300 Type-89I/II (mod Type-85), WZ-523, Type-92 (WZ-551), Type-86/86A (WZ-501), 100 BMD-3
TOWED ARTY 12,000: **100mm**: Type-59 (fd/ATK); **122mm**: Type-54-1, Type-60, Type-83; **130mm**: Type-59/-59-1; **152mm**: Type-54, Type-66, Type-83; **155mm**: 300+ Type-88 (WAC-21)
SP ARTY 122mm: ε1,200 incl Type-70/-70I, Type-89; **152mm**: Type-83
COMBINED GUN/MOR 100 2S23 *Nona-SVK*
MRL 2,500: **122mm**: Type-81, Type-89 SP; **130mm**: Type-70 SP, Type-82; **273mm**: Type-83; **320mm**: Type-96
MOR 82mm: Type-53/-67/-W87/-82 (incl SP); **100mm**: Type-71 reported; **120mm**: Type-55 (incl SP); **160mm**: Type-56
ATGW 7,000: HJ-73 (*Sagger*-type), HJ-8 (TOW/*Milan*-type)

RL 62mm: Type-70-1
RCL 75mm: Type-56; **82mm**: Type-65, Type-78; **105mm**: Type-75
ATK GUNS 100mm: Type-73, Type-86; **120mm**: 300+ Type-89 SP
AD GUNS 23mm: Type-80; **25mm**: Type-87; **35mm**: Type-90; **37mm**: Type-55/-65/-74/-89SP; **57mm**: Type-59, -80 SP; **85mm**: Type-56; **100mm**: Type-59
SAM HN-5A/-B/-C (SA-7 type), QW-1/-2, HQ-61A, HQ-7, PL-9C, 26 SA-15 (Tor-M1)
SURV *Cheetah* (arty), Type-378 (veh), RASIT (veh, arty)
AC 2 Y-8
HEL 24 Mi-17, 30 Mi-8TB, 30 Mi-171, 3 Mi-6, 4 Z-8A, 73 Z-9/-WZ-9, 8 SA-342 (with HOT), 20 S-70C2, 20 Z-11
UAV ASN-104/-105

RESERVES

(undergoing major re-org on provincial basis): some 500–600,000: 50 inf, arty and AD div, 100 indep inf, arty regt

DEPLOYMENT

(GA units only)
North-east Shenyang MR (Heilongjiang, Jilin, Liaoning MD): ε250,000: 4 GA, 2 tk, 10 inf div, 1 tk bde, 1 arty div
North Beijing MR (Beijing, Tianjin Garrison, Nei Mongol, Hebei, Shanxi MD): ε300,000: 5 GA, 2 tk, 12 inf div, 3 tk, 3 inf bde, 1 arty div
West Lanzhou MR (incl Ningxia, Shaanxi, Gansu, Qing-hai, Xinjiang MD): ε220,000: 2 GA, 2 tk, 4 inf div, 1 tk bde
South-west Chengdu MR (incl Chongqing Garrison, Sichuan, Guizhou, Yunnan, Xizang MD): ε180,000: 2 GA, 4 inf, 1 arty div plus 2 tk bde
South Guangzhou MR (Hubei, Hunan, Guangdong, Guangxi, Hainan MD): ε180,000: 2 GA, 4 inf plus 2 tk bde. Hong Kong: ε7,000: 1 inf bde (3 inf, 1 mech inf, 1 arty regt, 1 engr bn), 1 hel unit
Centre Jinan MR (Shandong, Henan MD): ε190,000: 3 GA, 3 tk, 7 inf div, 2 tk, 4 inf bde, 1 arty div
East Nanjing MR (Shanghai Garrison, Jiangsu, Zhejiang, Fujian, Jiangxi, Anhui MD): ε250,000: 3 GA, 1 tk, 5 inf div, 1 tk, 3 inf bde, 1 arty div

Navy ε220,000

(incl Coastal Regional Defence Forces, 26,000 Naval Air Force, some 5,000 Marines and some 40,000 conscripts)
SUBMARINES 65
STRATEGIC 1 SSBN
TACTICAL 64
SSN 5 *Han* (Type 091) with YJ 8-2 (C-801 derivative) ASSM, 6 × 533 TT
SSG 1 mod *Romeo* (Type S5G), with 6 C-801 (YJ-6, *Exocet* derivative) ASSM; 533mm TT (test platform)
SSK 57
1 *Song* with YJ 8-2 ASSM (C-802 derivative), 6 × 533mm TT
2 *Kilo*-class (Type EKM 877) with 533mm TT
3 *Kilo*-class (Type EKM 636) with 533mm TT
2 *Ming* (Type ES5C/D) with 533mm TT
15 imp *Ming* (Type ES5E) with 533mm TT
34 *Romeo* (Type ES3B)† with 533mm TT
OTHER ROLES 1 *Golf* (SLBM trials) SS
PRINCIPAL SURFACE COMBATANTS 60
DESTROYERS 20
DDG 20
1 *Sovremenny* with 2 × 4 SS-N-22 *Sunburn* SSM, 2 SA-N-7 *Gadfly* SAM, 2 × 2 130mm guns, 2 × 2 533mm ASTT, 2 ASW mor, 1 Ka-28 hel
1 *Luhai* with 4 × 4 CSS-N-4 SSM, 1 × 8 *Crotale* SAM, 1 × 2 100mm guns, 2 × 3 ASTT, 2 Ka-28 hel
2 *Luhu* with 4 × 2 YJ-8/CSS-N-4 SSM, 1 × 8 *Crotale* SAM, 2 × 100mm guns, 2 × 3 ASTT, 2 Z-9A (Fr *Panther*) hel
1 *Luda* III with 4 × 2 YJ-8/CSS-N-4 SSM, 2 × 2 130mm gun, 2 × 3 ASTT
2 mod *Luda* with 2 × 3 HY-1/CSS-N-2 SSM, 1 × 2 130mm guns, 2 × 3 ASTT, 2 Z-9C (Fr *Panther*) hel
13 *Luda* (Type-051) with 2 × 3 CSS-N-2 or CSS-N-4 SSM, 2 × 2 130mm guns, 6 × 324mm ASTT, 2 × 12 ASW RL (2 also with 1 × 8 *Crotale* SAM)
FRIGATES about 40 FFG
6 *Jiangwei II* with CSS-N-4 *Sardine* SSM, 1 × 8 *Croatale* SAM, 1 × 2 100mm guns, 2 × 6 ASW mor, 1 Z-9A (Fr *Dauphin*) hel
4 *Jiangwei* I with 2 × 3 C-801 SSM, 1 × 6 × HQ-61/CSA-N-1 SAM, 1 × 2 100mm guns, 2 × 6 ASW mor, 1 Z-9C (Fr *Panther*) hel
About 30 *Jianghu*; 3 variants:
About 26 Type I, with 2 × 2 *SY*-1/CSS-N-1 SSM, 2 × 100mm guns, 4 × 5 ASW mor
About 1 Type II, with 1 × 2 *SY*-1/CSS-N-1 SSM, 1 × 2 × 100mm guns, 2 × 5 ASW RL, 1 Z-9C (Fr *Panther*) hel
About 3 Type III, with 8 CSS-N-4 SSM, 2 × 2 100mm guns, 4 × 5 ASW RL
PATROL AND COASTAL COMBATANTS about 368
MISSILE CRAFT 93
5 *Huang* PFM with 6 YJ-8/CSS-N-4 SSM
20 *Houxin* PFM with 4 YJ-8/CSS-N-4 SSM
Some 38 *Huangfeng/Hola* (Sov *Osa* I-Type) PFM with 4 SY-1 SSM
30 *Houku* (*Komar*-Type) PFM with 2 SY-1 SSM
TORPEDO CRAFT about 16
16 *Huchuan* PHT
PATROL CRAFT about 259
COASTAL about 118
2 *Haijui* PCC with 3 × 5 ASW RL
About 96 *Hainan* PCC with 4 ASW RL
20 *Haiqing* PCC with 2 × 6 ASW mor
INSHORE about 111
100 *Shanghai* PCI<, 11 *Haizhui* PCI<

RIVERINE about 30<

MINE WARFARE about 39

MINELAYERS 1

1 *Wolei*

In addition, *Luda* class DDG, *Hainan*, *Shanghai* PC and T-43 MSO have minelaying capability

MINE COUNTERMEASURES about 38

27 Sov T-43 MSO

7 *Wosao* MSC

3 *Wochang* and 1 *Shanghai* II MSI

plus about 50 Lienyun aux MSC, 4 drone MSI and 42 reserve drone MSI

AMPHIBIOUS 59

7 *Yukan* LST, capacity about 200 tps, 10 tk

3 *Shan* (US LST-1) LST, capacity about 150 tps, 16 tk

8 *Yuting* LST, capacity 4 *Jingsah* ACV, 2 hel plus tps

28 *Yuliang*, 1 *Yudeng* LSM, capacity about 100 tps, 3 tk

12 *Yuhai* LSM capacity 250 tps 2tk

1 *Yudao* LSM

craft: 45 LCU, 10 LCAC plus over 230 LCU in reserve

SUPPORT AND MISCELLANEOUS about 159

1 *Nanchang* AO, 2 *Fuqing* AO, 33 AOT, 14 AF, 10 AS, 1 ASR, 2 AR; 2 *Qiongsha* AH, 30 tpt, 4 icebreakers, 25 AT/F, 1 hel trg, 1 trg; 33 AGOR/AGOS

MERCHANT FLEET

1,449 ocean-going ships over 1,000t (incl 252 AOT, 335 dry bulk, 94 container, 15 ro-ro, 4 pax, 749 other)

COASTAL REGIONAL DEFENCE FORCES

ε40 indep arty and ε10 SSM regt deployed to protect naval bases, offshore islands and other vulnerable points

SSM HY-2/C-201/CSS-C-3, HY-4/C-401/CSS-C-7

AD GUNS 37mm, 57mm

MARINES (some 5,000)

2 bde (3 marine, 1 mech inf, 1 lt tk, 1 arty bn); special recce units

3 Army div also have amph role

EQUIPMENT

LT TK Type-63 amph

APC Type-77-II

ARTY 122mm: Type-83

MRL 107mm: Type-63

ATGW HJ-8

SAM HN-5

NAVAL AIR FORCE (25,000)

507 shore-based cbt ac, 37 armed hel

BBR 7 H-6, 18 H-6D reported with 2 YJ-6/61 anti-ship ALCM; about 50 H-5 torpedo-carrying lt bbr

FGA some 30 Q-5, 20 JH-7

FTR some 250 J-6, 66 J-7, 18 J-8/8A, 12 J-8B, 12 J-8D

RECCE 7 HZ-5

MR/ASW 4* PS-5 (SH-5), 4 Y-8X

AEW 1 Y-8 (possibly 2 more delivered shortly)

HEL

ASW 9 SA-321, 12 Z-8, 12 Z-9C, 4 Ka-28

TPT 50 Y-5, 4 Y-7, 6 Y-8, 2 YAK-42, 6 An-26, 10 Mi-8

TRG 53 PT-6, 16* JJ-6, 4* JJ-7

MISSILES

ALCM YJ-6/C-601, YJ-61/C-611, YJ-81/C-801K

(Naval ftr integrated into national AD system)

DEPLOYMENT AND BASES

NORTH SEA FLEET

coastal defence from Korean border (Yalu River) to south of Lianyungang (approx 35°10′N); equates to Shenyang, Beijing and Jinan MR, and to seaward

BASES Qingdao (HQ), Dalian (Luda), Huludao, Weihai, Chengshan, Yuchi; 9 coastal defence districts

FORCES 2 SS, 3 escort, 1 MCM, 1 amph sqn; plus Bohai Gulf trg flotillas; about 300 patrol and coastal combatants

EAST SEA FLEET

coastal defence from south of Lianyungang to Dongshan (approx 35°10′N to 23°30′N); equates to Nanjing Military Region, and to seaward

BASES HQ Dongqian Lake (Ninbo), Shanghai Naval base, Dinghai, Hangzhou, Xiangshan; 7 coastal defence districts

FORCES 2 SS, 2 escort, 1 MCM, 1 amph sqn; about 250 patrol and coastal combatants

Marines 1 div (cadre)

Coastal Regional Defence Forces Nanjing Coastal District

SOUTH SEA FLEET

coastal defence from Dongshan (approx 23°30′N) to Vietnamese border; equates to Guangzhou MR, and to seaward (including Paracel and Spratly Islands)

BASE Hong Kong

PATROL AND COASTAL COMBATANTS 8

4 *Houjian* PCC with 6 YJ-8/C-801 SSM, 4 PCI

SUPPORT AND MISCELLANEOUS 5

2 *Wuhu* LSM capacity 250 tps 2 tk; 3 *Catamaran*

OTHER BASES Zhanjiang (HQ), Shantou, Guangzhou, Haikou, Dongguan City, Yulin, Beihai, Huangpu; plus outposts on Paracel and Spratly Islands; 9 coastal defence districts

FORCES 2 SS, 2 escort, 1 MCM, 1 amph sqn; about 300 patrol and coastal combatants

Marines 1 bde

Air Force 420,000

(incl strategic forces, 220,000 AD personnel and 160,000 conscripts); over 3,000 cbt ac, some armed hel **Flying hours** H-6: 80; J-7 and J-8: <100; Su-27: <100

HQ Beijing. 7 Mil Air Regions (Beijing, Chengdu, Guangzhou, Jinan, Lanzhou, Nanjing, Shenyang). 4 Air Army HQs. 33 air divs (27 ftr, 4 bbr, 2 tpt) and other indep regts. Up to 4 sqn, each with 10–15 ac, 1 maint unit, some tpt and trg ac, make up an air regt;

3 air regt form an air div. Varying numbers of air divs in the Air Regions – many in the south-east

BBR med 3 regt with 120 H-6E/F (some may be nuclear-capable/30 modified to carry YJ-6/C-601 ASUWM)

FTR 400 J-7II/IIA/IIH/ IIM, 100 J-7III, 200 J-7E, 100 J-8A/E, 150 J-8B/D, 65 Su-27SK/UBK Flanker (J-11)

FGA First of 40+ Su-30MKK delivered, but not yet entered service. 300 Q-5, some 60 regt with about 1,500 J-6/B/D/E

RECCE/ELINT ε290: ε40 HZ-5, 100 JZ-6, some JZ-7, 2 Tu-154M

TPT ε425: incl some 15 Tu-154M, 2 Il-18, 14 Il-76MD, 300 Y-5, 45 Y-7/An-24/An-26, 68 Y-8/An-12, 15 Y-11, 8 Y-12, 6 Boeing 737-200 (VIP), 2 CL-601 *Challenger*

TKR 6 HY-6

HEL some 170: incl 6 AS-332 (VIP), 4 Bell 214, 30 Mi-8, 100 Z-5, 30 Z-9

TRG ε200: incl HJ-5, JJ-6, JJ-7, JL-8, K-8, PT-6 (CJ-6)

MISSILES

AAM PL-2, PL-5, PL-8, PL-9, 250+ AA-10, 250+ AA-11, *Python* 3, 100 AA-12 on order for Su-30MKK

ASM YJ-6/C-601, YJ-61/C-611, HY-2/HY-4; YJ-81K/C-801K, YJ-81K

UAV Chang Hong 1

AD ARTY ε8,000: ε8 regt: 16,000 **85mm** and **100mm** guns; 28 indep AD bde/bn (100+ SAM units with 500+ HQ-2/2A/2B, 100+ HQ-7, 120 SA-10, 20+ HQ-15 FT-2000)

Forces Abroad

UN AND PEACEKEEPING

MIDDLE EAST (UNTSO): 4 obs **IRAQ/KUWAIT** (UNIKOM): 11 obs **SIERRA LEONE** (UNAMSIL): 6 obs **WESTERN SAHARA** (MINURSO): 16 obs

Paramilitary ε1,100,000 active

PEOPLE'S ARMED POLICE ε1,300,000

45 div (14 each with 4 regt, remainder no standard org; with 1–2 div per province) incl **Internal security** ε800,000 **Border defence** some 100,000 **Guards, Comms** ε69,000

Fiji

Fiji

Fijian dollar F$		1998	1999	2000	2001
GDP	F$	3.3bn	3.5bn		
	US$	1.7bn	1.8bn		
per capita	US$	5,600	6,100		
Growth	%	1.0	7.0		
Inflation	%	5.7	2.0		
Debt	US$	241m			
Def exp	F$	65m	68m		
	US$	33m	35m		
Def bdgt	F$	45m	54m	58m	
	US$	23m	27m	27m	
FMA (US)	US$			0.2m	0.2m
FMA (Aus)	US$	3m	3m		
US$1=F$		1.99	1.98	2.14	

Population **813,000**

(Fijian 51%, Indian 44%, European/other 5%)

Age	13–17	18–22	23–32
Men	46,000	45,000	66,000
Women	44,000	43,000	64,000

Total Armed Forces

ACTIVE some 3,500

(incl recalled reserves)

RESERVES some 6,000

(to age 45)

Army 3,200

(incl 300 recalled reserves)

7 inf bn (incl 4 cadre) • 1 engr bn • 1 arty bty • 1 special ops coy

EQUIPMENT

TOWED ARTY 88mm: 4 25-pdr (ceremonial)

MOR 81mm: 12

HEL 1 AS-355, 1 SA-365

Navy 300

BASES Walu Bay, Viti (trg)

PATROL AND COASTAL COMBATANTS 9

PATROL, COASTAL/INSHORE 9

3 *Kula* (*Pacific Forum*) PCC, 4 *Vai* (Il *Dabur*) PCI<, 2 *Levuka* PCI<

SUPPORT AND MISCELLANEOUS 2

1 *Cagi Donu* presidential yacht (trg), 1 *Tovutu* AGHS

Forces Abroad

UN AND PEACEKEEPING

EAST TIMOR (UNTAET): 188 **EGYPT** (MFO): 339; 1 inf bn(-) **IRAQ/KUWAIT** (UNIKOM): 7 obs **LEBANON** (UNIFIL): 588; 1 inf bn **PAPUA NEW GUINEA**: Bougainville Peace Monitoring Group

Indonesia Indo

rupiah Rp		1998	1999	2000	2001
GDP	Rp	1,253tr	1,107tr		
	US$	125bn	140bn		
per capita	US$	3,900	4,000		
Growth	%	-13.7	1.8		
Inflation	%	57.7	20.5		
Debt	US$	150bn	200bn		
Def exp	Rp	ε9.5tr	ε11.8tr		
	US$	950m	1.5bn		
Def bdgt	Rp	9.4tr	12.2tr	18.9tr	
	US$	939m	1,553m	2,271m	
FMA (US)	US$	0.1m	0.5m	0.6m	0.4m
FMA (Aus)	US$	4.5m	4.0m	5.2m	
US$1=Rp		10,014	7,855	8,320	

Population **206,213,000**

(Muslim 87%; Javanese 45%, Sundanese 14%, Madurese 8%, Malay 8%, Chinese 3%, other 22%)

Age	13–17	18–22	23–32
Men	11,053,000	10,994,000	17,815,000
Women	10,571,000	10,534,000	18,024,000

Total Armed Forces

ACTIVE 297,000

Terms of service 2 years selective conscription authorised

RESERVES 400,000

Army cadre units; numbers, str n.k., obligation to age 45 for officers

Army ε230,000

Strategic Reserve (KOSTRAD) (30,000)
2 inf div HQ • 3 inf bde (9 bn) • 3 AB bde (9 bn) • 2 fd arty regt (6 bn) • 1 AD arty regt (2 bn) • 2 armd bn • 2 engr bn
11 Mil Area Comd (KODAM) (150,000) (Provincial (KOREM) and District (KODIM) comd)
2 inf bde (6 bn) • 65 inf bn (incl 5 AB) • 8 cav bn • 11 fd arty, 10 AD bn • 8 engr bn • 1 composite avn sqn, 1 hel sqn
Special Forces (KOPASSUS) (ε6,000 incl 4,800 cbt); to be 5 SF gp (incl 2 para-cdo, 1 counter-terrorist, 1 int, 1 trg)

EQUIPMENT

LT TK some 275 AMX-13 (to be upgraded), 30 PT-76, 50 *Scorpion*-90
RECCE 69 *Saladin* (16 upgraded), 55 *Ferret* (13 upgraded), 18 BL
AIFV 11 BMP-2
APC 200 AMX-VCI, 45 *Saracen* (14 upgraded), 60 V-150 *Commando*, 22 *Commando Ranger*, 80 BTR-40, 14 BTR-50PK, 40 *Stormer* (incl variants)
TOWED ARTY 76mm: 100 M-48; **105mm**: 170 M-101, 10 M-56; **155mm**: 5 FH 2000
MOR 81mm: 800; **120mm**: 75 Brandt
RCL 90mm: 90 M-67; **106mm**: 45 M-40A1
RL 89mm: 700 LRAC
AD GUNS 20mm: 125; **40mm**: 90 L/70; **57mm**: 200 S-60
SAM 51 *Rapier*, 42 RBS-70
AC 1 BN-2 *Islander*, 4 NC-212, 2 *Commander* 680, 3 DHC-5
HEL 30 Bell 205A, 17 Bo-105, 28 NB-412, 15 Hughes 300C (trg)

Navy 40,000

(incl ε1,000 Naval Air and 13,000 Marines)

PRINCIPAL COMMAND

WESTERN FLEET HQ Teluk Ratai (Jakarta)
BASES Primary Teluk Ratai, Belawan **Other** 10 plus minor facilities
EASTERN FLEET HQ Surabaya
BASES Primary Surabaya, Ujung Pandang, Jayapura **Other** 13 plus minor facilities

MILITARY SEALIFT COMMAND (KOLINLAMIL)

controls some amph and tpt ships used for inter-island comms and log spt for Navy and Army (assets incl in Navy and Army listings)

SUBMARINES 2

SSK 2 *Cakra* (Ge *T-209*) with 8 × 533mm TT (Ge HWT)

FRIGATES 17

FFG 10
6 *Ahmad Yani* (Nl *Van Speijk*) with 2 × 4 *Harpoon* SSM, 2 × 2 *Mistral* SAM, 1 × 76mm gun, 2 × 3 ASTT, 1 *Wasp* hel
3 *Fatahillah* with 2 × 2 MM-38 *Exocet* SSM, 1 × 120mm gun, 2 × 3 ASTT (not *Nala*), 1 × 2 ASW mor, 1 *Wasp* hel (*Nala* only)
1 *Hajar Dewantara* (trg) with 2 × 2 MM-38 *Exocet* SSM, 2 × 533mm ASTT, 1 ASW mor
FF 7
4 *Samadikun* (US *Claud Jones*) with 1 × 76mm gun, 2 × 3 324mm ASTT
3 *M. K. Tiyahahu* (UK *Tribal*) with *Mistral* SAM, 2 × 114mm guns, 1 × 3 *Limbo* ASW mor, 1 *Wasp* hel

PATROL AND COASTAL COMBATANTS 36

CORVETTES 16 *Kapitan Patimura* (GDR *Parchim*) FS with SA-N-5 SAM (in some), 1 × 57mm gun, 4 × 400mm ASTT, 2 ASW RL
MISSILE CRAFT 4 *Mandau* (Ko *Dagger*) PFM with 4 MM-38 *Exocet* SSM
TORPEDO CRAFT 4 *Singa* (Ge Lürssen 57m) with 2 × 533mm TT
PATROL CRAFT 12
OFFSHORE 4
4 *Kakap* (Ge Lürssen 57m) PCO with hel deck
COASTAL/INSHORE 8
8 *Sibarau* (Aust *Attack*) PCC
plus 18 craft

MINE COUNTERMEASURES 12

2 *Pulau Rengat* (mod Nl *Tripartite*) MCC (sometimes used for coastal patrol)
2 *Pulau Rani* (Sov T-43) MCC (mainly used for coastal patrol)
8 *Palau Rote* (GDR *Kondor* II)† MSC (mainly used for coastal patrol, 7 non-op)

AMPHIBIOUS 26

6 *Teluk Semangka* (SK *Tacoma*) LST, capacity about 200 tps, 17 tk, 2 with 3 hel (1 fitted as hospital ship)
1 *Teluk Amboina* LST, capacity about 200 tps, 16 tk
7 *Teluk Langsa* (US *LST-512*) LST, capacity 200 tps, 16 tks
12 *Teluk Gilimanuk* (GDR *Frosch* I/II) LST
Plus about 65 LCM and LCVP

SUPPORT AND MISCELLANEOUS 15

1 *Sorong* AO, 1 *Arun* AO (UK *Rover*), 2 Sov *Khobi* AOT, 1 cmd/spt/replenish; 1 AR, 2 AT/F, 1 *Barakuda* (Ge *Lürsson Nav* IV) presidental yacht; 6 AGOR/AGOS

NAVAL AIR (ε1,000)

no cbt ac, 18 armed hel
ASW 6 *Wasp* HAS-1 (3 non-op)
MR 9 N-22 *Searchmaster* B, 6 *Searchmaster* L, 10 NC-212 (MR/ELINT), 14 N-22B, 6 N-24, 3 CN-235 MP
TPT 4 *Commander*, 10 NC-212, 2 DHC-5, 20 *Nomad* (6 VIP)
TRG 2 *Bonanza* F33, 6 PA-38
HEL 3* NAS-332F (2 non-op), 5* NBo-105, 4* Bell-412

MARINES (KORMAR) (13,000)

2 inf bde (6 bn) • 1 SF bn(-) • 1 cbt spt regt (arty, AD)

EQUIPMENT

LT TK 100 PT-76†
RECCE 14 BRDM
AIFV 10 AMX-10 PAC 90
APC 24 AMX-10P, 60 BTR-50P
TOWED ARTY 48: **105mm**: 20 LG-1 Mk II; **122mm**: 28 M-38
MOR 81mm
MRL 140mm: 15 BM-14
AD GUNS 40mm: 5 L60/70; **57mm**

Air Force 27,000

108 cbt ac, no armed hel; 2 operational cmds (East and West Indonesia) plus trg cmd
FGA 5 sqn
1 with 21 A-4 (18 -E, 1 TA-4H, 2 TA-4J)
1 with 10 F-16 (7 -A, 3 -B)
2 with 8 *Hawk* Mk 109 and 31 (1 more to be delivered) *Hawk* Mk 209 (FGA/ftr)
1 with 14 *Hawk* Mk 53 (FGA/trg)
FTR 1 sqn with 12 F-5 (8 -E, 4 -F)
RECCE 1 flt with 12* OV-10F (only a few operational)
MR 1 sqn with 3 Boeing 737-200
TKR 2 KC-130B
TPT 4 sqn with 19 C-130 (9 -B, 3 -H, 7 -H-30), 3 L100-30, 1 Boeing 707, 4 Cessna 207, 5 Cessna 401, 2 C-402, 6 F-27-400M, 1 F-28-1000, 2 F-28-3000, 10 NC-212, 1 *Skyvan* (survey), 23 CN-235-110
HEL 3 sqn with 10 S-58T, 10 Hughes 500, 11 NAS-330, 5 NAS-332L (VIP/CSAR), 4 NBO-105CD, 2 Bell 204B
TRG 3 sqn with 39 AS-202, 2 Cessna 172, 22 T-34C, 6 T-41D
MISSILES
AIM-9P *Sidewinder*, AGM-65G *Maverick*

Forces Abroad

UN AND PEACEKEEPING

CROATIA (UNMOP): 2 obs **GEORGIA** (UNOMIG): 4 obs **IRAQ/KUWAIT** (UNIKOM): 4 obs **SIERRA LEONE** (UNAMSIL): 10 obs

Paramilitary ε195,000 active

POLICE (*POLRI*) ε195,000

incl 14,000 police 'mobile bde' (BRIMOB) org in 56 coy, incl counter-terrorism unit (*Gegana*)
EQPT APC 34 *Tactica*; **ac** 1 *Commander*, 2 Beech 18, 1 PA-31T, 1 Cessna-U206, 2 NC- 212 **hel** 19 NBO-105, 3 Bell 206

MARINE POLICE (12,000)

about 10 PCC, 9 PCI and 6 PCI< (all armed)

KAMRA (People's Security) (R)

ε40,000 report for 3 weeks' basic trg each year; part-time police auxiliary

WANRA (People's Resistance) (R)

part-time local military auxiliary force under Regional Military Comd (KOREM)

CUSTOMS

about 72 PFI<, armed

SEA COMMUNICATIONS AGENCY (responsible to Department of Communications)

5 Kujang PCI, 4 Golok PCI (SAR), plus boats

Opposition

FREE PAPUA ORGANISATION (OPM) ε150 (100 armed)

FREE ACEH MOVEMENT (*Gerakan Aceh Merdeka*) armed wing (AGAM) ε2,000

Other Forces

Militia gps operating in some provinces incl:-
Laskar Jihad (Holy war soldiers): Java-based. With ε2,000 to Ambon in Maluku

Foreign Forces

UN (UNTAET): some 7,905 tps, 179 mil obs and 1,262 civ pol

Japan J

yen ¥		1998	1999	2000	2001
GDP	¥	498tr	495tr		
	US$	3.8tr	4.3tr		
per capita	US$	23,700	23,800		
Growth	%	-2.6	-1.4		
Inflation	%	0.7	-0.3		
Publ Debt	%	97.3	105.4		
Def exp	¥	4.9tr	4.6tr		
	US$	37.7bn	40.8bn		
Def bdgt	¥	4.9tr	4.9tr	4.9tr	
	US$	37.6bn	43.2bn	45.6bn	
US$1=¥		131	113	108	

Population		**126,840,000** (Korean <1%)
Age	13–17	18–22	23–32
Men	3,678,000	4,129,000	9,553,000
Women	3,504,000	3,930,000	9,121,000

Total Armed Forces

ACTIVE some 236,700

(incl 1,400 Central Staffs; some 9,800 women)

RESERVES some 49,200

READY RESERVE Army (GSDF) some 3,300

GENERAL RESERVE Army (GSDF) some 44,000 **Navy** (MSDF) some 1,100 **Air Force** (ASDF) some 800

Army (Ground Self-Defense Force) some 148,500

5 Army HQ (Regional Comds) • 1 armd div • 11 inf div (6 at 7,000, 5 at 9,000 each); 1 inf bde • 2 composite bde • 1 AB bde • 1 arty bde; 2 arty gp • 2 AD bde; 3 AD gp • 4 trg bde (incl 1 spt) • 5 engr bde •1 hel bde • 5 ATK hel sqn

EQUIPMENT

MBT some 20 Type-61 (retiring), some 860 Type-74, some 190 Type-90

RECCE some 90 Type-87

AIFV some 60 Type-89

APC some 220 Type-60, some 340 Type-73, some 230 Type-82

TOWED ARTY 155mm: some 470 FH-70

SP ARTY 155mm: some 200 Type-75; **203mm**: some 90 M-110A2

MRL 130mm: some 60 Type-75 SP; **227mm**: some 50 MLRS

MOR incl **81mm**: some 720; **107mm**: some 270; **120mm**: some 310 (some SP)

SSM some 90 Type-88 coastal

ATGW some 150 Type-64, some 240 Type-79, some 280 Type-87

RL 89mm: some 1,480

RCL 84mm: some 2,720 *Carl Gustav*; **106mm**: some 250 (incl Type 60 SP)

AD GUNS 35mm: some 30 twin, some 50 Type-87 SP

SAM some 320 *Stinger*, some 60 Type 81, some 110 Type 91, some 50 Type 93, some 200 I HAWK

AC some 10 LR-1, some LR-2

ATTACK HEL some 90 AH-1S

TPT HEL 3 AS-332L (VIP), some 40 CH-47J/JA, some V-107, some 160 OH-6D, some 150 UH-1H/J, some UH-60JA

SURV Type-92 (mor), J/MPQ-P7 (arty)

Maritime Self-Defence Force some 42,600

(incl some 12,000 Air Arm; and some 1,800 women)

BASES Yokosuka, Kure, Sasebo, Maizuru, Ominato

FLEET Surface units org into 4 escort flotillas of 8 DD/FF each **Bases** Yokosuka, Kure, Sasebo, Maizuru

SS org into 2 flotillas **Bases** Kure, Yokosuka

Remainder assigned to 5 regional districts

SUBMARINES 16

SSK 16

6 *Harushio* with *Harpoon* USGW, 6 × 533mm TT (J Type-89 HWT)

7 *Yuushio* with *Harpoon* USGW, 533mm TT (J Type-89 HWT)

3 *Oyashio* with *Harpoon* USGW, 6 × 533mm *Harpoon* TT

PRINCIPAL SURFACE COMBATANTS some 55

DESTROYERS 42

DDG 30

4 *Kongou* with 2 × 4 *Harpoon* SSM, 2 VLS for *Standard* SAM and ASROC SUGW, 1 × 127mm gun, 2 × 3 ASTT, hel deck

2 *Hatakaze* with 2 × 4 *Harpoon* SSM, 1 SM-1-MR SAM, 2 × 127mm guns, 2 × 3 ASTT, 1 × 8 ASROC SUGW

3 *Tachikaze* with 2 × 4 *Harpoon* SSM, 1 SM-1-MR SAM, 2 × 127mm guns, 2 × 3 ASTT, 1 × 8 ASROC SUGW

2 *Takatsuki* (Japanese DDA) with 2 × 4 *Harpoon* SSM, *Sea Sparrow* SAM, 1 × 127mm gun, 2 × 3 ASTT, 1 × 8 ASROC SUGW, 1 × 4 ASW RL

8 *Asagiri* (Japanese DD) with 2 × 4 *Harpoon* SSM, *Sea Sparrow* SAM, 2 × 3 ASTT, 1 × 8 ASROC SUGW, 1 SH-60J hel

11 *Hatsuyuki* (Japanese DD) with 2 × 4 *Harpoon* SSM, *Sea Sparrow* SAM, 2 × 3 ASTT, 1 × 8 ASROC SUGW, 1 SH-60J hel

DD 12

6 *Murasame* with 1 VLS *Sea Sparrow* SAM, 2 × 3 ASTT, 1 VLS ASROC SUGW, 1 SH-60J hel

2 *Shirane* (Japanese DDH) with *Sea Sparrow* SAM, 2 × 127mm guns, 2 × 3 ASTT, 1 × 8 ASROC SUGW, 3 SH-60J hel

2 *Haruna* (Japanese DDH) with 2 × 127mm guns, 2 × 3 ASTT, 1 × 8 ASROC SUGW, 3 SH-60J hel

2 *Yamagumo* (Japanese DDH) with 4 × 76mm gun, 2 × 3 ASTT, 1 × 8 ASROC SUGW, 1 × 4 ASW RL

FRIGATES some 13
FFG 9
6 *Abukuma* (Japanese DE) with 2 × 4 *Harpoon* SSM, 1 × 76mm gun, 2 × 3 ASTT, 1 × 8 ASROC SUGW
2 *Yubari* (Japanese DE) with 2 × 4 *Harpoon* SSM, 2 × 3 ASTT, 1 × 4 ASW RL
1 *Ishikari* (Japanese DE) with 2 × 4 *Harpoon* SSM, 2 × 3 ASTT, 1 × 4 ASW RL
FF 4
4 *Chikugo* (Japanese DE) with 2 × 76mm guns, 2 × 3 ASTT, 1 × 8 ASROC SUGW

PATROL AND COASTAL COMBATANTS 3
MISSILE CRAFT 3 *Ichi-Go* (Japanese PG) PHM with 4 SSM-1B

MINE COUNTERMEASURES 31
2 *Uraga* MCM spt (Japanese MST) with hel deck; can lay mines
3 *Yaeyama* MSO
13 *Hatsushima* MSC
9 *Uwajima* MSC
2 *Sugashima* MSC
2 *Nijma* coastal MCM spt

AMPHIBIOUS 9
1 *Osumi* LST, capacity 330 tps, 10 tk, 2 LCAC (large flight deck)
3 *Miura* LST, capacity 200 tps, 10 tk
1 *Atsumi* LST, capacity 130 tps, 5 tk
2 *Yura* and 2 *Ichi-Go* LSM
Plus craft: 2 LCAC, 11 LCM

SUPPORT AND MISCELLANEOUS 19
3 *Towada* AOE, 1 *Sagami* AOE (all with hel deck), 2 AS/ARS, 1 *Minegumo* trg, 1 *Kashima* (trg), 2 trg spt, 8 AGHS/AGOS, 1 icebreaker

AIR ARM (ε12,000)
some 80 cbt ac, some 80 armed hel
7 Air Groups
MR 10 sqn (1 trg) with some 80 P-3C
ASW 6 land-based hel sqn (1 trg) with some 30 HSS-2B, 4 shipboard sqn with some 50 SH-60J
MCM 1 hel sqn with some 10 MH-53E
EW 1 sqn with several EP-3
TPT 1 sqn with several YS-11M
SAR some 10 US-1A, several 10 S-61 hel, some 20 UH-60J
TRG 4 sqn with **ac** some 40 T-5, some 30 TC-90, some 10 YS-11T **hel** some 10 OH-6D, several OH-6DA

Air Self-Defence Force some 44,200

some 331 cbt ac, no armed hel, 7 cbt air wings
Flying hours 150
FGA 2 sqn with some 40 F-I, 1 sqn with some 20 F-4EJ
FTR 10 sqn
8 with some 160 F-15J/DJ
2 with some 50 F-4EJ
RECCE 1 sqn with some 20* RF-4E/EJ
AEW 1 sqn with some 10 E-2C, 4 Boeing E-767 (AWACS)
EW 2 sqn with 1 EC-1, some 10 YS-11 E
AGGRESSOR TRG 1 sqn with some 10 F-15DJ
TPT 4 sqn, 4 flt
3 with some 20 C-1, some 10 C-130H, a few YS-11
1 with a few 747-400 (VIP)
4 flt heavy-lift hel with some 10 CH-47J
SAR 1 wg (10 det) with **ac** some 11 MU-2, some 13 U-125 **hel** some 15 KV-107, some 15 UH-60J
CAL 1 sqn with a few YS-11, a few U-125-800
TRG 5 wg, 12 sqn with some 22 T-1A/B, some 41* T-2, some 43 T-3, some 60 T-4, some 10 T-400
LIAISON some 10 T-33, some 90 T-4, a few U-4
TEST 1 wg with a few F-15J, some 10 T-4

AIR DEFENCE
ac control and warning: 4 wg, 28 radar sites
6 SAM gp (24 sqn) with some 120 *Patriot*
Air Base Defence Gp with **20mm** *Vulcan* AA guns, Type 81 short-range SAM, Type 91 portable SAM, *Stinger* SAM
ASM ASM-1, ASM-2
AAM AAM-1, AAM-3, AIM-7 *Sparrow*, AIM-9 *Sidewinder*

Forces Abroad

UN AND PEACEKEEPING
SYRIA/ISRAEL (UNDOF): 30

Paramilitary 12,000

MARITIME SAFETY AGENCY (Coast Guard) 12,000 (Ministry of Transport, no cbt role)
PATROL VESSELS some 343
Offshore (over 1,000 tons) 52, incl 1 *Shikishima* with 2 *Super Puma* hel, 2 *Mizuho* with 2 Bell 212, 8 *Soya* with 1 Bell 212 hel, 2 *Izu*, 28 *Shiretok* and 1 Kojima (trg) **Coastal** (under 1,000 tons) 66 **Inshore** some 225 patrol craft most<
MISC 93: 12 AGHS, 60 nav tender, 14 fire fighting boats, 4 buoy tenders, 3 trg
AC 5 NAMC YS-11A, 2 Saab 340, 19 *King Air*, 1 Cessna U-206G
HEL 26 Bell 212, 4 Bell 206B, 6 Bell 412, 4 *Super Puma*, 4 Sikorsky S76C

Foreign Forces

US 39,750: **Army** 1,800; 1 Corps HQ **Navy** 5,200; bases at Yokosuka (HQ 7th Fleet) and Sasebo **Marines** 19,200; 1 MEF in Okinawa **Air Force** 13,550; 1 Air Force HQ (5th Air Force), 90 cbt ac, 1 ftr wg, 2 sqn with 36 F-16, 1 wg, 3 sqn with 54 F-15C/D, 1 sqn with 15 KC-135, 1 SAR sqn with 8 HH-60, 1 sqn with 2 E-3 AWACS; 1 airlift wg with 16 C-130E/H, 4 C-21, 3 C-9; 1 special ops gp with 4 MC-130P, 4 MC-130E

Korea, Democratic People's Republic of (North) DPRK

won		1998	1999	2000	2001
GNP[a]	US$	ε14bn	ε14.7bn		
per capita	US$	950	1,000		
Growth	%	5.0			
Inflation	%	ε5			
Debt	US$	12bn			
Def exp	US$	ε2.0bn	ε2.1bn		
Def bdgt	won	2.92bn	2.96bn	2.96bn	
	US$	1.3bn	1.3bn	1.3bn	
US$1=won		2.2	2.2	2.2	

[a] PPP est. GNP is larger than GDP because of remitted earnings of DPRK expatriates in Japan and ROK

Population			ε21,500,000
Age	13–17	18–22	23–32
Men	1,055,000	951,000	2,510,000
Women	1,092,000	1,033,000	2,135,000

Total Armed Forces

ACTIVE ε1,082,000

Terms of service **Army** 5–8 years **Navy** 5–10 years **Air Force** 3–4 years, followed by compulsory part-time service to age 40. Thereafter service in the Worker/Peasant Red Guard to age 60

RESERVES 4,700,000 of which

Army 600,000 **Navy** 65,000 are assigned to units (see also *Paramilitary*)

Army ε950,000

20 Corps (1 armd, 4 mech, 12 inf, 2 arty, 1 capital defence) • 27 inf div • 15 armd bde • 14 inf • 21 arty • 9 MRL bde

Special Purpose Forces Comd (88,000): 10 *Sniper* bde (incl 2 amph, 2 AB), 12 lt inf bde (incl 3 AB), 17 recce, 1 AB bn, 'Bureau of Reconnaissance SF' (8 bn)

Army tps: 6 hy arty bde (incl MRL), 1 *Scud* SSM bde, 1 FROG SSM regt

Corps tps: 14 arty bde incl 122mm, 152mm SP, MRL

RESERVES

40 inf div, 18 inf bde

EQUIPMENT

MBT some 3,500: T-34, T-54/-55, T-62, Type-59

LT TK 560 PT-76, M-1985

APC 2,500 BTR-40/-50/-60/-152, PRC Type-531, VTT-323 (M-1973), some BTR-80A

TOTAL ARTY (excl mor) 11,500

TOWED ARTY 3,500: **122mm**: M-1931/-37, D-74, D-30; **130mm**: M-46; **152mm**: M-1937, M-1938, M-1943

SP ARTY 4,400: **122mm**: M-1977, M-1981, M-1985, M-1991; **130mm**: M-1975, M-1981, M-1991; **152mm**: M-1974, M-1977; **170mm**: M-1978, M-1989

COMBINED GUN/MOR: **120mm** (reported)

MRL 2,500: **107mm**: Type-63; **122mm**: BM-21, BM-11, M-1977/-1985/-1992/-1993; **240mm**: M-1985/-1989/-1991

MOR 7,500: **82mm**: M-37; **120mm**: M-43 (some SP); **160mm**: M-43

SSM 24 FROG-3/-5/-7; some 30 *Scud*-C, *No-dong*

ATGW: AT-1 *Snapper*, AT-3 *Sagger* (some SP), AT-4 *Spigot*, AT-5 *Spandrel*

RCL 82mm: 1,700 B-10

AD GUNS 11,000: **14.5mm**: ZPU-1/-2/-4 SP, M-1984 SP; **23mm**: ZU-23, M-1992 SP; **37mm**: M-1939, M-1992; **57mm**: S-60, M-1985 SP; **85mm**: KS-12; **100mm**: KS-19

SAM ε10,000+ SA-7/-16

Navy ε46,000

BASES East Coast Toejo (HQ), Changjon, Munchon, Songjon-pardo, Mugye-po, Mayang-do, Chaho Nodongjagu, Puam-Dong, Najin **West Coast** Nampo (HQ), Pipa Got, Sagon-ni, Chodo-ri, Koampo, Tasa-ri 2 Fleet HQ

SUBMARINES 26

SSK 26

22 PRC Type-031/Sov *Romeo* with 533mm TT

4 Sov *Whiskey*† with 533mm and 406mm TT

(Plus some 45 SSI and 21 *Sang-O* SSC mainly used for SF ops, but some with 2 TT, all +)

FRIGATES 3

FF 3

1 *Soho* with 4 SS-N-2 *Styx* SSM, 1 × 100mm gun and hel deck, 4 ASW RL

2 *Najin* with 2 SS-N-2 *Styx* SSM, 2 × 100mm guns, 2 × 5 ASW RL

PATROL AND COASTAL COMBATANTS some 310

CORVETTES 6

4 *Sariwon* FS with 1 × 85mm gun

2 *Tral* FS with 1 × 85mm gun

MISSILE CRAFT 43

15 *Soju*, 8 Sov *Osa*, 4 PRC *Huangfeng* PFM with 4 SS-N-2 *Styx* SSM, 6 *Sohung*, 10 Sov *Komar* PFM with 2 SS-N-2 *Styx* SSM

TORPEDO CRAFT some 103

3 Sov *Shershen* PFT with 4 × 533mm TT

60 *Ku Song* PHT

40 *Sin Hung* PHT

PATROL CRAFT 158

COASTAL 25

6 *Hainan* PFC with 4 ASW RL, 13 *Taechong* PFC with 2 ASW RL, 6 *Chong-Ju* with 1 85mm gun, (2 ASW mor)

INSHORE some 133

18 SO-1<, 12 *Shanghai* II<, 3 *Chodo*<, some 100<

MINE COUNTERMEASURES about 23 MSI<

AMPHIBIOUS 10

10 *Hantae* LSM, capacity 350 tps, 3 tk
plus craft 15 LCM, 15 LCU, about 100 Nampo LCVP, plus about 130 hovercraft

SUPPORT AND MISCELLANEOUS 7

2 AT/F, 1 AS, 1 ocean and 3 inshore AGHS

COASTAL DEFENCE

2 SSM regt: *Silkworm* in 6 sites, and probably some mobile launchers
GUNS 122mm: M-1931/-37; **130mm**: SM-4-1, M-1992; **152mm**: M-1937

Air Force 86,000

6 air divs, one per military district:
3 bbr and ftr divs, 2 support ac divs, 1 trg div
Approx 70 full time/contingency air bases
621 cbt ac, ε24 armed hel
Flying hours 30 or less
BBR 3 lt regt with 80 H-5 (Il-28)
FGA/FTR 15 regt
3 with 107 J-5 (MiG-17), 4 with 159 J-6 (MiG-19), 4 with 130 J-7 (MiG-21), 1 with 46 MiG-23, 1 with 16 MiG-29, 1 with 18 Su-7, 1 with 35 Su-25, 30 MiG-29 (25 -As, 5 -Us), and 10 more being assembled, to start replacing J-5/J-6
TPT ac ε300 An-2/Y-5 (to infiltrate 2 air force sniper brigades deep into S Korean rear areas), 6 An-24, 2 Il-18, 4 Il-62M, 2 Tu-134, 4 Tu-154
HEL ε320. Large hel aslt force spearheaded by 24 Mi-24*. Tpt/utility: 80 Hughes 500D, 139 Mi-2, 15 Mi-8/-17, 48 Z-5
TRG incl 10 CJ-5, 7 CJ-6, 6 MiG-21, 170 Yak-18, 35 FT-2 (MiG-15UTI)
MISSILES
AAM AA-2 *Atoll*, AA-7 *Apex*
SAM ε45 SA-2 bty, 7 SA-3, 2 SA-5, many thousands of SA-7/14/16

Forces Abroad

advisers in some 12 African countries

Paramilitary 189,000 active

SECURITY TROOPS (Ministry of Public Security) 189,000 incl border guards, public safety personnel

WORKER/PEASANT RED GUARD some 3,500,000 (R)
Org on a provincial/town/village basis; comd structure is bde – bn – coy – pl; small arms with some mor and AD guns (but many units unarmed)

Korea, Republic of (South) ROK

won		1998	1999	2000	2001
GDP	won	450tr	484tr		
	US$	426bn	407bn		
per capita	US$	12,200	13,700		
Growth	%	-6.7	10.7		
Inflation	%	7.5	0.8		
Debt	US$	148bn	141bn		
Def exp	won	14.3tr	14.3tr		
	US$	10.2bn	12.0bn		
Def bdgt	won	13.9tr	13.7tr	14.4tr	
	US$	9.9bn	11.6bn	12.8bn	
US$1=won		1,401	1,186	1,129	

Population			47,500,000
Age	13–17	18–22	23–32
Men	1,815,000	1,964,000	4,365,000
Women	1,701,000	1,829,000	4,091,000

Total Armed Forces

ACTIVE 683,000
(incl ε159,000 conscripts)
Terms of service conscription **Army** 26 months **Navy** and **Air Force** 30 months; First Combat Forces (Mobilisation Reserve Forces) or Regional Combat Forces (Homeland Defence Forces) to age 33

RESERVES 4,500,000
being re-org

Army 560,000

(incl 140,000 conscripts)
HQ: 3 Army (1st and 3rd to merge), 11 Corps (two to be disbanded)
3 mech inf div (each 3 bde: 3 mech inf, 3 tk, 1 recce, 1 engr bn; 1 fd arty bde) • 19 inf div (each 3 inf regt, 1 recce, 1 tk, 1 engr bn; 1 arty regt (4 bn)) • 2 indep inf bde • 7 SF bde • 3 counter-infiltration bde • 3 SSM bn with NHK-I/-II (*Honest John*) • 3 AD arty bde • 3 I HAWK bn (24 sites), 2 *Nike Hercules* bn (10 sites) • 1 avn comd

RESERVES
1 Army HQ, 23 inf div

EQUIPMENT
MBT 1,000 Type 88, 80 T-80U, 400 M-47, 850 M-48
AIFV 40 BMP-3
APC incl 1,700 KIFV, 420 M-113, 140 M-577, 200 Fiat 6614/KM-900/-901, 20 BTR-80
TOWED ARTY some 3,500: **105mm**: 1,700 M-101, KH-178; **155mm**: M-53, M-114, KH-179; **203mm**: M-115
SP ARTY 155mm: 1,040 M-109A2, some K-9; **175mm**: M-107; **203mm**: 13 M-110

East Asia and Australasia

MRL 130mm: 156 *Kooryong* (36-tube); **227mm**: 29 MLRS (with ATACM)
MOR 6,000: **81mm**: KM-29; **107mm**: M-30
SSM 12 NHK-I/-II
ATGW TOW-2A, *Panzerfaust*, AT-7
RCL 57mm, 75mm, 90mm: M67; **106mm**: M40A2
ATK GUNS 58: **76mm**: 8 M-18; **90mm**: 50 M-36 SP
AD GUNS 600: **20mm**: incl KIFV (AD variant), 60 M-167 *Vulcan*; **30mm**: 20 B1 HO SP; **35mm**: 20 GDF-003; **40mm**: 80 L60/70, M-1
SAM 350 *Javelin*, 60 *Redeye*, 130 *Stinger*, 170 *Mistral*, SA-16, 110 I HAWK, 200 *Nike Hercules*, *Chun Ma* (reported)
SURV RASIT (veh, arty), AN/TPQ-36 (arty, mor), AN/TPQ-37 (arty)
AC 5 O-1A
HEL
ATTACK 60 AH-1F/-J, 45 Hughes 500 MD, 12 BO-105
TPT 18 CH-47D
UTL 130 Hughes 500, 20 UH-1H, 116 UH-60P, 3 AS-332L

Navy 60,000

(incl 25,000 Marines and ε19,000 conscripts)
BASES Chinhae (HQ), Cheju, Mokpo, Mukho, Pohang, Pusan, Pyongtaek, Tonghae
FLEET COMMANDS 3
1st: Tonghae (Sea of Japan); 2nd: Pyongtaek (Yellow Sea); 3rd: Chinhae (Korean Strait)

SUBMARINES 19
SSK 8 *Chang Bogo* (Ge T-209/1200) with 8 × 533 TT
SSI 11
3 KSS-1 *Dolgorae* (175t) with 2 × 406mm TT
8 *Dolphin* (175t) with 2 × 406mm TT

PRINCIPAL SURFACE COMBATANTS 39
DESTROYERS 6
DDG 6
3 *King Kwanggaeto* with 8 *Harpoon* SSM, 1 *Sea Sparrow* SAM, 1 × 127mm gun, 1 *Super Lynx* hel
3 *Kwang Ju* (US *Gearing*) with 2 × 4 *Harpoon* SSM, 2 × 2 × 127mm guns, 2 × 3 ASTT, 1 × 8 ASROC SUGW, 1 *Alouette* III hel
FRIGATES 9
FFG 9 *Ulsan* with 2 × 4 *Harpoon* SSM, 2 × 76mm gun, 2 × 3 ASTT (Mk 46 LWT)
CORVETTES 24
24 *Po Hang* FS with 2 × 3 ASTT; some with 2 × 1 MM-38 *Exocet* SSM

PATROL AND COASTAL COMBATANTS 84
CORVETTES 4 *Dong Hae* (ASW) FS with 2 × 3 ASTT
MISSILE CRAFT 5
5 *Pae Ku*-52 (US *Asheville*) PFM, 2 × 2 *Harpoon* SSM, 1 × 76mm gun
PATROL, INSHORE 75
75 *Kilurki*-11 (*Sea Dolphin*) 37m PFI

MINE WARFARE 15
MINELAYERS 1
1 *Won San* ML
MINE COUNTERMEASURES 14
6 *Kan Keong* (mod It *Lerici*) MHC
8 *Kum San* (US MSC-268/289) MSC

AMPHIBIOUS 14
4 *Alligator* (RF) LST, capacity 700
7 *Un Bong* (US LST-511) LST, capacity 200 tps, 16 tk
3 *Ko Mun* (US LSM-1) LSM, capacity 50 tps, 4 tk
Plus about 36 craft; 6 LCT, 10 LCM, about 20 LCVP

SUPPORT AND MISCELLANEOUS 13
2 AOE, 2 spt AK, 2 AT/F, 2 salv/div spt, 1 ASR, about 4 AGHS (civil-manned, Ministry of Transport-funded)

NAVAL AIR
23 cbt ac; 48 armed hel
ASW 3 sqn
2 **ac** 1 with 15 S-2E, 1 with 8 P-3C
1 **hel** with 25 Hughes 500MD
1 flt with 8 SA-316 hel, 12 *Lynx*, 3 *Super Lynx*

MARINES (25,000)
2 div, 1 bde • spt units
EQUIPMENT
MBT 60 M-47
AAV 60 LVTP-7, 3 AAV-7
TOWED ARTY 105mm, 155mm
SSM *Harpoon* (truck-mounted)

Air Force 63,000

3 Cmds (Ops, Logs, Trg), Tac Airlift Wg and Composite Wg are all responsible to ROK Air Force HQ. Ops Cmd controls Anti-Aircraft Artillery Cmd, Air Traffic Centre and tac ftr wgs.
555 cbt ac, no armed hel
FTR/FGA 7 tac ftr wgs
2 with 160 F-16C/D
3 with 195 F-5E/F
2 with 130 F-4D/E
CCT 1 wg with 22* A-37B
FAC 1 wg with 20 O-1A, 10 O-2A
RECCE 1 gp with 18* RF-4C, 5* RF-5A
SAR 1 hel sqn, 5 UH-1H, 4 Bell-212
TAC AIRLIFT WG ac 2 BAe 748 (VIP), 1 Boeing 737-300 (VIP), 1 C-118, 10 C-130H, 15 CN-235M **hel** 6 CH-47, 3 AS-332, 3 VH-60
TRG 25* F-5B, 50 T-37, 30 T-38, 25 T-41B, 18 *Hawk* Mk-67
UAV 3 *Searcher*, 100 *Harpy*
MISSILES
ASM AGM-65A *Maverick*, AGM-88 HARM, AGM-130, AGM-142
AAM AIM-7 *Sparrow*, AIM-9 *Sidewinder*, AIM-120B AMRAAM
SAM *Nike-Hercules*, I HAWK, *Javelin*, *Mistral*

Forces Abroad

UN AND PEACEKEEPING

EAST TIMOR (UNTAET): 444 **GEORGIA** (UNOMIG): 3 obs **INDIA/PAKISTAN** (UNMOGIP): 9 obs **WESTERN SAHARA** (MINURSO): 20

Paramilitary ε4,500 active

CIVILIAN DEFENCE CORPS3,500,000 (R) (to age 50)

MARITIME POLICE ε4,500

- **PATROL CRAFT** 81
  - **OFFSHORE** 10
    3 *Mazinger* (HDP-1000) (1 CG flagship), 1 *Han Kang* (HDC-1150), 6 *Sea Dragon/Whale* (HDP-600)
  - **COASTAL** 33
    22 *Sea Wolf/Shark*, 2 *Bukhansan*, 7 *Hyundai*-type, 2 *Bukhansan*
  - **INSHORE** 38
    18 *Seagull*, about 20<, plus numerous boats
- **SUPPORT AND MISCELLANEOUS** 3 salvage
- **HEL** 9 Hughes 500

Foreign Forces

US 36,630: **Army** 27,500; 1 Army HQ, 1 inf div **Navy** 300 **Air Force** 8,700: 1 HQ (7th Air Force); 90 cbt ac, 2 ftr wg; 3 sqn with 72 F-16, 1 sqn with 6 A-10, 12 OA-10, 1 special ops sqn with 5MH -53J **USMC** 130

Laos Lao

kip		1998	1999	2000	2001
GDP	kip	4.4tr	6.8tr		
	US$	1.3bn	1.0bn		
per capita	US$	2,500	2,600		
Growth	%	4.0	3.8		
Inflation	%	91	128		
Debt	US$	2.5bn			
Def exp	kip	ε110bn	ε156bn		
	US$	33m	22m		
Def bdgt	kip	ε110bn	ε110bn		
	US$	33m	15m		
FMA (US)	US$	3.5m	4.0m	1.5m	1.5m
US$1=kip		3,298	7,102	7,600	

Population 5,500,000

(*lowland* Lao Loum 68% *upland* Lao Theung 22% *highland* Lao Soung incl Hmong and Yao 9%, Chinese and Vietnamese 1%)

Age	13–17	18–22	23–32
Men	313,000	247,000	381,000
Women	308,000	243,000	379,000

Total Armed Forces

ACTIVE ε29,100

Terms of service conscription, 18 months minimum

Army 25,000

4 Mil Regions • 5 inf div • 7 indep inf regt • 1 armd, 5 arty, 9 AD arty bn • 3 engr (2 construction) regt • 65 indep inf coy • 1 lt ac liaison flt

EQUIPMENT

- **MBT** 30 T-54/-55, T-34/85
- **LT TK** 25 PT-76
- **APC** 30 BTR-40/-60, 40 BTR-152
- **TOWED ARTY 75mm**: M-116 pack; **105mm**: 25 M-101; **122mm**: 40 M-1938 and D-30; **130mm**: 10 M-46; **155mm**: M-114
- **MOR 81mm; 82mm; 107mm**: M-2A1, M-1938; **120mm**: M-43
- **RCL 57mm**: M-18/A1; **75mm**: M-20; **106mm**: M-40; **107mm**: B-11
- **AD GUNS 14.5mm**: ZPU-1/-4; **23mm**: ZU-23, ZSU-23-4 SP; **37mm**: M-1939; **57mm**: S-60
- **SAM** SA-3, SA-7

Navy (Army Marine Section) ε600

PATROL AND COASTAL COMBATANTS some 16

PATROL, RIVERINE some 16
some 12 PCR<, 4 LCM, plus about 40 boats

Air Force 3,500

14† cbt ac; no armed hel

FGA 2 sqn with some 12 MiG-21bis/2-UMs (serviceability in doubt)

TPT 1 sqn with 4 An-2, 5 An-24, 3 An-26, 1 Yak-40 (VIP), 1 An-74

HEL 1 sqn with 1 Mi-6, 9 Mi-8, 12 Mi-17, 3 SA-360, 1 Ka-32T (5 more on order), 1 Mi-26

TRG 8 Yak-18

AAM AA-2 *Atoll*†

Paramilitary

MILITIA SELF-DEFENCE FORCES 100,000+
village 'home-guard' org for local defence

Opposition

Numerous factions/groups; total armed str: ε2,000 **United Lao National Liberation Front (ULNLF)** largest group

Malaysia Mal

ringgit RM		1998	1999	2000	2001
GDP	RM	284bn	299bn		
	US$	72bn	78bn		
per capita	US$	10,000	10,600		
Growth	%	-7.5	5.4		
Inflation	%	5.3	2.8		
Debt	US$	44bn	48bn		
Def exp[a]	RM	7.3bn	12.0bn		
	US$	1.9bn	3.2bn		
Def bdgt[b]	RM	4.5bn	6.9bn	6.0bn	
	US$	1.2bn	1.8bn	1.6bn	
FMA (US)	US$	0.9m	0.7m	0.7m	0.7m
FMA (Aus)	US$	3.5m	4.2m		
US$1=RM		3.9	3.8	3.8	

[a] Incl procurement and def industry exp
[b] Excl procurement allocation in 1999 and 2000

Population **21,868,000**

(Muslim 54%; Malay and other indigenous 64%, Chinese 27%, Indian 9%; in *Sabah* and *Sarawak* non-Muslim Bumiputras form the majority of the population; 1,000,000+ Indonesian and Filipino illegal immigrants in 1997)

Age	13–17	18–22	23–32
Men	1,251,000	1,034,000	1,767,000
Women	1,191,000	987,000	1,712,000

Total Armed Forces

ACTIVE 96,000

RESERVES 49,800

Army 47,000 **Navy** 2,200 **Air Force** 600

Army 80,000

2 Mil Regions • 1 HQ fd comd, 4 area comd (div) • 1 mech inf, 11 inf bde • 1 AB bde (3 AB bn, 1 lt arty regt, 1 lt tk sqn – forms Rapid Deployment Force)

Summary of combat units

5 armd regt • 36 inf bn • 3 AB bn • 5 fd arty, 1 AD arty, 5 engr regt

1 SF regt (3 bn)

AVN 1 hel sqn

RESERVES

Territorial Army 1 bde HQ; 12 inf regt, 4 highway sy bn

EQUIPMENT

LT TK 26 *Scorpion* **(90mm)**

RECCE 162 SIBMAS, 140 AML-60/-90, 92 *Ferret* (60 mod)

APC 111 KIFV (incl variants), 184 V-100/-150 *Commando*, 25 *Stormer*, 459 *Condor* (150 to be upgraded), 37 M-3 Panhard

TOWED ARTY 105mm: 130 Model 56 pack, 40 M-102A1 († in store); **155mm**: 12 FH-70

MOR 81mm: 300

ATGW SS-11, *Eryx*

RL 89mm: M-20; **92mm**: FT5

RCL 84mm: *Carl Gustav*; **106mm**: 150 M-40

AD GUNS 35mm: 24 GDF-005; **40mm**: 36 L40/70

SAM 48 *Javelin*, *Starburst*, 12 *Rapier*

HEL 10 SA-316B

ASLT CRAFT 165 *Damen*

Navy 8,000

(incl 160 Naval Air)

Fleet Operations Comd (HQ Lumut)

Naval Area 1 Kuantan **Naval Area 2** Labuan plus trg base at Pengelih (new base being built at Sepanggar Bay, Sabah)

SUBMARINES 0

but personnel have trained in France, the Netherlands and UK

PRINCIPAL SURFACE COMBATANTS 4

FRIGATES 4

FFG 2 *Leiku* with 8 × MM-40 *Exocet* SSM, 1 × 16 VLS *Seawolf* SAM, 6 × 324mm ASTT

FF 2 (both used for training)

1 *Hang Tuah* (UK *Mermaid*) with 1 × 57mm gun, 1 × 3 *Limbo* ASW mor, hel deck

1 *Rahmat* with 1 × 114mm gun, 1 × 3 ASW mor, hel deck

PATROL AND COASTAL COMBATANTS 41

CORVETTES 6

4 *Laksamana* (It *Assad*) FSG with 6 OTO *Melara* SSM, 1 *Selenia* SAM, 1 × 76mm gun, 6 × 324mm ASTT

2 *Kasturi* (FS 1500) FS with 4 MM-38 *Exocet* SSM, 1 × 100mm gun, 2 × 2 ASW mor, hel deck

MISSILE CRAFT 8

4 *Handalan* (Swe *Spica*) PFM with 4 MM-38 *Exocet* SSM, 1 × 57mm gun

4 *Perdana* (Fr *Combattante* II) PFM with 2 *Exocet* SSM, 1 × 57mm gun

PATROL CRAFT 27

OFFSHORE 2 *Musytari* PCO with 1 × 100mm gun, hel deck

COASTAL/INSHORE 25

6 *Jerong* PFC, 4 *Sabah* PCC, 14 *Kris* PCC, 1 *Kedah* PCI<

MINE COUNTERMEASURES 4

4 *Mahamiru* (mod It *Lerici*) MCO

plus 1 diving tender (inshore)

AMPHIBIOUS 2

1 *Sri Banggi* (US LST-511) LST, capacity 200 tps, 16 tk (but usually employed as tenders to patrol craft)

1 *Sri Inderapura* (US *Newport*) LST, capacity 400 tps, 10 tk

Plus 115 craft: LCM/LCP/LCU

SUPPORT AND MISCELLANEOUS 4

2 log/fuel spt, 2 AGOR/AGOS

NAVAL AIR (160)
no cbt ac, 17 armed hel
HEL 17 *Wasp* HAS-1
SPECIAL FORCES
1 Naval Commando Unit

Air Force 8,000

84 cbt ac, no armed hel; 4 Air Div
Flying hours 60
FGA 4 sqn
3 with 8 *Hawk* 108, 17 *Hawk* 208, 9 MB-339
1 with 8 F/A-18D
FTR 3 sqn
2 with 15 MiG-29N, 2 MiG-29U
1 fighter lead-in/'aggressor'/recce with 4 F-5E/3 F-5F/2 RF-5E
MR 1 sqn with 4 Beech-200T
TRANSPORT 4 sqn
1 with 11 DHC-4, 6 CN-235
1 with 5 C-130H
1 with 6 C-130H-30, 1 C-130H-MP, 2 KC-130H (tkr), 9 Cessna 402B (2 modified for aerial survey)
1 with **ac** 1 *Falcon*-900 (VIP), 1 Bombardier Global Express, 1 F-28 **hel** 2 AS-61N, 1 Agusta-109, 2 S-70A
HEL 3 tpt/SAR sqn with 32 S-61A, 12 SA-316A/B, 2 Mi-17 (firefighting)
TRAINING
AC 20 MD3-160, 37 PC-7 (12* wpn trg)
HEL 12 SA-316
MISSILES
AAM AIM-9 *Sidewinder*, AA-10 *Alamo*, AA-11 *Archer*
ASM AGM-65 *Maverick*, AGM-84D *Harpoon*

AIRFIELD DEFENCE
1 field sqn
SAM 1 sqn with *Starburst*

Forces Abroad

UN AND PEACEKEEPING
EAST TIMOR (UNTAET): 32 incl 20 obs **DROC** (MONUC): 9 obs **IRAQ/KUWAIT** (UNIKOM): 5 obs **SIERRA LEONE** (UNAMSIL): 10 obs **WESTERN SAHARA** (MINURSO): 13 obs

Paramilitary ε20,100

POLICE-GENERAL OPS FORCE 18,000
5 bde HQ: 21 bn (incl 2 Aboriginal, 1 Special Ops Force), 4 indep coy
EQPT ε100 Shorland armd cars, 140 AT-105 *Saxon*, ε30 SB-301 APC
MARINE POLICE about 2,100
BASES Kuala, Kemaman, Penang, Tampoi, Kuching, Sandakan
PATROL CRAFT, INSHORE 30
15 *Lang Hitam* (38m) PFI, 6 *Sangitan* (29m) PFI, 9 improved PX PFI, plus 6 tpt, 2 tugs, 120 boats
POLICE AIR UNIT
ac 6 Cessna *Caravan* I, 4 Cessna 206, 7 PC-6 **hel** 1 Bell 206L, 2 AS-355F
AREA SECURITY UNITS (aux General Ops Force) 3,500
89 units
BORDER SCOUTS (in Sabah, Sarawak) 1,200
PEOPLE'S VOLUNTEER CORPS (RELA) 240,000
some 17,500 armed
CUSTOMS SERVICE
PATROL CRAFT, INSHORE 8
6 *Perak* (Vosper 32m) armed PFI, 2 *Combatboat 90H* PFI, plus about 36 craft

Foreign Forces

AUSTRALIA 148: **Army** 115; 1 inf coy **Air Force** 33; det with 2 P-3C **ac**

Mongolia Mgl

tugrik t		1998	1999	2000	2001
GDP	t	878bn	1.0tr		
	US$	1.0bn	980m		
per capita	US$	2,200	2,300		
Growth	%	3.5	3.5		
Inflation	%	9.5	7.3		
Debt	US$	670m			
Def exp	t	17.0bn	19.3bn		
	US$	20m	19m		
Def bdgt	t	18.7bn	19.7bn	25.1bn	
	US$	24m	21m	24.6m	
FMA (US)	US$	0.4m	0.4m	0.5m	0.5m
US$1=t		841	1,021	1,018	

Population			2,460,000
(Kazak 4%, Russian 2%, Chinese 2%)			
Age	13–17	18–22	23–32
Men	158,000	140,000	231,000
Women	152,000	134,000	223,000

Total Armed Forces

ACTIVE 9,100
(incl 300 construction tps and 500 Civil Defence – see *Paramilitary*; 4,000 conscripts)
Terms of service conscription: males 18–28 years, 1 year

RESERVES 140,000
Army 140,000

East Asia and Australasia

Army 7,500

(incl 4,000 conscripts)
7 MR bde (all under str) • 1 arty bde • 1 lt inf bn (rapid-deployment) • 1 AB bn

EQUIPMENT

MBT 650 T-54/-55/-62
RECCE 120 BRDM-2
AIFV 400+ BMP-1
APC 250+ BTR-60
TOTAL ARTY (incl ATK and AD Guns) 1,500
TOWED ARTY 122mm: M-1938/D-30; **130mm**: M-46; **152mm**: ML-20
MRL 122mm: BM-21
MOR 140: **82mm, 120mm, 160mm**
ATK GUNS 200 incl: **85mm**: D-44/D-48; **100mm**: BS-3, MT-12

Air Defence 800

9 cbt ac; 11 armed hel
Flying hours 22
2 AD regt
FTR 1 sqn with 8 MiG-21, 1 Mig-21U
ATTACK HEL 11 Mi-24
TPT (Civil Registration) 15 An-2, 12 An-24, 3 An-26, 1 An-30, 2 Boeing 727, 1 Airbus A310-300
AD GUNS: 150: **14.5mm**: ZPU-4; **23mm**: ZU-23, ZSU-23-4; **57mm**: S-60
SAM 250 SA-7

Paramilitary 7,200 active

BORDER GUARD 6,000 (incl 4,700 conscripts)

INTERNAL SECURITY TROOPS 1,200 (incl 800 conscripts) 4 gd units

CIVIL DEFENCE TROOPS (500)

CONSTRUCTION TROOPS (300)

Myanmar My

kyat K		1998	1999	2000	2001
GDP[a]	K	1,645bn	1,559bn		
	US$	31bn	29bn		
per capita	US$	1,100	1,200		
Growth	%	7.0	7.0		
Inflation	%	51.5	18.4		
Debt	US$	5.6bn	6.0bn		
Def exp[a]	K	33bn	31bn		
	US$	2.1bn	2.0bn		
Def bdgt[a]	K	24.5bn	31.8bn		
	US$	1.7bn	1.7bn		
US$1=K[b]		6.34	6.35	6.25	

[a] PPP est
[b] Market rate **1999** US$1=K250–300

Population			**48,500,000**
(Burmese 68%, Shan 9%, Karen 7%, Rakhine 4%, Chinese 3+%, *Other* Chin, Kachin, Kayan, Lahu, Mon, Palaung, Pao, Wa, 9%)			
Age	13–17	18–22	23–32
Men	2,707,000	2,410,000	4,321,000
Women	2,636,000	2,365,000	4,275,000

Total Armed Forces

ACTIVE some 429,000 reported (incl People's Police Force and People's Militia – see *Paramilitary*)

Army 325,000

10 lt inf div (each 3 tac op comd (TOC))
12 Regional Comd (each with 10 regt)
32 TOC with 145 garrison inf bn
Summary of cbt units
245 inf bn • 7 arty bn • 4 armd bn • 2 AA arty bn

EQUIPMENT†

MBT 100 PRC Type-69II
LT TK 105 Type-63 (ε60 serviceable)
RECCE 45 *Ferret*, 40 *Humber*, 30 *Mazda* (local manufacture)
APC 20 *Hino* (local manufacture), 250 Type-85
TOWED ARTY 76mm: 100 M-1948; **88mm**: 50 25-pdr; **105mm**: 96 M-101; **122mm**; **130mm**: 16 M-46; **140mm**: 5.5in; **155mm**: 16 Soltam
MRL 107mm: 30 Type-63
MOR 81mm; 82mm: Type-53; **120mm**: Type-53, 80 Soltam
RCL 84mm: 500 *Carl Gustav*; **106mm**: M40A1
ATK GUNS 60: **57mm**: 6-pdr; **76.2mm**: 17-pdr
AD GUNS 37mm: 24 Type-74; **40mm**: 10 M-1; **57mm**: 12 Type-80
SAM HN-5A (reported)

Navy† 10,000

(incl 800 Naval Infantry)
BASES Bassein, Mergui, Moulmein, Seikyi, Yangon (Monkey Point), Sittwe
PATROL AND COASTAL COMBATANTS 68
CORVETTES 2
1 *Yan Taing Aung* (US PCE-827) FS† with 1 × 76mm gun
1 *Yan Gyi Aung* (US *Admirable* MSF) FS† with 1 × 76mm gun
MISSILE CRAFT 6 *Houxin* PFM with 4 C-801 SSM
PATROL, OFFSHORE 3 *In Daw* (UK *Osprey*) PCO
PATROL, COASTAL 10 *Yan Sit Aung* (PRC *Hainan*) PCC
PATROL, INSHORE 18
12 US PGM-401/412, 3 FRY PB-90 PFI<, 3 *Swift* PCI 421

PATROL, RIVERINE about 29
2 *Nawarat*, 2 imp FRY Y-301 and 10 FRY Y-301, about 15<, plus some 25 boats

AMPHIBIOUS craft only
1 LCU, 10 LCM

SUPPORT 9
6 coastal tpt, 1 AOT, 1 diving spt, 1 buoy tender, plus 6 boats

NAVAL INFANTRY (800) 1 bn

Air Force 9,000

83 cbt ac, 29 armed hel
FTR 3 sqn with 25 F-7, 5 FT-7
FGA 2 sqn with 22 A-5M
CCT 2 sqn with 12 PC-7, 9 PC-9, 10 *Super Galeb* G4
TPT 1 sqn with 3 F-27, 4 FH-227, 5 PC-6A/-B, 2 Y-8D
LIAISON/TRG 4 Cessna 180, 1 Cessna *Citation* II, 12 K-8
HEL 4 sqn with 12 Bell 205, 6 Bell 206, 9 SA-316, 18* Mi-2, 11* Mi-17, 10 PZL W-3 *Sokol*

Paramilitary ε85,250

PEOPLE'S POLICE FORCE 50,000

PEOPLE'S MILITIA 35,000

PEOPLE'S PEARL AND FISHERY MINISTRY ε250
11 patrol boats (3 *Indaw* (Dk *Osprey*) PCC, 3 US *Swift* PGM PCI, 5 Aus *Carpentaria* PCI<)

Opposition and Former Opposition

GROUPS WITH CEASE-FIRE AGREEMENTS

UNITED WA STATE ARMY (UWSA) ε12,000 **Area** Wa hills between Salween river and Chinese border; formerly part of CPB
KACHIN INDEPENDENCE ARMY (KIA) some 8,000 **Area** northern Myanmar, incl Kuman range, the Triangle. Reached cease-fire agreement with government in October 1993
MONG TAI ARMY (MTA) (formerly Shan United Army) ε3,000+ **Area** along Thai border and between Lashio and Chinese border
SHAN STATE ARMY (SSA) ε3,000 **Area** Shan state
MYANMAR NATIONAL DEMOCRATIC ALLIANCE ARMY (MNDAA) 2,000 **Area** north-east Shan state
MON NATIONAL LIBERATION ARMY (MNLA) ε1,000 **Area** on Thai border in Mon state
NATIONAL DEMOCRATIC ALLIANCE ARMY (NDAA) ε1,000 **Area** eastern corner of Shan state on China–Laos border; formerly part of CPB
PALAUNG STATE LIBERATION ARMY (PSLA) ε700 **Area** hill tribesmen north of Hsipaw
NEW DEMOCRATIC ARMY (NDA) ε500 **Area** along Chinese border in Kachin state; former Communist Party of Burma (CPB)
DEMOCRATIC KAREN BUDDHIST ORGANISATION (DKBO) ε100–500 armed

GROUPS STILL IN OPPOSITION

KAREN NATIONAL LIBERATION ARMY (KNLA) ε4,000 **Area** based in Thai border area; political wg is Karen National Union (KNU)
ALL BURMA STUDENTS DEMOCRATIC FRONT ε2,000
KARENNI ARMY (KA) >1,000 **Area** Kayah state, Thai border

New Zealand NZ

New Zealand dollar NZ$

		1998	1999	2000	2001
GDP	NZ$	98bn	99bn		
	US$	58bn	51bn		
per capita	US$	18,100	19,300		
Growth	%	0.2	4.5		
Inflation	%	1.3	-0.1		
Publ debt	%	38.7	36.7		
Def exp	NZ$	1.6bn	1.6bn		
	US$	881m	824m		
Def bdgt	NZ$	1.6bn	1.6bn	1.6bn	
	US$	860m	824m	804m	
US$1=NZ$		1.86	1.92	1.99	

Population			3,860,000
(Maori 15%, Pacific Islander 6%)			
Age	13–17	18–22	23–32
Men	132,000	129,000	292,000
Women	124,000	122,000	278,000

Total Armed Forces

ACTIVE 9,230
(incl some 1,340 women)
RESERVES some 5,490
Regular some 2,410 **Army** 1,550 **Navy** 850 **Air Force** 10
Territorial 3,080 **Army** 2,650 **Navy** 390 **Air Force** 40

Army 4,450

(incl 550 women)
1 Land Force Comd HQ • 2 Land Force Gp HQ • 1 APC/Recce regt (-) • 2 inf bn • 1 arty regt (2 fd bty, 1 AD tp) • 1 engr regt (-) • 2 SF sqn (incl 1 reserve)

RESERVES
Territorial Force 6 Territorial Force Regional Training regt (each responsible for providing trained individuals for top-up and round-out of deployed forces)

EQUIPMENT
LT TK 8 *Scorpion* (for disposal)
APC 56 M-113 (plus 21 variants)

TOWED ARTY 105mm: 24 *Hamel*
MOR 81mm: 50
RL 94mm: LAW
RCL 84mm: 63 *Carl Gustav*
SAM 12 *Mistral*
SURV *Cymbeline* (mor)

Navy 1,980

(incl 360 women)
BASE Auckland (Fleet HQ)
FRIGATES 3
FF 3
2 *Anzac* with 8 *Sea Sparrow* VLS SAM, 1 × 127mm gun, 6 × 324mm TT, 1 SH-2F hel
1 *Canterbury* (UK *Leander*) with 2 × 114mm guns, 6 × 324mm ASTT, 1 SH-2F hel
PATROL AND COASTAL COMBATANTS 4
4 *Moa* PCI (reserve trg)
SUPPORT AND MISCELLANEOUS 7
1 *Endeavour* AO; 1 trg, 1 sail trg, 1 diving spt; 1 *Resolution* (US *Stalwart*) AGHS, 2 inshore AGHS

NAVAL AIR
no cbt ac, 3 armed hel
HEL 3 SH-2F *Sea Sprite* (see Air Force)

Air Force 2,800

(incl 430 women); 42 cbt ac, no armed hel
Flying hours A-4: 180

AIR COMMAND
FGA 2 sqn with 14 A-4K, 5 TA-4K
MR 1 sqn with 6* P-3K *Orion*
LIGHT ATTACK/TRG 1 sqn for *ab initio* and ftr lead-in trg with 17* MB-339C
ASW/ASUW 3 SH-2F (Navy-assigned)
TPT 2 sqn
ac 1 with 5 C-130H, 2 Boeing 727
hel 1 with 13 UH-1H, 5 Bell 47G (trg)
TRG 1 sqn with 13 CT-4E
MISSILES
ASM AGM-65B/G *Maverick*
AAM AIM-9L *Sidewinder*

Forces Abroad

AUSTRALIA 47; 3 A-4K, 3 TA-4K, 9 navigation trg
SINGAPORE 11; spt unit
UN AND PEACEKEEPING
BOSNIA (SFOR II): 27 **CAMBODIA** (CMAC): 2 **CROATIA** (UNMOP): 2 obs **EAST TIMOR** (UNTAET): 674 incl 9 obs **EGYPT** (MFO): 26 **MIDDLE EAST** (UNTSO): 6 obs **PAPUA NEW GUINEA**: 31 (Bougainville Peace Monitoring Group) **SIERRA LEONE** (UNAMSIL): 2 obs

Papua New Guinea PNG

kina K		1998	1999	2000	2001
GDP	K	7.9bn	8.7bn		
	US$	3.3bn	3.2bn		
per capita	US$	2,600	2,800		
Growth	%	3.8	6.1		
Inflation	%	13.5	14.9		
Debt	US$	2.7bn	2.3bn		
Def exp	K	115m	ε126m		
	US$	56m	46m		
Def bdgt	K	86m	80m	ε88m	
	US$	42m	29m	36m	
FMA (US)	US$	0.2m	0.2m	0.2m	0.2m
FMA (Aus)	US$	9.6m	6.7m		
US$1=K		2.06	2.73	2.41	

Population			**4,862,000**
Age	13–17	18–22	23–32
Men	273,000	244,000	419,000
Women	260,000	229,000	383,000

Total Armed Forces

ACTIVE ε4,400

Army ε3,800

2 inf bn • 1 engr bn
EQUIPMENT
MOR 81mm; 120mm: 3

Maritime Element 400

BASES Port Moresby (HQ), Lombrum (Manus Island) (patrol boat sqn); forward bases at Kieta and Alotau
PATROL AND COASTAL COMBATANTS 4
PATROL, INSHORE 4 *Tarangau* (Aust *Pacific Forum* 32-m) PCI
AMPHIBIOUS 2
2 *Salamaua* (Aust *Balikpapan*) LSM, plus 4 landing craft, manned and operated by the civil administration

Air Force 200

no cbt ac, no armed hel
TPT 2 CN-235, 3 IAI-201 *Arava*, 1 CN-212
HEL †4 UH-1H

Foreign Forces

AUSTRALIA 38; trg unit
BOUGAINVILLE PEACE MONITORING GROUP some 300 tps from Aus (ε150), NZ (31), Fiji, Tonga, Vanuatu

Philippines Pi

peso P		1998	1999	2000	2001
GDP	P	2.7tr	3.0tr		
	US$	65bn	78.5bn		
per capita	US$	3,200	3,300		
Growth	%	-0.5	3.2		
Inflation	%	9.7	6.7		
Debt	US$	47bn	52bn		
Def exp[a]	P	61bn	62bn		
	US$	1.5bn	1.6bn		
Def bdgt[b]	P	41bn	52bn	54bn	
	US$	1.0bn	1.4bn	1.3bn	
FMA (US)	US$	1.3m	1.4m	1.4m	1.4m
FMA (Aus)	US$	3.0m	3.8m		
US$1=P		40.1	38.1	42.5	

[a] Incl paramil exp

[b] A five-year supplementary procurement budget of P50bn (US$1.9bn) for 1996–2000 was approved in Dec 1996

Population 77,268,000

(Muslim 5–8%; *Mindanao provinces* Muslim 40–90%; Chinese 2%)

Age	13–17	18–22	23–32
Men	4,279,000	3,795,000	6,300,000
Women	4,133,000	3,661,000	6,088,000

Total Armed Forces

ACTIVE 106,000

RESERVES 131,000

Army 100,000 (some 75,000 more have commitments) **Navy** 15,000 **Air Force** 16,000 (to age 49)

Army 66,000

5 Area Unified Comd (joint service) • 8 inf div (each with 3 inf bde, 1 arty bn) • 1 special ops comd with 1 lt armd bde ('regt'), 1 scout ranger, 1 SF regt • 5 engr bn • 1 arty regt HQ • 1 Presidential Security Group

EQUIPMENT

LT TK 40 *Scorpion*
AIFV 85 YPR-765 PRI
APC 100 M-113, 20 *Chaimite*, 100 V-150, ε140 *Simba*
TOWED ARTY 105mm: 230 M-101, M-102, M-26 and M-56; **155mm**: 12 M-114 and M-68
MOR 81mm: M-29; **107mm**: 40 M-30
RCL 75mm: M-20; **90mm**: M-67; **106mm**: M-40 A1
AC 4 Cessna (-170, -172, P-206, U-206), 1 Beech 65

Navy† ε24,000

(incl 8,000 Marines and 2,000 Coast Guard)
6 Naval Districts
BASES Sangley Point/Cavite, Zamboanga, Cebu
FRIGATES FF 1 *Rajah Humabon* (US *Cannon*) with 3 × 76mm gun, ASW mor

PATROL AND COASTAL COMBATANTS 60

PATROL, OFFSHORE 13
2 *Rizal* (US *Auk*) PCO with 2 × 76mm gun, 3 × 2 ASTT, hel deck
3 *Emilio Jacinto* (ex-UK *Peacock*) PCO with 1 × 76mm gun
8 *Miguel Malvar* (US PCE-827) PCO with 1 × 76mm gun

PATROL, COASTAL 10
2 *Aguinaldo* PCC, 3 *Kagitingan* PCC, 5 *Thomas Batilo* (ROK *Sea Dolphin*) PCC

PATROL, INSHORE 37
22 *José Andrada* PCI< and about 15 other PCI<

AMPHIBIOUS some 9
2 US *F. S. Beeson*-class LST, capacity 32 tk plus 150 tps, hel deck
Some 7 *Zamboanga del Sur* (US LST-1/511/542) LST, capacity either 16tk or 10tk plus 200 tps
Plus about 39 craft: 30 LCM, 3 LCU, some 6 LCVP

SUPPORT AND MISCELLANEOUS 11
2 AOT (small), 1 AR, 3 spt, 2 AWT, 3 AGOR/AGOS

NAVAL AVIATION

no cbt ac, no armed hel
MR/SAR ac 2 BN-2A *Defender*, 1 *Islander* **hel** 4 Bo-105 (SAR)

MARINES (8,000)

3 bde (10 bn) to be 2 bde (6 bn)

EQUIPMENT

AAV 30 LVTP-5, 55 LVTP-7
LAV 24 LAV-300 (reported)
TOWED ARTY 105mm: 150 M-101
MOR 4.2in (**107mm**): M-30

Air Force ε16,000

47 cbt ac, some 97 armed hel
FTR 1 sqn with 11 F-5 (9 -5A, 2 -5B)
ARMED HEL 3 sqn with 60 Bell UH-1H/M, 16 AUH-76 (S-76 gunship conversion), 21 Hughes 500/520MD
MR 1 F-27M
RECCE 4 RT-33A, 21* OV-10 *Broncos*
SAR ac 4 HU-16 **hel** 10 Bo-105C
PRESIDENTIAL AC WG ac 1 F-27, 1 F-28 **hel** 2 Bell 212, 4 Bell-412, 2 S-70A, 2 SA-330
TPT 3 sqn
1 with 2 C-130B, 3 C-130H, 3 L-100-20, 5 C-47, 7 F-27
2 with 2 BN-2 *Islander*, 14 N-22B *Nomad Missionmaster*
HEL 2 sqn with 55 Bell 205, 16 UH-1H, 33 MD-520
LIAISON 10 Cessna (7 -180, 2 -210, 1 -310), 5 DHC-2, 12 U-17A/B
TRG 4 sqn
1 with 4 T-33A, 1 with 14 T-41D, 1 with 28 SF-260TP, 1 with 15* S-211
AAM AIM-9B *Sidewinder*

Forces Abroad

UN AND PEACEKEEPING

EAST TIMOR (UNTAET): 623 incl 20 obs

Paramilitary 42,500 active

PHILIPPINE NATIONAL POLICE 40,500 (Department of Interior and Local Government)

62,000 active aux; 15 Regional, 73 Provincial Comd

COAST GUARD 2,000

Part of Department of Transport; but mainly funded, manned and run by the Navy

EQPT 3 *De Haviland* PCI, 4 *Basilan* (US PGM-39/42) PCI, plus some 35 *Swift* PCI, 3 SAR hel (by 2000)

CITIZEN ARMED FORCE GEOGRAPHICAL UNITS (CAFGU) 40,000

Militia, 56 bn; part-time units which can be called up for extended periods

Opposition and Former Opposition

Groups with Peace Agreements

BANGSA MORO ARMY (armed wing of Moro National Liberation Front (MNLF); Muslim) ε5,700 integrated into national army

Groups Still in Opposition

NEW PEOPLE'S ARMY (NPA; communist) ε9,500

MORO ISLAMIC LIBERATION FRONT (breakaway from MNLF; Muslim) 10,000 (up to 15,000 reported)

MORO ISLAMIC REFORMIST GROUP (breakaway from MNLF; Muslim) 900

ABU SAYYAF GROUP ε1,500

Singapore Sgp

Singapore dollar S$		1998	1999	2000	2001
GDP	S$	138bn	144bn		
	US$	83bn	84bn		
per capita	US$	23,500	24,400		
Growth	%	0.3	5.4		
Inflation	%	-0.3	0.5		
Debt	US$	10.5bn			
Def exp	S$	8.1bn	8.1bn		
	US$	4.8bn	4.7bn		
Def bdgt	S$	7.3bn	7.3bn	7.4bn	
	US$	4.4bn	4.2bn	4.4bn	
FMA (Aus)	US$	0.5m	0.5m		
US$1=S$		1.67	1.72	1.72	

Population 4,130,000

(Chinese 76%, Malay 15%, Indian 6%)

Age	13–17	18–22	23–32
Men	119,000	108,000	244,000
Women	114,000	102,000	237,000

Total Armed Forces

ACTIVE 60,500

(incl 39,800 conscripts)

Terms of service conscription 24–30 months

RESERVES ε213,800

Army ε200,000; annual trg to age 40 for men, 50 for officers **Navy** ε6,300 **Air Force** ε7,500

Army 50,000

(35,000 conscripts)

3 combined arms div (mixed active/reserve formations) each with 2 inf bde (each 3 inf bn), 1 armd bde, 1 recce, 2 arty, 1 AD, 1 engr bn

1 Rapid Deployment div (mixed active/reserve formation) with 3 inf bde (incl 1 air mob, 1 amph)

1 mech bde

Summary of active units

9 inf bn • 4 lt armd/recce bn • 4 arty bn • 1 cdo (SF) bn • 4 engr bn

RESERVES

9 inf bde incl in mixed active/reserve formations listed above • 1 op reserve div with additional inf bde • 2 People's Defence Force cmd with 7+ bde gp • Total cbt units ε60 inf, ε8 lt armd/recce, ε8 arty, 1 cdo (SF), ε8 engr bn

EQUIPMENT

MBT 63 *Centurion* (trg only)

LT TK ε350 AMX-13SM1

RECCE 22 AMX-10 PAC 90

AIFV 22 AMX-10P, some IFV-25

APC 750+ M-113A1/A2 (some with 40mm AGL, some with 25mm gun), 30 V-100, 250 V-150/-200 *Commando*, some IFV-40/50

TOWED ARTY 105mm: 37 LG1; **155mm**: 38 Soltam M-71S, 22 M-114A1 (may be in store), 16 M-68 (may be in store), 52 FH-88, 36 FH-2000

MOR 81mm (some SP); **120mm**: 50 (some SP in M-113); **160mm**: 12 Tampella

ATGW 30+ *Milan*, *Spike*

RL *Armbrust*; **89mm**: 3.5in M-20

RCL 84mm: ε200 *Carl Gustav*; **106mm**: 90 M-40A1 (in store)

AD GUNS 20mm: 30 GAI-CO1 (some SP)

SAM 75: RBS-70 (some SP in V-200) (Air Force), *Mistral* (Air Force)

SURV AN/TPQ-36/-37 (arty, mor)

Navy 4,500

(incl 1,800 conscripts)

COMMANDS Fleet (1st and 3rd Flotillas) **Coastal** and **Naval Logistic** and **Training Command**

BASES Pulau Brani, Tuas (Jurong), Changi (building, ready end 2000)

SUBMARINES 1

1 *Challenger* (Swe *Sjoormen*) SSK with 4 × 533 TT (3 more in Sweden; personnel undergoing training in Sweden)

PATROL AND COASTAL COMBATANTS 24

CORVETTES 6 *Victory* (Ge Lürssen 62m) FSG with 8 *Harpoon* SSM, 1 × 2 *Barak* SAM, 1 × 76mm gun, 2 × 3 ASTT

MISSILE CRAFT 6

6 *Sea Wolf* (Ge Lürssen 45m) PFM with 2 × 4 *Harpoon* SSM, 4 × 2 *Gabriel* SSM, 1 × 2 *Mistral/Simbad* SAM, 1 × 57mm gun

PATOL CRAFT 12

12 *Fearless* PCO with 2 *Mistral / Sadral* SAM, 1 × 76mm gun (6 with 6 × 324mm TT)

MINE COUNTERMEASURES 4

4 *Bedok* (SW *Landsort*) MHC

AMPHIBIOUS 6

1 *Perseverance* (UK *Sir Lancelot*) LSL with 1 × 2 *Mistral/ Simbad* SAM, capacity 340 tps, 16 tk, hel deck

3 *Endurance* LST with 2 × 2 *Mistral/Simbad* SAM, 1 × 76mm gun; capacity: 350 tps, 18 tk, 4 LCVP, 2 hel plus 1 LST (ex US *County*) in reserve

2 *Excellence* (US 5II-II52) LST

Plus craft: 6 LCM, 30 LCU, and boats

SUPPORT AND MISCELLANEOUS 2

1 *Jupiter* diving spt and salvage, 1 trg

Air Force 6,000

(incl 3,000 conscripts); 136 cbt ac, 20 armed hel

FGA 6 sqn

2 with 50 A-4SU

1 sqn (plus another forming) with 3 F-16A, 4 F-16B, 22 F-16C, 20 F-16D (some fitted for SEAD). 8 F-16C, 10 F-16D in Singapore: 24 F-16C/D in US

2 with 28 F-5S, 9 F-5T (secondary GA role)

RECCE 1 sqn with 8 RF-5S

AEW 1 sqn with 4 E-2C

TKR 1 KC-135 in US (3 more on order – 1 KC-135 to be delivered to Singapore in 2000), 1 KC-130H

TPT/TKR/RECCE 2 sqn

1 with 4 KC-130B (tkr/tpt), 5 C-130H (1 ELINT)

1 with 5 F-50 *Enforcer* (tpt/MR)

ARMED HEL 2 sqn with 20 AS 550A2/C2 (8 AH-64D to be delivered from 2002)

HEL 4 sqn

1 with 19 UH-1H, 6 AB-205A, 2 with 22 AS-332M (incl 5 SAR), 6 AS-532UL

1 with CH-47D (first of 6 delivered)

TRG

1 sqn with 27 SIAI S-211

1 trg detachment with 18 TA-4SU

1 with 26 SF-260

UAV 1 sqn with 40 *Searcher*, 24 *Chukar* III

AIR DEFENCE SYSTEMS DIVISION

4 field def sqn

Air Defence Bde 1 sqn with **35mm** Oerlikon, 1 sqn with 18 I HAWK, 1 sqn with Blind Fire *Rapier*, 1 sqn with SA-18 IGLA

Air Force Systems Bde 1 sqn mobile radar, 1 sqn LORADS

Divisional Air Def Arty Bde (attached to Army divs) 1 bn with 36 *Mistral* (SAM), 3 bn with RBS 70 (SAM), 1 bn with SA-18 *Igla*

MISSILES

AAM AIM-7P *Sparrow*, AIM-9 N/P *Sidewinder*, *Python* 4 reported

ASM AGM-45 *Shrike*, AGM-65B *Maverick*, AGM-65G *Maverick*, AGM-84 *Harpoon*

Forces Abroad

AUSTRALIA 230; flying trg schools at Oakey (6 AS-532), and Pearce (27 S-211)

BRUNEI 500; trg school, incl hel det (with 5 UH-1H)

FRANCE 200; trg 18, mostly TA-4SU (Cazaux AFB)

SOUTH AFRICA *Searcher* UAV trg det Hoedspruit AFB

TAIWAN 3 trg camps (incl inf, arty and armd)

THAILAND 1 trg camp (arty)

US trg detachment CH-47D (ANG facility Grand Prairie, TX); 12 F-16C/D (leased from USAF at Luke AFB, AZ), 12 F-16C/D (at Cannon AFB, NM); KC-135 trg det at McConnell AFB, KS

UN AND PEACEKEEPING

EAST TIMOR (UNTAET): 24 **IRAQ/KUWAIT** (UNIKOM): 5 obs

Paramilitary ε108,000+ active

SINGAPORE POLICE FORCE

incl Police Coast Guard

12 *Swift* PCI< and about 60 boats

Singapore Gurkha Contingent (750)

CIVIL DEFENCE FORCE 120,000

(incl 1,500 regulars, 3,600 conscripts, ε60,000 former Army reservists, 60,000+ volunteers); 1 construction bde (2,500 conscripts)

Foreign Forces

US 150: **Air Force** 40 **Navy** 90 **USMC** 20

Taiwan (Republic of China) ROC

new Taiwan dollar NT$

		1998	1999	2000	2001
GNP	NT$	8.9tr	9.4tr		
	US$	295bn	288bn		
per capita	US$	14,700	15,600		
Growth	%	4.7	5.5		
Inflation	%	1.7	0.8		
Debt	US$	26bn			
Def exp[a]	NT$	425bn	490bn		
	US$	14.2bn	15.0bn		
Def bdgt[b]	NT$	275bn	357bn	395bn	
	US$	8.3bn	10.9bn	12.8bn	
US$1=NT$		30.0	32.7	30.8	

[a] Incl special appropriations for procurement and infrastructure amounting to NT$301bn (US$11bn) 1993–2001. Between 1993–98, NT$208bn (US$8bn) was spent out of NT$289bn (US$11bn) appropriated for these years.
[b] 1999 def bdgt covers 18-month period Jul 1999–Dec 2000.

Population			21,960,000
(Taiwanese 84%, mainland Chinese 14%)			
Age	13–17	18–22	23–32
Men	975,000	1,006,000	1,826,000
Women	934,000	952,000	1,725,000

Total Armed Forces

ACTIVE ε370,000 (to be 350,000)

Terms of service 2 years

RESERVES 1,657,500

Army 1,500,000 with some obligation to age 30 **Navy** 32,500 **Marines** 35,000 **Air Force** 90,000

Army ε240,000 (to be 200,000)

(incl mil police)

3 Army, 1 AB Special Ops HQ • 10 inf div • 2 mech inf div • 2 AB bde • 6 indep armd bde • 1 tk gp • 2 AD SAM gp with 6 SAM bn: 2 with *Nike Hercules*, 4 with I HAWK • 2 avn gp, 6 avn sqn

RESERVES

7 lt inf div

EQUIPMENT

MBT 100 M-48A5, 450+ M-48H, 189 M-60A3
LT TK 230 M-24 (**90mm** gun), 675 M-41/Type 64
AIFV 225 M-113 with **20–30mm** cannon
APC 650 M-113, 300 V-150 *Commando*
TOWED ARTY 105mm: 650 M-101 (T-64); **155mm**: M-44, 90 M-59, 250 M-114 (T-65); **203mm**: 70 M-115
SP ARTY 105mm: 100 M-108; **155mm**: 45 T-69, 110 M-109A2/A5; **203mm**: 60 M-110
COASTAL ARTY 127mm: US Mk 32 (reported)
MRL 117mm: KF VI; **126mm**: KF III/IV towed and SP
MOR 81mm: M-29 (some SP); **107mm**
SSM *Ching Feng*
ATGW 1,000 TOW (some SP)
RCL 90mm: M-67; **106mm**: 500 M-40A1, Type 51
AD GUNS 40mm: 400 (incl M-42 SP, Bofors)
SAM 40 Nike *Hercules* (to be retired), 100 HAWK, *Tien Kung (Sky Bow)* -1/-2, *Stinger*, 74 *Avenger*, 2 *Chaparral*, 25 *Patriot*
AC 20 O-1
HEL 110 UH-1H, 53 AH-1S, 30 TH-67, 26 OH-58D, 12 KH-4, 7 CH-47, 5 Hughes 500
UAV *Mastiff* III

DEPLOYMENT

Quemoy 35–40,000; 4 inf div **Matsu** 8–10,000; 1 inf div

Navy ε62,000

(incl 30,000 Marines)
3 Naval Districts
BASES Tsoying (HQ), Makung (Pescadores), Keelung, Hualien (ASW HQ) (New East Coast fleet set up and based at Suo; 6 *Chin Yang*-class FF)

SUBMARINES 4

SSK 4
2 *Hai Lung* (Nl mod *Zwaardvis*) with 533mm TT
2 *Hai Shih* (US *Guppy* II) with 533mm TT (trg only)

PRINCIPAL SURFACE COMBATANTS 33

DESTROYERS 12
DDG 12
7 *Chien Yang* (US *Gearing*) (*Wu Chin* III conversion) with 4 *Hsiung Feng* SSM, SM-1-MR SAM, 2 × 3 ASTT, 1 × 8 ASROC SUGW, 1 *Hughes* MD-500 hel
3 *Fu Yang* (US *Gearing*) with 5 *Hsiung Feng* I/*Gabriel* II SSM, 1 or 2 × 127mm guns, 2 × 3 ASTT, 1 *Hughes* MD-500 hel (1 also with 1 × 8 ASROC SUGW)
2 *Po Yang* (US *Sumner*)† with *Hsiung Feng* SSM, 1 or 2 × 127mm guns, 2 × 3 ASTT, 1 *Hughes* MD-500 hel

FRIGATES 21
FFG 21
7 *Cheng Kung* (US *Perry*) with 8 *Hsiung Feng* II SSM, 1 SM-1 MR SAM, 1 × 76mm gun, 2 × 3 ASTT, 2 S-70C hel
6 *Kang Ding* (Fr *La Fayette*) with 8 *Hsiung Feng* SSM, 4 *Sea Chaparral* SAM, 1 × 76mm gun, 6 × 324mm ASTT, 1 S-70C hel
8 *Chin Yang* (US *Knox*) with *Harpoon* SSM, 1 × 127mm gun, 4 ASTT, 1 × 8 ASROC SUGW, 1 SH-2F hel

PATROL AND COASTAL COMBATANTS 59

MISSILE CRAFT 59
2 *Lung Chiang*† PFM with 2 *Hsiung Feng I* SSM
9 *Jinn Chiang* PFM with 4 *Hsiung Feng I* SSM
48 *Hai Ou* (mod Il *Dvora*) PFM< with 2 *Hsiung Feng I* SSM

MINE COUNTERMEASURES 12
4 (ex-US) *Aggressive* MSO
4 *Yung Chou* (US *Adjutant*) MSC
4 *Yung Feng* MSC converted from oil-rig spt ships

AMPHIBIOUS 18

1 *Shiu Hai* (US *Anchorage*) LSD
2 *Chung Ho* (US *Newport*) LST capacity 400 troops, 500 tons vehicles, 4 LCVP
1 *Kao Hsiung* (US LST 511) LCC
10 *Chung Hai* (US LST 511) LST, capacity 16 tk, 200 tps
4 *Mei Lo* (US LSM-1) LSM, capacity about 4 tk
Plus about 325 craft; some 20 LCU, 205 LCM, 100 LCVP and assault LCVP

SUPPORT AND MISCELLANEOUS 20

3 AO, 2 AR, 1 *Wu Yi* combat spt with hel deck, 2 *Yuen Feng* and 2 *Wu Kang* attack tpt with hel deck, 2 tpt, 7 AT/F, 1 *Te Kuan* AGOR

COASTAL DEFENCE 1

1 SSM coastal def bn with *Hsiung Feng* (*Gabriel*-type)

NAVAL AIR

31 cbt ac; 21 armed hel
MR 1 sqn with 31 S-2 (24 -E, 7 -G)
HEL 12* Hughes 500MD, 9* S-70C ASW *Defender*, 9 S-70C(M)-1

MARINES (30,000)

2 div, spt elm

EQUIPMENT

AAV LVTP-4/-5
TOWED ARTY 105mm, 155mm
RCL 106mm

Air Force 68,000

570 cbt ac, no armed hel
Flying hours 180
FTR 3 sqn with 58 *Mirage* 2000-5 (47 -5EI, 11 -5DI)
FGA/FTR 20 sqn
6 with 200 F-5 (7 -B, 213 -E, 52 -F) (ε70 in store)
6 with 128 *Ching-Kuo*
7 with 126 F-16A/B (incl one sqn recce capable)
1 with 22 AT-3
RECCE 1 with 8 RF-5E
AEW 4 E-2T
EW 1 with 2 C-130HE, 2 CC-47
SAR 1 sqn with 17 S-70C
TPT 3 **ac** sqn
2 with 19 C-130H (1 EW)
1 VIP with 4 Boeing 727-100, 1 Boeing 737-800, 10 Beech 1900, 3 *Fokker* F-50
HEL 1 S-62A (VIP), 14 S-70, 9 CH-475D being delivered
TRG ac incl 36* AT-3A/B, 42 T-34C

MISSILES

ASM AGM-65A *Maverick*
AAM AIM-4D *Falcon*, AIM-9J/P *Sidewinder*, *Shafrir*, *Sky Sword* I and II, MATRA *Mica*, MATRA R550 *Magic* 2

Forces Abroad

US F-16 conversion unit at Luke AFB

Paramilitary ε26,650

SECURITY GROUPS 25,000

National Police Administration (Ministry of Interior); **Bureau of Investigation** (Ministry of Justice); **Military Police** (Ministry of Defence); **Coast Guard Administration**

MARITIME POLICE ε1,000

about 38 armed patrol boats

CUSTOMS SERVICE (Ministry of Finance) 650

5 PCO, 2 PCC, 1 PCI, 5 PCI<; most armed

COAST GUARD ADMINISTRATION 22,000 (all civilians)

responsible for guarding the Spratly and Pratas island groups, and to enforce law and order

Foreign Forces

SINGAPORE 3 trg camps

Thailand Th

baht b		1998	1999	2000	2001
GDP	b	4.9tr	5.0tr		
	US$	119bn	135bn		
per capita	US$	7,400	7,800		
Growth	%	-10.4	4.2		
Inflation	%	8.1	0.2		
Debt	US$	86bn	79bn		
Def exp	b	86bn	98bn		
	US$	2.1bn	2.6bn		
Def bdgt	b	81.0bn	77.4bn	77.3bn	88.4bn
	US$	2.0bn	2.1bn	2.0bn	2.3bn
FMA (US)	US$	3.9m	4.6m	$1.6m	$2.8m
FMA (Aus)	US$	2.5m	3.0m		
US$1=b		41.4	37.2	38.9	

Population 62,400,000
(Thai 75%, Chinese 14%, Muslim 4%)

Age	13–17	18–22	23–32
Men	3,147,000	3,178,000	6,144,000
Women	3,040,000	3,085,000	5,979,000

Total Armed Forces

ACTIVE 301,000
Terms of service 2 years

RESERVES 200,000

Army 190,000

(incl ε70,000 conscripts)
4 Regional Army HQ, 2 Corps HQ • 2 cav div • 3 armd inf div • 2 mech inf div • 1 lt inf div • 2 SF div • 1 arty div, 1 AD arty div (6 AD arty bn) • 1 engr div • 4 economic development div • 1 indep cav regt • 8

indep inf bn • 4 recce coy • armd air cav regt with 3 air-mobile coy • Some hel flt • Rapid Reaction Force (1 bn per region forming)

RESERVES

4 inf div HQ

EQUIPMENT

MBT 50 PRC Type-69 (trg/in store), 105 M-48A5, 127 M-60 (74 A3, 53 A1)
LT TK 154 *Scorpion* (ε50 in store), 200 M-41, 106 *Stingray*
RECCE 32 Shorland Mk 3, HMMWV
APC 340 M-113A1/A3, 162 V-150 *Commando*, 18 *Condor*, 450 PRC Type-85 (YW-531H)
TOWED ARTY 105mm: 24 LG1 Mk 2, 285 M-101/-101 mod, 12 M-102, 32 M-618A2 (local manufacture); **130mm**: 15 PRC Type-59; **155mm**: 56 M-114, 62 M-198, 32 M-71, 42 GHN-45A1
SP ARTY 155mm: 20 M-109A2
MOR 81mm (incl 21 M-125A3 SP), **107mm** incl M-106A1 SP; **120mm**: 12 M-1064A3 SP
ATGW TOW (incl 18 M-901A5), 300 *Dragon*
RL M-72 LAW
RCL 75mm: 30 M-20; **106mm**: 150 M-40
AD GUNS 20mm: 24 M-163 *Vulcan*, 24 M-167 *Vulcan*; **37mm**: 122 Type-74; **40mm**: 80 M-1/M-42 SP, 48 L/70; **57mm**: 24+ PRC Type-59 (ε6 op)
SAM *Redeye*, some *Aspide*, HN-5A
UAV *Searcher*
AIRCRAFT
TPT 2 C-212, 2 Beech 1900C-1, 4 C-47, 2 Short 330UTT, 2 *Beech King Air*, 2 *Jetstream* 41
LIAISON 10 O-1A, 5 T-41A, 5 U-17A
TRG 10 T-41D, 18 MX-7-235
HELICOPTERS
ATTACK 4 AH-1F
TPT 8 CH-47D, 50 Bell (incl -206, -212, -214), 69 UH-1H
TRG 36 Hughes 300C
SURV RASIT (veh, arty), AN-TPQ-36 (arty, mor)

Navy 68,000

(incl 1,700 Naval Air, 18,000 Marines, 7,000 Coastal Defence; incl 27,000 conscripts)
FLEETS 1st North Thai Gulf **2nd** South Thai Gulf **3rd** Andaman Sea
1 Naval Air Division
BASES Bangkok, Sattahip (Fleet HQ), Songkhla, Phang Nga, Nakhon Phanom (HQ Mekong River Operating Unit)

PRINCIPAL SURFACE COMBATANTS 15

AIRCRAFT CARRIER 1 *Chakri Naruebet* with 8 AV-8S *Matador (Harrier)*†, 6 S-70B *Seahawk* hel
FRIGATES 14
FFG 8
2 *Naresuan* with 2 × 4 *Harpoon* SSM, 8 cell *Sea Sparrow* SAM, 1 × 127mm gun, 6 × 324mm TT, 1 SH-2G hel
2 *Chao Phraya* (PRC *Jianghu* III) with 8 C-801 SSM, 2 × 2 × 100mm guns, 2 × 5 ASW RL, 1 Bell 212 hel
2 *Kraburi* (PRC *Jianghu* IV type) with 8 C-801 SSM, 1 × 2 100mm guns, 2 × 5 ASW RL and 1 Bell 212 hel
2 *Phutthayotfa Chulalok* (US *Knox*) (leased from US) with 8 *Harpoon* SSM, 1 × 127mm gun, 4 × 324 ASTT, 1 Bell 212 hel
FF 6
1 *Makut Rajakumarn* with 2 × 114mm guns, 2 × 3 ASTT
2 *Tapi* (US PF-103) with 1 × 76mm gun, 6 × 324mm ASTT (Mk 46 LWT)
2 *Tachin* (US *Tacoma*) with 3 × 76mm gun
1 *Pin Klao* (US *Cannon*) with 3 × 76mm gun, 6 × 324mm ASTT

PATROL AND COASTAL COMBATANTS 88

CORVETTES 5
2 *Rattanakosin* FSG with 2 × 4 *Harpoon* SSM, 8 *Aspide* SAM, 1 × 76mm gun, 2 × 3 ASTT
3 *Khamronsin* FS with 1 × 76mm gun, 2 × 3 ASTT
MISSILE CRAFT 6
3 *Ratcharit* (It Breda 50m) PFM with 4 MM-38 *Exocet* SSM
3 *Prabparapak* (Ge Lürssen 45m) PFM with 5 *Gabriel* SSM
PATROL CRAFT 77
OFFSHORE
1 *Kua Hin* PCO with 1 × 76mm gun
COASTAL 12
3 *Chon Buri* PFC, 6 *Sattahip*, 3 PCC
INSHORE 64
7 T-11 (US PGM-71), 9 T-91, about 33 PCF and 15 PCR plus boats

MINE COUNTERMEASURES 7

2 *Lat Ya* (It *Gaeta*) MCMV
2 *Bang Rachan* (Ge Lürssen T-48) MCC
2 *Bangkeo* (US *Bluebird*) MSC
1 *Thalang* MCM spt with minesweeping capability
(Plus some 12 MSB)

AMPHIBIOUS 9

2 *Sichang* (Fr PS-700) LST, capacity 14 tk, 300 tps with hel deck (trg)
5 *Angthong* (US LST-511) LST, capacity 16 tk, 200 tps
2 *Kut* (US LSM-1) LSM, capacity about 4 tk
Plus about 51 craft: 9 LCU, about 24 LCM, 1 LCG, 2 LSIL, 3 hovercraft, 12 LCVP

SUPPORT AND MISCELLANEOUS 16

1 *Similan* AO (1 hel) , 1 *Chula* AO, 5 AO, 3 AGHS, 6 trg

NAVAL AIR (1,700)

(incl 300 conscripts); 67 cbt ac; 5 armed hel
FTR 9 *Harrier* (7 AV-8, 2 TAV-8)
MR/ATTACK 5 Cessna T-337 *Skymasters*, 14 A-7E, 4 TA-7C, 5 O-1G, 4 U-17B
MR/ASW 3 P-3T *Orion* (plus 2 P-3A in store), 1 UP-3T, 6 Do-228, 3 F-27 MPA, 8 S-2F, 5 N-24A *Nomad*

ASW HEL 5 S-70B
SAR/UTILITY 2 CL-215, 8 Bell 212, 5 Bell 214, 4 UH-1H, 5 S-76N
ASM AGM-84 *Harpoon* (for F-27MPA, P-3T)

MARINES (18,000)

1 div HQ, 2 inf regt, 1 arty regt (3 fd, 1 AA bn); 1 amph aslt bn; recce bn

EQUIPMENT

AAV 33 LVTP-7
TOWED ARTY 155mm: 12 GC-45
ATGW TOW, *Dragon*

Air Force 43,000

4 air divs, one flying trg school
153 cbt ac, no armed hel
Flying hours 100
FGA 3 sqn
1 with 14 F-5A/B • 2 with 34 F-16 (26 -A, 8 -B)
FTR/AGGRESSOR 3 sqn with 33 F-5E, 5 -F
ARMED AC 5 sqn
1 with 4 AC-47 • 3 with 22 AU-23A • 1 with 19* N-22B *Missionmaster* (tpt/armed) • 1 with 19 OV-10C (coin/obs). To be replaced by 25 Alphajets (first 5 being delivered)
ELINT 1 sqn with 3 IAI-201
RECCE 3* RF-5A
SURVEY 2 *Learjet* 35A, 3 *Merlin* IVA, 3 GAF N-22B *Nomads*
TPT 3 sqn
1 with 6 C-130H, 6 C-130H-30, 3 DC-8-62F
1 with 3 C-123-K, 4 BAe-748
1 with 6 G-222
VIP Royal flight **ac** 1 Airbus A-310-324, 1 Boeing 737-200, 3 *King Air* 200, 2 BAe-748, 3 *Merlin* IV **hel** 2 Bell 412, 3 AS-532A2
TRG 24 CT-4, 29 *Fantrainer*-400, 13 *Fantrainer*-600, 10 SF-260, 15 T-33A/RT-33A, 22 PC-9, 6 -C, 12 T-37, 34 L-39ZA/MP
LIAISON 3 *Commander*, 1 *King Air* E90, 2 O-1 *Bird Dog*, 2 *Queen Air*, 3 *Basler Turbo*-67
HEL 2 sqn
1 with 17 S-58T • 1 with 25 UH-1H
AAM AIM-9B/J *Sidewinder*, *Python* 3

AIR DEFENCE

1 AA arty bty: 4 *Skyguard*, 1 *Flycatcher* radars, each with 4 fire units of 2 30mm Mauser/Kuka guns
SAM *Blowpipe*, *Aspide*, RBS NS-70, *Starburst*

Forces Abroad

UN AND PEACEKEEPING

EAST TIMOR (UNTAET): 920 incl 10 obs **IRAQ/KUWAIT** (UNIKOM): 5 obs **SIERRA LEONE** (UNAMSIL): 5 obs

Paramilitary 115,600 active

THAHAN PHRAN (Hunter Soldiers) 22,600
volunteer irregular force; 27 regt of some 200 coy
NATIONAL SECURITY VOLUNTEER CORPS 50,000
MARINE POLICE 2,500
3 PCO, 3 PCC, 8 PFI, some 110 PCI<
POLICE AVIATION 500
ac 1 *Airtourer*, 6 AU-23, 2 Cessna 310, 1 Fokker 50, 1 CT-4, 2 CN 235, 8 PC-6, 2 Short 330 **hel** 27 Bell 205A, 14 Bell 206, 3 Bell 212, 6 UH-12, 5 KH-4
BORDER PATROL POLICE 40,000
PROVINCIAL POLICE ε50,000
incl ε500 Special Action Force

Foreign Forces

SINGAPORE 1 trg camp (arty)
US Army 40 **Air Force** 30 **Navy** 10 **USMC** 40

Vietnam Vn

dong d		1998	1999	2000	2001
GDP	d	372tr	416tr		
	US$	28.7bn	30bn		
per capita	US$	1,200	1,200		
Growth	%	3.5	3.5		
Inflation	%	7.7	7.6		
Debt	US$	22bn	22bn		
Def exp	US$	ε925m	ε890m		
Def bdgt	US$	ε924m	ε891m	ε1.0bn	
US$1=d		12,980	13,893	14,081	

Population		**82,014,000** (Chinese 3%)
Age	13–17	18–22	23–32
Men	4,479,000	4,060,000	7,028,000
Women	4,327,000	3,927,000	6,857,000

Total Armed Forces

ACTIVE ε484,000
(referred to as 'Main Force')
Terms of service 2 years Army and Air Defence, 3 years Air Force and Navy, specialists 3 years, some ethnic minorities 2 years

RESERVES some 3–4,000,000
'Strategic Rear Force' (see also *Paramilitary*)

Army ε412,000

8 Mil Regions, 2 special areas • 14 Corps HQ • 58 inf div[a] • 3 mech inf div • 10 armd bde • 15 indep inf regt • SF incl AB bde, demolition engr regt • Some 10 fd arty bde • 8 engr div • 10–16 economic construction

div • 20 indep engr bde

EQUIPMENT

MBT 45 T-34, 850 T-54/-55, 70 T-62, 350 PRC Type-59
LT TK 300 PT-76, 320 PRC Type-62/63
RECCE 100 BRDM-1/-2
AIFV 300 BMP-1/-2
APC 1,100 BTR-40/-50/-60/-152, 80 YW-531, M-113
TOWED ARTY 2,300: **76mm; 85mm; 100mm**: M-1944, T-12; **105mm**: M-101/-102; **122mm**: Type-54, Type-60, M-1938, D-30, D-74; **130mm**: M-46; **152mm**: D-20; **155mm**: M-114
SP ARTY 152mm: 30 2S3; **175mm**: M-107
COMBINED GUN/MOR 120mm: 2S9 reported
ASLT GUNS 100mm: SU-100; **122mm**: ISU-122
MRL 107mm: 360 Type 63; **122mm**: 350 BM-21; 140mm: BM-14-16
MOR 82mm, 120mm: M-43; **160mm**: M-43
SSM *Scud* B/C (reported)
ATGW AT-3 *Sagger*
RCL 75mm: PRC Type-56; **82mm**: PRC Type-65, B-10; **87mm**: PRC Type-51
AD GUNS 12,000: **14.5mm; 23mm**: incl ZSU-23-4 SP; **30mm; 37mm; 57mm; 85mm; 100mm**
SAM SA-7/-16

[a] Inf div str varies from 5,000 to 12,500

Navy ε42,000

(incl 27,000 Naval Infantry)
Four Naval Regions
BASES Hanoi (HQ), Cam Ranh Bay, Da Nang, Haiphong, Ha Tou, Ho Chi Minh City, Can Tho, plus several smaller bases

SUBMARINES 2

SSI 2 DPRK *Yugo*

FRIGATES 6

FF 6
1 *Barnegat* (US *Cutter*) with 1 × 127mm gun
3 Sov *Petya* II with 4 × 76mm gun, 10 × 406mm ASTT, 2 ASW RL
2 Sov *Petya* III with 4 × 76mm gun, 3 × 533mm ASTT, 2 ASW RL

PATROL AND COASTAL COMBATANTS 42

CORVETTES 1 HO-A (Type 124A) FSG with 8 SS-N-25 *Zvezda* SSM, SA-N-5 SAM
MISSILE CRAFT 12
8 Sov *Osa* II with 4 SS-N-2 *Styx* SSM
4 Sov *Tarantul* with 4 SS-N-2D *Styx* SSM
TORPEDO CRAFT 10
5 Sov *Turya* PHT with 4 × 533mm TT (2 without TT)
5 Sov *Shershen* PFT with 4 × 533mm TT
PATROL, INSHORE 19
4 Sov SO-1, 3 US PGM-59/71, 10 *Zhuk*<, 2 Sov *Poluchat* PCI; plus large numbers of river patrol boats

MINE COUNTERMEASURES 10

2 *Yurka* MSC, 3 *Sonya* MSC, 2 PRC *Lienyun* MSC, 1 *Vanya* MSI, 2 *Yevgenya* MSI, plus 5 K-8 boats

AMPHIBIOUS 6

3 US LST-510-511 LST, capacity 200 tps, 16 tk
3 Sov *Polnocny* LSM, capacity 180 tps, 6 tk
Plus about 30 craft: 12 LCM, 18 LCU

SUPPORT AND MISCELLANEOUS 30+

incl 1 trg, 1 AGHS, 4 AO, about 12 small tpt, 2 ex-Sov floating docks and 3 div spt. Significant numbers of small merchant ships and trawlers are taken into naval service for patrol and resupply duties. Some of these may be lightly armed

NAVAL INFANTRY (27,000)

(amph, cdo)

People's Air Force 30,000

3 air divs (each with 3 regts), a tpt bde, an Air Force Academy
189 cbt ac, 26 armed hel
FGA 2 regt with 53 Su-22 M-3/M-4/MR (recce dedicated) and UM-3; 12 Su-27 SK/UBK
FTR 6 regt with 124 MiG-21bis/PF
ATTACK HEL 26 Mi-24
MR 4 Be-12
TPT 3 regt with ac: 12 An-2, 12 An-26, 4 Yak-40 (VIP) hel: 30 Mi-8/Mi-17, 4 Mi-6
ASW The PAF also maintains Vietnam's naval air arm, operating 3 Ka-25s, 10 Ka-28s and 2 Ka-32s.
TRG 10 Yak-18, 10 BT-6, 18 L-39, some MiG-21UM
AAM AA-2 *Atoll*, AA-8 *Aphid*, AA-10 *Allamo*
ASM AS-9 *Kyle*
SAM some 66 sites with SA-2/-3/-6/-7/-16
AD 4 arty bde: **37mm, 57mm, 85mm, 100mm, 130mm**
People's Regional Force: ε1,000 units, 6 radar bde: 100 sites

Paramilitary 40,000 active

LOCAL FORCES some 4–5,000,000

incl **People's Self-Defence Force** (urban units), **People's Militia** (rural units); these comprise static and mobile cbt units, log spt and village protection pl; some arty, mor and AD guns; acts as reserve

BORDER DEFENCE CORPS ε40,000

COAST GUARD

came into effect on 1 Sept 1998

Foreign Forces

RUSSIA 700: naval facilities; ELINT station

MILITARY DEVELOPMENTS

Colombia's long-running insurgency is now having increasingly significant effects on neighbouring countries. Increasing military pressure on the guerrillas and intensified anti-narcotics activity by the government are driving the coca growers, whose crops are a crucial source of finance for *Fuerzas Armadas Revolucionarias de Colombia* (FARC) and the *Ejército de Liberación Nacional de Colombia* (ELN), to move out of Colombia into Bolivia, Ecuador, Panama and Venezuela. Elsewhere in the region, territorial claims to marine and river waters are creating tensions involving Costa Rica, El Salvador, Guatemala, Guyana, Honduras, Nicaragua and Suriname, but none of the parties are currently likely to resort to military action. Civil–military relations continue to be a problem in some parts of the region, but there continues to be an overall trend for the military, at least those in uniform, to withdraw from political control. Apart from US military aid to Colombia, military spending in the region remains constrained, and acquisitions of major weapons systems are declining or being delayed.

Insurgency and Terrorism

There continues to be far more guerrilla activity in Colombia than anywhere else in Latin America. Peace talks between the government and the main rebel groups have failed to make real progress to date. Government security forces are suffering from having ceded control of territory to both FARC and the ELN in return for their participation in the talks. However, neither group has engaged in serious negotiation; both have concentrated on gaining full control of territory. They have even demanded that government forces withdraw from provinces adjacent to their enclaves as a condition for fully engaging in a peace process. This is in order to ease government pressure on their drug trafficking and other illegal activities, such as taking hostages for ransom, which provide their income. A major complication for government efforts to develop a peace process is the activity of the right-wing militias of the *Autodefensas Unidas de Colombia* (AUC). They have been held responsible in a report by the Ombudsman's Office in Santafé de Bogotá for one-third of the 1,000 or so unarmed civilians killed during the first half of 2000. FARC has also been held responsible for a substantial number of these atrocities although only about half as many as the AUC, according to the same report. Over the past year there has been an alarming increase in the number of civilians kidnapped by the guerrilla organisations. Abductions for ransom are reported to be running at 3,000 a year. On 11 July 2000, Colombian President Andrés Pastrana signed a law that officially outlaws genocide and 'disappearances'. The same law also created a commission to investigate the fate of the estimated 3,000 missing people.

Colombia's armed forces, which have a bad record of human-rights abuse, have implemented major reforms, punishing abuses by military personnel and reforming their training and command arrangements. This has resulted in misconduct by the armed forces falling sharply, but it has also had an adverse impact on morale: while military personnel are vigorously prosecuted, the guerrilla groups escape the legal process, mainly through fear of court officials and witnesses being intimidated or killed. One of the reasons why the armed forces introduced stringent reforms was to ensure that they met the conditions for receiving much-needed US military and financial support (Colombia is second only to Israel in the receipt of US military aid). On 30 June 2000, the US Senate gave final approval to an $11.3bn emergency aid measure whose most controversial element was $1.3bn earmarked for training the Colombian army and police in anti-narcotics operations. Most of it benefits the armed forces rather than the police. The most expensive part of the package is the provision of 30 *Blackhawk* and 33 *Huey* helicopters, to train and equip Colombian military and national police battalions and for intelligence activities. Plans for their use include retaking rebel-controlled portions of southern Colombia and destroying coca fields with chemicals. The increasing tempo of these operations is forcing coca growers out of the country into Ecuador and Bolivia.

The US aid is only part of the $7.5bn *Plan Colombia* to combat the drug trade and revitalise the Colombian economy, for which President Pastrana has been trying to drum up international support. On 7 July 2000, a conference in Madrid to promote the plan, involving 27 countries, pledged $871 million to assist Colombia. European Union countries withheld a decision on contributions as some of them sympathise with the FARC and ELN and fear that the money would be used for military operations against them.

US efforts to combat the drugs trade from Colombia reaches beyond the northern Andean region to Brazil and even to El Salvador. Washington has been trying to persuade Brazil to use its military to support anti-narcotics operations along its border with Colombia, in particular to block the thriving drugs-traffic down the Amazon. This 1,000km border is at present only lightly policed, and much of the Colombian side is controled by FARC. However, the Brazilian military is reluctant to take action, as the constitution does not allow it to become involved in civil policing. In any case, the Brazilian government does not perceive FARC as a security threat. US plans in El Salvador involve seeking government agreement to setting up a military base in order to intercept drugs moving up the Pacific coast from the northern Andes. A base there is important to US Southern Command after withdrawal of their forces from Panama in 1998. On 7 July 2000, a bill allowing an American base was passed by the Salvadorean parliament despite vigorous opposition from sympathisers with El Salvador's former left-wing guerrilla movements.

Mexico

In Mexico, there have been no peace talks between the government and the rebel groups in the Chiapas and Guerrero regions for three years. However, very little of the recent violence in these areas can be attributed to the *Ejército Zapatista de Liberación Nacional* (EZLN) and the smaller *Ejército Popular Revolucionario* (EPR). These groups were anxious to avoid an increased military presence in the region in the run-up to the June 2000 presidential election. Additionally, the government attempted to court popularity before the election by injecting more resources into new infrastructure projects in Chiapas such as roads, water supplies and schools. Many of the troops deployed in the Chiapas region were engineers and other specialists employed in building tunnels and bridges as part of the new Southern Border Highway project. The violence in that region seems to have been principally perpetuated by right-wing militias and criminal groups engaged in the drug trade. It remains to be seen whether the change in government following the defeat of the Institutional Revolutionary Party (PRI), after more than 70 years in power, will bring progress towards a lasting settlement with the EZLN and the EPR. An encouraging start has been made to rooting out corruption in central and local government and tackling organised crime, which could reduce the need for the military's prominent role in policing.

Maritime Disputes

There have been prominent disputes over territorial waters in the region. In June 2000, Suriname used its naval patrol vessels to force the removal of a Canadian-owned oil-drilling platform in territorial waters contested with Guyana. The area contains an oilfield that could yield millions of barrels of oil. Despite talks between the two countries' presidents, mediated by Jamaican Prime Minister Percival Patterson, the dispute remains unresolved. In May 2000, Nicaragua accused Honduras and Colombia of designs on 100,000km² and 30,000km² respectively of its territorial waters. Nicaragua claimed that a 1986 treaty between Honduras and Colombia, (ratified in 1999) encroaches on its maritime borders. Honduras has a similar dispute with Guatemala. Despite these tensions, with, in some cases, highly valuable assets involved, recourse to military action is unlikely. This surge in maritime border disputes is due largely to the entry-into-force in 1994 of the UN Convention on the Law of the Sea. It created Exclusive Economic Zones (EEZs), whereby coastal states have sovereign rights in waters out to 200 nautical miles from their coastline over natural resources and certain commercial activities, such as fishing. As states assert their rights under this new legal regime, tensions will inevitably arise over maritime borders. On land, border disputes between Guyana and Venezuela and between Belize and Guatemala remain unresolved, but are not causing military confrontation as they have done in the past.

DEFENCE SPENDING

Pressures on public spending and the strength of the dollar against local currencies have resulted in continuing constraints on regional defence spending in 1998–99. Nor did higher oil prices result in significantly more money for defence spending by oil-exporting countries. In Venezuela, for example, not only are there heavy demands for social spending, but the country is also still recovering from the 1999 storm in which 20,000 people are reported to have died, and essential infrastructure still needs to be rebuilt.

Regional defence expenditure declined by 9% to $35bn in 1999, measured in constant 1999 US dollars. However, this was a smaller fall than expected, as the budgets set for 1999 had suggested a drop of 19% over 1998. These budgets were planned during the 1998 international financial crisis and suffered as a result of general austerity measures. With the oil-price recovery and the relative stability of the major regional currencies against the US dollar in 2000, budgets for 2000 suggest a nominal increase in regional defence spending of 7%. This modest increase is unlikely to result in significantly more spending on procurement, as most of it will be needed for operations and maintenance and military pay and pensions.

The constraints on military spending were illustrated in May 2000 when the Brazilian Air Force commander was reported as saying that nearly 60% of Brazil's nearly 700 aircraft were not operational, mainly because of a lack of spares. In contrast, the Brazilian Navy has restarted its nuclear-submarine programme, spending $70m on uranium-enriching facilities, a research unit and a prototype nuclear-powered submarine (SSN). The Navy now has four *Tupi*-class German-designed diesel submarines in service. Two improved *Tupi*-class vessels, designated the *Tikuna*-class, also designed in Germany, were intended to be in service by 2005, acting as a 'bridge' to the SSNs. In July 2000, however, the Navy cancelled this order, deciding instead to start a project for the first completely Brazilian-designed and -built diesel submarine, planned to be in service in 2008. In this field Brazil is the regional leader, as Argentina cannot now refit its own submarines in its own dockyards – one is currently under refit in Rio de Janeiro. Venezuela has also expressed interest in refitting its submarines in Brazil. Should the Brazilian Navy carry through its SSN programme, it would become one of only six navies able to deploy submarines worldwide, but it is difficult to find a military rationale for this capability. Brazil's naval ambitions are further underlined by its reported negotiations to buy the French aircraft carrier *Foch* to replace the ageing *Minais Gerais*. It is unlikely that Brazil can afford both the SSNs and a new aircraft carrier in view of all the refitting that would be necessary, although possible co-operation with Argentina on naval aviation could lower the cost.

Argentina's naval activities provide a clue to why Brazil regards enhanced naval capabilities to be necessary. Argentina provided a frigate in the 1991 Gulf War, and in 2000 it provided a frigate to participate in the enforcement of the oil embargo against Iraq. This small contribution could reap important political benefits, particularly from the US. Argentina has long appreciated and valued the rewards of participating in UN and other multinational operations. Nonetheless, tight budgets have resulted in a contraction of its naval programmes. No other major conversions or acquisitions are planned and even the two *Meko*-class frigates currently under construction in the country will probably be sold on completion. Chile too is reducing its procurement, although it will still commission two *Scorpene* diesel submarines from France between 2006 and 2008. It has been confirmed that financial cutbacks will reduce the Chilean *Tridente* general-purpose frigate programme from eight to four ships.

The majority of arms deliveries to the region are likely to result from the two-year US$1.3bn military-aid package for Colombia, whose most costly elements are the *Black Hawk* and *Huey* helicopters for supporting counter-insurgency. Some of the aid is for police equipment and training and for activities such as crop-substitution.

Selected Defence Budgets and Expenditure

Brazil Brazil's initial defence budget for 1999 of R17.5bn was subsequently reduced to R16.6bn. In 2000, the budget rose to R17.9bn ($9.9bn). Brazil's military spending remains low relative to the size of the national economy, accounting for just 3.2% of government spending and 1.7% of gross domestic product (GDP) in 1999. Spending increases mainly result from higher military salaries and pensions. Otherwise, heavy limitations on procurement and operations remain following the 1998 international financial crisis that imposed severe fiscal limitations on government spending.

Table 26 **Brazil: defence expenditure, 1997–2000**

US$m	1997 Budget	1997 Revised	1998 Outlay	1998 Budget	1998 Revised	1999 Budget	1999 Revised	2000 Budget
Army	6,906	7,001	6,606	6,493	6,321	5,083	4,444	5,239
Navy	4,032	4,088	3,769	3,755	3,755	2,611	2,538	2,528
Air Force	3,697	3,774	3,439	3,354	3,353	2,347	1,976	1,925
MoD	132	132	132	250	250	205	193	234
Total	14,767	14,995	13,945	13,851	13,678	10,245	9,151	9,926

Argentina Of the major Latin American countries, Argentina has carried out the most thorough reforms, to direct military spending to the real tasks faced by the armed forces at home and abroad. Its defence spending remains less than 2% of GDP. This means that only a small fraction of its defence budget, just under 6%, is available for procurement. Nearly 83% of military expenditure is on personnel, pensions and operations.

Table 27 **Argentina: Defence, *Gendarmerie* and Coast Guard Budgets, 2000**

US$m	MoD	Army	Navy	Air Force	Total Defence	%	*Gendarmerie*	Coast Guard	Total	%
Personnel and operations	99	823	539	456	1,917	50.9	328	206	2,451	52.2
Procurement	15	129	23	57	224	5.9	28	23	275	5.9
Pensions					1,206	32.0	172	107	1,485	31.6
Other	109	76	30	208	423	11.2	25	33	481	10.3
Total	223	1,029	592	721	3,770	100.0	553	369	4,692	100.0

Chile Continuing low copper prices put pressure on all Chilean government spending, but military expenditure still remains at nearly 4% of GDP. The comparable figure is half as great in Argentina and significantly lower in Brazil. However, as for its regional neighbours, over three-quarters of Chile's defence budget is absorbed by pay, pensions, operations and maintenance.

Table 28 **Chile: defence and security funding by service, 1999–2000**

US$m	1999	%	2000	%
Army	476	18.2	492	16.7
Navy	437	16.7	458	15.5
Air Force	228	8.7	241	8.2
Sub-total	**1,141**	**43.7**	**1,190**	**40.4**
Military pensions	592	22.7	763	25.9
Other	130	5.0	165	5.6
Sub-total	**1,863**	**71.3**	**2118**	**71.8**
Paramilitary	749	28.7	830	28.2
Total	**2,612**	**100.0**	**2,947**	**100.0**

Mexico, Venezuela and Bolivia The Mexican defence budget rose to NP28.3bn ($3bn) in 2000 from NP23.2bn ($2.4bn) in 1999. The inauguration of a new president Vincente Fox in November 2000 is unlikely to change the defence-spending pattern in the short term. Even if the security situation in Chiapas and Guerrero improved, support for the police and major public infrastructure projects will continue to place heavy demands on the military in the foreseeable future.

President Hugo Chavez of Venezuela has said that he will cut defence expenditure 'to bare minimums' because of other demands on public spending. There is some unrest and great unease among those sections of the population which suffered severely in the 1999 floods. With the campaign in full swing for the election in July 2000, President Chavez felt compelled to make a clear response to the majority of voters' perceived priorities for government spending.

Partly because of the boliviano's 5% fall in value against the US dollar in 1999–2000, the Bolivian defence budget for 2000 is down 12% from the previous year. Bolivia demonstrates the regional trend of modest allocations for defence procurement.

Table 29 **Arms orders and deliveries, Caribbean and Latin America, 1998–2000**

Country	Country supplier	Classification	Designation	Quantity	Order date	Delivery date	Comment
Argentina	dom	arty	**155mm L45CALA**		1995	2000	Entering production
	US	hel	**UH-1H**	8	1996	1998	
	US	MPA	**P-3B**	8	1996	1997	Deliveries to 1999
	US	LAW	**M72**	900	1997	1999	
	US	FGA	**A-4M**	8	1997	1999	Further 11 for spares
	US	hel	**UH-1H**	8	1997	1998	
	US	tkr ac	**KC-135**	1	1998	2000	
	Fr	AO	***Durance***	1	1998	1999	
	dom	trg	**IA-63**	1	1999	1999	

Country	Country supplier	Classification	Designation	Quantity	Order date	Delivery date	Comment
	US	APC	**M113A2**	90	1999		Ex-US Army
Bahamas	US	tpt	**C-26**	2	1997	1998	
	US	PCO	***Bahamas***	2	1997	1999	Contract options for 4 more
Bolivia	dom	PCR	**PCR**	23	1997	1999	
	US	FGA	**TA-4J**	18	1997	1998	12 for op and 6 for spares
Brazil	dom	AAM	**MAA-1**	40	1976	1998	Under test since mid-1998
	col	FGA	**AM-X**	54	1980	1989	Deliveries continue: 2 in 1997
	Fr	hel	**AS-350**	77	1985	1988	Lic prod continues at low rate
	Ge	SSK	**Type 209**	4	1985	1989	Last delivered in 2000
	Ge	PCC	***Grauna***	12	1986	1993	Last 2 delivered 1999
	dom	MRL	***ASTROS* 2**	20	1994	1998	4 ordered 1996, 16 1998
	Be	MBT	***Leopard* 1**	87	1995	1997	55 delivered 1998–99
	dom	FF	***Niteroi***	6	1995	1999	Upgrade to 2001
	dom	AEW	**EMB-145**	8	1997	2001	5 AEW, 3 Remote Sensing
	Fr	tpt	**F-406**	5	1997	1999	For delivery 1999–2001
	dom	ATGW	**MSS-1.2**	40	1997	2001	Development
	col	FGA	**AMX**	13	1998	2001	3rd batch
	Kwt	FGA	**A-4**	23	1998	1998	Ex-Kwt Air Force; includes 3 TA-4
	Il	FGA	**F-5**	45	1998	2000	Upgrade
	UK	arty	**105mm**	18	1999	2001	
	Swe	HWT	**Tp-62**	50	1999	2000	For *Tupi* SSK
	US	MPA	**P-3A/B**	12	1999	2002	
Chile	Ge	FAC	**Type 148**	6	1995	1997	2 in 1997, 4 in 1998
	Be	APC	**M-113**	128	1995	1998	
	Fr	trg	**C-101**	2	1995	1997	1st upgrade to *Halcon* II delivered
	UK	ASSM	**MM-38 Exocet**	4	1996	1998	*Excalibur* ASSM; refurbished in Fr
	Fr	MBT	**AMX 30B**	60	1996	1998	
	US	recce	***Caravan* 1**	3	1996	1998	
	UK	arty	**M101**	100	1996	1998	Upgrade
	col	MRL	***Rayo***		1996	2000	Prototype trials
	US	tpt	**R-182**	8	1997	1998	
	RSA	arty	**M71**	24	1997	1998	
	Ge	PFM	***Tiger***	2	1997	1998	
	Fr	SSK	***Scorpene***	2	1997	2003	1st delivery 2003, 2nd 2005
	US	hel	**UH-60**	12	1998	1998	1st delivery Jul 1998
	dom	MPA	**P-3**	2	1998	1999	Upgrade for up to 8
	Nl	MBT	***Leopard* 1**	200	1998	1999	Deliveries completed in 2000
Colombia	Sp	tpt	**CN-235**	3	1996	1998	
	dom	utl	***Gavilan***	12	1997	1998	
	US	hel	**B-212**	6	1998	1998	First 3 in 1998
	US	hel	**UH-60L**	6	1998	1999	For delivery Sep 1999–Jan 2000
	US	hel	**UH-1H**	25	1998	1999	Delivered 1999
	US	hel	**MD-530F**	2	1998	1999	National Police

Country	Country supplier	Classification	Designation	Quantity	Order date	Delivery date	Comment
	US	hel	***Black Hawk***	30	2000	2001	For counter-drug operations
	US	hel	**UH-1H**	33	2000	2001	For counter-drug operations
Ecuador	Il	AAM	***Python* 3**	100	1996	1999	
	US	ASW hel	**Bell 412EP**	2	1996	1998	
	RF	hel	**Mi-17**	7	1997	1998	
	Il	FGA	***Kfir***	2	1998	1999	Ex-IAF; also upgrade of 11
El Salvador	US	hel	**MD-520N**	2	1997	1998	
Guatemala	dom	APC	***Danto***		1994	1998	For internal security duties
	Chl	trg	**T-35B**	10	1997	1998	Ex-Chl Air Force
Honduras	US	FGA	***Super Mystere***	11	1997	1998	
Jamaica	Fr	hel	**AS-555**	4	1997	1999	
Mexico	dom	PCO	***Holzinger* 2000**	8	1997	1997	Final delivery 2001
	Be	lt tk	**AMX 13**	136	1994	1995	Final 34 in 1998
	Ukr	hel	**Mi-17**	12	1995	1997	*Erint* delivered 1998
	US	FF	***Knox***	3	1996	1998	Third for delivery 1999
	US	LST	***Newport***	1	1998	1999	Excess Defense Articles (EDA)
	US	hel	**MD-520N**	8	1998	1999	
	RF	hel	**Mi-26**	1	2000	2000	
Paraguay	ROC	FGA	**F-5E**	4	1997	1998	Total of 12 in all
	ROC	PCI		2	1998	1999	Free transfer
Peru	Bel	FGA	**Su-25**	18	1995	1998	
	Bel	FGA	**MiG-29**	18	1995	1996	Deliveries 1996–97
	Fr	ASSM	***Exocet***	8	1995	1997	Deliveries to 1998
	It	ASSM	***Otomat***	12	1995	1997	Deliveries to 1998
	RF	FGA	**MiG-29**	3	1998	1998	Plus spares
	Cz	arty	**D-30**	6	1998	1998	
	Cz	trg	**ZLIN-242L**	18	1998	1998	
	RF	tpt	**Il-103**	6	1999	1999	
	US	PCI		6	2000	2000	For Coast Guard
Suriname	Sp	MPA	**C-212-400**	2	1997	1998	2nd delivered 1999
Uruguay	UK	tac hel	***Wessex***	6			Free transfer
	Il	MBT	**T-55**	11	1996	1997	Deliveries to 1998
	Cz	MRL	**RM-70**	1	1998	1998	
	Cz	SP arty	**2S1**	6	1998	1998	
Venezuela	Pl	tpt	**M-28**	12	1996	1996	Deliveries 1996–98
	Sp	MPA	**C-212**	3	1997	1998	Plus modernisation of existing C-212-200
	US	FGA	**F-16B**	2	1997	1999	
	US	hel	**B-212**	2	1997	1999	US grant aid for counter-drug op
	Fr	hel	**AS-532**	6	1997	2000	
	US	hel	**UH-1H**	5	1997	1999	
	Swe	ATGW	**AT-4**		1997	1999	

Country	Country supplier	Classification	Designation	Quantity	Order date	Delivery date	Comment
	US	FF	***Lupo***	2	1998	2001	Upgrade and modernisation
	Swe	radar	***Giraffe***	4	1998	1999	4 truck-mounted systems
	It	trg	**SF-260E**	12	1998	1999	Requirement for 12 more
	US	PCI	**PCI**	12	1998	1999	Aluminium 80 foot craft
	US	PCI	**PCI**	10	1998	1999	Aluminium 54 foot craft
	It	trg	**MB-339FD**	10	1998	2000	Req for up to 24; del to 2001
	col	FGA	**AMX**	8	1998	2001	Br and It; up to 24 req
	US	SAR hel	**AB-412EP**	4	1998	1999	Option for a further 2
	Il	SAM	***Barak*-1**	6	1999	2000	Upgrade
	Swe	SAM	**RBS-70**	500	1999	2000	Includes AT-4 ATGW
	Fr	radar	***Flycatcher***	3	1999	2000	Del to early 2002; part of *Guardian*

Dollar GDP figures for several countries in Latin America are based on Inter-American Development Bank estimates. In some cases, the dollar conversion rates are different from the average exchange rate values shown under the country entry. Dollar GDP figures may vary from those cited in *The Military Balance* in previous years. Defence budgets and expenditures have been converted at the dollar exchange rate used to calculate GDP.

Antigua and Barbuda AB

East Caribbean dollar EC$		1998	1999	2000	2001
GDP	EC$	1.7bn	1.8bn		
	US$	617m	653m		
per capita	US$	5,800	6,000		
Growth	%	3.9	3.5		
Inflation	%	3.4	3.0		
Ext Debt	US$	357m	350m		
Def exp	EC$	11m	11m		
	US$	4m	4m		
Def bdgt	EC$	11m	11m	12m	
	US$	4m	4m	4m	
FMA	US$	0.1m	0.1m	0.1m	0.1m
US$1=EC$		2.7	2.7	2.7	2.7

Population			**72,000**
Age	13–17	18–22	23–32
Men	5,000	5,000	6,000
Women	5,000	5,000	8,000

Total Armed Forces

ACTIVE 150 (all services form combined **Antigua and Barbuda Defence Force**)

RESERVES 75

Army 125

Navy 45

BASE St Johns

PATROL CRAFT 3

PATROL, INSHORE 3

1 *Swift* PCI< • 1 *Dauntless* PCI< • 1 *Point* PCI<

Argentina Arg

peso P		1998	1999	2000	2001
GDP	P	298bn	283bn		
	US$	298bn	283bn		
per capita	US$	10,400	10,100		
Growth	%	3.9	-3.0		
Inflation	%	0.9	-1.2		
Debt	US$	144bn	145bn		
Def exp	P	5.3bn	5.4bn		
	US$	5.3bn	5.4bn		
Def bdgt	P	3.7bn	3.5bn	3.8bn	
	US$	3.7bn	3.5bn	3.8bn	
FMA (US)	US$	0.6m	1.4m	0.7m	1.8m
US$1=P		1.0	1.0	1.0	

Population			**37,300,000**
Age	13–17	18–22	23–32
Men	1,631,000	1,625,000	2,767,000
Women	1,578,000	1,577,000	2,701,000

Total Armed Forces

ACTIVE 71,100

RESERVES none formally established or trained

Army 41,400

3 Corps
- 1 with 1 armd, 1 jungle bde, 1 trg bde
- 1 with 1 AB, 1 mech, 1 mtn bde
- 1 with 1 armd, 3 mech, 1 mtn bde

Army tps
- 1 mot inf bn (Army HQ Escort Regt), 1 mot cav regt (Presidential Escort), 1 AD arty, 3 avn, 2 engr bn, 1 SF coy

EQUIPMENT

MBT 200 TAM
LT TK 50 AMX-13, 100 SK-105 *Kuerassier*
RECCE 75 AML-90
AIFV 160 VCTP (incl variants)
APC 126 M-5 half-track, 323 M-113
TOWED ARTY 105mm: 100 M 56 *Oto Melara;* **155mm**: 100 CITEFA Models 77/-81
SP ARTY 155mm: 20 Mk F3, 15 VCA (*Palmaria*)
MRL 105mm: 5 SLAM *Pampero;* **127mm**: 5 SLAM SAPBA-1
MOR 81mm: 1,100; **120mm**: 360 Brandt (37 SP in VCTM AIFV)
ATGW 600 SS-11/-12, *Cobra (Mamba)*, 2,100 *Mathogo*
RL 66mm: M-72
RCL 75mm: 75 M-20; **90mm**: 100 M-67; **105mm**: 930 M-1968
AD GUNS 30mm: 21; **40mm**: 76 L/60/-70
SAM <40† *Tigercat*, <40† *Blowpipe*
SURV RASIT also RATRAS (veh, arty), *Green Archer* (mor), *Skyguard*
AC 1 C212-200, 3 Cessna 207, 2 *Commander* 690, 3 DHC-6, 2 G-222, 1 *Merlin* IIIA, 5 *Merlin* IV, 1 *Queen Air*, 1 *Sabreliner*, 5 T-41, 23 OV-1D
HEL 6 A-109, 3 AS-332B, 1 Bell 212, 4 FH-1100, 5 SA-315B, 1 SA-330, 13 UH-1H, 8 UH-12

Navy 17,200

(incl 2,000 Naval Aviation and 2,800 Marines)
NAVAL AREAS Centre from River Plate to 42°45'S **South** from 42°45'S to Cape Horn **Antarctica**
BASES Buenos Aires, Puerto Belgrano (HQ Centre), Ushuaio (HQ South), Mar del Plata (submarines), Trelew, Punta Indio (naval air training), Rio Santiago (shipbuilding), Caleta Paula (fisheries protection)

SUBMARINES 3
SSK 3
2 *Santa Cruz* (Ge TR-1700) with 6 × 533mm TT (SST-4 HWT)
1 *Salta* (Ge T-209/1200) with 8 × 533mm TT (SST-4 HWT)

PRINCIPAL SURFACE COMBATANTS 13
DESTROYERS 6
DDG 6
2 *Hercules* (UK Type 42) with 4 MM-38 *Exocet* SSM, 1 × 2 *Sea Dart* SAM, 1 × 114mm gun, 2 × 3 ASTT, 1 AS-319B hel (one in deep refit)
4 *Almirante Brown* (Ge MEKO 360) with 8 MM-40 *Exocet* SSM, 1 × 127mm gun, 2 × 3 ASTT, 1 AS-555 hel
FRIGATES 7
FFG 7
4 *Espora* (Ge MEKO 140) with 4 MM-38 *Exocet* SSM, 1 × 76mm gun, 2 × 3 ASTT, 1 SA 319B hel
3 *Drummond* (Fr A-69) with 4 MM-38 *Exocet* SSM, 1 × 100mm gun, 2 × 3 ASTT

PATROL AND COASTAL COMBATANTS 15
TORPEDO CRAFT 2 *Intrepida* (Ge Lürssen 45m) PFT with 2 × 533mm TT (SST-4 HWT) (one with 2 MM-38 SSM)
PATROL, OFFSHORE 8
1 *Teniente Olivieri* (ex-US oilfield tug)
3 *Irigoyen* (US *Cherokee* AT)
2 *King* (trg) with 3 × 105mm guns
2 *Sobral* (US *Sotoyomo* AT)
PATROL, INSHORE 5
4 *Baradero* (*Dabur*) PCI<
1 *Point* PCI<

MINE COUNTERMEASURES 2
2 *Neuquen* (Uk *Ton*) MHC

AMPHIBIOUS craft only
4 LCM, 16 LCVP

SUPPORT AND MISCELLANEOUS 11
1 *Durance* AO, 3 *Costa* tpt; 3 *Red* buoy tenders, 1 icebreaker, 1 sail trg, 1 AGOR, 1 AGHS (plus 2 craft)

NAVAL AVIATION (2,000)
21 cbt ac, 14 armed hel
Carrier air crew training on Brazilian CV *Minas Gerais*
ATTACK 1 sqn with 11 *Super Etendard*
MR/ASW 1 sqn with 5 S-2T, 4 P-3B, 5 BE-200M/G
EW 1 L-188E
HEL 2 sqn
1 ASW/tpt with 5 ASH-3H (ASW) *Sea King* and 2 AS-61D (tpt), 4 AS-555 *Fenec*
1 spt with 5 SA-319B *Alouette III*†
TPT 1 sqn with 3 F-28-3000
SURVEY 2 B-200F, 1 PL-6A
TRG 2 sqn with 10 T-34C, 8 MC-32
MISSILES
ASM AM-39 *Exocet*, AS-12, *Martín Pescador*
AAM R-550 *Magic*

MARINES (2,800)
FLEET FORCES 2
1 with 1 marine inf, 1 AAV, 1 arty, 1 AAA bn, 1 cdo gp
1 with 2 marine inf bn, 2 naval det
AMPH SPT FORCE 1 marine inf bn
6 marine sy coy

EQUIPMENT
RECCE 12 ERC-90 *Lynx*
AAV 21 LVTP-7, 13 LARC-5
APC 6 MOWAG *Grenadier*, 36 Panhard VCR
TOWED ARTY 105mm: 6 M-101, 12 Model 56; **155mm**: 4 M-114
MOR 81mm: 70; **120mm**: 12
ATGW 50 *Bantam*, *Cobra (Mamba)*
RL 89mm: 60 M-20
RCL 105mm: 30 1974 FMK1
AD GUNS 30mm: 12 HS-816; **35mm**: GDF-001
SAM 6 RBS-70

Air Force 12,500

133 cbt ac, 27 armed hel, 4 Major Cmds – Air Operations, Personnel, Air Regions, Logistics
AIR OPERATIONS COMMAND (8 bde, 2 Air Mil Bases, 1 Airspace Surveillance and Control Gp, 1 EW Gp)
STRATEGIC AIR 5 sqn
2 with 23 *Dagger Nesher*
1 with 6 *Mirage* V Mara
2 with 36 A-4AR *Fightinghawk*
AIRSPACE DEFENCE 1 sqn with 14 *Mirage* III/EA, 6 TPS-43 field radars, SAM -3 *Roland*
AD GUNS 35mm: 1; 200mm: 86
TAC AIR 2 sqn
2 with 30 IA-58 *Pucara*
SURVEY/RECCE 1 sqn with 1 Boeing 707, 3 *Learjet* 35A, 2 IA-50
TPT/TKR 6 sqn
1 with 4 Boeing 707
2 with 13 C-130 *Hercules* (5 -B, 5 -H, 2 KC-H, 1 L-100-30)
1 with 8 F-27
1 with 4 F-28
1 with 6 DHC-6 *Twin Otter*
plus 3 IA-50 for miscellaneous communications
SAR
4 Bell 212, 10* UH-1H, 17* MD-500 hel

PERSONNEL COMMAND

TRG

30 *Mentor* B-45 (basic), 27 *Tucano* EMB-312 (primary), 14* *Pampa* IA-63, 10* MS-760 (advanced), 8 Su-29AR **hel** 3 Hughes MD-500

MISSILES

ASM ASM-2 *Martín Pescador*

AAM AIM-9B *Sidewinder*, R-530, R-550, *Shafrir*

Forces Abroad

UN AND PEACEKEEPING

CROATIA (UNMOP): 1 obs **CYPRUS** (UNFICYP) 412: 1 inf bn **IRAQ/KUWAIT** (UNIKOM): 84 engr, 4 obs **MIDDLE EAST** (UNTSO): 3 obs **WESTERN SAHARA** (MINURSO): 1 obs **YUGOSLAVIA** (KFOR): 113

Paramilitary 31,240

GENDARMERIE (Ministry of Interior) 18,000

5 Regional Comd, 16 bn

EQPT Shorland recce, 40 UR-416, 47 MOWAG *Grenadier*; **81mm** mor; **ac** 3 *Piper*, 5 PC-6 **hel** 5 SA-315

PREFECTURA NAVAL (Coast Guard) 13,240

7 comd

SERVICEABILITY better than Navy

EQPT 5 *Mantilla*, 1 *Delfin* PCO, 1 *Mandubi* PCO; 4 PCI, 21 PCI< plus boats; **ac** 5 C-212 **hel** 1 AS-330L, 2 AS-365, 4 AS-565MA, 2 Bell-47, 2 Schweizer-300C

Bahamas — Bs

Bahamian dollar B$		1998	1999	2000	2001
GDP	B$	3.6bn	3.8bn		
	US$	3.6bn	3.8bn		
per capita	US$	13,900	14,500		
Growth	%	2.5	4.0		
Inflation	%	1.4	1.5		
Debt	US$	325m	315m		
Def exp	B$	25m	26m		
	US$	25m	26m		
Def bdgt	B$	25m	26m	26m	
	US$	25m	26m	26m	
FMA (US)	US$	1.1m	1.3m	1.2m	1.5m
US$1=B$		1.0	1.0	1.0	

Population			**300,000**
Age	13–17	18–22	23–32
Men	14,000	15,000	32,000
Women	13,000	14,000	30,000

Total Armed Forces

ACTIVE 860

Navy (Royal Bahamian Defence Force) 860

(incl 70 women)

BASE Coral Harbour, New Providence Island

MILITARY OPERATIONS PLATOON 1

ε120; Marines with internal and base sy duties

PATROL AND COASTAL COMBATANTS 7

PATROL, OFFSHORE 2 *Bahamas* PCO

PATROL, INSHORE 5

3 *Protector* PFI<, 1 *Cape* PCI<, 1 *Kieth Nelson* PCI<

SUPPORT AND MISCELLANEOUS 3

1 *Fort Montague* (AG), 2 *Dauntless* (AG)

HARBOUR PATROL UNITS 4

4 *Boston* whaler

AIRCRAFT 4

1 Cessna 404, 1 Cessna 421C, 2 C-26

Barbados — Bds

Barbadian dollar B$		1998	1999	2000	2001
GDP	B$	4.7bn	5.0bn		
	US$	2.4bn	2.5bn		
per capita	US$	6,700	7,000		
Growth	%	4.2	3.5		
Inflation	%	1.5	1.6		
Debt	US$	607m	490m		
Def exp	B$	23m	24m		
	US$	12m	12m		
Def bdgt	B$	25m	24m	26m	
	US$	13m	12m	13m	
FMA	US$	0.1m	0.1m	0.1m	0.1m
US$1=B$		2.0	2.0	2.0	2.0

Population			**266,000**
Age	13–17	18–22	23–32
Men	11,000	11,000	23,000
Women	11,000	10,000	22,000

Total Armed Forces

ACTIVE 610

RESERVES 430

Army 500

Navy 110

BASES St Ann's Fort Garrison (HQ), Bridgetown

PATROL AND COASTAL COMBATANTS 5

PATROL, OFFSHORE 1

1 *Kebir* PCO

PATROL, INSHORE 4

1 *Dauntless* PCI< • 3 *Guardian* PCI< • plus boats

Belize — Bze

Belize dollar BZ$		1998	1999	2000	2001
GDP	BZ$	1.2bn	1.3bn		
	US$	624m	674m		
per capita	US$	2,700	2,800		
Growth	%	1.7	4.6		
Inflation	%	-0.7	-1.2		
Debt	US$	337m	260m		
Def exp	BZ$	33m	34m		
	US$	17m	17m		
Def bdgt	BZ$	17m	17m	17m	15m
	US$	8m	8m	9m	8m
FMA (US)	US$	0.3m	0.3m	0.4m	0.5m
US$1=BZ$		2.0	2.0	2.0	2.0

Population			**242,000**
Age	13–17	18–22	23–32
Men	14,000	13,000	20,000
Women	14,000	13,000	20,000

Total Armed Forces

ACTIVE ε1,050

RESERVES 700

Army ε1,050

3 inf bn (each 3 inf coy), 1 spt gp, 3 Reserve coy

EQUIPMENT

MOR 81mm: 6
RCL 84mm: 8 *Carl Gustav*

MARITIME WING

PATROL CRAFT some 14 armed boats and 3 LCU

AIR WING

No cbt ac or armed hel
MR/TPT 1 BN-2B *Defender*
TRG 1 T67-200 *Firefly*, 1 Cessna 182

Foreign Forces

UK Army 180

Bolivia — Bol

boliviano B		1998	1999	2000	2001
GDP	B	47bn	51bn		
	US$	8.5bn	8.8bn		
per capita	US$	3,100	3,100		
Growth	%	4.6	1.0		
Inflation	%	7.7	2.1		
Debt	US$	6.1bn	5.8bn		
Def exp	B	1,128m	864m		
	US$	205m	149m		
Def bdgt	B	1,184	942m	796m	
	US$	215m	162m	130m	
FMA[a] (US)	US$	36m	55m	49m	53m
US$1=B		5.5	5.8	6.1	

[a] Excl Plan Colombia allocation for 2001

Population			**8,140,000**
Age	13–17	18–22	23–32
Men	418,000	337,000	546,000
Women	424,000	360,000	581,000

Total Armed Forces

ACTIVE 32,500 (to be 35,000)
(incl some 21,800 conscripts)
Terms of service 12 months, selective

Army 25,000

(incl some 18,000 conscripts)
HQ: 6 Mil Regions
Army HQ direct control
2 armd bn • 1 mech cav regt • 1 Presidential Guard inf regt
10 'div'; org, composition varies; comprise
8 cav gp (5 horsed, 2 mot, 1 aslt) • 1 mot inf 'regt' with 2 bn • 22 inf bn (incl 5 inf aslt bn) • 10 arty 'regt' (bn) • 1 AB 'regt' (bn) • 6 engr bn

EQUIPMENT

LT TK 36 SK-105 *Kuerassier*
RECCE 24 EE-9 *Cascavel*
APC 18 M-113, 10 V-100 *Commando*, 20 MOWAG *Roland*, 24 EE-11 *Urutu*
TOWED ARTY 75mm: 70 incl M-116 pack, ε10 Bofors M-1935; **105mm**: 30 incl M-101, FH-18; **122mm**: 18 PRC Type-54
MOR 81mm: 50; **107mm**: M-30
AC 1 C-212, 1 *King Air* B90, 1 *Cheyenne* II, 1 *Seneca* III, 5 Cessna (4 -206, 1 -421B)

Navy 3,500

(incl 1,700 Marines)
NAVAL AREAS 3 (Strategic Logistic Support)
NAVAL DISTRICTS 6, covering Lake Titicaca and the rivers; each 1 flotilla
BASES Riberalta (HQ), Tiquina (HQ), Puerto Busch, Puerto Guayaramerín (HQ), Puerto Villaroel, Trinidad (HQ), Puerto Suárez (HQ), Cobija (HQ), Santa Cruz (HQ), Bermejo (HQ), Cochabamba (HQ), Puerto Villarroel

PATROL CRAFT, RIVERINE some 60 riverine craft/boats, all<

SUPPORT AND MISCELLANEOUS some 18 logistic support and patrol craft plus 27 *Rodman* boats<

MARINES (1,700)
6 bn (1 in each District)

Air Force 3,000

(incl perhaps 2,000 conscripts); 62 cbt ac, 10 armed hel
FGA 18 AT-33AN
ARMED HEL 1 sqn with 10 Hughes 500M hel
SAR 1 hel sqn with 4 HB-315B, 2 SA-315B, 1 UH-1
SURVEY 1 sqn with 5 Cessna 206, 1 C-210, 1 C-402, 3 *Learjet* 25/35
TPT 3 sqn
1 VIP tpt with 1 L-188, 1 *Sabreliner*, 2 *Super King Air*
2 tpt with 14 C-130A/B/H, 4 F-27-400, 1 IAI-201, 2 *King Air*, 2 C-47, 4 *Convair* 580
LIAISON ac 9 Cessna 152, 1 C-185, 13 C-206, 1 C-208, 2 C-402, 2 Beech *Bonanza*, 2 Beech *Barons*, PA-31, 4 PA-34 **hel** 2 Bell 212, 22 UH-1H
TRG 1 Cessna 152, 2 C-172, 20* PC-7, 4 SF-260CB, 15 T-23, 12* T-33A, 12* T-34A, 1 *Lancair* 320
AD 1 air-base def regt† (Oerlikon twin **20mm**, 18 PRC Type-65 **37mm**, some truck-mounted guns)

Forces Abroad

UN AND PEACEKEEPING
DROC (MONUC): 2 obs **EAST TIMOR** (UNTAET): 2 obs **SIERRA LEONE** (UNAMSIL): 4 obs

Paramilitary 37,100

NATIONAL POLICE some 31,100
9 bde, 2 rapid action regt, 27 frontier units

NARCOTICS POLICE some 6,000

Brazil Br

real R		1998	1999	2000	2001
GDP	R	902bn	1,089bn		
	US$	583bn	600bn		
per capita	US$	6,200	6,300		
Growth	%	0.2	0.5		
Inflation	%	3.2	4.9		
Debt	US$	232bn	240bn		
Def exp[a]	R	ε21.4bn	ε29.0bn		
	US$	18.4bn	16.0bn		
Def bdgt[b]	R	16.1bn	16.6bn	17.9bn	
	US$	13.8bn	9.1bn	9.9bn	
FMA (US)	US$	0.7m	1.4m	1.7m	2.3m
US$1=R		1.16	1.82	1.80	

[a] Incl spending on paramilitary forces
[b] 1999 defence bdgt revised down to R16.6bn (US$9.2bn)

Population			164,000,000
Age	13–17	18–22	23–32
Men	8,798,000	8,355,000	14,533,000
Women	8,707,000	8,348,000	14,667,000

Total Armed Forces

ACTIVE 287,600
(incl 48,200 conscripts)
Terms of service 12 months (can be extended to 18)

RESERVES
Trained first-line 1,115,000; 400,000 subject to immediate recall **Second-line** 225,000

Army 189,000

(incl 40,000 conscripts)
HQ: 7 Mil Comd, 12 Mil Regions; 8 div (3 with Regional HQ)
1 armd cav bde (2 armd cav, 1 armd, 1 arty bn), 3 armd inf bde (each 2 armd inf, 1 armd cav, 1 arty bn), 4 mech cav bde (each 2 mech cav, 1 armd cav, 1 arty bn) • 10 motor inf bde (26 bn) • 1 lt inf bde (3 bn) • 4 jungle bde • 1 frontier bde (6 bn) • 1 AB bde (3 AB, 1 arty bn) • 1 coast and AD arty bde (6 bn) • 3 cav guard regt • 10 arty gp (4 SP, 6 med) • 2 engr gp (9 bn) • 10 engr bn (incl 2 railway) (to be increased to 34 bn)
AVN 1 hel bde (1 bn of 4 sqn)
EQUIPMENT
MBT 87 *Leopard* 1, 91 M-60A3
LT TK 286 M-41B/C
RECCE 409 EE-9 *Cascavel*
APC 219 EE-11 *Urutu*, 584 M-113
TOWED ARTY 105mm: 319 M-101/-102, 56 pack, 22 L118; **155mm**: 92 M-114
SP ARTY 105mm: 72 M-7/-108
MRL 108mm: SS-06; 16 ASTROS II
MOR 81mm: 707; **107mm**: 236 M-30; **120mm**: 77 K6A3
ATGW 4 *Milan*, 18 *Eryx*
RL 84mm: 115 AT-4
RCL 84mm: 127 *Carl Gustav*; **106mm**: 163 M-40A1
AD GUNS 134 incl **35mm**: GDF-001; **40mm**: L-60/-70 (some with BOFI)
SAM 4 *Roland* II, 40 SA-18
HEL 3 S-70A, 35 SA-365, 19 AS-550 *Fennec*, 16 AS-350 (armed)

Navy 48,600

(incl 1,150 Naval Aviation, 13,900 Marines and 3,200 conscripts)
OCEANIC NAVAL DISTRICTS 5 plus 1 Riverine; 1 Comd
BASES Ocean Rio de Janeiro (*HQ I Naval District*), Salvador (*HQ II District*), Recife (*HQ III District*), Belém

(*HQ IV District*), Floriancholis (*HQ V District*) **River** Ladario (*HQ VI District*)

SUBMARINES 5

SSK 5

4 *Tupi* (Ge T-209/1400) with 8 × 533mm TT (UK *Tigerfish* HWT)

1 *Tonelero* (UK *Oberon*) with 533mm TT (UK *Tigerfish* HWT)

PRINCIPAL SURFACE COMBATANTS 19

CARRIERS 1 *Minas Gerais* (UK *Colossus*) CV, typically ASW **hel** 4–6 ASH-3H, 3 AS-332 and 2 AS-355; has been used by Argentina for embarked aircraft training

FRIGATES 14

FFG 10

4 *Greenhaigh* (ex-UK *Broadsword*) with 4 MM-38 *Exocet* SSM, GWS 25 *Seawolf* SAM, 6 × 324mm ASTT (Mk 46 LWT), 2 *Super Lynx* hel

6 *Niteroi* with 2 × 2 MM 40 *Exocet* SSM, 2 × 3 *Seacat* SAM, 1 × 115mm gun, 6 × 324mm ASTT (Mk 46 LWT), 1 × 2 ASW mor, 1 *Super Lynx* hel

FF 4

4 *Para* (US *Garcia*) with 2 × 127mm guns, 2 × 3 ASTT, 1 × 8 ASROC SUGW, 1 *Super Lynx* hel

CORVETTES 4

4 *Inhauma* FSG, with 4 MM-40 *Exocet* SSM, 1 × 114mm gun, 2 × 3 ASTT, 1 *Super Lynx* hel

PATROL AND COASTAL COMBATANTS 50

PATROL, OFFSHORE 19

7 *Imperial Marinheiro* PCO with 1 × 76mm gun, 12 *Grajaü* PCO

PATROL, COASTAL 10

6 *Piratini* (US PGM) PCC, 4 *Bracui* (UK *River*) PCC

PATROL, INSHORE 16

16 *Tracker* PCI<

PATROL, RIVERINE 5

3 *Roraima* PCR and 2 *Pedro Teixeira* PCR

MINE WARFARE

MINELAYERS 0 but both SSK classes can lay mines

MINE COUNTERMEASURES 6

6 *Aratü* (Ge *Schütze*) MSC

AMPHIBIOUS 4

2 *Ceara* (US *Thomaston*) LSD capacity 345 tps, 21 LCM or 6 LCM and 3 LCUs

1 *Duque de Caxais* (US *de Soto County*) LST, capacity 575 tps, 75 tons veh, 4 LCVPs

1 *Mattoso Maia* (US *Newport* LST) capacity 400 tps, 500 tons veh, 3 LCVP, 1 LCPL

Plus some 48 craft: 3 LCU, 10 LCM, 35 LCVP

SUPPORT AND MISCELLANEOUS 25

1 AO; 1 river gp of 1 AOT, 1 AK, 1 AF; 1 AK, 3 trp tpt; 2 AH, 1 ASR, 5 ATF, 4 AG; 2 polar AGOR, 2 AGOR, 1 AGHS plus 6 craft

NAVAL AVIATION (1,150)

22 cbt ac, 54 armed hel

FGA 22 A-4/TA-4 (being delivered)

ASW 6 SH-3B, 7 SH-3D, 6 SH-3G/H

ATTACK 14 *Lynx* MK-21A

UTL 2 sqn with 5 AS-332, 12 AS-350 (armed), 9 AS-355 (armed)

TRG 1 hel sqn with 13 TH-57

ASM AS-11, AS-12, *Sea Skua*

MARINES (13,900)

FLEET FORCE 1 amph div (1 comd, 3 inf bn, 1 arty gp)

REINFORCEMENT COMD 5 bn incl 1 engr, 1 SF

INTERNAL SECURITY FORCE 8+ regional gp

EQUIPMENT

RECCE 6 EE-9 Mk IV *Cascavel*

AAV 11 LVTP-7A1, 13 AAV-7A1

APC 28 M-113, 5 EE-11 *Urutu*

TOWED ARTY 105mm: 15 M-101, 18 L-118; **155mm**: 6 M-114

MOR 81mm; 120mm: 8 K 6A3

ATGW RB-56 *Bill*

RL 89mm: 3.5in M-20

RCL 106mm: 8 M-40A1

AD GUNS 40mm: 6 L/70 with BOFI

Air Force 50,000

(incl 5,000 conscripts); 268 cbt ac, 29 armed hel

AIR DEFENCE COMMAND 1 gp

FTR 2 sqn with 18 *Mirage* F-103E/D (14 *Mirage* IIIE/4 DBR)

TACTICAL COMMAND 10 gp

FGA 3 sqn with 47 F-5E/-B/-F, 37 AMX

CCT 2 sqn with 53 AT-26 (EMB-326) - 33 to be upgraded

RECCE 2 sqn with 4 RC-95, 10 RT-26, 12 *Learjet* 35 recce/VIP, 3 RC-130E

SURVEILLANCE/CALIBRATION First of 4 *Hawker* 800XP delivered for Amazon inspection/ATC calibration

LIAISON/OBS 7 sqn

1 with **ac** 8 T-27

5 with **ac** 31 U-7

1 with **hel** 29 UH-1H (armed)

MARITIME COMMAND 4 gp

MR/SAR 3 sqn with 10 EMB-110B, 20 EMB-111

TRANSPORT COMMAND

6 gp (6 sqn)

1 with 9 C-130H, 2 KC-130H • 1 with 4 KC-137 (tpt/tkr) • 1 with 12 C-91 • 1 with 17 C-95A/B/C • 1 with 17 C-115 • 1 (VIP) with **ac** 1 VC-91, 12 VC/VU-93, 2 VC-96, 5 VC-97, 5 VU-9, 2 Boeing 737-200 **hel** 3 VH-4

7 regional sqn with 7 C-115, 86 C-95A/B/C, 6 EC-9 (VU-9)

HEL 6 AS-332, 8 AS-355, 4 Bell 206, 27 HB-350B

LIAISON 50 C-42, 3 Cessna 208, 30 U-42

TRAINING COMMAND

AC 38* AT-26, 97 C-95 A/B/C, 25 T-23, 98 T-25, 61* T-27 (*Tucano*), 14* AMX-T
HEL 4 OH-6A, 25 OH-13
CAL 1 unit with 2 C-95, 1 EC-93, 4 EC-95, 1 U-93

MISSILES

AAM AIM-9B *Sidewinder*, R-530, *Magic* 2, MAA-1 *Piranha*

Forces Abroad

UN AND PEACEKEEPING

CROATIA (UNMOP): 1 obs **EAST TIMOR** (UNTAET): 10 obs, 71 tps

Paramilitary

PUBLIC SECURITY FORCES (R) some 385,600

in state mil pol org (state militias) under Army control and considered Army Reserve

Chile — Chl

Chilean peso pCh		1998	1999	2000	2001
GDP	pCh	33.5tr	34.3tr		
	US$	79bn	67bn		
per capita	US$	12,200	12,200		
Growth	%	3.3	-1.1		
Inflation	%	5.3	3.3		
Debt	US$	36bn	34bn		
Def exp[a]	pCh	1,386bn	1,371bn		
	US$	3.0bn	2.7bn		
Def bdgt	pCh	969bn	1,033bn	1,096bn	
	US$	2.1bn	2.0bn	2.1bn	
FMA (US)	US$	0.5m	0.5m	0.5m	0.5m
US$1=pCh		460	509	515	

[a] Incl spending on paramilitary forces

Population			15,088,000
Age	13–17	18–22	23–32
Men	712,000	636,000	1,221,000
Women	685,000	614,000	1,192,000

Total Armed Forces

ACTIVE 87,000

(incl 30,600 conscripts)

Terms of service **Army** 1 year **Navy** and **Air Force** 22 months

RESERVES 50,000

Army 50,000

Army 51,000

(incl 27,000 conscripts)
7 Mil Regions, 2 Corps HQ
7 div; org, composition varies; comprise
23 inf (incl 10 mtn, 13 mot), 10 armd cav, 8 arty, 6 engr regt
Army tps: 1 avn bde, 1 engr, 1 AB regt (1 AB, 1 SF bn)

EQUIPMENT

MBT 51 AMX-30, 200 *Leopard* 1
RECCE 50 EE-9 *Cascavel*
AIFV 20 MOWAG *Piranha* with **90mm** gun, some M-113C/-R
APC 355 M-113, 180 Cardoen/MOWAG *Piranha*, 30 EE-11 *Urutu*
TOWED ARTY 105mm: 66 M-101, 54 Model 56; **155mm**: 12 M-71
SP ARTY 155mm: 12 Mk F3
MOR 81mm: 300 M-29; **107mm**: 15 M-30; **120mm**: 125 FAMAE (incl 50 SP)
ATGW *Milan/Mamba, Mapats*
RL 89mm: 3.5in M-20
RCL 150 incl: **57mm**: M-18; **106mm**: M-40A1
AD GUNS 20mm: 60 incl some SP (Cardoen/MOWAG)
SAM 50 *Blowpipe, Javelin*, 12 *Mistral*

AIRCRAFT

TPT 6 C-212, 1 *Citation* (VIP), 5 CN-235, 4 DHC-6, 4 PA-31, 8 PA-28 Piper *Dakota*, 3 Cessna-208 *Caravan*
TRG 16 Cessna R-172, 8 Cessna R-182
HEL 2 AB-206, 3 AS-332, 15 Enstrom 280 FX, some Hughes MD-530F (armed trg), 10 SA-315, 12 SA-330

Navy 24,000

(incl 600 Naval Aviation, 2,700 Marines, 1,300 Coast Guard and 2,100 conscripts)

DEPLOYMENT AND BASES

MAIN COMMAND Fleet (includes DD and FF), SS flotilla, tpt. Remaining forces allocated to 4 Naval Zones **1st** 26°S–36°S approx: Valparaiso (HQ) **2nd** 36°S–43°S approx: Talcahuano (HQ), Puerto Montt **3rd** 43°S to Antarctica: Punta Arenas (HQ), Puerto Williams **4th** north of 26°S approx: Iquique (HQ)

SUBMARINES 3

SSK 3
1 *O'Brien* (UK *Oberon*) with 8 × 533mm TT (Ge HWT)
2 *Thompson* (Ge T-209/1300) with 8 × 533mm TT (HWT)

PRINCIPAL SURFACE COMBATANTS 5

DESTROYERS 2

DDG 2
1 *Prat* (UK *Norfolk*) with 4 MM-38 *Exocet* SSM, 1 × 2 *Seaslug* SAM, 1 × 2 × 114mm guns, 2 × 3 ASTT (Mk 44 LWT), 1 AB-206B hel
1 *Blanco Encalada* (UK *Norfolk*) with 2 × 8 *Barak 1* SAM, 2 × 114mm guns, 2 × 3 ASTT (Mk 44 LWT), 2

AS-332F hel

FRIGATES 3

FFG 3 *Condell* (mod UK *Leander*), with 2 × 114mm guns, 2 × 3 ASTT (Mk 44 LWT); 1 with 2 × 2 MM 38 *Exocet* SSM and 1 AB 206B hel; 2 with 2 × 2 MM 40 *Exocet* SSM and AS-332F hel

PATROL AND COASTAL COMBATANTS 26

MISSILE CRAFT 9

3 *Casma* (Il *Sa'ar* 4) PFM with 4 *Gabriel* SSM, 2 × 76mm gun

2 *Iquique* (Il *Sa'ar* 3) PFM with 6 *Gabriel* SSM, 1 × 76mm gun

4 *Tiger* (Ge Type 148) PFM with 4 *Exocet* SSM, 1 × 76mm gun

PATROL, OFFSHORE 3

3 *Micalvi* PCO

PATROL, COASTAL 4

4 *Guacolda* (Ge Lürssen 36m) PCC

PATROL, INSHORE 10

10 *Grumete Diaz* (Il *Dabur*) PCI<

AMPHIBIOUS 3

2 *Maipo* (Fr *Batral*) LST, capacity 140 tps, 7 tk

1 *Valdivia*† (US *Newport*) LST, capacity 400 tps, 500t vehicles (non-op)

Plus craft: 2 *Elicura* LSM, 1 *Pisagua* LCU

SUPPORT AND MISCELLANEOUS 12

1 *Araucano* AO, 1 AK; 1 tpt, 2 AG; 1 training ship, 4 ATF; 1 AGOR, 1 AGHS

NAVAL AVIATION (600)

no cbt ac, 20 armed hel

MR 1 sqn with 1* EMB-110, 2* P-3A *Orion*, 8 Cessna *Skymaster* (plus 2 in store)

ASW HEL 1 sqn with 6 AS-532 (4 with AM-39 *Exocet*, 2 with torp)

LIAISON 1 sqn with 3 C-212A

TPT 1 sqn with 2 P-3A *Orion*

HEL 1 sqn with 8 BO-105, 6 UH-57

TRG 1 sqn with 10 PC-7

ASM AM-39 *Exocet*

MARINES (2,700)

4 gp: 4 inf, 2 trg bn, 4 cdo coy, 4 fd arty, 1 SSM bty, 4 AD arty bty • 1 amph bn

EQUIPMENT

LT TK 30 *Scorpion*

APC 40 MOWAG *Roland*

TOWED ARTY 105mm: 16 KH-178, **155mm**: 28 G-5

MOR 81mm: 50

SSM *Excalibur*

RCL 106mm: ε30 M-40A1

SAM *Blowpipe*

COAST GUARD (1,300)

(integral part of the Navy)

PATROL CRAFT 23

2 *Alacalufe* PCC, 15 *Rodman* PCI, 6 PCI, plus about 30 boats

Air Force 12,000

(incl 1,500 conscripts); 88 cbt ac, no armed hel

Flying hours: 100

5 Air Bde, 5 wg

FGA 2 sqn

1 with 15 *Mirage* 5BA (MRIS), 6 *Mirage* BD (MRIS)

1 with 16 F-5 (13 -E, 3 -F)

CCT 2 sqn with 24 A-37B (12 being progresively retired), 8 A-36

FTR/RECCE 1 sqn with 15 *Mirage* 50 (8 -FCH, 6 -CH, 1 -DCH), 4 *Mirage* 5-BR

RECCE 2 photo units with 1 *King Air* A-100, 2 *Learjet* 35A

AEW 1 IAI-707 *Phalcon* ('*Condor*')

TPT ac 3 Boeing 707(2 tpt, 1 tkr), 1 Boeing 737-500 (VIP), 2 C-130H, 4 C-130B, 4 C-212, 9 Beech 99 (ELINT, tpt, trg), 14 DHC-6 (5 -100, 9 -300), 1 *Gulfstream* III (VIP), 1 *Beechcraft* 200 (VIP), 1 Cessna 206 (amph)

HEL 1 S-70A (likely to be sold as unsuited to Air Force requirements), 9 UH-1H (5 of which abandoned in Iraq), 2 Bell 412 (first of 10–12 planned to replace UH-1H), 12 UH-60, 8 Bo-105, 5 SA-315B

TRG 1 wg, 3 flying schools **ac** 12 PA-28, 19 T-35A/B, 11 T-36, 12 T-37B/C, 5 *Extra* 300 **hel** 2 Bell 206A

MISSILES

ASM AS-11/-12

AAM AIM-9B *Sidewinder*, *Shafrir*, *Python* III

AD 1 regt (5 gp) with **20mm**: S-639/-665, GAI-CO1 twin; **35mm**: Oerlikon GDF-005, MATRA *Mistral*, *Mygalle*

Forces Abroad

UN AND PEACEKEEPING

EAST TIMOR (UNTAET): 33 **INDIA/PAKISTAN** (UNMOGIP): 4 obs **MIDDLE EAST** (UNTSO): 3 obs

Paramilitary 29,500

CARABINEROS (Ministry of Defence) 29,500

13 zones, 39 districts

APC 20 MOWAG *Roland*

MOR 60mm, 81mm

AC 22 Cessna (6 C-150, 10 C-182, 6 C-206), 1 *Metro*

HEL 2 Bell 206, 12 Bo-105

Opposition

FRENTE PATRIOTICO MANUEL RODRIGUEZ – AUTONOMOUS FACTION (FPMR-A) ε800

leftist

Colombia Co

Colombian peso pC		1998	1999	2000	2001
GDP	pC	146tr	160tr		
	US$	79bn	77bn		
per capita	US$	6,100	5,800		
Growth	%	2.3	-5.0		
Inflation	%	19.0	11.2		
Debt	US$	33.2bn	34.4bn		
Def exp	pC	3.6tr	3.8tr		
	US$	2.5bn	2.2bn		
Def bdgt	pC	3.6tr	3.7tr	4.0tr	
	US$	2.5bn	2.1bn	2.0bn	
FMA (US)	US$	44m	210m	820m	265m
US$1=pC		1,426	1,756	2,005	

Population			42,400,000
Age	13–17	18–22	23–32
Men	1,956,000	1,882,000	3,313,000
Women	1,866,000	1,809,000	3,257,000

Total Armed Forces

ACTIVE 153,000

(incl some 74,700 conscripts)
Terms of service 12–18 months, varies (all services)

RESERVES 60,700

(incl 2,000 first-line) **Army** 54,700 **Navy** 4,800 **Air Force** 1,200

Army 130,000

(incl 63,800 conscripts)
5 div HQ
17 bde
6 mech each with 3 inf, 1 mech cav, 1 arty, 1 engr bn
2 air-portable each with 2 inf bn
9 inf (8 with 2 inf bn, 1 with 4 inf bn)
2 arty bn
Army tps
3 Mobile Counter Guerrilla Force (bde) (each with 1 cdo unit, 4 bn) – 2 more forming
2 trg bde with 1 Presidential Guard, 1 SF, 1 AB, 1 mech, 1 arty, 1 engr bn
1 AD arty bn
1 army avn 'bde'

EQUIPMENT

LT TK 30 M-3A1 (in store)
RECCE 12 M-8, 8 M-20, 127 EE-9 *Cascavel*
APC 80 M-113, 76 EE-11 *Urutu*, 4 RG-31 *Nyala*
TOWED ARTY 75mm: 30 M-116; **105mm**: 130 M-101
MOR 81mm: 125 M-1; **107mm**: ε100 M-30; **120mm**: 120 Brandt
ATGW 20 TOW (incl 8 SP)
RCL 106mm: 65 M-40A1
AD GUNS 40mm: 30 Bofors
HEL 6 OH-6A, 12 UH-60

Navy (incl Coast Guard) 15,000

(incl 8,500 Marines, 100 Naval Aviation and 7,000 conscripts)
BASES Ocean Cartagena (main), Buenaventura, Málaga (Pacific) **River** Puerto Leguízamo, Barrancabermeja, Puerto Carreño (tri-Service Unified Eastern Command HQ), Leticia, Puerto Orocue, Puerto Inirida

SUBMARINES 4

SSK 2 *Pijao* (Ge T-209/1200) with 8 × 533mm TT (Ge HWT)
SSI 2 *Intrepido* (It SX-506) (SF delivery)

CORVETTES 4

4 *Almirante Padilla* with 8 MM-40 *Exocet* SSM, 1 × 76mm gun, 2 × 3 ASTT, 1 Bo-105 hel

PATROL AND COASTAL COMBATANTS 27

PATROL, OFFSHORE 5
2 *Pedro de Heredia* (ex-US tugs) PCO with 1 × 76mm gun, 2 *Lazaga* PCO, 1 *Esperanta* (Sp *Cormoran*) PFO
PATROL, COASTAL/INSHORE 9
1 *Quito Sueno* (US *Asheville*) PFC with 1 × 76mm gun, 2 *Castillo Y Rada* PCC, 2 *José Garcia* PCC, 2 *José Palas* PCI, 2 *Jaime Gomez* PCI
PATROL, RIVERINE 13
3 *Arauca* PCR, 10 *Diligente* PCR, plus 76 craft: 9 *Tenerife*, 5 *Rio Magdalena*, 20 *Delfin*, 42 *Pirana*

SUPPORT AND MISCELLANEOUS 7

1 tpt; 1 AH, 1 sail trg; 2 AGOR, 2 AGHS

MARINES (8,500)

2 bde (each of 2 bn), 1 amph aslt, 1 river ops (15 amph patrol units), 1 SF, 1 sy bn
No hy eqpt (to get EE-9 *Cascavel* recce, EE-11 *Urutu* APC)

NAVAL AVIATION (100)

AC 2 *Commander*, 2 PA-28, 2 PA-31, 2 *Cessna* 206
HEL 2 Bo-105, 2 AS 555SN *Fenec*

Air Force 8,000

(some 3,900 conscripts); 72 cbt ac, 60 armed hel

AIR COMBAT COMMAND

FGA 2 sqn
1 with 12 *Mirage* 5, 1 with 13 *Kfir* (11 -C2, 2 -TC2)

TACTICAL AIR SUPPORT COMMAND

CBT ac 1 AC-47, 2 AC-47T, 3 IA-58A, 22 A-37B, 6 AT-27
UTILITY/ARMED HEL 12 Bell 205, 5 Bell 212, 2 Bell 412, 2 UH-1B, 13 UH-60A/L, 7 S-70 being delivered, 11 MD-500ME, 2 MD-500D, 3 MD-530F, 10 Mi-17
RECCE 8 *Schweizer* SA 2-37A, 13* OV-10, 3 C-26

MILITARY AIR TRANSPORT COMMAND

AC 1 Boeing 707, 2 Boeing 727, 14 C-130B, 2 C-130H, 1 C-117, 2 C-47, 2 CASA 212, 2 *Bandeirante*, 1 F-28, 3 CN-235
HEL 17 UH-1H

AIR TRAINING COMMAND

AC 14 T-27 (*Tucano*), 3 T-34M, 13 T-37, 8 T-41
HEL 2 UH-1B, 4 UH-1H, 12 F-28F

MISSILES

AAM AIM-9 *Sidewinder*, R-530

Forces Abroad

UN AND PEACEKEEPING

EGYPT (MFO) 358: 1 inf bn

Paramilitary 95,000

NATIONAL POLICE FORCE 95,000

ac 5 OV-10A, 12 Gavilan, 11 *Turbo Thrush* **hel** 11 Bell-206L, 7 Bell-212, 2 Hughes 500D, 49 UH-1H, 3 UH-60L

COAST GUARD

integral part of Navy

Opposition

***COORDINADORA NACIONAL GUERRILLERA SIMON BOLIVAR* (CNGSB)** loose coalition of guerrilla gp incl **Revolutionary Armed Forces of Colombia (FARC)** up to 17,000 reported active; **National Liberation Army (ELN)** ε5,000, pro-Cuban; **People's Liberation Army (EPL)** ε500

Other Forces

UNITED SELF DEFENCE FORCE OF COLOMBIA (AUC): ε5,000 right-wing paramilitary group

Costa Rica CR

colon C		1998	1999	2000	2001
GDP	C	2.7tr	3.2tr		
	US$	10.6bn	11.3bn		
per capita	US$	6,900	7,500		
Growth	%	6.8	8.3		
Inflation	%	11.6	10.0		
Debt	US$	3.6bn	3.8bn		
Sy exp[a]	C	17.6bn	19.8bn		
	US$	69m	69m)		
Sy bdgt[a]	C	17.6bn	19.8bn	25.6bn	
	US$	69m	69m	86m	
FMA (US)	US$	0.2m	0.2m	0.2m	0.2m
US$1=C		57	86	297	

[a] No defence forces. Budgetary data are for border and maritime policing and internal security.

Population			3,722,000
Age	13–17	18–22	23–32
Men	195,000	176,000	300,000
Women	187,000	170,000	291,000

Total Armed Forces

ACTIVE Nil

Paramilitary 8,400

CIVIL GUARD 4,400

7 urban *comisaria* (reinforced coy) • 1 tac police *comisaria* • 1 special ops unit • 6 provincial *comisaria*

BORDER SECURITY POLICE 2,000

2 Border Sy Comd (8 *comisaria*)

MARITIME SURVEILLANCE UNIT (300)

BASES Pacific Golfito, Punta Arenas, Cuajiniquil, Quepos **Atlantic** Limon, Moin

PATROL CRAFT, COASTAL/INSHORE 10

1 *Isla del Coco* (US *Swift* 32m) PFC
1 *Astronauta* (US *Cape*) PCC
2 *Point* PCI<
6 PCI<; plus about 10 boats

AIR SURVEILLANCE UNIT (300)

No cbt ac
ac 1 Cessna O-2A, 1 DHC-4, 1 PA-31, 1 PA-34, 4 U206G **hel** 2 MD-500E, 1 Mi-17

RURAL GUARD (Ministry of Government and Police) 2,000

8 comd; small arms only

Cuba C

Cuban peso P		1998	1999	2000	2001
GDP	US$	ε14bn	ε15bn		
per capita	US$	2,300	2,400		
Growth	%	1.2	6.2		
Inflation	%	2.7	7.0		
Debt	US$	12bn	12bn		
Def exp	US$	ε750m	ε750m		
Def bdgt	P	630m	650m		
	US$	27m	31m		
US$1=P		23	23	21	

Population			11,320,000
Age	13–17	18–22	23–32
Men	414,000	381,000	1,047,000
Women	388,000	357,000	985,000

Total Armed Forces

ACTIVE ε58,000

Terms of service 2 years

RESERVES

Army 39,000 **Ready Reserves** (serve 45 days per year) to fill out Active and Reserve units; see also *Paramilitary*

Army ε45,000

(incl conscripts and Ready Reserves)
HQ: 3 Regional Comd, 3 Army
4–5 armd bde • 9 mech inf bde (3 mech inf, 1 armd, 1 arty, 1 AD arty regt) • 1 AB bde • 14 reserve bde • 1 frontier bde
AD arty regt and SAM bde
EQUIPMENT † (some 75% in store)
MBT ε900 incl: T-34, T-54/-55, T-62
LT TK some PT-76
RECCE some BRDM-1/-2
AIFV some BMP-1
APC ε700 BTR-40/-50/-60/-152
TOWED ARTY 500: **76mm**: ZIS-3; **122mm**: M-1938, D-30; **130mm**: M-46; **152mm**: M-1937, D-1
SP ARTY 40: **122mm**: 2S1; **152mm**: 2S3
MRL 175: **122mm**: BM-21; **140mm**: BM-14
MOR 1,000: **82mm**: M-41/-43; **120mm**: M-38/-43
STATIC DEF ARTY JS-2 (**122mm**) hy tk, T-34 (**85mm**)
ATGW AT-1 *Snapper*, AT-3 *Sagger*
ATK GUNS 85mm: D-44; **100mm**: SU-100 SP, T-12
AD GUNS 400 incl: **23mm**: ZU-23, ZSU-23-4 SP; **30mm**: M-53 (twin)/BTR-60P SP; **37mm**: M-1939; **57mm**: S-60 towed, ZSU-57-2 SP; **85mm**: KS-12; **100mm**: KS-19
SAM SA-6/-7/-8/-9/-13/-14/-16

Navy ε3,000

(incl 550+ Naval Infantry)
NAVAL DISTRICTS Western HQ Cabanas **Eastern** HQ Holquin
BASES Cienfuegos, Cabanas, Havana, Mariel, Punta Movida, Nicaro
PATROL AND COASTAL COMBATANTS 5†
MISSILE CRAFT 4 Sov *Osa* II
PATROL, COASTAL 1 Sov *Pauk* II PFC with 1 × 76mm gun, 4 ASTT, 2 ASW RL
MINE COUNTERMEASURES 6†
2 Sov *Sonya* MSC, 4 Sov *Yevgenya* MHC
SUPPORT AND MISCELLANEOUS 1
1 AGHS†
NAVAL INFANTRY (550+)
2 amph aslt bn
COASTAL DEFENCE
ARTY 122mm: M-1931/37; **130mm**: M-46; **152mm**: M-1937
SSM 2 SS-C-3 systems, some mobile *Bandera* IV (reported)

Air Force ε10,000

(incl AD and conscripts); 130† cbt ac of which only some 25 are operational, 45 armed hel
Flying hours less than 50
FGA 2 sqn with 10 MiG-23BN
FTR 4 sqn
2 with 30 MiG-21F, 1 with 50 MiG-21bis, 1 with 20 MiG-23MF, 6 MiG-29
(Probably only some 3 MiG-29, 10 MiG-23, 5 MiG-21bis in operation)
ATTACK HEL 45 Mi-8/-17, Mi-25/35
ASW 5 Mi-14 hel
TPT 4 sqn with 8 An-2, 1 An-24, 15 An-26, 1 An-30, 2 An-32, 4 Yak-40, 2 Il-76 (Air Force ac in civilian markings)
HEL 40 Mi-8/-17
TRG 25 L-39, 8* MiG-21U, 4* MiG-23U, 2* MiG-29UB, 20 Z-326
MISSILES
ASM AS-7
AAM AA-2, AA-7, AA-8, AA-10, AA-11
SAM 13 active SA-2, SA-3 sites
CIVIL AIRLINE
10 Il-62, 7 Tu-154, 12 Yak-42, 1 An-30 used as troop tpt

Paramilitary 26,500 active

YOUTH LABOUR ARMY 65,000
CIVIL DEFENCE FORCE 50,000
TERRITORIAL MILITIA (R) ε1,000,000
STATE SECURITY (Ministry of Interior) 20,000
BORDER GUARDS (Ministry of Interior) 6,500
about 20 Sov *Zhuk* and 3 Sov *Stenka* PFI<, plus boats

Foreign Forces

US 1,080: **Navy** 590 **Marines** 490
RUSSIA 810: 800 SIGINT, ε10 mil advisers

Dominican Republic DR

peso República Dominican pRD

		1998	1999	2000	2001
GDP	pRD	242bn	279bn		
	US$	11.1bn	12.2bn		
per capita	US$	5,100	5,500		
Growth	%	7.3	8.2		
Inflation	%	4.6	6.6		
Debt	US$	3.8bn	3.5bn		
Def exp	pRD	ε2.4bn	ε2.5bn		
	US$	113m	114m		
Def bdgt	pRD	1.8bn	2.1bn	2.4bn	
	US$	84m	92m	105m	
FMA (US)	US$	0.6m	0.9m	1.0m	1.1m
US$1=pRD		14.7	16.0	16.3	

Population			**8,269,000**
Age	13–17	18–22	23–32
Men	458,000	414,000	720,000
Women	447,000	404,000	710,000

Total Armed Forces

ACTIVE 24,500

Army 15,000

3 Defence Zones • 4 inf bde (with 8 inf, 1 arty bn, 2 recce sqn) • 1 armd, 1 Presidential Guard, 1 SF, 1 arty, 1 engr bn

EQUIPMENT

LT TK 12 AMX-13 (**75mm**), 12 M-41A1 (**76mm**)
RECCE 8 V-150 *Commando*
APC 20 M-2/M-3 half-track
TOWED ARTY 105mm: 22 M-101
MOR 81mm: M-1; **120mm**: 24 ECIA

Navy 4,000

(incl marine security unit and 1 SEAL unit)
BASES Santo Domingo (HQ), Las Calderas

PATROL AND COASTAL COMBATANTS 12

PATROL, OFFSHORE 5
2 *Cohoes* PCO with 2 × 76mm gun, 1 *Prestol* (US *Admirable*) with 1 × 76mm gun, 1 *Sotoyoma* PCO with 1 × 76mm gun, 1 *Balsam* PCO
PATROL, COASTAL/INSHORE 7
1 *Betelgeuse* (US PGM-71) PCC, 2 *Canopus* PCI<, 4 PCI<

SUPPORT AND MISCELLANEOUS 4
1 AOT (small harbour), 3 AT

Air Force 5,500

10 cbt ac, no armed hel
Flying hours probably less than 60
CCT 1 sqn with 8 A-37B
TPT 1 sqn with 3 C-47, 1 *Commander* 680, 2 C-212-400
LIAISON 1 Cessna 210, 2 PA-31, 3 *Queen Air* 80, 1 *King Air*
HEL 8 Bell 205, 2 SA-318C, 1 SA-365 (VIP), 10 AS 550
TRG 2* AT-6, 6 T-34B, 3 T-41D, 8 T-35B
AB 1 SF (AB) bn
AD 1 bn with 4 **20mm** guns

Paramilitary 15,000

NATIONAL POLICE 15,000

Ecuador Ec

Ecuadorean sucre ES		1998	1999	2000	2001
GDP	ES	107tr	170tr		
	US$	20bn	15bn		
per capita	US$	4,600	4,300		
Growth	%	0.9	-7.0		
Inflation	%	36.1	52.3		
Debt	US$	15.1bn	16.1bn		
Def exp[a]	ES	ε2.9tr	ε4.0tr		
	US$	532m	339m		
Def bdgt[a]	ES	ε2.0tr	ε4.0tr	ε10.0tr	
	US$	367m	339m	400m	
FMA[b] (US)	US$	1.0m	2.8m	2.7m	4.0m
US$1=ES		5,447	11,787	25,000	

[a] incl extra-budgetary funding
[b] MOMEP **1998** εUS$15m

Population			12,884,000
Age	13–17	18–22	23–32
Men	712,000	657,000	1,122,000
Women	691,000	640,000	1,099,000

Total Armed Forces

ACTIVE 57,500

Terms of service conscription 1 year, selective

RESERVES 100,000

Ages 18–55

Army 50,000

4 Defence Zones
1 div with 2 inf bde (each 3 inf, 1 armd, 1 arty bn) • 2 armd bde (3 armd, 1 mech inf, 1 SP arty bn) • 2 inf bde (5 inf, 3 mech inf, 2 arty bn) • 3 jungle bde (2 with 3, 1 with 4 jungle bn)
Army tps: 1 SF (AB) bde (4 bn), 1 AD arty gp, 1 avn gp (4 bn), 3 engr bn

EQUIPMENT

MBT 3 T-55
LT TK 108 AMX-13
RECCE 27 AML-60/-90, 30 EE-9 *Cascavel*, 10 EE-3 *Jararaca*
APC 20 M-113, 80 AMX-VCI, 30 EE-11 *Urutu*
TOWED ARTY 105mm: 50 M2A2, 30 M-101, 24 Model 56; **155mm**: 12 M-198, 12 M-114
SP ARTY 155mm: 10 Mk F3
MRL 122mm: 6 RM-70
MOR 81mm: M-29; **107mm**: 4.2in M-30; **160mm**: 12 Soltam
RCL 90mm: 380 M-67; **106mm**: 24 M-40A1
AD GUNS 14.5mm: 128 ZPU-1/-2; **20mm**: 20 M-1935; **23mm**: 34 ZU-23; **35mm**: 30 GDF-002 twin; **37mm**: 18 Ch; **40mm**: 30 L/70
SAM 75 *Blowpipe*, 90 SA-18 (reported), SA-8

AIRCRAFT

SURVEY 1 Cessna 206, 1 *Learjet* 24D
TPT 1 CN-235, 1 DHC-5, 3 IAI-201, 1 *King Air* 200, 2 PC-6
LIAISON/TRG/OBS 1 Cessna 172, 1 -182

HELICOPTERS

SURVEY 3 SA-315B
TPT/LIAISON 9 AS-332, 4 AS-350B, 1 Bell 214B, 3 SA-315B, 3 SA-330, 30 SA-342

Navy 4,500

(incl 250 Naval Aviation and 1,500 Marines)
BASES Guayaquil (main base), Jaramijo, Galápagos Islands

SUBMARINES 2

SSK 2 *Shyri* (Ge T-209/1300) with 8 × 533mm TT (Ge SUT HWT)

FRIGATES 2

FFG 2 *Presidente Eloy Alfaro* (ex-UK *Leander Batch* II) with 4 MM-38 *Exocet* SSM, 1 206B hel

PATROL AND COASTAL COMBATANTS 11

CORVETTES 6 *Esmeraldas* FSG with 2 × 3 MM-40 *Exocet* SSM, 1 × 4 *Albatros* SAM, 1 × 76mm gun, 6 × 324mm ASTT, hel deck
MISSILE CRAFT 5
3 *Quito* (Ge Lürssen 45m) PFM with 4 MM-38 *Exocet* SSM, 1 × 76mm gun
2 *Manta*† (Ge Lürssen 36m) PFM with 4 *Gabriel* II SSM

AMPHIBIOUS 1

1 *Hualcopo* (US LST-512-1152) LST, capacity 150 tps

SUPPORT AND MISCELLANEOUS 7

2 AOT (small); 1 AE; 2 ATF, 1 sail trg; 1 AGOR

NAVAL AVIATION (250)

LIAISON 1 *Super King Air* 200, 1 *Super King Air* 300, 1 CN-235
TRG 3 T-34C
HEL 4 Bell 206, 2 Bell 412 EP

MARINES (1,500)

3 bn: 2 on garrison duties, 1 cdo (no hy weapons/veh)

Air Force 3,000

78 cbt ac, no armed hel

OPERATIONAL COMMAND

2 wg, 5 sqn
FGA 3 sqn
1 with 8 *Jaguar* S (6 -A(E), 2 -B(E))
1 with 10 *Kfir* C-2 (being modernised to CE standard), 2 TC-2
1 with 20 A-37B
FTR 1 sqn with 13 *Mirage* F-1JE, 1 F-1JB
CCT 4 *Strikemaster* Mk 89A

MILITARY AIR TRANSPORT GROUP

2 civil/military airlines:
TAME 6 Boeing 727, 2 BAe-748, 2 C-130B, 1 C-130H, 1 DHC-6, 1 F-28, 1 L-100-30
ECUATORIANA 3 Boeing 707-320, 1 DC-10-30, 2 A-310
LIAISON 1 *King Air* E90, 1 *Sabreliner*
LIAISON/SAR hel 2 AS-332, 1 Bell 212, 6 Bell-206B, 5 SA-316B, 1 SA-330, 2 UH-1B, 24 UH-1H
TRG incl 22 AT-33*, 20 Cessna 150, 5 C-172, 17 T-34C, 1 T-41

MISSILES

AAM R-550 *Magic*, *Super* 530, *Shafrir*, *Python* 3, *Python* 4
AB 1 AB sqn

Paramilitary 270

COAST GUARD 270

PATROL, COASTAL/INSHORE 4
2 5 *De Agosto* PCC, 1 PGM-71 PCI, 1 *Point* PCI plus some 8 boats

El Salvador — EIS

colon C		1998	1999	2000	2001
GDP	C	135bn	139bn		
	US$	15.4bn	15.9bn		
per capita	US$	2,900	2,900		
Growth	%	3.5	2.6		
Inflation	%	2.5	0.5		
Debt	US$	2.6bn	2.7bn		
Def exp	C	ε1.4bn	ε1.5bn		
	US$	160m	171m		
Def bdgt	C	986m	983m	980m	
	US$	113m	112m	112m	
FMA (US)	US$	0.5m	0.5m	0.5m	0.5m
US$1=C		8.76	8.76	8.76	

Population			**6,179,000**
Age	13–17	18–22	23–32
Men	369,000	354,000	536,000
Women	357,000	343,000	553,000

Total Armed Forces

ACTIVE 16,800

Terms of service selective conscription, 1 year

RESERVES

Ex-soldiers registered

Army ε15,000

(incl 4,000 conscripts)
6 Mil Zones • 6 inf bde (each of 2 inf bn) • 1 special sy bde (4 MP, 2 border gd bn) • 8 inf det (bn) • 1 engr comd (2 engr bn) • 1 arty bde (2 fd, 1 AD bn) • 1 mech

cav regt (2 bn) • 1 special ops gp (1 para bn, 1 naval inf, 1 SF coy)

EQUIPMENT

RECCE 10 AML-90 (2 in store)
APC 40 M-37B1 (mod), 8 UR-416
TOWED ARTY 105mm: 24 M-101 (in store), 36 M-102, 18 M-56
MOR 81mm: incl 300 M-29; **120mm**: 60 UB-M52, M-74 (in store)
RL 94mm: LAW; **82mm**: B-300
RCL 90mm: 400 M-67; **106mm**: 20+ M-40A1 (incl 16 SP)
AD GUNS 20mm: 36 FRY M-55, 4 TCM-20

Navy 700

(incl some 90 Naval Infantry and spt forces)
BASES La Uníon, La Libertad, Acajutla, El Triunfo, Guija Lake

PATROL AND COASTAL COMBATANTS 5

PATROL, COASTAL/INSHORE 5
3 *Camcraft* 30m PCC, 2 PCI<, plus 22 river boats

NAVAL INFANTRY (Marines) (some 90)

1 sy coy

Air Force 1,100

(incl AD and ε200 conscripts); 23 cbt ac, 10 armed hel
Flying hours A-37: 90
CBT AC 1 sqn with 5 A-37B, 4 OA-37B, 1 *Ouragan*, 9 O-2A, 2 O-2B (psyops), 2 CM-170 in store
ARMED HEL 1 sqn with 1 MD-500D, 6 MD-500E, 3 UH-1M, (11 UH-1H in store)
TPT 1 sqn with **ac** 2 C-47, 6 Basler Turbo-67 (3 capable of being converted back to AC-47 gunships), 1 T-41D, 1 Cessna 337G, 1 *Merlin* IIIB, (1 C-123K and 1 OC-6B in store) **hel** 1 sqn with 18 UH-1H tpt hel (incl 4 SAR), (15 UH-1H in store)
TRG 5 *Rallye*, 5 T-35 *Pillan*, **hel** 6 Hughes 269A (of which 4 stored)
AAM *Shafrir*

Forces Abroad

UN AND PEACEKEEPING

WESTERN SAHARA (MINURSO): 2 obs

Paramilitary 12,000

NATIONAL CIVILIAN POLICE (Ministry of Public Security) some 12,000 (to be 16,000)

small arms; **ac** 1 Cessna O-2A **hel** 1 UH-1H, 2 Hughes-520N, 1 MD-500D
10 river boats

Guatemala Gua

quetzal q		1998	1999	2000	2001
GDP	q	121bn	133bn		
	US$	13.3bn	14.1bn		
per capita	US$	3,900	4,000		
Growth	%	4.7	3.5		
Inflation	%	7.0	4.9		
Debt	US$	4.0bn	4.0bn		
Def exp	q	ε1.0bn	ε1.1bn		
	US$	156m	149m		
Def bdgt	q	874m	845m	950m	
	US$	137m	114m	123m	
FMA (US)	US$	2.2m	3.3m	3.2m	3.3m
US$1=q		6.39	7.39	7.71	

Population			**12,224,000**
Age	13–17	18–22	23–32
Men	732,000	628,000	942,000
Women	711,000	612,000	929,000

Total Armed Forces

(National Armed Forces are combined; the Army provides log spt for Navy and Air Force)

ACTIVE ε31,400

(ε23,000 conscripts)
Terms of service conscription; selective, 30 months

RESERVES

Army ε35,000 (trained) **Navy** (some) **Air Force** 200

Army 29,200

(incl ε23,000 conscripts)
15 Mil Zones (22 inf, 1 trg bn, 6 armd sqn) • 2 strategic bde (4 inf, 1 lt armd bn, 1 recce sqn, 2 arty bty) • 1 SF gp (3 coy incl 1 trg) • 2 AB bn • 5 inf bn gp (each 1 inf bn, 1 recce sqn, 1 arty bty) • 1 Presidential Guard bn • 1 engr bn • 1 Frontier Detachment
RESERVES ε19 inf bn

EQUIPMENT

RECCE 7 M-8 (in store), 9 RBY-1
APC 10 M-113 (plus 5 in store), 7 V-100 *Commando*, 30 *Armadillo*
TOWED ARTY 105mm: 12 M-101, 8 M-102, 56 M-56
MOR 81mm: 55 M-1; **107mm**: 12 M-30 (in store); **120mm**: 18 ECIA
RL 89mm: 3.5in M-20 (in store)
RCL 57mm: M-20; **105mm**: 64 Arg M-1974 FMK-1; **106mm**: 56 M-40A1
AD GUNS 20mm: 16 M-55, 16 GAI-DO1

Navy ε1,500

(incl some 650 Marines)
BASES Atlantic Santo Tomás de Castilla **Pacific** Puerto Quetzal

PATROL CRAFT, COASTAL/INSHORE 9

1 *Kukulkan* (US *Broadsword* 32m) PCC, 2 *Stewart* PCI<, 6 *Cutlas* PCI<, plus 6 *Vigilante* boats

PATROL CRAFT, RIVERINE 20 boats

MARINES (some 650)

2 bn (-)

Air Force 700

10† cbt ac, 12 armed hel. Serviceability of ac is less than 50%.

CBT AC 1 sqn with 4 Cessna A-37B, 1 sqn with 6 PC-7
TPT 1 sqn with 4 T-67 (mod C-47 *Turbo*), 2 F-27, 1 *Super King Air* (VIP), 1 PA 301 *Navajo*, 4 Arava 201
LIAISON 1 sqn with 2 Cessna 206, 1 Cessna 310
HEL 1 sqn with 12 armed hel (9 Bell 212, 3 Bell 412), 9 Bell 206, 3 UH-1H, 3 S-76
TRG 6 T-41, 5 T-35B, 5 Cessna R172K

TACTICAL SECURITY GROUP (Air Military Police)

3 CCT coy, 1 armd sqn, 1 AD bty (Army units for air-base sy)

Paramilitary 19,000 active

NATIONAL POLICE 19,000

21 departments, 1 SF bn, 1 integrated task force (incl mil and treasury police)

TREASURY POLICE (2,500)

Guyana Guy

Guyanan dollar G$		1998	1999	2000	2001
GDP	G$	111bn	119bn		
	US$	740m	774m		
per capita	US$	3,200	3,300		
Growth	%	-3.0	2.0		
Inflation	%	5.0	5.0		
Debt	US$	1.5bn	1.5bn		
Def exp	G$	1.2bn	1.2bn		
	US$	8m	7m		
Def bdgt	G$	ε850m	ε900m	ε950m	
	US$	6m	5m	5m	
FMA (US)	US$	0.2m	0.3m	0.3m	0.3m
US$1=G$		151	178	181	

Population			846,000
Age	13–17	18–22	23–32
Men	43,000	39,000	78,000
Women	41,000	37,000	74,000

Total Armed Forces

ACTIVE (combined **Guyana Defence Force**) some 1,600

RESERVES some 1,500
People's Militia (see *Paramilitary*)

Army 1,400

(incl 500 Reserves)
1 inf bn, 1 SF, 1 spt wpn, 1 engr coy

EQUIPMENT

RECCE 3 Shorland, 6 EE-9 *Cascavel* (reported)
TOWED ARTY 130mm: 6 M-46
MOR 81mm: 12 L16A1; **82mm**: 18 M-43; **120mm**: 18 M-43

Navy 100

(plus 170 reserves)
BASES Georgetown, New Amsterdam
2 boats

Air Force 100

no cbt ac, no armed hel
TPT ac 1 BN-2A, 1 *Skyvan* 3M **hel** 1 Bell 206, 1 Bell 412

Paramilitary

GUYANA PEOPLE'S MILITIA (GPM) some 1,500

Haiti RH

gourde G		1998	1999	2000	2001
GDP	G	59bn	66bn		
	US$	3.5bn	3.9bn		
per capita	US$	1,100	1,100		
Growth	%	3.1	2.5		
Inflation	%	12.7	8.7		
Debt	US$	1,086m	1,195m		
Sy exp	G	ε800m	ε850m		
	US$	48m	50m		
Sy bdgt	G	ε800m	ε850m	ε900m	
	US$	48m	50m	49m	
FMA[a] (US)	US$	0.3m	4.0m	7.3m	4.8m
US$1=G		16.8	16.9	18.3	

[a] UN **1998** US$18m **1999** US$19m

Population			7,682,000
Age	13–17	18–22	23–32
Men	424,000	375,000	615,000
Women	414,000	369,000	614,000

Total Armed Forces

ACTIVE Nil

Caribbean and Latin America

Paramilitary

In 1994, the military government of Haiti was replaced by a civilian administration. The former armed forces and police were disbanded and an Interim Public Security Force (IPSF) of 3,000 formed. A National Police Force of ε5,300 personnel has now been formed. All Army equipment has been destroyed.

The United Nations Civilian Police Mission in Haiti (MIPONUH) maintains some 285 civ pol to assist the government of Haiti by supporting and contributing to the professionalisation of the National Police Force.

NAVY (Coast Guard) 30

BASE Port-au-Prince

PATROL CRAFT boats only

AIR FORCE (disbanded in 1995)

Foreign Forces

US USMC 230

Honduras Hr

lempira L		1998	1999	2000	2001
GDP	L	70.2bn	76.6bn		
	US$	5.3bn	5.4bn		
per capita	US$	2,200	2,200		
Growth	%	3.0	-1.9		
Inflation	%	13.7	11.6		
Debt	US$	4.4bn	4.5bn		
Def exp	L	ε1,300m	ε1,350m		
	US$	97m	95m		
Def bdgt	L	ε479m	ε500m	ε520m	
	US$	36m	35m	35m	
FMA (US)	US$	0.5m	0.5m	0.5m	0.5m
US$1=L		13.40	14.10	14.68	

Population			6,795,000
Age	13–17	18–22	23–32
Men	402,000	345,000	556,000
Women	389,000	335,000	545,000

Total Armed Forces

ACTIVE 8,300

RESERVES 60,000

Ex-servicemen registered

Army 5,500

6 Mil Zones

4 inf bde

3 with 3 inf, 1 arty bn • 1 with 3 inf bn

1 special tac gp with 1 inf (AB), 1 SF bn

1 armd cav regt (2 mech bn, 1 lt tk, 1 recce sqn, 1 arty, 1 AD arty bty)

1 engr bn

1 Presidential Guard coy

RESERVES

1 inf bde

EQUIPMENT

LT TK 12 *Scorpion*

RECCE 3 *Scimitar*, 1 *Sultan*, 50 *Saladin*, 13 RBY-1

TOWED ARTY 105mm: 24 M-102; **155mm**: 4 M-198

MOR 60mm; **81mm**; **120mm**: 60 FMK; **160mm**: 30 *Soltam*

RL 84mm: 120 *Carl Gustav*

RCL 106mm: 80 M-40A1

AD Guns 20mm: 24 M-55A2, 24 TCM-20

Navy 1,000

(incl 400 Marines)

BASES Atlantic Puerto Cortés, Puerto Castilla **Pacific** Amapala

PATROL CRAFT, COASTAL/INSHORE 10

3 *Guaymuras* (US *Swiftship* 31m) PFC

2 *Copan* (US *Guardian* 32m) PFI<

5 PCI<, plus 28 riverine boats

AMPHIBIOUS craft only

1 *Punta Caxinas* LCT

MARINES (400)

3 indep coy (-)

Air Force 1,800

50 cbt ac, no armed hel

FGA 2 sqn

1 with 13 A-37B (3 C-101 in store)

1 with 11 F-5E/F

FTR 11 *Super Mystère* B2

TPT 5 C-47, 3 C-130A, 1 IAI-201, 2 IAI-1123

LIAISON 1 sqn with 3 Cessna 172, 2 C-180, 2 C-185, 3 *Commander*, 1 PA-31, 1 PA-34

HEL 9 Bell 412, 3 Hughes 500, 6 UH-1B/H, 1 S-76

TRG 4* C-101CC, 6 U-17A, 11* EMB-312, 5 T-41A

AAM *Shafrir*

Forces Abroad

UN AND PEACEKEEPING

WESTERN SAHARA (MINURSO): 12 obs

Paramilitary 6,000

PUBLIC SECURITY FORCES (Ministry of Public Security and Defence) 6,000

11 regional comd

Foreign Forces

US 410: **Army** 160 **Marines** 70 **Air Force** 180

Jamaica Ja

Jamaican dollar J$		1998	1999	2000	2001
GDP	J$	251bn	259bn		
	US$	6.9bn	6.6bn		
per capita	US$	3,500	3,500		
Growth	%	-1.9	-1.0		
Inflation	%	8.6	5.9		
Debt	US$	4.1bn	3.8bn		
Def exp	J$	1.6bn	2.0bn		
	US$	44m	51m		
Def bdgt	J$	1.6bn	2.0bn	2.1bn	
	US$	44m	51m	50m	
FMA (US)	US$	1.1m	1.7m	1.7m	2.3m
US$1=J$		36.6	39.0	41.8	

Population			**2,512,000**
Age	13–17	18–22	23–32
Men	121,000	119,000	226,000
Women	121,000	116,000	226,000

Total Armed Forces

ACTIVE (combined **Jamaican Defence Force**) some 2,830

RESERVES some 953

Army 877 **Coast Guard** 60 **Air Wing** 16

Army 2,500

2 inf, 1 spt bn, 1 engr regt (4 sqn)

EQUIPMENT

APC 13 V-150 *Commando* (all reported non-op)

MOR 81mm: 12 L16A1

RESERVES

1 inf bn

Coast Guard 190

BASE Port Royal, out stations at Discovery Bay and Pedro Cays

PATROL CRAFT 4

COASTAL/INSHORE 4

1 *Fort Charles* (US 34m) PFC, 1 *Paul Bogle* (US-31m) PFI<, 2 *Point* PCI<

plus 4 craft and boats

Air Wing 140

no cbt ac, no armed hel

AC 2 BN-2A, 1 Cessna 210, 1 *King Air*

HEL 4 Bell 206, 3 Bell 212, 3 UH-1H, 4 AS-355

Mexico Mex

new peso NP		1998	1999	2000	2001
GDP	NP	3.8tr	4.6tr		
	US$	415bn	484bn		
per capita	US$	7,900	8,200		
Growth	%	4.8	3.7		
Inflation	%	15.9	16.6		
Debt	US$	161bn	162bn		
Def exp[a]	NP	35bn	41bn		
	US$	3.8bn	4.3bn		
Def bdgt	NP	20.1bn	23.2bn	28.4bn	
	US$	2.2bn	2.4bn	3.0bn	
FMA (US)	US$	6m	6m	9m	11m
US$1=NP		9.14	9.56	9.41	

[a] Incl spending on paramilitary forces.

Population	**104,000,000** (Chiapas region 4%)		
Age	13–17	18–22	23–32
Men	5,287,000	4,893,000	8,971,000
Women	5,136,000	4,791,000	8,975,000

Total Armed Forces

ACTIVE 192,770

(60,000 conscripts)

Terms of service 1 year conscription (4 hours per week) by lottery

RESERVES 300,000

Army 144,000

(incl ε60,000 conscripts)

12 Mil Regions

44 Zonal Garrisons with 81 inf bn (1 mech), 19 mot cav, 3 arty regt plus 1 air-mobile SF unit per Garrison

3 Corps HQ each with 3 inf bde

STRATEGIC RESERVE

4 armd bde (each 2 armd recce, 1 arty regt, 1 mech inf bn, 1 ATK gp)

1 AB bde (3 bn)

1 MP bde (3 MP bn, 1 mech cav regt)

EQUIPMENT

RECCE 40 M-8, 119 ERC-90F *Lynx*, 40 VBL, 25 MOWAG, 40 MAC-1

APC 40 HWK-11, 32 M-2A1 half-track, 40 VCR/TT, 24 DN-3, 40 DN-4 *Caballo*, 70 DN-5 *Toro*, 495 AMX-VCI, 95 BDX, 26 LAV-150 ST, some BTR-60 (reported)

TOWED ARTY 75mm: 18 M-116 pack; **105mm**: 16 M-2A1/M-3, 80 M-101, 80 M-56

SP ARTY 75mm: 5 DN-5 *Bufalo*

MOR 81mm: 1,500; **120mm**: 75 Brandt

ATGW *Milan* (incl 8 VBL)
RL 82mm: B-300
ATK GUNS 37mm: 30 M-3
AD GUNS 12.7mm: 40 M-55; **20mm**: 40 GAI-BO1
SAM RBS-70

Navy 37,000

(incl 1,100 Naval Aviation and 8,600 Marines)
NAVAL COMMANDS: Gulf, Pacific
NAVAL ZONES Gulf 6 **Pacific** 11
BASES Gulf Vera Cruz (HQ), Tampico, Chetumal, Ciudad del Carmen, Yukalpetén, Lerna, Frontera, Coatzacoalcos, Isla Mujéres **Pacific** Acapulco (HQ), Ensenada, La Paz, San Blas, Guaymas, Mazatlán, Manzanillo, Salina Cruz, Puerto Madero, Lázaro Cárdenas, Puerto Vallarta

PRINCIPAL SURFACE COMBATANTS 11
DESTROYERS 3
DD 3
2 *Ilhuicamina (*ex-*Quetzalcoatl*) (US *Gearing*) with 2 × 2 127mm guns, 1 Bo-105 hel
1 *Cuitlahuac* (US *Fletcher*) with 5 × 127mm guns, 5 × 533mm ASTT
FRIGATES 8
FF 8
2 *Knox* with 1 × 127mm gun, 4 × 324mm ASTT, 2 × 8 ASROC SUGW, 1 × Bo 105 hel
2 *H. Galeana* (US *Bronstein*) with 6 × 324mm ASTT, ASROC SUGW
3 *Hidalgo* (US *Lawrence/Crosley*) with 1 × 127mm gun
1 *Comodoro Manuel Azueta* (US *Edsall*) (trg) with 2 × 76mm gun

PATROL AND COASTAL COMBATANTS 109
PATROL, OFFSHORE 44
4 *Holzinger 2000* PCO with MD 902 hel
4 *S. J. Holzinger* (ex-*Uxmal*) (imp *Uribe*) PCO with Bo-105 hel
6 *Uribe* (Sp *'Halcon'*) PCO with Bo-105 hel
11 *Negrete* (US *Admirable* MSF) PCO with 1 Bo-105 hel
17 *Leandro Valle* (US *Auk* MSF) PCO
1 *Guanajuato* PCO with 2 × 102mm gun
1 *Centenario* PCO
PATROL, COASTAL 41
31 *Azteca* PCC
3 *Cabo* (US *Cape Higgon*) PCC
7 *Tamiahua* (US *Polimar*) PCC
PATROL, INSHORE 6
4 *Isla* (US *Halter*) XFPCI<
2 *Punta* (US *Point*) PCI<
PATROL, RIVERINE 18<, plus boats

AMPHIBIOUS 3
2 *Panuco* (US-511) LST
1 *Grijalva* (US-511) LST

SUPPORT AND MISCELLANEOUS 19
1 AOT; 4 AK, 2 log spt; 6 AT/F, 1 sail trg; 2 AGHS, 3 AGOR

NAVAL AVIATION (1,100)
9 cbt ac, no armed hel
MR 1 sqn with 9* C-212-200M
MR HEL 12 Bo-105 (8 afloat), 10 MD Explorer
TPT 1 C-212, 2 C-180, 3 C-310, 1 DHC-5, 1 FH-227, 1 *King Air* 90, 1 *Learjet* 24, 1 *Commander*, 2 C-337, 2 C-402, 3 An-32
HEL 3 Bell 47, 4 SA-319, 20 UH-1H, 20 Mi-8/17, 4 AS-555
TRG ac 8 Cessna 152, 10 F-33C *Bonanza*, 10 L-90 *Redigo* **hel** 4 MD-500E

MARINES (8,600)
3 marine bde (each 3 bn), 1 AB bde (3 bn) • 1 Presidential Guard bn • 14 regional bn • 1 Coast def gp: 2 coast arty bn • 1 indep sy coy
EQUIPMENT
AAV 25 VAP-3550
TOWED ARTY 105mm: 16 M-56
MRL 51mm: 6 *Firos*
MOR 100 incl **60mm, 81mm**
RCL 106mm: M-40A1
AD GUNS 20mm: Mk 38; **40mm**: Bofors

Air Force 11,770

107 cbt ac, 71 armed hel
FTR 1 sqn with 8 F-5E, 2 -F
CCT 9 sqn
7 with 70 PC-7
2 with 17 AT-33
ARMED HEL 1 sqn with 1 Bell 205A, 15 Bell 206B, 7 Bell 206L-3, 24 Bell 212
RECCE 1 photo sqn with 10* *Commander* 500S, 2 SA 2-37A, 4 C-26
TPT 5 sqn with 1 Convair CV-580, 1 Lockheed L-1329 *Jetstar*, 1 Cessna 500 *Citation*, 1 C-118, 7 C-130A, 1 L-100 *Hercules*, 10 *Commander* 500S, 1 sqn with 9 IAI-201 (tpt/SAR)
HEL 6 S-70A, 1 Mi-2, 11 Mi-8, 24 Mi-17, 1 Mi-26T
PRESIDENTIAL TPT ac 1 Boeing 757, 3 Boeing 727-100
LIAISON/UTL 9 IAI *Arava*, 1 *King Air* A90, 3 *King Air* C90, 1 *Super King* 300, 1 *Musketeer*, 29 Beech *Bonanza* F-33C, 73 Cessna 182S, 11 Cessna 206, 11 Cessna 210, 4 PC-6, 6 Turbo Commander
TRG ac 6 Maule M-7, 21 Maule MXT-7-180, 12 PT-17 Stearman, 30 SF-260 **hel** 24* MD 530F (SAR/paramilitary/trg)

Paramilitary

RURAL DEFENCE MILITIA (R) 14,000

COAST GUARD
4 *Mako* 295 PCI

Opposition

ZAPATISTA ARMY OF NATIONAL LIBERATION str n.k.

POPULAR INSURGENT REVOLUTIONARY ARMY str n.k.

MEXICAN PEASANT WORKERS FRONT OF THE SOUTH EAST str n.k.

POPULAR MOVEMENT OF NATIONAL LIBERATION str n.k.

REVOLUTIONARY INSURGENT ARMY OF THE SOUTH EAST str n.k.

Nicaragua Nic

Cordoba oro Co		1998	1999	2000	2001
GDP	Co	22.5bn	28.0bn		
	US$	2.7bn	2.9bn		
per capita	US$	2,100	2,200		
Growth	%	4.0	6.0		
Inflation	%	17.0	10.9		
Debt	US$	6.3bn	6.7bn		
Def exp	Co	314m	294m		
	US$	30m	25m		
Def bdgt	Co	268m	294m	329m	
	US$	25m	25m	26m	
FMA (US)	US$	0.1m	0.2m	0.2m	0.2m
US$1=Co		10.58	11.81	12.48	

Population			**4,816,000**
Age	13–17	18–22	23–32
Men	329,000	274,000	350,000
Women	291,000	247,000	385,000

Total Armed Forces

ACTIVE ε16,000

Terms of service voluntary, 18–36 months

Army 14,000

Reorganisation in progress

5 Regional Comd (10 inf, 1 tk coy) • 2 mil det (2 inf bn) • 1 lt mech bde (1 mech inf, 1 tk, 1 recce bn, 1 fd arty gp (2 bn), 1 atk gp) • 1 comd regt (1 inf, 1 sy bn) • 1 SF bde (3 SF bn) • 1 tpt regt (incl 1 APC bn) • 1 engr bn

EQUIPMENT

MBT some 127 T-55 (42 op remainder in store)
LT TK 10 PT-76 (in store)
RECCE 20 BRDM-2
APC 102 BTR-152 (in store), 64 BTR-60
TOWED ARTY 122mm: 12 D-30, 100 *Grad* 1P (single-tube rocket launcher); **152mm**: 30 D-20 (in store)
MRL 107mm: 33 Type-63; **122mm**: 18 BM-21
MOR 82mm: 579; **120mm**: 24 M-43; **160mm**: 4 M-160 (in store)
ATGW AT-3 *Sagger* (12 on BRDM-2)
RCL 82mm: B-10
ATK GUNS 57mm: 354 ZIS-2 (90 in store); **76mm**: 83 Z1S-3; **100mm**: 24 M-1944
SAM 200+ SA-7/-14/-16

Navy ε800

BASES Corinto, Puerto Cabezzas, El Bluff

PATROL AND COASTAL COMBATANTS 5

PATROL, INSHORE 5

2 Sov *Zhuk* PFI<, 3 *Dabur* PCI<, plus boats

MINE COUNTERMEASURES 2

2 *Yevgenya* MHI

Air Force 1,200

no cbt ac, 15 armed hel
TPT 1 An-2, 4 An-26, 1 Cessna 404 Titan (VIP)
HEL 15 Mi-17 (tpt/armed) (3 serviceable), 1 Mi-17 (VIP)
UTL/TRG ac 1 Cessna T-41D
ASM AT-2 *Swatter* ATGW
AD GUNS 1 air def gp, 18 ZU-23, 18 C3-*Morigla* M1

Panama Pan

balboa B		1998	1999	2000	2001
GDP	B	9.1bn	9.7bn		
	US$	9.1bn	9.7bn		
per capita	US$	6,600	6,900		
Growth	%	3.9	3.2		
Inflation	%	0.7	1.3		
Debt	US$	5.4bn	5.4bn		
Sy bdgt	B	120m	128m	135m	
	US$	120m	128m	135m	
FMA (US)	US$		0.7m	0.1m	0.1m
US$1=B		1.0	1.0	1.0	

Population			**2,880,000**
Age	13–17	18–22	23–32
Men	145,000	137,000	257,000
Women	139,000	131,000	249,000

Total Armed Forces

ACTIVE Nil

Paramilitary ε11,800

NATIONAL POLICE FORCE 11,000

Presidential Guard bn (-), 1 MP bn plus 8 coys, 18 Police coy, 1 SF unit (reported); no hy mil eqpt, small arms only

NATIONAL MARITIME SERVICE ε400

BASES Amador (HQ), Balboa, Colón

PATROL AND COASTAL COMBATANTS 14

PATROL CRAFT, COASTAL 5

2 *Panquiaco* (UK *Vosper* 31.5m) PCC, 3 other PCC

PATROL CRAFT, INSHORE 9

3 *Tres de Noviembre* (ex-US *Point*) PCI, 1 *Swiftships* 65ft PCI, 1 ex-US MSB 5 class, 1 *Negrita* PCI, 3 ex-US PCI (plus some 25 boats)

NATIONAL AIR SERVICE 400

TPT 1 CN-235-2A, 1 BN-2B, 1 PA-34, 3 CASA-212M *Aviocar*

TRG 6 T-35D

HEL 2 Bell 205, 6 Bell 212, 13 UH-1H

Paraguay Py

Paraguayan guarani Pg

		1998	1999	2000	2001
GDP	Pg	23.4tr	28.9tr		
	US$	9.0bn	9.3bn		
per capita	US$	3,700	3,700		
Growth	%	-0.5	0.0		
Inflation	%	11.5	6.8		
Debt	US$	2.2bn	2.1bn		
Def exp	Pg	ε360bn	ε400bn		
	US$	132m	128m		
Def bdgt	Pg	280bn	262bn	290bn	
	US$	103m	84m	83m	
FMA (US)	US$	0.2m	0.2m	0.2m	0.2m
US$1=Pg		2,756	3,119	3,495	

Population			5,639,000
Age	13–17	18–22	23–32
Men	307,000	265,000	439,000
Women	296,000	256,000	424,000

Total Armed Forces

ACTIVE 20,200 (to reduce)

(incl 12,900 conscripts)

Terms of service 12 months **Navy** 2 years

RESERVES some 164,500

Army 14,900

(incl 10,400 conscripts)

3 corps HQ • 9 div HQ (6 inf, 3 cav) • 9 inf regt (bn) • 3 cav regt (horse) • 3 mech cav regt • Presidential Guard (1 inf, 1 MP bn, 1 arty bty) • 20 frontier det • 3 arty gp (bn) • 1 AD arty gp • 4 engr bn

RESERVES

14 inf, 4 cav regt

EQUIPMENT

MBT 12 M-4A3

RECCE 8 M-8, 5 M-3, 30 EE-9 *Cascavel*

APC 10 EE-11 *Urutu*

TOWED ARTY 75mm: 20 Model 1927/1934; **105mm**: 15 M-101; **152mm**: 6 Vickers 6in (coast)

MOR 81mm: 80

RCL 75mm: M-20

AD GUNS 30: **20mm**: 20 Bofors; **40mm**: 10 M-1A1

Navy 3,600

(incl 900 Marines, 800 Naval Aviation)

BASES Asunción (Puerto Sajonia), Bahía Negra, Ciudad Del Este

PATROL AND COASTAL COMBATANTS 10

PATROL, RIVERINE 10

2 *Paraguais* PCR with 4 × 120mm guns†

2 *Nanawa* PCR

1 *Itapu* PCR

1 *Capitan Cabral* PCR

2 *Capitan Ortiz* PCR (ROC *Hai Ou*) PCR<

2 ROC PCR

plus some 20 craft

SUPPORT AND MISCELLANEOUS 5

1 tpt, 1 trg/tpt, 1 AGHS<, 2 LCT

MARINES (900)

(incl 200 conscripts); 4 bn(-)

NAVAL AVIATION (800)

2 cbt ac, no armed hel

CCT 2 AT-6G

LIAISON 2 Cessna 150, 2 C-206, 1 C-210

HEL 2 HB-350, 1 OH-13

Air Force 1,700

(incl 600 conscripts); 28 cbt ac, no armed hel

FTR/FGA 8 F-5E, 4 F-5F

CCT 6 AT-33, 6 EMB-326, 4 T-27

LIAISON 1 Cessna 185, 4 C-206, 2 C-402, 2 T-41

HEL 3 HB-350, 1 UH-1B, 2 UH-1H, 4 UH-12, 4 Bell 47G

TPT 1 sqn with 5 C-47, 4 C-212, 3 DC-6B, 1 DHC-6 (VIP), 1 C-131D

TRG 6 T-6, 10 T-23, 5 T-25, 10 T-35, 1 T-41

Paramilitary 14,800

SPECIAL POLICE SERVICE 14,800

(incl 4,000 conscripts)

Peru Pe

new sol NS		1998	1999	2000	2001
GDP	NS	184bn	193bn		
	US$	58bn	57bn		
per capita	US$	4,300	4,500		
Growth	%	0.3	3.8		
Inflation	%	7.3	3.5		
Debt	US$	32bn	29bn		
Def exp	NS	ε2.9bn	ε3.0bn		
	US$	990m	888m		
Def bdgt	NS	2.7bn	2.7bn	2.9bn	
	US$	913m	820m	825m	
FMA[a] (US)	US$	32m	79m	50m	50m
US$1=NS		2.93	3.38	3.48	

[a] MOMEP **1998** εUS$15m

Population			25,916,000
Age	13–17	18–22	23–32
Men	1,358,000	1,293,000	2,267,000
Women	1,346,000	1,284,000	2,258,000

Total Armed Forces

ACTIVE 115,000

(incl 64,000 conscripts)

Terms of service 2 years, selective

RESERVES 188,000

Army only

Army 75,000

(incl 52,000 conscripts)

6 Mil Regions

Army tps

1 AB div (3 cdo, 1 para bn, 1 arty gp) • 1 Presidential Escort regt • 1 AD arty gp

Regional tps

3 armd div (each 2 tk, 1 armd inf bn,1 arty gp, 1 engr bn) • 1 armd gp (3 indep armd cav, 1 fd arty, 1 AD arty, 1 engr bn) • 1 cav div (3 mech regt, 1 arty gp) • 7 inf div (each 3 inf bn, 1 arty gp) • 1 jungle div • 2 med arty gp • 2 fd arty gp • 1 indep inf bn • 1 indep engr bn • 3 hel sqn

EQUIPMENT

MBT 300 T-54/-55 (ε50 serviceable)

LT TK 110 AMX-13 (ε30 serviceable)

RECCE 60 M-8/-20, 10 M-3A1, 50 M-9A1, 15 Fiat 6616, 30 BRDM-2

APC 130 M-113, 12 BTR-60, 130 UR-416, Fiat 6614, *Casspir*, 4 *Repontec*

TOWED ARTY 105mm: 20 Model 56 pack, 130 M-101; **122mm**: 42 D-30; **130mm**: 36 M-46; **155mm**: 36 M-114

SP ARTY 155mm: 12 M-109A2, 12 Mk F3

MRL 122mm: 14 BM-21

MOR 81mm: incl some SP; **107mm**: incl some SP; **120mm**: 300 Brandt, ECIA

ATGW 400 SS-11

RCL 106mm: M40A1

AD GUNS 23mm: 80 ZSU-23-2, 35 ZSU-23-4 SP; **30mm**: 10 2S6 SP; **40mm**: 45 M-1, 80 L60/70

SAM SA-7, 236 SA-16, *Javelin*

AC 13 Cessna incl 1 C-337, 1 *Queen Air* 65, 5 U-10, 3 U-17, 1 U-150, 2 U-206, 4 AN-32B

HEL 2 Bell 47G, 2 Mi-6, 26 Mi-8, 13 Mi-17, 6 SA-315, 5 SA-316, 3 SA-318, 2 *Agusta* A-109

Navy 25,000

(incl some 800 Naval Aviation, 3,000 Marines, 1,000 Coast Guard and 10,000 conscripts)

NAVAL AREAS Pacific, Lake Titicaca, Amazon River

BASES Ocean Callao, San Lorenzo Island, Paita, Talara **Lake** Puno **River** Iquitos, Puerto Maldonado

SUBMARINES 8

SSK 6 *Casma* (Ge T-209/1200) with 533mm TT (It A184 HWT) (2 in refit)

SSC 2 *Abato* (US *Mackerel*) with 533mm TT, 1 × 127mm gun

PRINCIPAL SURFACE COMBATANTS 5

CRUISERS 1

1 *Almirante Grau* (Nl *De Ruyter*) CG with 8 *Otomat* SSM, 4 × 2 152mm guns

FRIGATES 4

FFG 4 *Carvajal* (mod It *Lupo*) CG with 8 *Otomat* SSM, *Albatros* SAM, 1 × 127mm gun, 2 × 3 324mm ASTT (Mk 32 HWT), 1 AB-212 or SH-3D hel

PATROL AND COASTAL COMBATANTS 10

MISSILE CRAFT 6 *Velarde* PFM (Fr PR-72 64m) with 4 MM-38 *Exocet* SSM, 1 × 76mm gun

PATROL CRAFT, RIVERINE 4

2 *Marañon* PCR

2 *Amazonas* PCR

(plus 4 craft for lake patrol)

AMPHIBIOUS 3

3 *Paita* (US *Terrebonne Parish*) LST, capacity 395 tps, 2000t

SUPPORT AND MISCELLANEOUS 9

3 AO, 1 AOT, 1 tpt; 1 AT/F (SAR); 1 AGOR, 2 AGHS

NAVAL AVIATION (some 800)

9 cbt ac, 9 armed hel

ASW/MR 4 sqn with **ac** 5 *Super King Air* B 200T, 3 EMB-111A, 1 F-27 **hel** 5 AB-212 ASW, 4 ASH-3D (ASW)

TPT 2 An-32B, 1 Y-12

LIAISON 4 Bell 206B, 6 UH-1D hel, 3 Mi-8

TRG 1 Cessna 150, 5 T-34C

ASM *Exocet* AM-39 (on SH-3 hel)

MARINES (3,000)

1 Marine bde (5 bn, 1 recce, 1 special ops gp)

EQUIPMENT

RECCE V-100
APC 15 V-200 *Chaimite*, 20 BMR-600
MOR 81mm; 120mm ε18
RCL 84mm: *Carl Gustav*; **106mm**: M-40A1
AD GUNS twin 20mm SP

COASTAL DEFENCE 3 bty with 18 **155mm** how

Air Force 15,000

(incl 2,000 conscripts); 121 cbt ac†, 19 armed hel
BBR 8 *Canberra*
FGA 2 gp, 6 sqn
3 with 28 Su-22 (incl 4* Su-22U), 18 Su-25 (incl 8* Su-25UB)
3 with 23 Cessna A-37B
FTR 3 sqn
1 with 10 *Mirage* 2000P, 2 -DP
2 with 9 *Mirage* 5P, 2 -DP
1 with 21 MiG-29 (incl 2 MiG-29UB)
ATTACK/ASSAULT HEL 1 sqn with 10 Mi-24/-25, 8 Mi-17TM, 1 Ka-50 (under evaluation)
RECCE 3 MiG-25RB, 1 photo-survey unit with 2 *Learjet* 25B, 2 -36A
TKR 1 Boeing KC 707-323C
TPT 3 gp, 7 sqn
ac 17 An-32, 3 AN-72, 4 C-130A, 6 -D, 5 L-100-20, 2 DC-8-62F, 12 DHC-5, 8 DHC-6, 1 FH-227, 9 PC-6, 6 Y-12 (II), 1 Boeing 737 **hel** 3 sqn with 8 Bell 206, 14 B-212, 5 B-214, 1 B-412, 10 Bo-105C, 5 Mi-6, 3 Mi-8, 35 Mi-17, 5 SA-316
PRESIDENTIAL FLT 1 F-28, 1 *Falcon* 20F
LIAISON ac 2 Beech 99, 3 Cessna 185, 1 Cessna 320, 15 *Queen Air* 80, 3 *King Air* 90, 1 PA-31T **hel** 8 UH-1D
TRG ac 2 Cessna 150, 25 EMB-312, 6 Il-103, 13 MB-339A, 20 T-37B/C, 15 T-41A/-D **hel** 12 Bell 47G
MISSILES
ASM AS-30
AAM AA-2 *Atoll*, AA-8 *Aphid*, AA-10 *Alemo*, R-550 *Magic*, AA-12 *Adder*
AD 3 SA-2, 6 SA-3 bn

Forces Abroad

UN AND PEACEKEEPING
DROC (MONUC): 30 obs

Paramilitary 77,000

NATIONAL POLICE 77,000
General Police 43,000 **Security Police** 21,000 **Technical Police** 13,000
100+ MOWAG *Roland* APC

COAST GUARD (1,000)
5 *Rio Nepena* PCC, 3 PCI, 10 riverine PCI<

RONDAS CAMPESINAS (peasant self-defence force)
perhaps 2,000 *rondas* 'gp', up to pl strength, some with small arms. Deployed mainly in emergency zone.

Opposition

SENDERO LUMINOSO (Shining Path) ε1,000
Maoist
MOVIMIENTO REVOLUCIONARIO TUPAC AMARU (MRTA) ε600
mainly urban gp

Suriname Sme

guilder gld		1998	1999	2000	2001
GDP	gld	328bn	344bn		
	US$	383m	409m		
per capita	US$	4,800	5,100		
Growth	%	11.4	4.0		
Inflation	%	21.1			
Debt	US$	158m	160m		
Def exp	gld	6.0bn	9.0bn		
	US$	15m	11m		
Def bdgt	gld	4.5bn	9.0bn	9.0bn	
	US$	11m	11m	11m	
FMA (US)	US$	0.1m	0.1m	0.1m	0.1m
US$1=gld		401	810	810	

Population			**418,000**
Age	13–17	18–22	23–32
Men	22,000	18,000	34,000
Women	22,000	18,000	34,000

Total Armed Forces

ACTIVE ε2,040
(all services form part of the Army)

Army 1,600

1 inf bn (4 inf coy) • 1 mech cav sqn • 1 MP 'bde' (coy)
EQUIPMENT
RECCE 6 EE-9 *Cascavel*
APC 15 EE-11 *Urutu*
MOR 81mm: 6
RCL 106mm: M-40A1

Navy 240

BASE Paramaribo

PATROL CRAFT, INSHORE 3
3 *Rodman* 100 PCI<, plus 5 boats

Air Force ε200

7 cbt ac, no armed hel
MPA 2 C-212-400
TPT/TRG 4* BN-2 *Defender*, 1* PC-7
LIAISON 1 Cessna U206
HEL 2 SA-316, 1 AB-205

Trinidad and Tobago TT

Trinidad and Tobago dollar TT$

		1998	1999	2000	2001
GDP	TT$	38.2bn	42.5bn		
	US$	6.2bn	6.8bn		
per capita	US$	10,500	11,400		
Growth	%	3.2	7.0		
Inflation	%	5.6	2.6		
Debt	US$	3.2bn	3.1bn		
Def exp	TT$	273m	392m		
	US$	43m	62m		
Def bdgt	TT$	250m	372m	390m	
	US$	40m	59m	62m	
FMA (US)	US$	0.1m	0.4m	0.4m	0.4m
US$1=TT$		6.3	6.3	6.3	

Population			**1,359,000**
Age	13–17	18–22	23–32
Men	73,000	65,000	103,000
Women	72,000	64,000	107,000

Total Armed Forces

ACTIVE ε2,700 (all services form part of the **Trinidad and Tobago Defence Force**)

Army ε2,000

2 inf bn • 1 spt bn
EQUIPMENT
MOR 60mm: ε40; **81mm**: 6 L16A1
RL 82mm: 13 B-300
RCL 82mm: B-300

Coast Guard 700

(incl 50 Air Wing)
BASE Staubles Bay (HQ), Hart's Cut, Point Fortin, Tobago, Galeota
PATROL COASTAL/INSHORE 10 (some non-op)
2 *Barracuda* PFC (Sw *Karlskrona* 40m) (non-op)
4 *Plymouth* PCI<
2 *Point* PCI<
2 *Wasp* PCI<
plus 10 boats and 2 aux vessels

AIR WING
2 C-26, 1 Cessna 310, 1 C-402, 1 C-172, 2 *Navajos* on order

Uruguay Ury

Uruguyan peso pU		1998	1999	2000	2001
GDP	pU	232bn	239bn		
	US$	13.7bn	13.7bn		
per capita	US$	9,200	9,100		
Growth	%	2.2	-2.5		
Inflation	%	10.8	5.6		
Debt	US$	6.9bn	5.5bn		
Def exp	pU	ε3.3bn	ε3.6bn		
	US$	315m	318m		
Def bdgt	pU	2.9bn	2.6bn	2.7bn	
	US$	279m	232m	227m	
FMA (US)	US$	0.3m	0.3m	0.3m	0.3m
US$1=pU		9.40	10.47	11.34	11.91

Population			**3,264,000**
Age	13–17	18–22	23–32
Men	131,000	136,000	250,000
Women	126,000	131,000	244,000

Total Armed Forces

ACTIVE 23,700

Army 15,200

4 Mil Regions/div HQ • 5 inf bde (4 of 3 inf bn, 1 of 1 mech, 1 mot, 1 para bn) • 3 cav bde (10 cav bn (4 horsed, 3 mech, 2 mot, 1 armd)) • 1 arty bde (2 arty, 1 AD arty bn) • 1 engr bde (3 bn) • 3 arty, 4 cbt engr bn
EQUIPMENT
MBT 15 T-55
LT TK 17 M-24, 29 M-3A1, 22 M-41A1
RECCE 18 EE-3 *Jararaca*, 15 EE-9 *Cascavel*
AIFV 10 BMP-1
APC 18 M-113, 50 *Condor*, 60 OT-64 SKOT
TOWED ARTY 75mm: 12 Bofors M-1902; **105mm**: 48 M-101A/M-102; **155mm**: 5 M-114A1
SP ARTY 122mm: 6 2S1
MRL 122mm: 3 RM-70
MOR 81mm: 97; **107mm**: 8 M-30; **120mm**: 44
ATGW 5 *Milan*
RCL 57mm: 30 M-18; **106mm**: 30 M-40A1
AD GUNS 20mm: 6 M-167 *Vulcan*; **40mm**: 8 L/60

Navy 5,500

(incl 310 Naval Aviation, 400 Naval Infantry, 1,950 *Prefectura Naval* (Coast Guard))
BASES Montevideo (HQ), Paysando (river), La Paloma (naval air)

FRIGATES 3

FFG 3 *General Artigas* (Fr *Cdt Rivière*) with 4 MM-38 *Exocet* SSM, 2 × 100mm guns, 2 × 3 ASTT, 1 × 2 ASW mor

PATROL AND COASTAL COMBATANTS 10

PATROL, COASTAL/INSHORE 10

3 *15 de Noviembre* PCC (Fr *Vigilante* 42m), 1 *Salto* PCC, 2 *Colonia* PCI< (US *Cape*), 1 *Paysandu* PCI<, 3 other PCI<

MINE COUNTERMEASURES 4

4 *Temerario* MSC (Ge *Kondor* II)

AMPHIBIOUS craft only

3 LCM, 2 LCVP

SUPPORT AND MISCELLANEOUS 7

1 *Presidente Rivera* AOT, 1 *Vanguardia* ARS, 1 *Campbell* (US *Auk* MSF) PCO (Antarctic patrol/research), 1 AT (ex-GDR *Elbe*-Class), 1 trg, 1 AGHS, 1 AGOR

NAVAL AVIATION (310)

1 cbt ac, no armed hel

ASW 1 *Super King Air* 200T

TRG/LIAISON 2 T-28, 2 T-34B, 2 T-34C, 2 PA-34-200T, 3 C-182

HEL 3 Wessex Mk60, 5 Wessex HC2, 2 Bell 47G, 2 SH-34J

NAVAL INFANTRY (400)

1 bn

Air Force 3,000

21 cbt ac, no armed hel

Flying hours 120

CBT AC 2 sqn

1 with 10 A-37B, 1 with 5 IA-58B

SURVEY 1 EMB-110B1

HEL 1 sqn with 2 Bell 212, 6 UH-1H, 6 *Wessex* HC2

TPT 3 sqn with 3 C-212 (tpt/SAR), 3 EMB-110C, 1 F-27, 3 C-130B, 1 Cessna 310 (VIP), 1 Cessna 206

LIAISON 2 Cessna 182, 2 *Queen Air* 80, 5 U-17, 1 T-34A

TRG 6 SF-260EU*, 5 T-41D, 5 PC-7U

Forces Abroad

UN AND PEACEKEEPING

DROC (MONUC): 19 obs **EAST TIMOR** (UNTAET): 5 obs **EGYPT** (MFO): 60 **GEORGIA** (UNOMIG): 3 obs **INDIA/PAKISTAN** (UNMOGIP): 4 obs **IRAQ/KUWAIT** (UNIKOM): 6 obs **SIERRA LEONE** (UNAMSIL): 11 obs **WESTERN SAHARA** (MINURSO): 13 obs

Paramilitary 920

GUARDIA DE GRANADEROS 450

GUARDIA DE CORACEROS 470

COAST GUARD (1,950)

Prefectura Naval (PNN) is part of the Navy operates 3 PCC, 2 LCMs plus 9 boats

Venezuela Ve

bolivar Bs		1998	1999	2000	2001
GDP	Bs	52bn	55bn		
	US$	90bn	85bn		
per capita	US$	8,600	8,000		
Growth	%	-0.2	-7.2		
Inflation	%	35.8	23.6		
Debt	US$	37bn			
Def exp	Bs	716bn	805bn		
	US$	1,307m	1,329m		
Def bdgt	Bs	716bn	805bn	949bn	
	US$	1,307m	1,329m	1,404m	
FMA (US)	US$	1.0m	1.1m	1.1m	1.6m
US$1=Bs		548	605	677	

Population			**24,238,000**
Age	13–17	18–22	23–32
Men	1,243,000	1,179,000	2,032,000
Women	1,197,000	1,139,000	1,977,000

Total Armed Forces

ACTIVE 79,000

(incl National Guard and ε31,000 conscripts)

Terms of service 30 months selective, varies by region for all services

RESERVES

Army ε8,000

Army 34,000

(incl 27,000 conscripts)

6 inf div HQ • 1 armd bde • 1 cav bde • 7 inf bde (18 inf, 1 mech inf, 4 fd arty bn) • 1 AB bde • 2 Ranger bde (1 with 4 bn, 1 with 2 bn) • 1 avn regt

RESERVES ε6 inf, 1 armd, 1 arty bn

EQUIPMENT

MBT 81 AMX-30

LT TK 75 M-18, 36 AMX-13, 80 *Scorpion* 90

RECCE 30 M-8

APC 25 AMX-VCI, 100 V-100, 30 V-150, 100 *Dragoon* (some with **90mm** gun), 35 EE-11 *Urutu*

TOWED ARTY 105mm: 40 Model 56, 40 M-101; **155mm**: 12 M-114

SP ARTY 155mm: 10 Mk F3

MRL 160mm: 20 LAR SP

MOR 81mm: 165; **120mm**: 60 Brandt

ATGW AS-11, 24 *Mapats*

RL 84mm: AT-4

RCL 84mm: *Carl Gustav*; **106mm**: 175 M-40A1

SURV RASIT (veh, arty)
AC 3 IAI-202, 2 Cessna 182, 2 C-206, 2 C-207
ATTACK HEL 5 A-109 (ATK)
TPT HEL 4 AS-61A, 3 Bell 205, 6 UH-1H
LIAISON 2 Bell 206

Navy 15,000

(incl 1,000 Naval Aviation, 5,000 Marines, 1,000 Coast Guard and ε4,000 conscripts)
NAVAL COMMANDS Fleet, Marines, Naval Avn, Coast Guard, Fluvial (River Forces)
NAVAL FLEET SQN submarine, frigate, patrol, amph, service
BASES Main bases Caracas (HQ), Puerto Cabello (submarine, frigate, amph and service sqn), Punto Fijo (patrol sqn) **Minor bases** Puerto de Hierro, Puerto La Cruz, El Amparo (HQ Arauca River), Maracaibo, La Guaira, Ciudad Bolivar (HQ Fluvial Forces)

SUBMARINES 2

SSK 2 *Sabalo* (Ge T-209/1300) with 8 × 533mm TT (SST-4 HWT)

FRIGATES 6

FFG 6 *Mariscal Sucre* (It *Lupo*) with 8 *Teseo* SSM, *Albatros* SAM, 1 × 127mm gun, 2 × 3 ASTT (A-244S LWT), 1 AB-212 hel

PATROL AND COASTAL COMBATANTS 6

MISSILE CRAFT 3
3 *Constitución* PFM (UK Vosper 37m), with 2 *Teseo* SSM
PATROL CRAFT, OFFSHORE 3
3 *Constitución* PCO with 1 × 76mm gun

AMPHIBIOUS 4

4 *Capana* LST (Sov *Alligator*), capacity 200 tps, 12 tk
Plus craft: 2 LCU (river comd), 12 LCVP

SUPPORT AND MISCELLANEOUS 5

1 log spt; 1 *Punta Brava* AGOR, 2 AGHS; 1 sail trg

NAVAL AVIATION (1,000)

7 cbt ac, 8 armed hel
ASW 1 hel sqn (afloat) with 8 AB-212
MR 1 sqn with 4 C-212-200 MPA, 3 C-212-400
TPT 2 C-212, 1 DHC-7, 1 *Rockwell Commander* 680
LIAISON 1 Cessna 310, 1 C-402, 1 *King Air* 90, 3 C-212-400
HEL 4 Bell 412-EP

MARINES (5,000)

4 inf bn • 1 arty bn (3 fd, 1 AD bty) • 1 amph veh bn • 1 river patrol, 1 engr, 2 para/cdo unit

EQUIPMENT
AAV 11 LVTP-7 (to be mod to -7A1)
APC 25 EE-11 *Urutu*, 10 *Fuchs/Transportpanzer* 1
TOWED ARTY 105mm: 18 Model 56
AD GUNS 40mm: 6 M-42 twin SP

COAST GUARD (1,000)

BASE La Guaira; operates under Naval Command and Control, but organisationally separate
PATROL, OFFSHORE 2
2 *Almirante Clemente* corvettes with 2 × 76mm guns, 3 × 2 ASTT
PATROL, INSHORE 4
4 *Petrel* (USCG *Point*-class) PCI
plus 27 river patrol craft and boats
plus 1 support ship

Air Force 7,000

(some conscripts); 124 cbt ac, 31 armed hel
Flying hours 155
FTR/FGA 6 air gp
1 with 16 CF-5A/B (12 A, 4 B), 7 NF-5A/B
1 with 16 *Mirage* 50EV/DV
2 with 21 F-16A/B (14 A, 7 B)
2 with 20 EMB-312
RECCE 15* OV-10A
ECM 3 *Falcon* 20DC
ARMED HEL 1 air gp with 10 SA-316, 12 UH-1D, 5 UH-1H, 4 AS-532
TPT ac 7 C-123, 5 C-130H, 8 G-222, 2 HS-748, 2 B-707 (tkr) **hel** 2 Bell 214, 4 Bell 412, 8 AS-332B, 2 UH-1N, 18 Mi-8/17
PRESIDENTIAL FLT 1 Boeing 737, 1 *Gulfstream* III, 1 *Gulfstream* IV, 1 *Learjet* 24D **hel** 1 Bell 412
LIAISON 9 Cessna 182, 1 *Citation* I, 1 *Citation* II, 2 *Queen Air* 65, 5 *Queen Air* 80, 5 *Super King Air* 200, 9 SA-316B *Alouette* III
TRG 1 air gp: 12* EMB-312, 20 T-34, 17* T-2D, 12 SF-260E

MISSILES

AAM R-530 *Magic*, AIM-9L *Sidewinder*, AIM-9P *Sidewinder*
ASM *Exocet*
AD GUNS 20mm: some IAI TC-20; **35mm**; **40mm**: 114: Bofors L/70 towed, Otobreda 40L70 towed
SAM 10 *Roland*, RBS-70

National Guard (*Fuerzas Armadas de Cooperación*) 23,000

(internal sy, customs)
8 regional comd

EQUIPMENT

20 UR-416 AIFV, 24 Fiat-6614 APC, 100 **60mm** mor, 50 **81mm** mor **ac** 1 *Baron*, 1 BN-2A, 2 Cessna 185, 5 -U206, 4 IAI-201, 1 *King Air* 90, 1 *King Air* 200C, 2 *Queen Air* 80, 6 M-28 *Skytruck* **hel** 4 A-109, 20 Bell 206, 2 Bell 212
PATROL CRAFT, INSHORE 52 craft/boats

Forces Abroad

UN AND PEACEKEEPING

IRAQ/KUWAIT (UNIKOM): 2 obs

Sub-Saharan Africa

MILITARY DEVELOPMENTS

Sub-Saharan Africa accounted for about two-thirds of the 100,000 people worldwide killed as a direct result of armed conflict over the year to August 2000. There has been armed conflict of some form in three-quarters of all countries in the region. Both military and non-military measures have been used in attempts to stabilise these situations. While the arms flow from outside and within the region continues unabated, efforts are being made to restrict funds used to buy weapons and logistic support and to pay militias and other private armed groups. An effort to curb the illegal trade in diamonds, a major source of income for some rebel groups particularly in Sierra Leone and Angola, is being led by Western governments, notably the UK, in partnership with diamond-trading companies. This is unlikely to have a short-term impact on the flow of mainly cheap small arms, but it could significantly influence the calculations of leaders of armed groups who profit most from this trade. The establishment of war crimes tribunals in Rwanda and most recently in Sierra Leone is another important non-military means of influencing leaders of armed groups.

Efforts to mediate peace, whether by international organisations such as the UN and the Organisation of African Unity (OAU) or by individuals such as Nelson Mandela in Burundi, seem at best only to bring a brief respite from all-out fighting. The suffering and loss of life in the region from armed conflict is serious enough but disease, in particular AIDS, is having an even larger impact on the region. AIDS alone accounts for an estimated 6,000 deaths each day and will harm long-term regional economic and political stability.

Horn of Africa

In May 2000, major Ethiopian military operations drove Eritrean forces from the disputed border areas they had occupied in fighting in early 1998. The Ethiopian–Eritrean conflict has been the largest in the region in terms of the numbers of troops engaged. Both sides fielded 14 divisions at its peak. With the mediation of the OAU chairman President Abdelaziz Bouteflika of Algeria, supported by EU and US representatives, a Cessation of Hostilities Agreement (CHA) was accepted by Eritrea on 9 June and by Ethiopia on 14 June 2000. The agreement calls for the two sides' withdrawal to positions held before the May 1998 hostilities and for a UN observer force to be deployed in a 25-kilometre-wide buffer zone along their 1,000km common border until international arbitrators demarcate it. However, Ethiopian forces will continue to patrol disputed areas, such as Badame, that they occupied before May 1998. The CHA envisaged an observer force of 2,000, but it will probably be much smaller, mainly because of the reluctance of governments to contribute personnel.

Hopes for peace in the 17-year civil war in Sudan were dashed in 2000 when fighting between the Sudanese People's Liberation Army (SPLA) and government forces resumed. By July 2000, when the government's forces had suffered a series of defeats that compelled it to declare a 'state of mobilisation', the SPLA had overrun several towns, including Gogrial, the third largest town in the southern Bar-e-Ghazal region.

In Somalia, warlords continue to battle for dominance in the Mogadishu area while in the north-east an orderly regime has emerged with all the trappings of a state. In 1999, the self-styled Puntland government established a Maritime Security Force (MSF), run by a Bermuda-based security firm. It has one UK-built fisheries protection vessel and 70, mainly British ex-service, personnel. In August 2000, President Ismael Omar Gelleh of Djibouti hosted a conference whose aim was to recreate a central government in Somalia and, more specifically, to form a transitional 225-seat National Assembly. But there were vigorous disagreements among the competing Somali clans over the allocation of seats, and the failure of 12 previous similar attempts to end the nine-year civil war is not encouraging. While a provisional national assembly for Somalia has been set up in neighbouring Djibouti, it does not have the support of all the clan leaders in the south of the country.

Central Africa

In the Democratic Republic of Congo (DROC), there has been fighting not only between government forces and the rebels, but also between Rwandan and Ugandan troops deployed to support the rebels, who clashed in the Kisangani area over the control of territory and resources. In response to domestic and international pressure, Ugandan troops withdrew from Kivu province in north-east DROC in July 2000, only to return after a few weeks. The area became unstable when the rebel *Interahamwe* moved in after the Ugandan withdrawal and threatened the local inhabitants. The situation remains very unstable. There is an understandable reluctance to deploy to this troubled area the necessary personnel for the relatively small UN Organization Mission in the Democratic Republic of Congo (MONUC), without the full cooperation of the parties involved, particularly after the UN's casualties in Sierra Leone. This observer force of 5,500 personnel is meant to oversee the shaky cease-fire agreed at Lusaka in July 1999. The UN had intended to deploy four battalions – two to the government-held cities of Gbandanka and Kanga held by government forces and two to the rebel-controlled cities of Kisangani and Kindu. But in late July 2000, President Laurent Kabila of DROC was still objecting to the presence of UN forces in areas he holds. Only around 100 soldiers had arrived, mainly headquarters personnel and liaison officers dealing with the governments party to the cease-fire.

One of Kabila's allies, the Angolan government, remains locked in a civil war with the forces of *União Nacional para a Independência Total de Angola* (UNITA) led by Jonas Savimbi. Its own *Forcas Armadas Angolanas* (FAA) have reversed the situation of mid-1999 when UNITA forces were threatening the country's oil-producing areas and even carrying out attacks within 100km of the capital Luanda. By July 2000, the FAA had taken control of 11 of the 13 districts in the diamond-rich Lunda North and South provinces and were likely soon to control the two remaining districts, Lubalo and Cuilo. The UN estimates that UNITA has obtained as much as $4 billion from selling diamonds over the past eight years, despite longstanding sanctions against the trade. In July 2000 the Angolan government installed a diamond-certification scheme for distinguishing legitimate from illegitimate trading. The FAA's military successes indicate that the most effective way to cut off the source of supply is to occupy the diamond-producing areas. Although now much more dispersed than in 1999, UNITA is still a fighting force, and the Angolan people still suffer from brutal attacks by both sides in this 25-year war. An estimated 6,000 people, mostly non-combatants, have been killed in the fighting over the year to August 2000.

Another principal Kabila supporter, Zimbabwean President Robert Mugabe was preoccupied in mid-2000 with an election campaign and domestic economic and political problems. The advantages to Zimbabwe of engagement in the war in the DROC are unclear, particularly as the country faces its worse economic crisis since its independence in 1980. There is no common border, and Zimbabwe is under no direct military threat from the DROC or any of the combatants. The war has become increasingly unpopular with Zimbabweans and could be one reason why the opposition Mugabe gained more support during the year. Mugabe admitted publicly for the first time, on 4 August 2000, that there were 11,000–12,000 Zimbabwean troops in DROC. At home, at least 30 lives were lost in the violence when 'war veterans' occupied white-owned farms in a campaign that was condoned and largely orchestrated by the government.

By July 2000, refugees from the conflict in the Republic of the Congo began to return home. An uneasy peace had settled after the December 1999 cease-fire signed between President Denis Sassou-Nguesso's government and the opposition forces. Military tension continued until the joint deployment of the two sides' forces in the major towns in the rebel-held south of the country. The Brazzaville government's ability to maintain the cease-fire is uncertain. There are many weapons still in circulation and tensions remain between the president, whose power base is primarily in the less populous north, and the rebels in the more densely populated south. President Nguesso can draw on the support of Angolan government forces in the adjacent oil-rich Cabinda province if necessary, but while this support may maintain him in power, conflict could resume at any time.

The Burundi peace process faltered, even with the full weight of Nelson Mandela's mediation, when rebel forces killed 50 civilians in the south-eastern province of Ryugi in July 2000. Government forces immediately launched an offensive there and also against rebels operating in the hills around the capital Bujumbura. Much of the violence is attributed to the *Front National de Liberation* (FNL), also known as the *Palipehutu* movement. The FNL has stayed away from the peace talks in Arusha, Tanzania as it considers the Burundian government representatives at the talks to be hostile to the country's Hutu population, from which the rebels draw their support. Further talks have been postponed until September 2000. The UN estimates that, as a result of the fighting in Burundi, there are some 800,000 people internally displaced and over 500,000 refugees in Tanzania.

West Africa

Hopes for peace in West Africa were shattered in May 2000 when the rebels of the Revolutionary United Front (RUF) launched an offensive in Sierra Leone, even though their leader Foday Sankoh was a member of the government with responsibility for the diamond trade. The offensive caught the UN force unprepared and at one point 500 of its members were taken hostage. In response, a 1,000-strong British force intervened, supported by a naval force off the coast by the Sierra Leone capital, Freetown. The British also provided a training team for the government forces. By July 2000, this effort had helped to obtain the release of the UN hostages and to stem the RUF offensive. The UN force was increased from the originally planned 5,000 personnel to 13,000 to enable it to conduct offensive operations against the RUF rather than simply to oversee a peace accord. As of August 2000, the fighting was still far from over and international efforts focused on trying to block the illegal diamond trade from which the RUF and its leaders and backers benefit. There is clear evidence of collusion between the Liberian government of President Charles Taylor and the RUF in trading diamonds on the world market. In July 2000, the UN Security Council's Sanctions Committee on Sierra Leone endorsed a ban on diamond trading other than through an official certification scheme run by governments in cooperation with the private companies involved. Its success seems likely to be limited given the indifferent result of the similar restriction in Angola. This has shown that the only successful way to choke off diamond-derived income for guerrilla movements is to occupy the producing areas.

In the effort to bolster the Sierra Leone government forces, British instructors will have given basic training to a brigade-equivalent of some 4,000 soldiers by the end of 2000. The UK has supplied sufficient rifles and ammunition to equip this force. But to gain a clear advantage over the RUF, the government forces also need combat support such as armed and support helicopters; in the most recent conflict they could field only two Mi-24 armed helicopters and one qualified pilot.

In December 1999, the Côte d'Ivoire government fell to a military coup led by General Robert Guei. The ousted government had mismanaged and embezzled the wealth of a country with the third-largest economy in Sub-Saharan Africa, resulting in the suspension of aid from the EU and the International Monetary Fund (IMF). A referendum was held in July 2000 on a new constitution that is intended to pave the way for elections and a return to civilian government. However, there have been divisions within the military, and mutinous actions by soldiers indicate the return to civilian rule could be difficult. General Guei has political ambitions and has said that he will run for the presidency as a civilian.

At a conference of the Economic Community of West African States (ECOWAS) Commission on Security and Defence held in the Ghanaian capital Accra in July 2000, a new role was announced for the Economic Community of West African States' Cease-Fire Monitoring Group (ECOMOG). It is planned that the sub-regional peacekeeping force will turn into a permanent stand-by force for responding to conflict and that there should be a common training programme for the troops involved, with two training bases, one to be set up in Ghana and the other in Côte d'Ivoire. The difficult political situation in the latter country, however, may delay these plans. The US is also providing training and support for West African peacekeeping forces. A 40-man training team from the US Special Forces arrived in Nigeria in July; several hundred more were due to come in August and September 2000. They will train and equip five Nigerian battalions of 800 soldiers each, one battalion from Ghana and one from a French-speaking West African country, possibly Mali or Senegal.

Southern Africa

Namibia has felt the consequences of the continuous fighting in southern Angola over the past two years. Up to 15,000 Angolans have fled the conflict by crossing the Kavango River into Namibian territory. Windhoek faced problems on another front in mid-1999 when the separatist Caprivi Liberation Army (CLA), led by exiled former opposition leader, Mishake Muyongo, attacked government installations near the north-eastern town of Katimo Mulilo in the Caprivi Strip. The government captured 100 CLA members while others escaped into Botswana. The trial of the captives has been repeatedly postponed and is unlikely to take place until 2001.

The reform of the South African National Defence Force (SANDF) continues. Its re-organisation has taken place slowly, with the regular troops now consisting of 'type' formations, responsible for producing combat-ready units. Two brigade headquarters have been retained but no units have been permanently assigned to them. There have been allegations of continued racism in the SANDF. Over 70% of its officers were in place during the apartheid era, and the stresses of the period apparently still prevail. This was illustrated by a shooting incident at Phalaborwa in which a major of the 7 SA Infantry Battalion was killed by a fellow officer. Evidently, some officers from the apartheid era are not adjusting readily to the politics of integration, particularly in relations with former members of the guerrilla forces. This incident was the second of its kind and a four-man ministerial commission has been set up to examine the circumstances leading to both episodes and to recommend improvements to be made in training and military discipline. Morale has also been lowered by financial constraints. In particular, the higher oil price over the past year has severely restricted training. For example, flying hours have been curtailed and parachute-training jumps for new recruits to airborne units were stopped in 2000; only operational units have been able to maintain their skills. Morale has been lowered further by doubts about the economic feasibility of the ambitious eight-year 'Defence Renewal' arms procurement plan for combat aircraft and ships.

The SANDF has begun its first deployments abroad. Its logistic units are on standby for peacekeeping operations with MONUC in the DROC, and naval patrol craft are in Comoran territorial waters as part of an OAU force blockading Anjouan Island, where rebels are seeking autonomy. In support of the Comoros government, the OAU voted in July 2000 to blockade the island and to enforce a trade embargo that began in April 2000.

DEFENCE SPENDING

Regional military expenditure fell slightly from $10bn in 1998 to $9.8bn (in constant 1999 dollars) in 1999. This reflected the strength of the dollar and many countries' serious economic difficulties. Estimating regional arms spending accurately is difficult as the many guerrilla movements account for a significant portion of it. Although this spending is low compared to that in other regions, it is still a major drain on impoverished countries' resources, and it sustains conflicts whose heavy toll on human and material resources cannot be quantified.

In the Horn of Africa, the Sudanese defence budget for 1999 was set at $425m. This figure probably bears little relationship to real spending as in 1999 the civil war cost the government $650m, according to some estimates, half of its entire $1.3bn budget. Ethiopia declared a defence budget of birr3.5bn ($432m) in 1999, rising to birr3.7bn ($456m) in 2000. Its spending in 1999 is estimated to have been slightly above budget, at birr3.6bn ($444m). No official budgets are available for Eritrea, but its spending in 1999 is estimated to have increased to $308m from $291m in 1998. Despite their enormous economic difficulties, both sides in the border war have used more advanced aircraft, such as Su-27 and MiG-29 combat aircraft and Ka-50 helicopters, during the recent conflicts. For this they have needed outside help. In Eritrea, Ukrainians provide support and maintenance for the MiG-29s, while Eritrean pilots fly the aircraft. In Ethiopia, Russians on contract provide support and maintenance and also pilot the Ethiopian Su-27 combat aircraft.

In Central Africa, Uganda continued to spend well over its budget on defence. The official budget for 1999 was Ush200bn ($137m) rising to Ush210bn ($132m) in 2000, but official outlays in 1999 were quoted at Ush290bn ($199m). As in previous years, the real outlay is probably higher still. Uganda was the first country to benefit from the new World Bank debt-relief initiative for developing countries; its annual debt repayments have fallen from $120m to $40m. The conflict in DROC continues to cost Uganda about $60m a year. One of the more substantial regional arms acquisitions in 2000 was Angola's purchase of 12 Su-22 combat aircraft from Slovakia and an undisclosed number of Su-24s from Russia.

The IMF suspended an aid package to Zimbabwe in 2000 because of differences with the government over the conditions for a $200m loan. Harare had reported to the Fund that the war in DROC was costing it $3m a month. However, official government documents leaked to the public domain indicated that its military involvement was costing at least $25m a month in January–June 2000. Reports indicate that Zimbabwean forces have lost equipment worth $200m since entering the DROC in 1998. In August 2000, the country's annual inflation rate was 60%; reductions in government expenditure were essential and it was questionable whether military spending at recent levels could be maintained. Zimbabwe's defence spending was an estimated $417m in 1999, representing 6.8% of gross domestic product (GDP), against a budgeted figure of $168m. Clearly the government, or elements of it, hope to reap financial rewards from the involvement in DROC, having set up companies to deal in its diamonds and gold. In addition, 500,000 hectares of land in the DROC have been granted to the Zimbabwe Rural Development Authority for Zimbabwean companies to use. The military appear to be deeply involved in these commercial operations, calling into question their motives and military competence if senior officers' attention is focused on such activities.

Nigeria's defence budget, the largest in West Africa, was N34.2bn ($340m) for 2000, up from N25.7bn ($278m) for 1999. The figure does not include items such as procurement, funding for military construction, military pensions, state-level funding for military governors or funding for paramilitary forces. When taken into account, these items bring the real level of Nigerian military spending in 1999 to just over $2.2bn.

Table 30 **Nigeria: defence expenditure, 1992–2000**

	1992	1993	1994	1995	1996	1997	1998	1999	2000
Total (Nm)	11,300	13,500	13,800	14,000	15,500	17,500	22,000	25,700	34,181
Total (US$m)	519	617	630	639	708	799	1,005	278	340
Real Spending (1999 US$m)	1,390	1,477	1,558	1,635	1,990	2,044	2,143	2,282	2,387
Defence as % of GDP	3.0	3.2	3.4	3.5	4.1	4.1	4.3	4.3	4.3
Exchange rate	21.9	21.9	21.9	21.9	21.9	21.9	21.9	92.3	100

Table 31 **South Africa: defence budget by programme, 1995–2001**

(Rm/*US$m*)	1995		1996		1997		1998		1999		2000		2001	
Administration and General Support	745	*205*	1,095	*255*	1,104	*240*	1,089	*197*	1,456	*235*	2,123	*342*	n.a.	*n.a.*
Army	3,980	*1,097*	4,214	*980*	4,288	*931*	3,924	*710*	3,619	*584*	3,210	*518*	n.a.	*n.a.*
Air Force	1,753	*483*	2,104	*489*	2,083	*452*	1,903	*344*	1,944	*314*	1,850	*298*	n.a.	*n.a.*
Navy	778	*215*	781	*182*	802	*174*	833	*151*	842	*136*	884	*143*	n.a.	*n.a.*
Medical Support	739	*204*	873	*203*	887	*192*	910	*165*	939	*151*	973	*157*	n.a.	*n.a.*
Special Defence Account and other	3,525	*972*	1,854	*431*	1,942	*421*	1,591	*288*	1,829	*295*	4,720	*761*	n.a.	*n.a.*
Total Defence Budget														
Rm	11,521		10,922		11,106		10,250		10,628		13,760		15,270	
US$m		*3,176*		*2,540*		*2,410*		*1,853*		*1,714*		*2,219*		*2,463*

South Africa

The South African defence budget rose from R10.7bn ($1.7bn) in 1999 to R13.4bn ($1.9bn) in 2000. The SANDF's principal intended acquisitions for its R32bn ($5.08bn) eight-year defence-procurement plan remain as outlined in 1999. For the naval forces it plans to purchase three German Type-209 diesel submarines and four German *Meko* corvettes for active service by 2006–07 and 2004–05 respectively. The SANDF also plans to acquire four helicopters for active service with the *Meko* corvettes by 2005. There are two contenders for this contract: Britain's Westland *Super Lynx*, probably costing R800m ($127m), and the Russian Kaman *Sea Sprite*, whose price is undisclosed, but likely to be less than that of its competitor. Additional major orders include combat aircraft and trainers – 28 JAS-39 *Gripens* from Sweden and 24 *Hawks* from the UK – and 40 A-119 helicopters from Italy. It is not certain that the orders for these and other major weapons will be confirmed or that the weapons will be delivered in the number and to the schedule specified by the procurement plan.

In its annual report on 4 August 2000, Armscor, the South African defence procurement agency, recorded a surplus of R2.3m ($365,000), down from R7.9m ($1.25m) in the previous year. This surplus will be reversed over the next decade as the SANDF's eight-year procurement plan gathers pace, even if some of the orders are not fully taken up. The report stated that Armscor had spent R2.6bn ($413m) on arms procurement over the past year. The South African arms industry experienced an increase in export orders, particularly from the UK, in 2000. These orders illustrate the wide range of items that the industry offers at all technological levels. For example, Barlow

Ltd has concluded its biggest contract ever: for R530m, it will service and maintain logistic vehicles for the UK Ministry of Defence at 583 facilities in 17 countries for ten years. At the other end of the technology scale, Paradigm Systems Technology has become the first African supplier to the Eurofighter programme. Paradigm will supply software to the UK and throughout Europe for British Aerospace Systems, a vital element in the Eurofighter's support infrastructure when it begins service in 2002. Paradigm already has major contracts with the UK and French Ministries of Defence as well as the SANDF.

Table 32 **Arms orders and deliveries, Sub-Saharan Africa, 1998–2000**

Country	Country supplier	Classification	Designation	Quantity	Order date	Delivery date	Comment
Angola	RF	MBT	**T-72**	n.k.	1997	1999	
	RF	FGA	**MiG-23**	18	1997	1997	Deliveries into 1998
	RF	FGA	**SU-24**	n.k.	1999	2000	
	RF	FGA	**SU-27**	8	1999	2000	
	Slvk	FGA	**SU-22M4**	12	1999	2000	
	Kaz	MRL	**BM-21**	4	1997	1998	
	Bel	APC	**BMP-1**	7	1998	1999	
	Bel	MRL	**BM-21**	n.k.	1998	1999	
Botswana	Ca	FGA	**F-5**	13	1996	1996	Final 10 in 1997
	A	lt tk	**SK-105**	30	1997	1998	Option on further 20
Burundi	RSA	APC	**RG-31**	12	1997	1998	
Cameroon	Il	arty	**155mm**	8	1996	1997	4 in 1997, 4 in 1998
Côte d'Ivoire	PRC	AF	***Atchan***	1	1994	1998	Logistic support ship
DR Congo	Pl	mor	**120mm**	18	1997	1998	
Eritrea	Il	tpt	**IAI-1125**	1	1997	1998	
	RF	FGA	**MiG-29**	6	1998	1998	
	SF	trg	***Rodrigo***	8	1998	1999	
	It	cbt hel	***Augusta***	n.k.	1998	1998	
	Bg	MRL	**BM-21**	n.k.	1998	1998	
	RF	hel	**Mi-17**	4	1998	1999	
	RF	SAM	**SA-18**	200	1999	1999	
	Mol	FGA	**MiG-21**	6	1999	1999	
	Ga	FGA	**Su-25**	8	1999	1999	
Ethiopia	US	tpt	**C-130B**	4	1995	1998	Ex-USAF
	RF	cbt hel	**Mi-24**	4	1998	1998	
	RF	hel	**Mi-17**	8	1998	1998	
	Bg	MBT	**T-55**	140	1998	1995	50 in 1998; deliveries to 1999
	R	FGA	**Mig-21/23**	10	1998	1999	
	RF	FGA	**Su-27**	8	1998	1998	
	RF	SPA	**152mm**	10	1999	1999	
Kenya	Fr	LACV		4	1997	1998	Riot control
Namibia	Br	PCI		n.k.	1996	1999	

Country	Country supplier	Classification	Designation	Quantity	Order date	Delivery date	Comment
	RSA	arty	**140mm**	24	1997	1998	Free transfer
	Ge	APC	***Wolf***	30	1998	1999	
Rwanda	RSA	APC	**RG-31**	14	1995	1997	4 in 1997, 10 in 1998
Senegal	Fr	LACV		10	1997	1998	Free transfer
Sierra Leone	Ukr	cbt hel	**Mi-24**	2	1996	1999	
South Africa	dom	AAM	**R-*Darter***	n.k.	1988	1998	Development
	dom	APC	***Mamba* Mk2**	586	1993	1995	Deliveries completed in 1998
	US	tpt	**C-130**	12	1995	1997	Upgrades to 2002
	dom	cbt hel	***Rooivalk***	12	1996	1999	Deliveries to 2000
	dom	arty	**155mm**		1997	2006	Development
	Ge	SSK	**Type 209**	3	1998	2004	Deliveries 2004–6
	Ge	FSG	***Meko A-200***	4	1998	2002	Deliveries to 2004
	dom	arty	**LIW 35 DPG**		1998		Development
	It	hel	**A-119**	40	1998	2002	
	Swe	FGA	**JAS-39**	28	1998	2009	Deliveries 2009–14
	UK	FGA	***Hawk***	24	1998	2004	
	UK	cbt hel	***Lynx***	4	1998	2002	
	dom	SSK	***Daphne***	2	1998	1999	Upgrade 1999–2000
Tanzania	RSA	hel	**SA-316**	4	1998	1998	Free transfer
Uganda	RF	FGA	**MiG-21/23**	28	1998	1998	
	Bg	MBT	**T-54**	90	1998	1998	All delivered in 1998
	RSA	APC	***Chubby***		1998		Mine Clearing veh
Zimbabwe	Fr	ACV	**ACMAT**	23	1992	1999	
	It	trg	**SF-260F**	6	1997	1999	

Dollar GDP figures in Sub-Saharan Africa are usually based on African Development Bank estimates. In several cases, the dollar GDP values do not reflect the exchange rates shown in the country entry.

Angola Ang

kwanza		1998	1999	2000	2001
GDP	US$	ε7.0bn	ε6.1bn		
per capita	US$	1,500	1,200		
Growth	%	-1.1	-19.0		
Inflation	%	74.7	124.9		
Debt	US$	12.0bn	12.6bn		
Def exp	US$	ε955m	ε1,005m		
Def bdgt	US$	458m	574m	ε542m	
FMA[a] (Fr)	US$	0.03m	0.5m	0.1m	
FMA (US)	US$	0.0m	3.6m	3.4m	
US$1=kwanza		392,000	2,790,706	7.4	

[a] UNOMA **1998** US$46m

Population			12,410,000
(Ovimbundu 37%, Kimbundu 25%, Bakongo 13%)			
Age	13–17	18–22	23–32
Men	672,000	568,000	864,000
Women	674,000	572,000	882,000

Total Armed Forces

ACTIVE ε107,500

A unified national army, including UNITA troops, was to have been formed with a str of ε90,000. The integration process has been abandoned and civil war between government and UNITA forces resumed in December 1998.

Army ε100,000

35 regts/dets/gps (armd and inf – str vary)

EQUIPMENT†

MBT 300 T-54/-55, ε230 T-62, ε30 T-72
RECCE some 40+ BRDM-2
AIFV ε400 BMP-1/-2
APC 100 BTR-60/-80/-152
TOWED ARTY 300: incl **76mm**: M-1942 (ZIS-3); **85mm**: D-44; **122mm**: D-30; **130mm**: M-46
ASLT GUNS 100mm: SU-100
MRL 122mm: 50 BM-21, 40 RM-70; **240mm**: some BM-24
MOR 82mm: 250; **120mm**: 40+ M-43
ATGW AT-3 *Sagger*
RCL 500: **82mm**: B-10; **107mm**: B-11
AD GUNS 200+: **14.5mm**: ZPU-4; **23mm**: ZU-23-2, 20 ZSU-23-4 SP; **37mm**: M-1939; **57mm**: S-60 towed, 40 ZSU-57-2 SP
SAM SA-7/-14

Navy ε1,500†

BASE Luanda (HQ)

PATROL, INSHORE 7†

4 *Mandume* Type 31.6m PCI<, 3 *Patrulheiro* PCI< (all non-op)

COASTAL DEFENCE†

SS-C-l *Sepal* at Luanda (non-op)

Air Force/Air Defence 6,000

104 cbt ac, 40 armed hel
FGA 30 MiG-23, 12 Su-22 (a further 9 Su-22M4 being delivered), 22 Su-25, 2 Su-27
FTR 20 MiG-21 MF/bis
CCT/RECCE 9* PC-7/9
MR 2 EMB-111, 1 F-27MPA, 1 *King Air* B-200B
ATTACK HEL 15 Mi-25/35, 5 SA-365M (guns), 6 SA-342 (HOT), 14 Mi-24B
TPT 2 An-2, 9 An-26, 6 BN-2, 2 C-212, 4 PC-6B, 2 L-100-20, 2 C-130, 8 An-12 and Il-76 leased from Ukraine
HEL 8 AS-565, 30 IAR-316, 25 Mi-8/17
TRG 3 Cessna 172, 6 Yak-11, Emb-312
AD 5 SAM bn, 10 bty with 40 SA-2, 12 SA-3, 25 SA-6, 15 SA-8, 20 SA-9, 10 SA-13 (mostly unserviceable)
MISSILES
ASM HOT, AT-2 *Swatter*
AAM AA-2 *Atoll*

Forces Abroad

DROC: 1,000 **CONGO**: 500 reported

Paramilitary 10,000

RAPID-REACTION POLICE 10,000

Opposition

UNITA (Union for the Total Independence of Angola)

ε20,000 fully equipped tps plus 30,000 spt militia reported

EQPT T-34/-85, T-55, T-62 MBT; BMP-1, BMP-2 AIFV; misc APC; **75mm, 76mm, 100mm, 122mm, 130mm, 155mm** fd guns; BM-21 **122mm** MRL; **81mm, 82mm, 120mm** mor; **85mm** RPG-7 RL; **75mm** RCL; **12.7mm** hy machine guns; **14.5mm, 20mm**, ZU-23-2 **23mm** AA guns; SAM-7 (much eqpt is unserviceable)

No cbt ac or armed hel

FLEC (Front for the Liberation of the Cabinda Enclave)

ε600 (claims 5,000)

Small arms only

Benin — Bn

CFA fr		1998	1999	2000	2001
GDP	fr	1.3tr	1.5tr		
	US$	2.3bn	2.4bn		
per capita	US$	1,900	2,000		
Growth	%	4.4	5.0		
Inflation	%	5.7	0.3		
Debt	US$	1.3bn	1.4bn		
Def exp	fr	ε19bn	ε21bn		
	US$	32m	34m		
Def bdgt	fr	ε17bn	ε21bn	ε26bn	
	US$	29m	34m	37m	
FMA (Fr)	US$	5m	4m	4m	
FMA (US)	US$	0.4m	0.4m	0.4m	0.4m
US$1=fr		584	616	708	

Population			6,295,000
Age	13–17	18–22	23–32
Men	375,000	308,000	435,000
Women	383,000	322,000	472,000

Total Armed Forces

ACTIVE **e4,750**

Terms of service conscription (selective), 18 months

Army 4,500

3 inf, 1 AB/cdo, 1 engr bn, 1 armd sqn, 1 arty bty

EQUIPMENT

LT TK 20 PT-76 (op status uncertain)
RECCE 9 M-8, 14 BRDM-2, 10 VBL
TOWED ARTY 105mm: 4 M-101, 12 L-118
MOR 81mm
RL 89mm: LRAC

Navy† ε100

BASE Cotonou

PATROL, INSHORE 1

1 *Patriote* PFI (Fr 38m)<

Air Force† 150

no cbt ac

AC 2 An-26, 2 C-47, 1 *Commander* 500B, 2 Do-128, 1 Boeing 707-320 (VIP), 1 F-28 (VIP), 1 DHC-6

HEL 2 AS-350B, 1 SE-3130

Forces Abroad

UN AND PEACEKEEPING

DROC (MONUC): 8 obs

Paramilitary 2,500

GENDARMERIE 2,500

4 mobile coy

Botswana — Btwa

pula P		1998	1999	2000	2001
GDP	P	20.4bn	23.3bn		
	US$	4.8bn	5.0bn		
per capita	US$	6,000	6,200		
Growth	%	4.5	4.0		
Inflation	%	6.8	7.1		
Debt	US$	589m	583m		
Def exp	P	1,080m	ε1,200m		
	US$	256m	260m		
Def bdgt	P	868m	990m	1,243m	
	US$	205m	214m	234m	
FMA (US)	US$	0.5m	0.5m	0.5m	0.5m
US$1=P		4.23	4.62	5.31	

Population			1,701,000
Age	13–17	18–22	23–32
Men	104,000	88,000	136,000
Women	106,000	90,000	139,000

Total Armed Forces

ACTIVE 9,000

Army 8,500 (to be 10,000)

2 inf bde: 4 inf bn, 1 armd recce, 2 AD arty, 1 engr regt, 1 cdo unit • 1 arty bde

EQUIPMENT

LT TK 36 *Scorpion* (incl variants), 30 SK-105 *Kuerassier*
RECCE 12 V-150 *Commando* (some with **90mm** gun), RAM-V
APC 30 BTR-60, 6 *Spartan*
TOWED ARTY 105mm: 12 L-118, 6 Model 56 pack; **155mm:** Soltam (reported)
MOR 81mm: 12; **120mm:** 6 M-43
ATGW 6 TOW (some SP on V-150)
RCL 84mm: 50 *Carl Gustav*
AD GUNS 20mm: 7 M-167
SAM 12 SA-7, 10 SA-16, 5 *Javelin*

Air Wing 500

32 cbt ac, no armed hel

FTR/FGA 10 F-5A, 3 F-5B

TPT 2 CN-235, 2 *Skyvan* 3M, 1 BAe 125-800, 3 C-130, 2 CN-212 (VIP), 1 *Gulfstream* IV, 12* BN-2 *Defender*

TRG 2 sqn 2 Cessna 152, 7* PC-7

HEL 5 AS-350B, 5 Bell 412

Paramilitary 1,000

POLICE MOBILE UNIT 1,000
(org in territorial coy)

Burkina Faso BF

CFA fr		1998	1999	2000	2001
GDP	fr	1.5tr	1.6tr		
	US$	3.2bn	3.5bn		
per capita	US$	1,000	1,000		
Growth	%	6.0	5.3		
Inflation	%	5.7	-1.1		
Debt	US$	1.4bn	1.6bn		
Def exp	fr	ε47bn	ε46bn		
	US$	80m	75m		
Def bdgt	fr	ε45bn	ε46bn	ε49bn	
	US$	76m	74m	69m	
FMA (Fr)	US$	5m	4m	3m	
FMA (US)	US$				0.1m
US$1=fr		590	616	708	

Population			**12,152,000**
Age	13–17	18–22	23–32
Men	706,000	576,000	842,000
Women	681,000	559,000	868,000

Total Armed Forces

ACTIVE 10,000
(incl *Gendarmerie*)

Army 5,600

6 Mil Regions • 5 inf 'regt': HQ, 3 'bn' (each 1 coy of 5 pl) • 1 AB 'regt': HQ, 1 'bn', 2 coy • 1 tk 'bn': 2 pl • 1 arty 'bn': 2 tp • 1 engr 'bn'

EQUIPMENT

RECCE 15 AML-60/-90, 24 EE-9 *Cascavel*, 10 M-8, 4 M-20, 30 *Ferret*
APC 13 M-3
TOWED ARTY 105mm: 8 M-101; **122mm**: 6
MRL 107mm: PRC Type-63
MOR 81mm: Brandt
RL 89mm: LRAC, M-20
RCL 75mm: PRC Type-52
AD GUNS 14.5mm: 30 ZPU
SAM SA-7

Air Force 200

5 cbt ac, no armed hel
TPT 1 *Beech Super King*, 1 *Commander* 500B, 2 HS-748, 2 N-262, 1 Boeing 727 (VIP)
LIAISON 2 Cessna 150/172, 1 SA-316B, 1 AS-350, 3 Mi-8/17
TRG 5* SF-260W/WL

Forces Abroad

UN AND PEACEKEEPING
DROC (MONUC): 2 obs

Paramilitary

GENDARMERIE 4,200
SECURITY COMPANY (CRG) 250
PEOPLE'S MILITIA (R) 45,000 trained

Burundi Bu

franc fr		1998	1999	2000	2001
GDP	fr	506bn	552bn		
	US$	1.1bn	1.1bn		
per capita	US$	600	600		
Growth	%	4.6	-4.7		
Inflation	%	12.5	3.4		
Debt	US$	1.2bn	1.1bn		
Def exp	fr	ε36bn	ε39bn		
	US$	80m	69m		
Def bdgt	fr	27bn	35bn	ε42bn	
	US$	60m	62m	62m	
FMA (US)	US$	0.1m			0.1m
US$1=fr		448	564	674	

Population	ε**7,232,000** (Hutu 85%, Tutsi 14%)		
Age	13–17	18–22	23–32
Men	440,000	354,000	532,000
Women	401,000	325,000	494,000

Total Armed Forces

ACTIVE 45,500
(incl *Gendarmerie*)

Army ε40,000

7 inf bn • 2 lt armd 'bn' (sqn), 1 arty bn • 1 engr bn • some indep inf coy • 1 AD bty

RESERVES

10 bn (reported)

EQUIPMENT

RECCE 85 incl 18 AML (6-60, 12-90), 7 Shorland, 30 BRDM-2
APC 9 Panhard M-3, 20 BTR-40
TOWED ARTY 122mm: 18 D-30
MRL 122mm: 12 BM-21
MOR 100+ incl **82mm**: M-43; **120mm**
RL 83mm: *Blindicide*
RCL 75mm: 15 PRC Type-52

AD GUNS 375: **14.5mm**: 15 ZPU-4; **23mm**: ZU-23; **37mm**: Type-54
SAM SA-7

AIR WING (200)
4 cbt ac, no armed hel
TRG 4* SF-260W/TP
TP 2 DC-3
HEL 3 SA-316B, 2 Mi-8

Forces Abroad

DROC: ε1,000 reported

Paramilitary

GENDARMERIE ε5,500 (incl ε50 Marine Police): 16 territorial districts
BASE Bujumbura
3 *Huchan* (PRC Type 026) PHT† plus 1 LCT, 1 spt, 4 boats

GENERAL ADMINISTRATION OF STATE SECURITY ε1,000

Opposition

FORCES FOR THE DEFENCE OF DEMOCRACY (FDD) up to 10,000 reported
HUTU PEOPLE'S LIBERATION PARTY (PALIPEHUTU) armed wing (FNL) ε2–3,000
NATIONAL COUNCIL FOR THE DEFENCE OF DEMOCRACY (CNDD) ε1,000

Cameroon Crn

CFA fr		1998	1999	2000	2001
GDP	fr	5.6bn	6.3bn		
	US$	9.5bn	10.2bn		
per capita	US$	2,300	2,300		
Growth	%	5.0	4.4		
Inflation	%	2.1	2.0		
Debt	US$	7.5bn	7.9bn		
Def exp	fr	86bn	ε95bn		
	US$	146m	154m		
Def bdgt	fr	86bn	95bn	ε111bn	
	US$	146m	154m	155m	
FMA (US)	US$	0.1m	0.2m	0.2m	0.2m
FMA (Fr)	US$	10m	9m	8m	
US$1=fr		590	616	708	

Population			**15,549,000**
Age	13–17	18–22	23–32
Men	866,000	745,000	1,121,000
Women	863,000	747,000	1,139,000

Total Armed Forces

ACTIVE ε22,100
(incl *Gendarmerie*)

Army 11,500

8 Mil Regions each 1 inf bn under comd • Presidential Guard: 1 guard, 1 armd recce bn, 3 inf coy • 1 AB/cdo bn • 1 arty bn (5 bty) • 5 inf bn (1 trg) • 1 AA bn (6 bty) • 1 engr bn

EQUIPMENT
RECCE 8 M-8, *Ferret*, 8 V-150 *Commando* (**20mm** gun), 5 VBL
AIFV 14 V-150 *Commando* (**90mm** gun)
APC 21 V-150 *Commando*, 12 M-3 half-track
TOWED ARTY 75mm: 6 M-116 pack; **105mm**: 16 M-101; **130mm**: 12 Type-59; **155mm**: 4 I1
MRL 122mm: 20 BM-21
MOR 81mm (some SP); **120mm**: 16 Brandt
ATGW *Milan*
RL 89mm: LRAC
RCL 57mm: 13 PRC Type-52; **106mm**: 40 M-40A2
AD GUNS 14.5mm: 18 PRC Type-58; **35mm**: 18 GDF-002; **37mm**: 18 PRC Type-63

Navy ε1,300

BASES Douala (HQ), Limbe, Kribi
PATROL AND COASTAL COMBATANTS 3
PATROL, COASTAL 2
1 *Bakassi* (Fr P-48) PCC, 1 *L'Audacieux* (Fr P-48) PCC
PATROL, INSHORE 1
1 *Quartier* PCI<
PATROL, RIVERINE craft only
6 US *Swift*-38†, 6 *Simonneau*†

Air Force 300

15 cbt ac, 4 armed hel
1 composite sqn, 1 Presidential Fleet
FGA 4† *Alpha Jet*, 5 CM-170, 6 MB-326
MR 2 Do-128D-6
ATTACK HEL 4 SA-342L (with HOT)
TPT ac 3 C-130H/-H-30, 1 DHC-4, 4 DHC-5D, 1 IAI-201, 2 PA-23, 1 *Gulfstream* III, 1 Do-128, 1 Boeing 707 **hel** 3 Bell 206, 3 SE-3130, 1 SA-318, 3 SA-319, 2 AS-332, 1 SA-365

Paramilitary

GENDARMERIE 9,000
10 regional groups; about 10 US *Swift*-38 (see Navy)

Cape Verde CV

escudo E		1998	1999	2000	2001
GDP	E	23bn	26bn		
	US$	237m	257m		
per capita	US$	2,200	2,300		
Growth	%	4.1	6.0		
Inflation	%	4.5	4.0		
Debt	US$	243m	261m		
Def exp	E	380m	700m		
	US$	4m	7m		
Def bdgt	E	380m	700m	925m	
	US$	4m	7m	8m	
FMA (US)	US$	0.1m	0.1m	0.1m	0.1m
FMA (Fr)	US$	0.1m			
US$1=E		98	103	122	

Population			**470,000**
Age	13–17	18–22	23–32
Men	27,000	24,000	37,000
Women	28,000	25,000	41,000

Total Armed Forces

ACTIVE ε1,150

Terms of service conscription (selective)

Army 1,000

2 inf bn gp

EQUIPMENT

RECCE 10 BRDM-2
TOWED ARTY 75mm: 12; **76mm**: 12
MOR 82mm: 12; **120mm**: 6 M-1943
RL 89mm: 3.5in
AD GUNS 14.5mm: 18 ZPU-1; **23mm**: 12 ZU-23
SAM 50 SA-7

Coast Guard ε50

1 *Kondor* I PCC
1 *Zhuk* PCI<, 1 *Espadarte* PCI<

Air Force under 100

no cbt ac
MR 1 Do-228

Central African Republic CAR

CFA fr		1998	1999	2000	2001
GDP	fr	621bn	696bn		
	US$	1.1bn	1.1bn		
per capita	US$	1,300	1,300		
Growth	%	5.6	5.0		
Inflation	%	2.2	2.4		
Debt	US$	923m	835m		
Def exp	fr	ε29bn	ε28bn		
	US$	49m	46m		
Def bdgt	fr	ε23bn	ε28bn	ε31bn	
	US$	39m	46m	44m	
FMA[a] (US)	US$	0.2m	0.1m	0.1m	0.1m
FMA (Fr)	US$	6.0m	5.0m	4.0m	
US$1=fr		590	616	708	

[a] MISAB **1997–98** US$102m; MINURCA **1998** US$52m **1999** US$34m

Population			**3,867,000**
Age	13–17	18–22	23–32
Men	210,000	170,000	296,000
Women	209,000	175,000	294,000

Total Armed Forces

ACTIVE ε4,150

(incl *Gendarmerie*)

Terms of service conscription (selective), 2 years; reserve obligation thereafter, term n.k.

Army ε3,000

1 territorial defence regt (bn) • 1 combined arms regt (1 mech, 1 inf bn) • 1 spt/HQ regty

EQUIPMENT†

MBT 4 T-55
RECCE 10 *Ferret*
APC 4 BTR-152, some 10 VAB, 25+ ACMAT
MOR 81mm; 120mm: 12 M-1943
RL 89mm: LRAC
RCL 106mm: 14 M-40
RIVER PATROL CRAFT 9<

Air Force 150

no cbt ac, no armed hel
TPT 1 Cessna 337, 1 *Mystère Falcon* 20, 1 *Caravelle*
LIAISON 6 AL-60, 6 MH-1521
HEL 1 AS-350, 1 SE-3130

Paramilitary

GENDARMERIE ε1,000

3 regional legions, 8 'bde'

Chad Cha

CFA fr		1998	1999	2000	2001
GDP	fr	989bn	1,024bn		
	US$	1.64bn	1.66bn		
per capita	US$	900	900		
Growth	%	6.4	-1.1		
Inflation	%	4.5	-6.8		
Debt	US$	1,034m	1,127m		
Def exp	fr	ε37bn	ε29bn		
	US$	63m	47m		
Def bdgt	fr	ε27bn	ε29bn	ε34bn	
	US$	45m	47m	48m	
FMA (Fr)	US$	10m	10m	8m	
FMA (US)	US$	0.1m	0.8m	0.7m	0.7m
US$1=fr		590	616	708	

Population			7,353,000
Age	13–17	18–22	23–32
Men	396,000	324,000	508,000
Women	395,000	325,000	517,000

Total Armed Forces

ACTIVE ε30,350
(incl Republican Guard)
Terms of service conscription authorised

Army ε25,000

(being re-organised)
7 Mil Regions

EQUIPMENT

MBT 60 T-55
AFV 4 ERC-90, some 50 AML-60/-90, 9 V-150 with **90mm**, some EE-9 *Cascavel*
TOWED ARTY 105mm: 5 M-2
MOR 81mm; 120mm: AM-50
ATGW *Milan*
RL 89mm: LRAC
RCL 106mm: M-40A1; **112mm:** APILAS
AD GUNS 20mm, 30mm

Air Force 350

4 cbt ac, no armed hel
TPT ac 3 C-130, 1 C-212, 1 An-26 **hel** 2 SA-316
LIAISON 2 PC-6B, 5 Reims-Cessna FTB 337
TRG 2* PC-7, 2* SF-260W

Paramilitary 4,500 active

REPUBLICAN GUARD 5,000

GENDARMERIE 4,500

Opposition

WESTERN ARMED FORCES str n.k.

Foreign Forces

FRANCE 990: 2 inf coy; 1 AML sqn(-); 1 C-160, 1 C-130, 3 F-ICT, 2 F-ICR, 3 SA-330 hel

Congo RC

CFA fr		1998	1999	2000	2001
GDP	fr	1.2tr	1.3tr		
	US$	2.1bn	2.2bn		
per capita	US$	1,800	1,800		
Growth	%	-8.0	0.7		
Inflation	%	5.5	2.4		
Debt	US$	9.2bn	8.8bn		
Def exp	fr	ε48bn	ε45bn		
	US$	81m	73m		
Def bdgt	fr	ε37bn	ε45bn	ε52bn	
	US$	63m	73m	73m	
FMA (US)	US$				0.1m
FMA (Fr)	US$	2.0m	1.0m	1.0m	
US$1=fr		590	616	708	

Population 3,138,000
(Kongo 48%, Sangha 20%, Teke 17%, M'Bochi 12%, European mostly French 3%)

Age	13–17	18–22	23–32
Men	181,000	145,000	229,000
Women	171,000	138,000	221,000

Total Armed Forces

ACTIVE ε10,000

Army 8,000

2 armd bn • 2 inf bn gp (each with lt tk tp, 76mm gun bty) • 1 inf bn • 1 arty gp (how, MRL) • 1 engr bn • 1 AB/cdo bn

EQUIPMENT†

MBT 25 T-54/-55, 15 PRC Type-59 (some T-34 in store)
LT TK 10 PRC Type-62, 3 PT-76
RECCE 25 BRDM-1/-2
APC M-3, 50 BTR (30 -60, 20 -152), 18 Mamba
TOWED ARTY 76mm: M-1942; **100mm:** 10 M-1944; **122mm:** 10 D-30; **130mm:** 5 M-46; **152mm:** some D-20
MRL 122mm: 8 BM-21; **140mm:** BM-14-16
MOR 82mm; 120mm: 28 M-43
RCL 57mm: M-18
ATK GUNS 57mm: 5 M-1943
AD GUNS 14.5mm: ZPU-2/-4; **23mm:** ZSU-23-4 SP; **37mm:** 28 M-1939; **57mm:** S-60; **100mm:** KS-19

Navy† ε800

BASE Pointe Noire

PATROL AND COASTAL COMBATANTS 3†

PATROL, INSHORE 3†

3 Sov *Zhuk* PFI< (all non-op)

PATROL, RIVERINE n.k.

boats only

Air Force† 1,200

12 cbt ac, no armed hel

FGA 12 MiG-21

TPT 5 An-24, 1 An-26, 1 Boeing 727, 1 N-2501

TRG 4 L-39

HEL 2 SA-316, 2 SA-318, 1 SA-365, 2 Mi-8

MISSILES

AAM AA-2 *Atoll*

Paramilitary 2,000 active

GENDARMERIE 2,000

20 coy

PEOPLE'S MILITIA 3,000

being absorbed into national Army

PRESIDENTIAL GUARD

(forming)

Foreign Forces

ANGOLA: 500 reported

Côte D'Ivoire CI

CFA fr		1998	1999	2000	2001
GDP	fr	6.9tr	8.0tr		
	US$	12.6bn	13.1bn		
per capita	US$	1,700	1,700		
Growth	%	6.0	1.4		
Inflation	%	4.7	0.8		
Debt	US$	14.6bn	15.1bn		
Def exp	fr	ε70bn	ε80bn		
	US$	119m	130m		
Def bdgt	fr	ε70bn	ε80bn	ε95bn	
	US$	119m	130m	134m	
FMA (US)	US$	0.2m	0.2m	0.2m	0.1m
FMA (Fr)	US$	7.0m	6.0m	5.0m	
US$1=fr		590	616	708	

Population			17,050,000
Age	13–17	18–22	23–32
Men	1,026,000	810,000	1,173,000
Women	1,023,000	813,000	1,167,000

Total Armed Forces

ACTIVE ε13,900

(incl Presidential Guard, *Gendarmerie*)

Terms of service conscription (selective), 6 months

RESERVES 12,000

Army 6,800

4 Mil Regions • 1 armd, 3 inf bn, 1 arty gp • 1 AB, 1 AAA, 1 engr coy

EQUIPMENT

LT TK 5 AMX-13

RECCE 7 ERC-90 *Sagaie*, 16 AML-60/-90, 10 *Mamba*

APC 16 M-3, 13 VAB

TOWED ARTY 105mm: 4 M-1950

MOR 81mm; 120mm: 16 AM-50

RL 89mm: LRAC

RCL 106mm: M-40A1

AD GUNS 20mm: 16, incl 6 M-3 VDA SP; **40mm**: 5 L/60

Navy ε900

BASE Locodjo (Abidjan)

PATROL AND COASTAL COMBATANTS 3

PATROL, COASTAL 1 *Le Valereux* (Fr SFCN 47m) PCC

PATROL, INSHORE 2 *L'Ardent* (Fr *Patra*) PCI†

AMPHIBIOUS 1

1 *L'Eléphant* (Fr *Batral*) LST, capacity 140 tps, 7 tk, hel deck, plus some 8 craft†

Air Force 700

5† cbt ac, no armed hel

FGA 1 sqn with 5† *Alpha Jet*

TPT 1 hel sqn with 1 SA-318, 1 SA-319, 1 SA-330, 4 SA 365C

PRESIDENTIAL FLT ac 1 F-28, 1 *Gulfstream* IV, 3 Fokker 100 **hel** 1 SA-330

TRG 3 Beech F-33C, 2 Reims Cessna 150H

LIAISON 1 Cessna 421, 1 *Super King Air* 200

Paramilitary

PRESIDENTIAL GUARD 1,100

GENDARMERIE 4,400

VAB APC, 4 patrol boats

MILITIA 1,500

Foreign Forces

FRANCE 500: 1 marine inf bn (18 AML 601/90); 1 AS-555 hel

Democratic Republic of Congo DROC

congolese franc fr		1998	1999	2000	2001
GDP	US$	5.5bn	5.3bn		
per capita	US$	400	400		
Growth	%	-6.0	-2.0		
Inflation	%	5.0	12.0		
Debt	US$	12.9bn	16bn		
Def exp	US$	ε364m	ε411m		
Def bdgt	US$	ε250m	ε400m	ε400m	
FMA (US)	US$			0.04m	0.1m
US$1=fr[a]		ε138,000	4.5	9.0	

[a] Congolese franc became sole legal tender in July 1999

Population ε49,000,000

(Bantu and Hamitic 45%; minority groups include Hutus and Tutsis)

Age	13–17	18–22	23–32
Men	3,041,000	2,427,000	3,516,000
Women	3,007,000	2,420,000	3,547,000

Total Armed Forces

ACTIVE ε55,900

Army ε55,000

10+ inf, 1 Presidential Guard bde
1 mech inf bde, 1 cdo bde (reported)

EQUIPMENT†

MBT 20 PRC Type-59 (being refurbished), some 40 PRC Type-62
APC M-113, YW-531, Panhard M-3, some *Casspir*, *Wolf* Turbo 2, *Fahd*
TOWED ARTY 75mm: M-116 pack; **85mm**: Type-56; **122mm**: M-1938/D-30, Type-60; **130mm**: Type-59
MRL 107mm: Type 63; **122mm**: BM-21
MOR 81mm; **107mm**: M-30; **120mm**: Brandt
RCL 57mm: M-18; **75mm**: M-20; **106mm**: M-40A1
AD GUNS 14.5mm: ZPU-4; **37mm**: M-1939/Type; **40mm**: L/60
SAM SA-7

Navy ε900

BASES Coast Banana **River** Boma, Matadi, Kinshasa **Lake Tanganyika** (4 boats)

PATROL AND COASTAL COMBATANTS 6†

PATROL, COASTAL/INSHORE 6†
4 PRC *Shanghai* II PCC (most non-op)
2 *Swiftships* PCI<, plus about 6 armed boats (most non-op)

Air Force

Only a handful of utility and communications ac remain serviceable. 10 Su-25 reported on order.

Paramilitary

NATIONAL POLICE incl Rapid Intervention Police (National and Provincial forces)

PEOPLE'S DEFENCE FORCE

Opposition

THE RALLY FOR CONGOLESE DEMOCRACY
ε30,000; split into two factions:
a. **Congolese Rally for Democracy – Liberation Movement** (RCD-ML)
b. **Congolese Rally for Democracy – Goma** (RCD-Goma)

MOVEMENT FOR THE LIBERATION OF THE CONGO
ε18,000

Foreign Forces

In support of government:
ANGOLA: 1,000 **NAMIBIA**: 1,000 **ZIMBABWE**: up to 12,000 reported
In support of opposition:
ANGOLA (UNITA): reported **BURUNDI**: ε1,000 reported **RWANDA**: 15–20,000 reported **UGANDA**: ε15,000
UN (MONUC): 264 obs from 37 countries

Djibouti Dj

franc fr		1998	1999	2000	2001
GDP	fr	74bn	79bn		
	US$	416m	442m		
per capita	US$	900	900		
Growth	US$	1.9	3.9		
Inflation	US$	2.0	2.0		
Debt	US$	291m	309m		
Def exp	fr	ε3.8bn	ε3.9bn		
	US$	21m	22m		
Def bdgt	fr	ε3.5bn	ε3.9bn	ε4.0bn	
	US$	20m	22m	23m	
FMA (US)	US$	0.1m	0.1m	0.4m	0.9m
FMA (Fr)	US$	7.0m	6.0m	5.0m	
US$1=fr		178	178	178	

Population 758,000 (Somali 60%, Afar 35%)

Age	13–17	18–22	23–32
Men	41,000	34,000	55,000
Women	40,000	35,000	60,000

Total Armed Forces

ACTIVE ε9,600
(incl *Gendarmerie*)

Army ε8,000

3 Comd (North, Central, South) • 1 inf bn, incl mor, ATK pl • 1 arty bty • 1 armd sqn • 1 border cdo bn • 1 AB coy • 1 spt bn

EQUIPMENT

RECCE 15 VBL, 4 AML-60†
APC 12 BTR-60 (op status uncertain)
TOWED ARTY 122mm: 6 D-30
MOR 81mm: 25; **120mm**: 20 Brandt
RL 73mm; 89mm: LRAC
RCL 106mm: 16 M-40A1
AD GUNS 20mm: 5 M-693 SP; **23mm**: 5 ZU-23; **40mm**: 5 L/70

Navy ε200

BASE Djibouti
PATROL CRAFT, INSHORE 7
5 *Sawari* PCI<, 2 *Moussa Ali* PCI<, plus boats

Air Force 200

no cbt ac or armed hel
TPT 2 C-212, 2 N-2501F, 2 Cessna U206G, 1 *Socata* 235GT
HEL 3 AS-355, 1 AS-350; Mi-8, Mi-24 hel from **Eth**

Paramilitary ε3,000 active

GENDARMERIE (Ministry of Defence) 1,200
1 bn, 1 patrol boat
NATIONAL SECURITY FORCE (Ministry of Interior) ε3,000

Foreign Forces

FRANCE 3,200: incl 2 inf coy, 2 AMX sqn, 26 ERC90 recce, 6 155mm arty, 16 AA arty, 3 amph craft: 1 sqn: **ac** 6 *Mirage* F-1C (plus 4 in store), 1 C-160 **hel** 2 SA-330, 1 AS-555

Opposition

FRONT FOR THE RESTORATION OF UNITY AND DEMOCRACY (FRUD) str n.k.

Equatorial Guinea EG

CFA fr		1998	1999	2000	2001
GDP	fr	360bn	415bn		
	US$	447m	527m		
per capita	US$	2,700	3,100		
Growth	%	14.5	15.2		
Inflation	%	3.0	3.0		
Debt	US$	235m	215m		
Def exp	fr	ε5bn	ε6bn		
	US$	8m	10m		
Def bdgt	fr	ε3bn	ε5bn	ε8bn	
	US$	5m	8m	11m	
FMA (Fr)	US$	1.0m	1.0m	1.0m	
US$1=fr		590	616	708	

Population			**524,000**
Age	13–17	18–22	23–32
Men	28,000	22,000	37,000
Women	28,000	23,000	37,000

Total Armed Forces

ACTIVE 1,320

Army 1,100

3 inf bn

EQUIPMENT

RECCE 6 BRDM-2
APC 10 BTR-152

Navy† 120

BASES Malabo (Santa Isabel), Bata
PATROL CRAFT, INSHORE 2 PCI<†

Air Force 100

no cbt ac or armed hel
TPT ac 1 Yak-40, 3 C-212, 1 Cessna-337 **hel** 2 SA-316

Paramilitary

GUARDIA CIVIL
2 coy
COAST GUARD
1 PCI<

Eritrea Er

nakfa		1998	1999	2000	2001
GDP	US$	ε700m	ε700m		
per capita	US$	491	466		
Growth	%	ε-1.0	ε-2.0		
Inflation	%	8.3			
Debt	US$	450m			
Def exp	US$	ε292m	ε309m		
Def bdgt	US$	ε208m	ε210m	ε263m	
FMA (US)	US$	0.4m	0.4m	1.4m	1.4m
US$1=nakfa		7.2	ε8.1	ε9.5	

Population			ε4,099,000
(Tigrinya 50%, Tigre and Kunama 40%, Afar 4%, Saho 3%)			
Age	13–17	18–22	23–32
Men	246,000	205,000	311,000
Women	243,000	204,000	310,000

Total Armed Forces

ACTIVE ε200–250,000 (incl ε150–200,000 conscripts)
Terms of service 16 months (4 month mil trg)
RESERVES ε120,000 (reported)
Total holdings of army assets n.k.

Army ε200–250,000 mobilised

6 inf div (incl 1 reserve)
1 cdo div
1 mech bde

EQUIPMENT

MBT 80 T-54/-55
RECCE BRDM-2
AIFV/APC BMP-1, BTR-60
TOWED ARTY 130: **85mm**: D-44; **122mm**: D-30; **130mm**: M-46
MRL 122mm: 30 BM-21
MOR 120mm; 160mm
RL 73mm: RPG-7
ATGW AT-3 *Sagger*
SAM SA-18 (reported)

Navy 1,100

BASES Massawa (HQ), Assab, Dahlak
PATROL AND COASTAL COMBATANTS 10
MISSILE CRAFT 1
1 *Osa* II PFM with 4 SS-N-2B *Styx* SSM
PATROL, INSHORE 9
2 Sov Zhuk PCI<, 4 *Super Dvora* PFI<, 3 *Swiftships* PCI<
AMPHIBIOUS 3
1 *Edic* LCT
2 *Chamo* LST (Ministry of Transport)
plus 2 *Soviet* LCU†

Air Force ε1,000

17† cbt ac
Current types and numbers are assessed as follows:
FTR/FGA 3† MiG-23, 5† MiG-21, 4 MiG-29 (1-UB)
TPT 3 Y-12(II), 1 IAI-1125
TRG 6 L-90 *Redigo*, 5* MB-339CE
HEL 2 Mi-17, 1 Mi-35

Opposition

ALLIANCE OF ERITREAN NATIONAL FORCES
str ε3,000 incl **Eritrean Liberation Front of Abdullah Idris (ELF-AI)** and **Eritrean Liberation Front – National Congress (ELF-NC)**: str n.k.

AFAR RED SEA FRONT str n.k.

Ethiopia Eth

birr EB		1998	1999	2000	2001
GDP	EB	45bn	51bn		
	US$	6.3bn	6.2bn		
per capita	US$	500	500		
Growth	%	-0.5	-3.5		
Inflation	%	6.2	3.6		
Debt	US$	10bn	11bn		
Def exp	EB	2,700m	ε3,600m		
	US$	379m	444m		
Def bdgt	EB	995m	3,500m	3,700m	
	US$	140m	432m	457m	
FMA (US)	US$	0.3m	0.5m	1.5m	1.4m
FMA (Fr)	US$		0.5m	0.5m	
US$1=EB		7.1	8.1	8.1	

Population			ε59,000,000
(Oromo 40%, Amhara and Tigrean 32%, Sidamo 9%, Shankella 6%, Somali 6%, Afar 4%)			
Age	13–17	18–22	23–32
Men	3,842,000	3,083,000	4,647,000
Women	3,724,000	2,944,000	4,477,000

Total Armed Forces

ACTIVE ε352,500

The Ethiopian armed forces were formed following Eritrea's declaration of independence in April 1993. Extensive demobilisation of former members of the Tigray People's Liberation Front (TPLF) has taken place. Ethiopia auctioned off its naval assets in Sep 1996. Currently 17 div reported. Peacetime re-org outlined below.

Army ε350,000 mobilised

Re-org to consist of 3 Mil Regions each with corps HQ (each corps 2 divs, 1 reinforced mech bde); strategic reserve div of 6 bde will be located at Addis Ababa.

MBT 160 T-54/-55, T-62
RECCE/AIFV/APC ε200, incl BRDM, BMP, BTR-60/-152
TOWED ARTY 76mm: ZIS-3; **85mm**: D-44; **122mm**: D-30/M-30; **130mm**: M-46
SP ARTY 122mm: 2S1; **152mm**: 2S19
MRL BM-21
MOR 81mm: M-1/M-29; **82mm**: M-1937; **120mm**: M-1944
ATGW AT-3 *Sagger*
RCL 82mm: B-10; **107mm**: B-11
AD GUNS 23mm: ZU-23, ZSU-23-4 SP; **37mm**: M-1939; **57mm**: S-60
SAM 65: SA-2, SA-3, SA-7

Air Force ε2,500

53 cbt ac, 16-18 armed hel
Air Force operability improved as it played an active role in the war with Eritrea. Types and numbers of ac are assessed as follows:
FGA 24 MiG-21MF, 17 MiG-23BN, 4 Su-25 (2 -25T, 2 -25UB), 8 Su-27
TPT 4 C-130B, 7 An-12, 2 DH-6, 1 Yak-40 (VIP), 2 Y-12
TRG 10 L-39, 10 SF-260
ATTACK HEL 16 Mi-24 (possibly 2 Ka-50)
TPT HEL 22 Mi-8/17

Opposition

THE OROMO LIBERATION FRONT str several hundred
OGADEN NATIONAL LIBERATION FRONT str n.k.

Gabon Gbn

CFA fr		1998	1999	2000	2001
GDP	fr	3.8tr	4.0tr		
	US$	6.1bn	6.4bn		
per capita	US$	5,500	5,600		
Growth	%	1.7	2.5		
Inflation	%	1.7	2.0		
Debt	US$	4.3bn	4.4bn		
Def exp	fr	ε78bn	ε83bn		
	US$	132m	135m		
Def bdgt	fr	ε73bn	ε77bn	ε89bn	
	US$	124m	125m	126m	
FMA (Fr)	US$	8.0m	7.0m	6.0m	
FMA (US)	US$		0.05m	0.1m	0.1m
US$1=fr		590	616	708	

Population			1,515,000
Age	13–17	18–22	23–32
Men	75,000	59,000	96,000
Women	75,000	60,000	101,000

Total Armed Forces

ACTIVE ε4,700

Army 3,200

Presidential Guard bn gp (1 recce/armd, 3 inf coy, arty, AA bty), under direct presidential control
8 inf, 1 AB/cdo, 1 engr coy
EQUIPMENT
RECCE 14 EE-9 *Cascavel*, 24 AML-60/-90, 6 ERC-90 *Sagaie*, 12 EE-3 *Jararaca*, 14 VBL
AIFV 12 EE-11 *Urutu* with **20mm** gun
APC 9 V-150 *Commando*, Panhard M-3, 12 VXB-170
TOWED ARTY 105mm: 4 M-101
MRL 140mm: 8 *Teruel*
MORS 81mm: 35; **120mm**: 4 Brandt
ATGW 4 *Milan*
RL 89mm: LRAC
RCL 106mm: M40A1
AD GUNS 20mm: 4 ERC-20 SP; **23mm**: 24 ZU-23-2; **37mm**: 10 M-1939; **40mm**: 3 L/70

Navy ε500

BASE Port Gentil (HQ)
PATROL AND COASTAL COMBATANTS 2
PATROL, COASTAL 2 *General Ba'Oumar* (Fr P-400 55m) PCC
AMPHIBIOUS 1
1 *President Omar Bongo* (Fr *Batral*) LST, capacity 140 tps, 7 tk; plus craft 1 LCM

Air Force 1,000

10 cbt ac, 5 armed hel
FGA 9 *Mirage* 5 (2 -G, 4 -GII, 3 -DG)
MR 1 EMB-111
TPT 1 C-130H, 3 L-100-30, 1 EMB-110, 2 YS-11A, 1 CN-235
HELICOPTERS 5 SA-342*, 3 SA-330C/-H, 3 SA-316/-319
PRESIDENTIAL GUARD
CCT 4 CM-170, 3 T-34
TPT ac 1 ATR-42F, 1 EMB-110, 1 *Falcon* 900 **hel** 1 AS-332

Paramilitary 2,000

GENDARMERIE 2,000
3 'bde', 11 coy, 2 armd sqn, air unit with 1 AS-355, 2 AS-350

Foreign Forces

FRANCE 680: 1 marine inf bn (4 AML 60) **ac** 2 C-160 **hel** 1 AS-555, 13 AS-532

The Gambia Gam

dalasi D		1998	1999	2000	2001
GDP	D	4.4bn	5.1bn		
	US$	417m	446m		
per capita	US$	1,200	1,200		
Growth	%	3.9	4.2		
Inflation	%	1.2	3.8		
Debt	US$	428m	441m		
Def exp	D	ε160m	ε180m		
	US$	15m	16m		
Def bdgt	D	ε156m	ε180m	ε190m	
	US$	15m	16m	15m	
US$1=D		10.6	11.4	12.6	

Population			1,232,000
Age	13–17	18–22	23–32
Men	66,000	55,000	83,000
Women	67,000	54,000	82,000

Total Armed Forces

ACTIVE 800

Gambian National Army 800

Presidential Guard (reported) • 2 inf bn • engr sqn

MARINE UNIT (about 70)

BASE Banjul

PATROL CRAFT, INSHORE 3

3 PCI<, boats

Forces Abroad

UN AND PEACEKEEPING

SIERRA LEONE (UNAMSIL): 26 obs

Ghana Gha

cedi C		1998	1999	2000	2001
GDP	C	21.8tr	26.7tr		
	US$	9.4bn	10.1bn		
per capita	US$	2,300	2,400		
Growth	%	4.5	4.5		
Inflation	%	14.5	10.0		
Debt	US$	6.4bn	6.6bn		
Def exp[a]	C	ε310bn	ε320bn		
	US$	135m	121m		
Def bdgt	C	133bn	150bn	ε210bn	
	US$	57m	57m	45m	
FMA (US)	US$	0.3m	0.4m	0.4m	$0.4m
US$1=C		2,300	2,647	4,660	

[a] Defence and security budget including police

Population			20,008,000
Age	13–17	18–22	23–32
Men	1,190,000	984,000	1,466,000
Women	1,184,000	982,000	1,477,000

Total Armed Forces

ACTIVE 7,000

Army 5,000

2 Comd HQ • 2 bde (6 inf bn (incl 1 UNIFIL, 1 ECOMOG), spt unit) • 1 Presidential Guard, 1 trg bn • 1 recce regt (3 sqn) • 1 arty 'regt' (1 arty, 2 mor bty) • 1 AB force (incl 1 para coy) • 1 SF bn • 1 fd engr regt (bn)

EQUIPMENT

RECCE 3 EE-9 *Cascavel*
AIFV 50 MOWAG *Piranha*
TOWED ARTY 122mm: 6 D-30
MOR 81mm: 50; **120mm**: 28 Tampella
RCL 84mm: 50 *Carl Gustav*
AD GUNS 14.5mm: 4 ZPU-2, ZPU-4; **23mm**: 4 ZU-23-2
SAM SA-7

Navy 1,000

COMMANDS Western and **Eastern**

BASES HQ Western Sekondi **HQ Eastern** Tema

PATROL AND COASTAL COMBATANTS 4

PATROL, COASTAL 4

2 *Achimota* (Ge *Lürssen* 57m) PFC
2 *Dzata* (Ge *Lürssen* 45m) PFC

Air Force 1,000

19 cbt ac, no armed hel

TPT 5 Fokker (4 F-27, 1 F-28 (VIP)); 1 C-212, 6 *Skyvan*, 1 *Gulfstream*

HEL 4 AB-212 (1 VIP, 3 utl), 2 Mi-2, 4 SA-319

TRG 12* L-29, 2* L-39, 2* MB 339F, 3* MB-326K

Forces Abroad

UN AND PEACEKEEPING

CROATIA (UNMOP): 2 obs **DROC** (MONUC): 8 obs **IRAQ/KUWAIT** (UNIKOM): 5 obs **LEBANON** (UNIFIL): 784; 1 inf bn **SIERRA LEONE** (UNAMSIL): 775 plus 4 obs **WESTERN SAHARA** (MINURSO): 13 incl 6 obs

Guinea Gui

franc fr		1998	1999	2000	2001
GDP	fr	4.2tr	4.5tr		
	US$	3.4bn	3.6bn		
per capita	US$	900	900		
Growth	%	4.9	3.7		
Inflation	%	4.0	4.0		
Debt	US$	3.3bn	3.0bn		
Def exp	fr	ε72bn	ε87bn		
	US$	58m	59m		
Def bdg	fr	ε65bn	ε75bn	ε90bn	
	US$	55m	57m	55m	
FMA (US)	US$	0.2m	0.2m	0.2m	0.2m
FMA (Fr)	US$	6m	5m	4m	
US$1=fr		1,237	1,458	1,645	

Population			7,628,000
Age	13–17	18–22	23–32
Men	436,000	361,000	539,000
Women	445,000	365,000	545,000

Total Armed Forces

ACTIVE 9,700

(perhaps 7,500 conscripts)

Terms of service conscription, 2 years

Army 8,500

1 armd bn • 1 arty bn • 1 cdo bn • 1 engr bn • 5 inf bn • 1 AD bn • 1 SF bn

EQUIPMENT†

MBT 30 T-34, 8 T-54

LT TK 20 PT-76

RECCE 29 BRDM-1/-2, 2 AML-90

APC 40 BTR (16 -40, 10 -50, 8 -60, 6 -152)

TOWED ARTY 76mm: 8 M-1942; **85mm**: 6 D-44; **122mm**: 12 M-1931/37

MOR 82mm: M-43; **120mm**: 20 M-1938/43

RCL 82mm: B-10

ATK GUNS 57mm: M-1943

AD GUNS 30mm: twin M-53; **37mm**: 8 M-1939; **57mm**: 12 S-60, PRC Type-59; **100mm**: 4 KS-19

SAM SA-7

Navy† 400

BASES Conakry, Kakanda

PATROL AND COASTAL COMBATANTS 2†

PATROL, INSHORE 2†

2 US *Swiftships* 77 PCI<

Air Force† 800

8 cbt ac, no armed hel

FGA 4 MiG-17F, 4 MiG-21

TPT 4 An-14, 1 An-24

TRG 2 MiG-15UTI

HEL 1 IAR-330, 1 Mi-8, 1 SA-316B, 1 SA-330, 1 SA-342K

MISSILES

AAM AA-2 *Atoll*

Forces Abroad

UN AND PEACEKEEPING

SIERRA LEONE (UNAMSIL): 778 plus 12 obs

WESTERN SAHARA (MINURSO): 3 obs

Paramilitary 2,600 active

GENDARMERIE 1,000

REPUBLICAN GUARD 1,600

PEOPLE'S MILITIA 7,000

Guinea-Bissau GuB

CFA fr		1998	1999	2000	2001
GDP	fr	164bn	186bn		
	US$	271m	303m)		
per capita	US$	1,000	1,000		
Growth	%	0.6	8.9		
Inflation	%	13.9	6.0		
Debt	US$	780m	790m		
Def exp	US$	5m	6m		
Def bdgt	US$	3m	3m	3m	
FMA (Fr)	US$	0.1m			
FMA (US)	US$	0.1m	–	0.4m	0.6m
US$1=fr		590	616	708	

Population			1,197,000
Age	13–17	18–22	23–32
Men	67,000	59,000	94,000
Women	66,000	55,000	87,000

Total Armed Forces

ACTIVE ε9,250 (all services, incl *Gendarmerie*, form part of the armed forces)

Terms of service conscription (selective)

As a result of the 1998 revolt by dissident army tps, manpower and equipment totals should be treated with caution.

Army 6,800

1 armd 'bn' (sqn) • 5 inf, 1 arty bn • 1 recce, 1 engr coy

EQUIPMENT

MBT 10 T-34

LT TK 20 PT-76

RECCE 10 BRDM-2

APC 35 BTR-40/-60/-152, 20 PRC Type-56

TOWED ARTY 85mm: 8 D-44; 122mm: 18 M-1938/ D-30
MOR 82mm: M-43; 120mm: 8 M-1943
RL 89mm: M-20
RCL 75mm: PRC Type-52; 82mm: B-10
AD GUNS 23mm: 18 ZU-23; 37mm: 6 M-1939; 57mm: 10 S-60
SAM SA-7

Navy ε350

BASE Bissau
PATROL AND COASTAL COMBATANTS 3
PATROL, INSHORE 3
2 *Alfeite* PCI<, 1 PCI<

Air Force 100

3 cbt ac, no armed hel
FTR/FGA 3 MiG-17
HEL 1 SA-318, 2 SA-319

Paramilitary

GENDARMERIE 2,000

Kenya Kya

shilling sh		1998	1999	2000	2001
GDP	sh	699bn	737bn		
	US$	10.1bn	10.5bn		
per capita	US$	1,400	1,500		
Growth	%	1.8	1.3		
Inflation	%	5.8	2.6		
Debt	US$	6.0bn	5.8bn		
Def exp	sh	19bn	ε23bn		
	US$	315m	327m		
Def bdgt	sh	13.0bn	ε16bn	ε18bn	
	US$	195m	228m	235m	
FMA (US)[a]	US$	0.4m	0.5m	0.4m	0.4m
US$1=sh		60.4	70.3	76.7	

[a] Excl ACRI and East Africa Regional funding

Population	**31,409,000** (Kikuyu ε22–32%)		
Age	13–17	18–22	23–32
Men	2,022,000	1,732,000	2,494,000
Women	2,015,000	1,735,000	2,520,000

Total Armed Forces

ACTIVE 22,200

Army ε18,200

1 armd bde (3 armd bn) • 2 inf bde (1 with 2, 1 with 3 inf bn) • 1 indep inf bn • 1 arty bde (2 bn) • 1 AD arty bn • 1 engr bde • 2 engr bn • 1 AB bn • 1 indep air cav bn

EQUIPMENT
MBT 76 Vickers Mk 3
RECCE 72 AML-60/-90, 12 *Ferret*, 8 Shorland
APC 52 UR-416, 10 Panhard M-3 (in store)
TOWED ARTY 105mm: 40 lt, 8 pack
MOR 81mm: 50; **120mm**: 12 Brandt
ATGW 40 *Milan*, 14 *Swingfire*
RCL 84mm: 80 *Carl Gustav*
AD GUNS 20mm: 50 TCM-20, 11 Oerlikon; **40mm**: 13 L/70

Navy 1,000

BASE Mombasa
PATROL AND COASTAL COMBATANTS 4
MISSILE CRAFT 4
2 *Nyayo* (UK Vosper 57m) PFM with 4 *Ottomat* SSM, 1 × 76mm gun, 1 *Mamba* PFM with *Gabriel* SSM, 1 *Madaraka* (UK *Brooke Marine* 37m) with *Gabriel* SSM
AMPHIBIOUS craft only
2 *Galana* LCM
SUPPORT AND MISCELLANEOUS 1
1 AT

Air Force 3,000

30 cbt ac, 34 armed hel
FGA 10 F-5 (8 -E, 2 -F)
TPT 7 DHC-5D, 12 Y-12 (II), 1 PA-31, 3 DHC-8, 1 Fokker 70 (VIP) (6 Do-28D-2 in store)
ATTACK HEL 11 Hughes 500MD (with TOW), 8 Hughes 500ME, 15 Hughes 500M
TPT HEL 9 IAR-330, 3 SA-330, 1 SA-342
TRG 12 *Bulldog* 103/127, 8* *Hawk* Mk 52, 12* *Tucano*, **hel** 2 Hughes 500D
MISSILES
ASM AGM-65 *Maverick*, TOW
AAM AIM-9 *Sidewinder*

Forces Abroad

UN AND PEACEKEEPING
CROATIA (UNMOP): 1 obs **DROC** (MONUC): 2 obs **EAST TIMOR** (UNTAET): 252 **IRAQ/KUWAIT** (UNIKOM): 4 obs **SIERRA LEONE** (UNAMSIL): 878 plus 11 obs **WESTERN SAHARA** (MINURSO): 8 obs

Paramilitary 5,000

POLICE GENERAL SERVICE UNIT 5,000 (may have increased to 9,000)
AIR WING ac 7 Cessna lt **hel** 3 Bell (1 206L, 2 47G)
POLICE NAVAL SQN/CUSTOMS about 5 PCI< (2 Lake Victoria), some 12 boats

Lesotho Ls

maloti M		1998	1999	2000	2001
GDP	M	4.8bn	5.0bn		
	US$	880m	820m		
per capita	US$	2,300	2,300		
Growth	%	1.5	0.5		
Inflation	%	8.4	7.2		
Debt	US$	800m	830m		
Def exp	M	230m	210m		
	US$	42m	34m		
Def bdgt	M	170m	210m	170m	
	US$	31m	34m	26m	
FMA (US)	US$	0.1m	0.1m	0.1m	0.1m
US$1=M		5.5	6.1	7.1	

Population			**2,229,000**
Age	13–17	18–22	23–32
Men	129,000	112,000	167,000
Women	127,000	111,000	169,000

Total Armed Forces

ACTIVE 2,000

A mutiny on 22 September 1998 was quelled by military intervention by South Africa and Botswana. Tps from these countries withdrew on 15 May 1999. Manpower and equipment str should be treated with caution.

Army 2,000

7 inf coy • 1 spt coy (incl recce/AB, 81mm mor) • 1 air sqn

EQUIPMENT

RECCE 10 Il *Ramta*, 8 Shorland, AML-90
MOR 81mm: some
RCL 106mm: M-40
AC 3 C-212 *Aviocar* 300, 1 Cessna 182Q
HEL 2 Bo-105 CBS, 1 Bell 47G, 1 Bell 412 SP, 1 Bell 412EP

Liberia Lb

Liberian dollar L$		1998	1999	2000	2001
GDP	US$	ε370m	ε450m		
per capita	US$	ε600	ε600		
Growth	%	ε-4.0	ε15.0		
Inflation	%	ε11.0	ε1.4		
Debt	US$	2.1bn	2.0bn		
Def exp	US$	ε45m	ε25m		
Def bdgt	US$	6m	13m	15m	
FMA[a] (US)	US$				0.1m
US$1=L$[b]		1.0	1.0	1.0	

[a] ECOMOG **1990–98** εUS$525m
[b] Market rate **1999** US$1=L$41

Population	**ε3,000,000** (Americo-Liberians 5%)		
Age	13–17	18–22	23–32
Men	172,000	142,000	196,000
Women	167,000	137,000	184,000

Total Armed Forces

ACTIVE ε11–15,000 mobilised

Total includes militias supporting government forces. No further details. Plans for a new unified armed foces to be implemented at a future date provide for:

Army 4,000 • **Navy** 1,000 • **Air Force** 300

Madagascar Mdg

franc fr		1998	1999	2000	2001
GDP	fr	20.4tr	23.1tr		
	US$	4.8bn	5.2bn		
per capita	US$	700	700		
Growth	%	3.5	4.5		
Inflation	%	6.2	9.9		
Debt	US$	3.0bn	2.9bn		
Def exp	fr	ε240bn	ε273bn		
	US$	44m	43m		
Def bdgt	fr	ε220bn	ε273bn	ε295bn	
	US$	40m	43m	42m	
FMA (US)	US$	0.1m	0.1m	0.1m	0.1m
FMA (Fr)	US$	6m	5m	5m	
US$1=fr		5,441	6,300	7,000	

Population			**15,500,000**
Age	13–17	18–22	23–32
Men	901,000	747,000	1,122,000
Women	879,000	730,000	1,112,000

Total Armed Forces

ACTIVE some 21,000

Terms of service conscription (incl for civil purposes), 18 months

Army some 20,000

2 bn gp • 1 engr regt

EQUIPMENT

LT TK 12 PT-76
RECCE 8 M-8, ε20 M-3A1, 10 *Ferret*, ε35 BRDM-2
APC ε30 M-3A1 half-track
TOWED ARTY 76mm: 12 ZIS-3; **105mm**: some M-101; **122mm**: 12 D-30
MOR 82mm: M-37; **120mm**: 8 M-43
RL 89mm: LRAC
RCL 106mm: M-40A1
AD GUNS 14.5mm: 50 ZPU-4; **37mm**: 20 Type-55

Navy† 500

(incl some 100 Marines)
BASES Diégo-Suarez, Tamatave, Fort Dauphin, Tuléar, Majunga
AMPHIBIOUS craft only
1 LCT (Fr *Edic*)
SUPPORT AND MISCELLANEOUS 1
1 tpt/trg

Air Force 500

12 cbt ac, no armed hel
FGA 1 sqn with 4 MiG-17F, 8 MiG-21FL
TPT 4 An-26, 1 BN-2, 2 C-212, 2 Yak-40 (VIP)
HEL 1 sqn with 6 Mi-8
LIAISON 1 Cessna 310, 2 Cessna 337, 1 PA-23
TRG 4 Cessna 172

Paramilitary 7,500

GENDARMERIE 7,500
incl maritime police with some 5 PCI<

Malawi Mlw

kwacha K		1998	1999	2000	2001
GDP	K	52bn	64bn		
	US$	1.7bn	1.5bn		
per capita	US$	900	900		
Growth	%	3.6	4.2		
Inflation	%	29.8	44.9		
Debt	US$	2.5bn	2.6bn		
Def exp	K	ε800m	ε1,170m		
	US$	26m	27m		
Def bdgt	K	ε650m	ε1,170m	ε1,300m	
	US$	21m	27m	26m	
FMA (US)	US$	0.3m	0.3m	0.3m	0.4m
FMA (ROC)	US$	2.0m			
US$1=K		31.1	44.1	48.7	

Population			11,092,000
Age	13–17	18–22	23–32
Men	656,000	523,000	780,000
Women	650,000	519,000	811,000

Total Armed Forces

ACTIVE 5,000 (all services form part of the Army)

Army 5,000

2 inf bde each with 3 inf bn • 1 indep para bn • 1 general spt bn (incl arty, engr)
EQUIPMENT (less than 20% serviceability)
RECCE 20 *Fox*, 8 *Ferret*, 12 *Eland*
TOWED ARTY 105mm: 9 lt
MOR 81mm: 8 L16
SAM 15 *Blowpipe*

MARITIME WING (220)
BASE Monkey Bay (Lake Nyasa)
PATROL CRAFT 2
1 *Kasungu* PCI<†, 1 *Namacurra* PCI<, some boats
AMPHIBIOUS craft only
1 LCU

AIR WING (80)
no cbt ac, no armed hel
TPT AC 1 sqn with 2 Basler T-67, 3 Do-228, 1 HS-125-800 (VIP)
TPT HEL 3 SA-330F, 1 AS-350L, 1 *Super Puma* (VIP)

Paramilitary 1,000

MOBILE POLICE FORCE (MPF) 1,000
8 Shorland armd car **ac** 3 BN-2T *Defender* (border patrol), 1 *Skyvan* 3M, 4 Cessna **hel** 2 AS-365

Mali RMM

CFA fr		1998	1999	2000	2001
GDP	fr	1.6tr	1.7tr		
	US$	2.7bn	2.9bn		
per capita	US$	600	600		
Growth	%	5.4	6.4		
Inflation	%	3.5	-0.8		
Debt	US$	3.2bn	3.2bn		
Def exp	fr	ε21bn	ε21bn		
	US$	36m	34m		
Def bdgt	fr	21bn	21bn	21bn	
	US$	36m	34m	30m	
FMA (US)	US$	0.3m	0.3m	0.3m	0.3m
FMA (Fr)	US$	5m	4m	4m	
US$1=fr		590	616	708	

Population		11,542,000 (Tuareg 6–10%)	
Age	13–17	18–22	23–32
Men	642,000	519,000	773,000
Women	667,000	542,000	815,000

Total Armed Forces

ACTIVE about 7,350 (all services form part of the Army)
Terms of service conscription (incl for civil purposes), 2 years (selective)

Army about 7,350

2 tk • 4 inf • 1 AB, 2 arty, 1 engr, 1 SF bn • 2 AD, 1 SAM bty
EQUIPMENT†

MBT 21 T-34, T-54/-55 reported
LT TK 18 Type-62
RECCE 20 BRDM-2
APC 30 BTR-40, 10 BTR-60, 10 BTR-152
TOWED ARTY 85mm: 6 D-44; **100mm**: 6 M-1944; **122mm**: 8 D-30; **130mm**: M-46 reported
MRL 122mm: 2 BM-21
MOR 82mm: M-43; **120mm**: 30 M-43
AD GUNS 37mm: 6 M-1939; **57mm**: 6 S-60
SAM 12 SA-3

NAVY† (about 50)

BASES Bamako, Mopti, Segou, Timbuktu
PATROL CRAFT, RIVERINE 3<

AIR FORCE (400)

16† cbt ac, no armed hel
FGA 5 MiG-17F
FTR 11 MiG-21
TPT 2 An-24, 1 An-26
HEL 1 Mi-8, 1 AS-350, 2 Z-9
TRG 6 L-29, 1 MiG-15UTI, 4 Yak-11, 2 Yak-18

Forces Abroad

UN AND PEACEKEEPING

DROC (MONUC): 3 obs **SIERRA LEONE** (UNAMSIL): 8 obs

Paramilitary 4,800 active

GENDARMERIE 1,800

8 coy

REPUBLICAN GUARD 2,000

NATIONAL POLICE 1,000

MILITIA 3,000

Mauritius Ms

rupee R		1998	1999	2000	2001
GDP	R	97bn	106bn		
	US$	4.4bn	4.6bn		
per capita	US$	16,000	16,700		
Growth	%	5.3	3.2		
Inflation	%	6.1	6.9		
Debt	US$	1.2bn	1.2bn		
Def exp	R	ε2.1bn	ε2.3bn		
	US$	88m	91m		
Def bdgt	R	189m	218m	235m	
	US$	8m	8m	9m	
FMA (US)	US$	0.01	0.1m	0.1m	0.1m
US$1=R		24.0	25.2	25.9	

Population			1,192,000
Age	13–17	18–22	23–32
Men	51,000	55,000	100,000
Women	50,000	54,000	100,000

Total Armed Forces

ACTIVE Nil

Paramilitary ε1,500

SPECIAL MOBILE FORCE ε1,000

6 rifle, 2 mob, 1 engr coy, spt tp
APC 10 VAB
MOR 81mm: 2
RL 89mm: 4 LRAC

COAST GUARD ε500

PATROL CRAFT 4
PATROL, OFFSHORE 1
1 *Vigilant* (Ca *Guardian* design) PCO, capability for 1 hel
PATROL, COASTAL 1
1 SDB-3 PCC
PATROL, INSHORE 2
2 Sov *Zhuk* PCI<, plus 26 boats
MR 1 Do-228-101, 1 BN-2T *Defender*, 3 SA-316B

POLICE AIR WING

2 *Alouette* III

Mozambique Moz

metical M		1998	1999	2000	2001
GDP	M	24tr	29tr		
	US$	2.0bn	2.0bn		
per capita	US$	1,100	1,200		
Growth	%	8.0	9.7		
Inflation	%	2.2	1.5		
Debt	US$	8.2bn	7.3bn		
Def exp	M	ε950bn	ε1,200bn		
	US$	80m	94m		
Def bdgt	M	ε950bn	ε1,200bn	ε1,400bn	
	US$	80m	94m	87m	
FMA (US)	US$	0.2m	2.1m	2.7m	2.2m
US$1=M		11,875	12,775	16,100	

Population			16,700,000
Age	13–17	18–22	23–32
Men	1,151,000	963,000	1,450,000
Women	1,162,000	978,000	1,487,000

Total Armed Forces

ACTIVE ε5,100–6,100

Terms of service conscription, 2–3 years

Army ε4–5,000 (to be 5–6,000)

5 inf, 3 SF, 1 log bn • 1 engr coy

EQUIPMENT† (ε10% or less serviceability)

MBT some 80 T-54/-55 (300+ T-34, T-54/-55 non-op)

RECCE 30 BRDM-1/-2
AIFV 40 BMP-1
APC 150+ BTR-60, 100 BTR-152, 5 *Casspir*
TOWED ARTY 100+: **76mm**: M-1942; **85mm**: D-44, D-48, Type-56; **100mm**: M-1944; **105mm**: M-101; **122mm**: M-1938, D-30; **130mm**: M-46; **152mm**: D-1
MRL 122mm: BM-21
MOR 82mm: M-43; **120mm**: M-43
RCL 75mm; 82mm: B-10; **107mm**: B-11
AD GUNS 20mm: M-55; **23mm**: ZU-23-2; **37mm**: M-1939; **57mm**: S-60 towed, ZSU-57-2 SP
SAM SA-7

Navy† 100

BASES Monkey Bay, Lake Malawi
PATROL AND COASTAL COMBATANTS 3†
PATROL, INSHORE 3 PCI< (non-op)

Air Force 1,000

(incl AD units); no cbt ac, 4† armed hel
TPT 1 sqn with 5 An-26, 2 C-212, 4 PA-32 *Cherokee*
TRG 1 Cessna 182, 7 ZLIN-326
HEL 4† Mi-24*, 5 Mi-8 (most non-op)
AD SAM †SA-2, 10 SA-3 (all non-op)

Forces Abroad

UN AND PEACEKEEPING
EAST TIMOR (UNTAET): 12 incl 2 obs

Namibia Nba

Namibian dollar N$		1998	1999	2000	2001
GDP	N$	17bn	18bn		
	US$	2.6bn	2.7bn		
per capita	US$	4,700	4,800		
Growth	%	1.5	2.9		
Inflation	%	5.0	8.5		
Debt	US$	77m	85m		
Def exp	N$	510m	732m		
	US$	92m	120m		
Def bdgt	N$	443m	559m	617m	
	US$	80m	92m	96m	
FMA (US)	US$	0.2m	1.2m	0.5m	0.3m
US$1=N$		5.5	6.1	7.1	

Population			**1,881,000**
Age	13–17	18–22	23–32
Men	112,000	92,000	140,000
Women	110,000	91,000	139,000

Total Armed Forces

ACTIVE 9,000

Army 9,000

6 inf bn • 1 cbt spt bde with 1 arty, 1 AD, 1 ATK regt

EQUIPMENT
MBT some T-34, T-54/-55 (serviceability doubtful)
RECCE BRDM-2
APC 20 *Casspir*, 30 *Wolf*, 10 BTR-60
TOWED ARTY 140mm: 24 G2
MRL 122mm: 5 BM-21
MOR 81mm; 82mm
RCL 82mm: B-10
ATK GUNS 57mm; 76mm: M-1942 (ZIS-3)
AD GUNS 14.5mm: 50 ZPU-4; **23mm**: 15 *Zumlac* (ZU-23-2) SP
SAM ε50 SA-7

AIR WING
ac 1 *Falcon* 900, 1 *Learjet* 36, 5 Cessna 337/02-A, 2 Y-12
hel 2 SA-319 *Alouette*

Coast Guard ε100

(fishery protection, part of the Ministry of Fisheries)
BASE Walvis Bay
PATROL, OFFSHORE/COASTAL 2
1 *Oryx* PCO, 1 *Osprey* PCC
AIRCRAFT
1 F406 Caravan ac, 1 hel

Forces Abroad

DROC: 1,000 reported

Paramilitary

SPECIAL FIELD FORCE 6,000

Niger Ngr

CFA fr		1998	1999	2000	2001
GDP	fr	993bn	1,022bn		
	US$	1.6bn	1.7bn		
per capita	US$	800	800		
Growth	%	2.7	2.0		
Inflation	%	4.6	2.9		
Debt	US$	1.6bn	1.6bn		
Def exp	fr	ε15bn	ε17bn		
	US$	25m	28m		
Def bdgt	fr	ε13bn	ε17bn	ε19bn	
	US$	22m	28m	27m	
FMA (US)	US$				0.1m
FMA (Fr)	US$	7m	7m	2m	
US$1=fr		590	616	708	

Population		**10,662,000** (Tuareg 8–10%)	
Age	13–17	18–22	23–32
Men	598,000	482,000	698,000
Women	602,000	490,000	726,000

Total Armed Forces

ACTIVE 5,300

Terms of service selective conscription (2 years)

Army 5,200

3 Mil Districts • 4 armd recce sqn • 7 inf, 2 AB, 1 engr coy

EQUIPMENT

RECCE 90 AML-90, 35 AML-60/20, 7 VBL
APC 22 M-3
MOR 81mm: 19 Brandt; **82mm**: 17; **120mm**: 4 Brandt
RL 89mm: 36 LRAC
RCL 75mm: 6 M-20; **106mm**: 8 M-40
ATK GUNS 85mm; 90mm
AD GUNS 20mm: 39 incl 10 M-3 VDA SP

Air Force 100

no cbt ac or armed hel
TPT 1 C-130H, 1 Do-28, 1 Do-228, 1 Boeing 737-200 (VIP), 1 An-26
LIAISON 2 Cessna 337D

Forces Abroad

UN AND PEACEKEEPING

DROC (MONUC): 3 obs

Paramilitary 5,400

GENDARMERIE 1,400

REPUBLICAN GUARD 2,500

NATIONAL POLICE 1,500

Nigeria Nga

naira N		1998	1999	2000	2001
GDP	N	ε4.1tr	ε4.6tr		
	US$	ε49bn	ε50bn		
per capita	US$	1,300	1,300		
Growth	%	-0.1	1.8		
Inflation	%	10.3	9.5		
Debt	US$	30bn	33bn		
Def exp	US$	ε2.1bn	ε2.2bn		
Def bdgt	N	22bn	26bn	34bn	
	US$	1,000m	340m	340m	
FMA (US)	US$		0.1m	10.6m	0.7m
US$1=N		21.9	92.3	102.4	

Population ε116,000,000

(***North*** Hausa and Fulani ***South-west*** Yoruba ***South-east*** Ibo; these tribes make up ε65% of population)

Age	13–17	18–22	23–32
Men	7,479,000	6,492,000	9,733,000
Women	7,468,000	6,552,000	10,117,000

Total Armed Forces

ACTIVE 76,500

RESERVES

planned, none org

Army 62,000

1 armd div (2 armd bde) • 1 composite div (1 mot inf, 1 amph bde, 1 AB bn) • 2 mech div (each 1 mech, 1 mot inf bde) • 1 Presidential Guard bde (2 bn) • 1 AD bde • each div 1 arty, 1 engr bde, 1 recce bn

EQUIPMENT

MBT 50 T-55†, 150 Vickers Mk 3
LT TK 140 *Scorpion*
RECCE ε120 AML-60, 60 AML-90, 55 *Fox*, 75 EE-9 *Cascavel*, 72 VBL (reported)
APC 10 *Saracen*, 300 *Steyr* 4K-7FA, 70 MOWAG *Piranha*, EE-11 *Urutu* (reported)
TOWED ARTY 105mm: 200 M-56; **122mm**: 200 D-30/-74; **130mm**: 7 M-46; **155mm**: 24 FH-77B (in store)
SP ARTY 155mm: 27 *Palmaria*
MRL 122mm: 25 APR-21
MOR 81mm: 200; **82mm**: 100; **120mm**: 30+
RCL 84mm: *Carl Gustav*; **106mm**: M-40A1
AD GUNS 20mm: some 60; **23mm**: ZU-23, 30 ZSU-23-4 SP; **40mm**: L/60
SAM 48 *Blowpipe*, 16 *Roland*
SURV RASIT (veh, arty)

Navy 5,000

(incl Coast Guard)
BASES Lagos **HQ Western Comd** Apapa **HQ Eastern Comd** Calabar **Akwa Ibom state** Warri, Port Harcourt, Ibaka

FRIGATES 1†

FFG 1 *Aradu* (Ge MEKO 360)† with 8 *Otomat* SSM, *Albatros* SAM, 1 × 127mm gun, 2 × 3 ASTT, 1 *Lynx* hel

PATROL AND COASTAL COMBATANTS 6

CORVETTES 1† *Erinomi* (UK Vosper Mk 9) FS with 1 × 3 *Seacat* SAM, 1 × 76mm gun, 1 × 2 ASW mor
MISSILE CRAFT 3
3† *Ayam* (Fr *Combattante*) PFM with 2 × 2 MM-38 *Exocet* SSM, 1 × 76mm gun

PATROL, COASTAL 2

2 *Ekpe* (Ge *Lurssen* 57m) PCC with 1 × 76mm gun

MINE COUNTERMEASURES 2†

2 *Ohue* (mod It *Lerici*) MCC (both non-op)

AMPHIBIOUS 1

1 *Ambe* (Ge) LST, capacity 220 tps, 5 tk

SUPPORT AND MISCELLANEOUS 5

3 AT, 1 nav trg, 1 AGHS

NAVAL AVIATION

HEL 2† *Lynx* Mk 89 MR/SAR

Air Force 9,500

91† cbt ac, 15† armed hel (only 50% serviceability)

FGA/FTR 3 sqn
- 1 with 19 *Alpha Jet* (FGA/trg)
- 1 with 6† MiG-21MF, 4† MiG-21U, 12† MiG-21B/FR
- 1 with 15† *Jaguar* (12 -SN, 3 -BN)

ARMED HEL †15 Bo-105D

TPT 2 sqn with 5 C-130H, 3 -H-30, 17 Do-128-6, 2 Do-228 (VIP), 5 G-222 **hel** 4 AS-332, 2 SA-330

PRESIDENTIAL FLT ac 1 Boeing 727, 2 *Gulfstream*, 2 *Falcon* 900, 1 BAe 125-1000

TRG ac† 23* L-39MS, 12* MB-339AN, 59 Air *Beetle* **hel** 14 Hughes 300

AAM AA-2 *Atoll*

Forces Abroad

UN AND PEACEKEEPING

CROATIA (UNMOP): 1 obs **DROC**(MONUC): 6 obs **IRAQ/KUWAIT** (UNIKOM): 5 obs **SIERRA LEONE** (UNAMSIL): 3,214 plus 4 obs **WESTERN SAHARA** (MINURSO): 5 obs

Paramilitary

COAST GUARD

incl in Navy

PORT SECURITY POLICE ε2,000

about 60 boats and some 5 hovercraft

SECURITY AND CIVIL DEFENCE CORPS (Ministry of Internal Affairs)

POLICE UR-416, 70 AT-105 *Saxon*† APC **ac** 1 Cessna 500, 3 Piper (2 *Navajo*, 1 *Chieftain*) **hel** 4 Bell (2 -212, 2 -222)

Rwanda Rwa

franc fr		1998	1999	2000	2001
GDP	fr	632bn	727bn		
	US$	2.0bn	2.2bn		
per capita	US$	500	500		
Growth	%	10.5	5.0		
Inflation	%	6.2	-2.4		
Debt	US$	1.2bn	1.2bn		
Def exp	fr	ε44bn	ε45bn		
	US$	141m	135m		
Def bdgt	fr	ε39bn	ε45bn	ε45bn	
	US$	112m	135m	125m	
FMA (US)	US$	0.5m	1.1m	0.5m	0.5m
US$1=fr		312	333	359	

Population	ε7,200,000 (Hutu 80%, Tutsi 19%)		
Age	13–17	18–22	23–32
Men	563,000	454,000	651,000
Women	579,000	471,000	682,000

Total Armed Forces

ACTIVE ε55–70,000 (all services, incl *Gendarmerie*; up to 90,000 reported)

Army ε49–64,000

6 inf bde, 1 mech inf regt

EQUIPMENT

MBT 12 T-54/-55

RECCE AML-245, 15 AML-60, AML-90, 16 VBL

APC some BTR, Panhard, 16 RG-31 *Nyala*

TOWED ARTY 35: **105mm**†; **122mm**: 6

MRL 122mm: 5 RM-70

MOR 250: **81mm**; **120mm**

AD GUNS ε150: **14.5mm**; **23mm**; **37mm**

SAM SA-7

Air Force

TPT 1 Bn-2A Islander

HEL 4 Mi-17MD, 2+ Mi-24

Forces Abroad

DROC: 15–20,000 reported

Paramilitary 6,000

GENDARMERIE 6,000

Opposition

ε7,000 former govt tps dispersed in DROC supported by ε55,000 *Interahamwe*. Some have returned to Rwanda. Equipped with small arms and lt mor only.

Sub-Saharan Africa

Senegal Sen

CFA fr		1998	1999	2000	2001
GDP	fr	2.9tr	3.2tr		
	US$	4.8bn	5.2bn		
per capita	US$	1,900	2,000		
Growth	%	4.8	5.1		
Inflation	%	1.1	0.9		
Debt	US$	3.3bn	3.2bn		
Def exp	fr	ε48bn	ε50bn		
	US$	81m	81m		
Def bdgt	fr	40bn	43bn	ε44bn	
	US$	68m	70m	62m	
FMA (US)	US$	0.8m	0.8m	0.7m	0.8m
FMA (Fr)	US$	8m	7m	6m	
US$1=fr		590	616	708	

Population 9,686,000

(Wolof 36%, Fulani 17%, Serer 17%, Toucouleur 9%, Mandingo 9%, Diola 9%, of which 30–60% in Casamance)

Age	13–17	18–22	23–32
Men	598,000	486,000	708,000
Women	592,000	480,000	713,000

Total Armed Forces

ACTIVE 9,400

Terms of service conscription, 2 years selective

RESERVES n.k.

Army 8,000 (3,500 conscripts)

7 Mil Zone HQ • 4 armd bn • 1 engr bn • 6 inf bn • 1 Presidential Guard (horsed) • 1 arty bn • 3 construction coy • 1 cdo bn • 1 AB bn • 1 engr bn

EQUIPMENT

RECCE 10 M-8, 4 M-20, 30 AML-60, 27 AML-90
APC some 16 Panhard M-3, 12 M-3 half-track
TOWED ARTY 18: **75mm**: 6 M-116 pack; **105mm**: 6 M-101/HM-2; **155mm**: ε6 Fr Model-50
MOR 81mm: 8 Brandt; **120mm**: 8 Brandt
ATGW 4 *Milan*
RL 89mm: 31 LRAC
AD GUNS 20mm: 21 M-693; **40mm**: 12 L/60

Navy 600

BASES Dakar, Casamance

PATROL AND COASTAL COMBATANTS 10

PATROL, COASTAL 5
1 *Fouta* (Dk *Osprey*) PCC
1 *Njambuur* (Fr SFCN 59m) PCC
3 *Saint Louis* (Fr 48m) PCC

PATROL, INSHORE 5
3 *Senegal* II PFI<, 2 *Alioune Samb* PCI<

AMPHIBIOUS craft only
2 *Edic* 700 LCT

Air Force 800

8 cbt ac, no armed hel
MR/SAR 1 EMB-111
TPT 1 sqn with 6 F-27-400M, 1 Boeing 727-200 (VIP), 1 DHC-6 *Twin Otter*
HEL 2 SA-318C, 2 SA-330, 1 SA-341H
TRG 4* CM-170, 4* R-235 *Guerrier*, 2 *Rallye* 160, 2 R-235A

Forces Abroad

UN AND PEACEKEEPING

DROC (MONUC): 30 obs **IRAQ/KUWAIT** (UNIKOM): 6 obs

Paramilitary ε5,800

GENDARMERIE ε5,800
12 VXB-170 APC

CUSTOMS
2 PCI<, boats

Opposition

CASAMANCE MOVEMENT OF DEMOCRATIC FORCES
2–3,000 eqpt with lt wpns

Foreign Forces

FRANCE 1,170: 1 marine inf bn (14 AML 60/90); **ac** 1 *Atlantic*, 1 C-160 **hel** 1 SA-319

Seychelles Sey

rupee SR		1998	1999	2000	2001
GDP	SR	3.1bn	3.3bn		
	US$	594m	618m		
per capita	US$	4,600	4,700		
Growth	%	2.0	1.4		
Inflation	%	2.6	2.7		
Debt	US$	181m	163m		
Def exp	SR	55m	61m		
	US$	11m	11m		
Def bdgt	SR	54m	56m	62m	
	US$	10m	11m	11m	
FMA (US)	US$	0.1m	0.1m	0.1m	0.1m
US$1=SR		5.26	5.34	5.75	

Population 74,000

Age	13–17	18–22	23–32
Men	4,000	4,000	6,000
Women	4,000	4,000	6,000

Total Armed Forces

ACTIVE 450 (all services, incl Coast Guard, form part of the Army)

Army 200

1 inf coy
1 sy unit
EQUIPMENT†
RECCE 6 BRDM-2
MOR 82mm: 6 M-43
RL RPG-7
AD GUNS 14.5mm: ZPU-2/-4; **37mm**: M-1939
SAM 10 SA-7

Paramilitary 250 active

NATIONAL GUARD 250

COAST GUARD (200)
(incl 20 Air Wing and ε80 Marines)
BASE Port Victoria
PATROL, COASTAL/INSHORE 2
1 *Andromache* (It *Pichiotti* 42m) PCC, 1 *Zhuk* PCI<
plus 1 *Cinq Juin* LCT (govt owned but civilian op)

AIR WING (20)
No cbt ac, no armed hel
MR 1 BN-2 *Defender*
TPT 1 Reims-Cessna F-406/*Caravan* 11
TRG 1 Cessna 152

Sierra Leone SL

leone L		1998	1999	2000	2001
GDP	L	1,035bn	1,227bn		
	US$	769m	724m		
per capita	US$	700	700		
Growth	%	0.7	-8.1		
Inflation	%	14.5	34		
Debt	US$	1.2bn	1.2bn		
Def exp	US$	ε26m	ε11m		
Def bdgt	US$	5m	11m	9m	
FMA (US)	US$			0.1m	0.1m
FMA (UK)	US$		7.3m		
US$1=L		1,564	1,804	2,232	

Population			**ε4,500,000**
Age	13–17	18–22	23–32
Men	290,000	242,000	374,000
Women	290,000	240,000	379,000

Total Armed Forces

ACTIVE ε3,000

The Lome Peace Agreement between the government and RUF rebels broke down in May 2000. Fighting is on-going with government forces supported by UNAMSIL. A new national army is forming with an initial strength of 4,000 by Dec 2000.

EQUIPMENT (in store)
MOR 81mm: 3; **82mm**: 2; **120mm**: 2
RCL 84mm: *Carl Gustav*
AD GUNS 12.7mm: 4; **14.5mm**: 3
SAM SA-7
HEL 3 Mi-24 (only 2 operational), 3† Mi-8/17 (contract flown and maintained)

Navy† ε200

BASE Freetown
PATROL AND COASTAL COMBATANTS 3†
1 PRC *Shanghai* II PFI<, 1 *Swiftship* 32m† PFI<, 1 *Fairy Marine Tracker* II (all non-op)<

Foreign Forces

UK ε160: trg unit **RUSSIA** 110: 4 Mi-24

UN AND PEACEKEEPING
UN (UNAMSIL): some 12,447 to be 13,500 incl
BANGLADESH: 792 **NIGERIA**: 3,214 **INDIA**: 3,147 **JORDAN**: 1,826 **KENYA**: 878 **GHANA**: 775 **GUINEA**: 778 **ZAMBIA**: 777

Opposition

REVOLUTIONARY UNITED FRONT (RUF): ε15,000

Somali Republic SR

shilling sh		1998	1999	2000	2001
GDP	US$	ε853m	ε874m		
per capita	US$	1,200	1,200		
Growth	%	0.0			
Inflation	%	ε16	ε16		
Debt	US$	3.0bn	3.2bn		
Def exp	US$	ε40m	ε40m		
Def bdgt	US$		13m	15m	
FMA (US)	US$		1.1m	1.3m	1.6m
US$1=sh[a]		2,620	2,620	2,620	

[a] Market rate June **1997** US$1=sh8,000

Population		**ε6,600,000** (Somali 85%)	
Age	13–17	18–22	23–32
Men	607,000	494,000	707,000
Women	605,000	491,000	709,000

Total Armed Forces

ACTIVE Nil
Following the 1991 revolution, no national armed forces

have yet been formed. The Somali National Movement has declared northern Somalia the independent 'Republic of Somaliland', while insurgent groups compete for local supremacy in the south. Heavy military equipment is in poor repair or inoperable.

Clan/Movement Groupings

'SOMALILAND' (northern Somalia) Total armed forces reported to be some 12,900

UNITED SOMALI FRONT str n.k. **clan** Issa **leader** Abdurahman Dualeh Ali

SOMALI DEMOCRATIC ALLIANCE str n.k. **clan** Gadabursi

SOMALI NATIONAL MOVEMENT 5–6,000 **clan** Issaq, 3 factions (Tur, Dhegaweyne, Kahin)

UNITED SOMALI PARTY str n.k. **clan** Midigan/Tumaal **leader** Ahmed Guure Adan

SOMALIA

SOMALI SALVATION DEMOCRATIC FRONT 3,000 **clan** Darod **leader** Abdullah Yusuf Ahmed

UNITED SOMALI CONGRESS str n.k. **clan** Hawiye **sub-clan** Habr Gidir **leaders** Hussein Mohammed Aideed/Osman Atto

ALI MAHDI FACTION 10,000(-) **clan** Abgal **leader** Mohammed Ali Mahdi

SOMALI NATIONAL FRONT 2–3,000 **clan** Darod sub-clan Marehan **leader** General Omar Hagi Mohammed Hersi

SOMALI DEMOCRATIC MOVEMENT str n.k. **clan** Rahenwein/Dighil

SOMALI PATRIOTIC MOVEMENT 2–3,000 **clan** Darod **leader** Ahmed Omar Jess

MARITIME SECURITY FORCE (70 civilians, based at Bosaso under Puntland govt control)
  1 PCO for fisheries protection

South Africa RSA

rand R		1998	1999	2000	2001
GDP	R	724bn	780bn		
	US$	131bn	128bn		
per capita	US$	5,700	5,800		
Growt	%	0.1	1.2		
Inflation	%	6.9	5.1		
Debt	US$	39bn	41bn		
Def exp	R	11.8bn	10.7bn		
	US$	2.1bn	1.8bn		
Def bdgt	R	10.3bn	10.7bn	13.8bn	15.3bn
	US$	1.9bn	1.8bn	1.9bn	2.1bn
FMA (US)	US$	0.8m	1.0m	0.8m	0.8m
US$1=R		5.5	6.1	7.14	

Population			**40,300,000**
Age	13–17	18–22	23–32
Men	2,527,000	2,284,000	3,765,000
Women	2,494,000	2,267,000	3,774,000

Total Armed Forces

ACTIVE 63,389

(incl 569 MoD staff, 5,500 South African Military Health Service; 8,758 women; excluding 17,749 civilians)
Terms of service voluntary service in 4 categories (full career, up to 10 yrs, up to 6 yrs, 1 yr voluntary military service)
Racial breakdown 38,976 black, 17,135 white, 6,419 coloured, 859 Asian

RESERVES 87,392

Army 85,228 **Navy** 1,070 **Air Force** 442 **Military Health Service** (SAMHS) 652

Army 42,490

(incl 7,671 women)

PERMANENT FORCE

8 'type' formations
Formations under direct command and control of Chief of Joint Operations:
  5 regional joint task forces (each consists of HQ and a number of group HQ, but no tps which are provided when necessary by permanent and reserve force units from 'type' formations)
  1 SF formation (2 bn)
1 bde HQ:
  1 tk, 1 armd car bn
  20 inf bn (incl 2 mech, 3 mot, 13 lt inf, 2 AB)
  2 arty (incl 1 AA), 3 engr bn, 7 engr sqn

RESERVE FORCE

1 bde HQ:
  cadre units comprising 9 armd, 26 inf, 7 arty, 7 AD, 4 engr bn
  some 183 'cdo' (bn) home defence units

EQUIPMENT

MBT some 168 *Olifant* 1A/-B (125 in store)
RECCE 242 *Rooikat-76* (94 in store)
AIFV 1,240 *Ratel*-20/-60/-90 (666 in store)
APC 429 *Casspir*, 538 *Mamba*
TOWED ARTY 140mm: 75 G-2 (in store); **155mm**: 72 G-5 (51 in store)
SP ARTY 155mm: 43 G-6 (31 in store)
MRL 127mm: 25 *Bataleur* (40 tube) (4 in store), 26 *Valkiri* (24 tube) (in store)
MOR 81mm: 1,190 (incl some SP); **120mm**: 36
ATGW 52 ZT-3 *Swift* (36 in store)
RL 92mm: FT-5
RCL 106mm: 100 M-40A1 (some SP)
AD GUNS 23mm: 36 *Zumlac* (ZU-23-2) SP; **35mm**: 99 GDF Mk1/3 (in store), 48 GDF Mk5 (32 in store)
SURV *Green Archer* (mor), *Cymbeline* (mor)

Navy 5,190

(incl 1,353 women)

FLOTILLAS submarine, strike, MCM
BASES Simon's Town (HQ), Durban (Salisbury Island)
SUBMARINES 2
SSK 2 *Spear* (Mod Fr *Daphné*) with 550mm TT
PATROL AND COASTAL COMBATANTS 9
MISSILE CRAFT 6 *Warrior* (Il *Reshef*) PFM with 6 *Skerpioen* (Il *Gabriel*) SSM (incl 2 in refit)
PATROL, INSHORE 3
3 T craft PCI
MINE COUNTERMEASURES 8
4 *Kimberley* (UK *Ton*) MSC (incl 2 in reserve)
4 *River* (Ge *Navors*) MHC (incl 2 in refit)
SUPPORT AND MISCELLANEOUS 35
1 *Drakensberg* AO with 2 hel and extempore amph capability (perhaps 60 tps and 2 small LCU)
1 *Outeniqua* AO with similar capability to *Drakensberg*
1 diving spt
3 AT
28 harbour patrol PCI<
1 AGHS (UK *Hecla*)
1 Antarctic tpt with 2 hel (operated by private co for Ministry of Environment)
plus craft: 8 LCU

Air Force 9,640

(incl 2,470 women); 87 cbt ac, ε3 attack and several extempore armed hel
Air Force office, Pretoria, and 5 type formations
FTR/FGA 2 sqn
1 sqn with 29 *Cheetah* C, 10 *Cheetah* D
1 sqn with 27 *Impala* Mk2, 21 *Impala* Mk1
TPT/TKR/EW 1 sqn with 5 Boeing 707-320 (EW/tkr)
TPT 5 sqn
1 with 3 *King Air* 200, 1 *King Air* 300, 11 Cessna-208 *Caravan*, 1 PC-12
1 (VIP) with 2 *Citation* II, 2 *Falcon* 50, 1 *Falcon* 900
1 with 11 C-47 TP (5 maritime, 4 tpt, 1 PR, 1 EW trg)
1 with 12 C-130
1 with 4 CASA-212, 1 CASA-235, 11 Cessna 185
HEL 1 cbt spt with 3* *Rooivalk*. 4 tpt, 1 flying school with 51 *Oryx*, 10 BK-117, 46 SA-316/319
TRG 1 flying school with 58 PC-7
UAV 3 *Seeker* with 1 control station
MISSILES
ASM *Raptor*
AAM V-3C
GROUND DEFENCE
RADAR 2 Air Control Sectors (Hoedspruit and Bushveld), 3 fixed and 6 mob radars (2 long-range (Ellisras and Mariepskop), 4 tactical)
SAAF Regt: 12 security sqn

South African Military Health Service (SAMHS) 5,290

(incl ε2,500 women); a separate service within the SANDF; 3 Type, 1 spt, 1 trg formation

Forces Abroad

UN AND PEACEKEEPING
DROC (MONUC): 1 obs

Foreign Forces

SINGAPORE: *Searcher* UAV trg det

Sudan — Sdn

pound S£		1998	1999	2000	2001
GDP	US$	ε8.0bn	ε8.7bn		
per capita	US$	1,400	1,500		
Growth	%	5.0	5.5		
Inflation	%	19.3	14.0		
Debt	US$	16bn	18bn		
Def exp	US$	ε373m	ε424m		
Def bdgt	US$	ε248m	ε424m	ε425m	
US$1=S£		2,008	2,526	2,588	

Population ε33,194,000
(Muslim 70%, *mainly in North;* Christian 10%, *mainly in South;* African 52%, *mainly in South;* Arab 39%, *mainly in North*)

Age	13–17	18–22	23–32
Men	1,940,000	1,644,000	2,471,000
Women	1,854,000	1,573,000	2,374,000

Total Armed Forces

ACTIVE 104,500
(incl ε20,000 conscripts)
Terms of service conscription (males 18–30), 3 years

Army ε100,000

(incl ε20,000 conscripts)
1 armd div • 1 recce bde • 6 inf div (regional comd) • 10+ arty bde (incl AD) • 1 AB div (incl 1 SF bde) • 3 arty regt • 1 mech inf bde • 1 engr div • 1 border gd div • 24 inf bde
EQUIPMENT
MBT 150 T-54/-55, 20 M-60A3
LT TK 20 PRC Type-62
RECCE 6 AML-90, 30 *Saladin*, 80 *Ferret*, 60 BRDM-1/-2
AIFV 6 BMP-2
APC 63 BTR-50/-152, 42 OT-62/-64, 42 M-113, 19 V-100/-150, 120 *Walid*
TOWED ARTY 400 incl: **85mm**: D-44; **105mm**: Model 56 pack; **122mm**: D-74, M-30, Type-54/D-30; **130mm**: M-46/PRC Type 59-1
SP ARTY 155mm: M-114A1

MRL 600: **107mm**: Type-63; **122mm**: BM-21, Type-81
MOR 81mm; **82mm**; **120mm**: M-43, AM-49
ATGW 4 *Swingfire*
RCL 106mm: 40 M-40A1
ATK GUNS 40 incl: **76mm**: M-1942; **100mm**: M-1944
AD GUNS 1,000+ incl: **14.5mm**; **20mm**: M-167 towed, M-163 SP; **23mm**: ZU-23-2; **37mm**: M-1939/Type-63, Type-55; **57mm**: Type-59; **85mm**: M-1944
SAM 54 SA-7
SURV RASIT (veh, arty)

Navy ε1,500

BASES Port Sudan (HQ), Flamingo Bay (Red Sea), Khartoum (Nile)

PATROL AND COASTAL COMBATANTS 6
PATROL, INSHORE 2 *Kadir* PCI<
PATROL, RIVERINE 4 PCI<, about 12 armed boats

AMPHIBIOUS craft only
some 2 *Sobat* (FRY DTK-221) LCT (used for transporting stores)

Air Force 3,000

(incl Air Defence); 30† cbt ac, 5 armed hel
FGA 9 F-5 (7 -E, 2 -F), 5 PRC J-5 (MiG-17), 7 PRC J-6 (MiG-19), 4 (non-op) F-7 (MiG-21), 3 BAC-167
FTR 2 MiG-23
TPT 4 C-130H, 1 DHC-5D, 1 F-27, 2 *Falcon* 20/50
HEL 8 AB-212, 8 IAR/SA-330, 7 (1 op) Mi-8, 5* Mi-24V
TRG 8 PT-6A
AD 5 bty SA-2 SAM (18 launchers)
AAM AA-2 *Atoll*

Paramilitary 15,000

POPULAR DEFENCE FORCE 15,000 active
85,000 reserve; mil wg of National Islamic Front; org in bn of 1,000

Opposition

NATIONAL DEMOCRATIC ALLIANCE
coalition of many groups, of which the main forces are:

SUDANESE PEOPLE'S LIBERATION ARMY (SPLA) 20–30,000
four factions, each org in bn, operating mainly in southern Sudan; some captured T-54/-55 tks, BM-21 MRL and arty pieces, but mainly small arms plus **60mm** and **120mm** mor, **14.5mm** AA, SA-7 SAM

SUDAN ALLIANCE FORCES ε500
based in Eritrea, operate in border area

BEJA CONGRESS FORCES ε500
operates on Eritrean border

NEW SUDAN BRIGADE ε2,000
operates on Ethiopian and Eritrean borders

Foreign Forces

CHINA, IRAN, IRAQ: some mil advisers

Tanzania — Tz

shilling sh		1998	1999	2000	2001
GDP	sh	5.6tr	6.3tr		
	US$	8.1bn	8.5bn		
per capita	US$	700	700		
Growth	%	4.0	3.9		
Inflation	%	12.8	8.2		
Debt	US$	7.6bn	7.8bn		
Def exp	sh	ε95bn	ε105bn		
	US$	143m	141m		
Def bdgt	sh	71bn	102bn	ε115bn	
	US$	106m	137m	144m	
FMA (US)[a]	US$	0.2m	0.2m	0.2m	0.2m
US$1=sh		665	745	799	

[a] Excl ACRI and East Africa Regional funding

Population			32,855,000
Age	13–17	18–22	23–32
Men	1,924,000	1,550,000	2,308,000
Women	1,976,000	1,639,000	2,455,000

Total Armed Forces

ACTIVE ε34,000
Terms of service incl civil duties, 2 years

RESERVES 80,000

Army 30,000+

5 inf bde • 1 tk bde • 2 arty bn • 2 AD arty bn • 2 mor bn • 2 ATK bn • 1 engr regt (bn)

EQUIPMENT†
MBT 30 PRC Type-59 (15 op), 35 T-54 (all non-op)
LT TK 30 PRC Type-62, 40 *Scorpion*
RECCE 40 BRDM-2
APC 66 BTR-40/-152, 30 PRC Type-56
TOWED ARTY 76mm: 45 ZIS-3; **85mm**: 80 PRC Type-56; **122mm**: 20 D-30, 100 PRC Type-54-1; **130mm**: 40 PRC Type-59-1
MRL 122mm: 58 BM-21
MOR 82mm: 350 M-43; **120mm**: 50 M-43
RCL 75mm: 540 PRC Type-52

Navy† ε1,000

BASES Dar es Salaam, Zanzibar, Mwanza (Lake Victoria – 4 boats)

PATROL AND COASTAL COMBATANTS 7

TORPEDO CRAFT 2 PRC *Huchuan* PHT< with 2 533mm TT
PATROL, COASTAL 5
2 PRC *Shanghai* II PFC
3 *Thornycroft* PCC
AMPHIBIOUS craft only
2 *Yunnan* LCU

Air Defence Command 3,000

(incl ε2,000 AD tps); 19 cbt ac†, no armed hel
FTR 3 sqn with 3 PRC J-5 (MiG-17), 10 J-6 (MiG-19), 6 J-7 (MiG-21)
TPT 1 sqn with 3 DHC-5D, 1 PRC Y-5, 2 Y-12(II), 3 HS-748, 2 F-28, 1 HS-125-700
HEL 4 AB-205
LIAISON ac 5 Cessna 310, 2 Cessna 404, 1 Cessna 206 **hel** 6 Bell 206B
TRG 2 MiG-15UTI, 5 PA-28
AD GUNS 14.5mm: 40† ZPU-2/-4; **23mm**: 40 ZU-23; **37mm**: 120 PRC Type-55
SAM† 20 SA-3, 20 SA-6, 120 SA-7

Forces Abroad

UN AND PEACEKEEPING
DROC (MONUC): 7 obs **SIERRA LEONE** (UNAMSIL): 12 obs

Paramilitary 1,400 active

POLICE FIELD FORCE 1,400
18 sub-units incl Police Marine Unit
MARINE UNIT (100)
boats only
AIR WING
ac 1 Cessna U-206 **hel** 2 AB-206A, 2 Bell 206L, 2 Bell 47G

Togo Tg

CFA fr		1998	1999	2000	2001
GDP	fr	860bn	929bn		
	US$	1.4bn	1.5bn		
per capita	US$	1,300	1,300		
Growth	%	0.1	3.5		
Inflation	%	6.1	-1.6		
Debt	US$	1.4bn	1.3bn		
Def exp	fr	ε20bn	ε21bn		
	US$	34m	34m		
Def bdgt	fr	ε19bn	ε21bn	ε22bn	
	US$	32m	35m	31m	
FMA (Fr)	US$	4m	4m	4m	
FMA (US)	US$				0.1m
US$1=fr		590	616	708	

Population			5,001,000
Age	13–17	18–22	23–32
Men	303,000	231,000	337,000
Women	303,000	241,000	364,000

Total Armed Forces

ACTIVE some 6,950 (up to 10,000 reported)
Terms of service conscription, 2 years (selective)

Army 6,500

2 inf regt
1 with 1 mech bn, 1 mot bn
1 with 2 armd sqn, 3 inf coy; spt units (trg)
1 Presidential Guard regt: 2 bn (1 cdo), 2 coy
1 para cdo regt: 3 coy
1 spt regt: 1 fd arty, 2 AD arty bty; 1 log/tpt/engr bn
EQUIPMENT
MBT 2 T-54/-55
LT TK 9 *Scorpion*
RECCE 6 M-8, 3 M-20, 10 AML (3 -60, 7 -90), 36 EE-9 *Cascavel*, 2 VBL
AIFV 20 BMP-2 (reported)
APC 4 M-3A1 half-track, 30 UR-416
TOWED ARTY 105mm: 4 HM-2
SP ARTY 122mm: 6 reported
MOR 82mm: 20 M-43
RCL 57mm: 5 ZIS-2; **75mm**: 12 PRC Type-52/-56; **82mm**: 10 PRC Type-65
AD GUNS 14.5mm: 38 ZPU-4; **37mm**: 5 M-39

Navy ε200

(incl Marine Infantry unit)
BASE Lomé
PATROL CRAFT, COASTAL 2
2 *Kara* (Fr *Esterel*) PFC

Air Force †250

16 cbt ac, no armed hel
FGA 5 *Alpha Jet*, 4 EMB-326G
TPT 2 *Baron*, 2 DHC-5D, 1 Do-27, 1 F-28-1000 (VIP), 1 Boeing 707 (VIP), 2 Reims-Cessna 337
HEL 1 AS-332, 2 SA-315, 1 SA-319, 1 SA-330
TRG 4* CM-170, 3* TB-30

Paramilitary 750

GENDARMERIE (Ministry of Interior) 750
1 trg school, 2 reg sections, 1 mob sqn

Uganda — Uga

shilling Ush		1998	1999	2000	2001
GDP	Ush	8.1tr	9.0tr		
	US$	7.3bn	8.0bn		
per capita	US$	1,800	1,900		
Growth	%	5.5	7.0		
Inflation	%	0.0	3.4		
Debt	US$	3.9bn	3.8bn		
Def exp	Ush	280bn	290bn		
	US$	226m	199m		
Def bdgt	Ush	165bn	200bn	210bn	
	US$	133m	138m	132m	
FMA (US)[a]	US$	0.4m	0.4m	0.4m	0.4m
US$1=Ush		1,240	1,455	1,588	

[a] Excl ACRI and East Africa Regional funding

Population			22,261,000
Age	13–17	18–22	23–32
Men	1,224,000	1,078,000	1,543,000
Women	1,246,000	1,061,000	1,666,000

Total Armed Forces

ACTIVE ε50–60,000

Ugandan People's Defence Force ε50–60,000

4 div (2 with 3, 2 with 4)

EQUIPMENT†

MBT ε140 T-54/-55
LT TK ε20 PT-76
RECCE 40 *Eland*, 60 *Ferret* (reported)
APC 12 BTR-60, 4 OT-64 SKOT, 20 *Mamba*, 20 *Buffel*
TOWED ARTY 76mm: 60 M-1942; **122mm**: 20 M-1938; **130mm**: ε12; **155mm**: 4 G5
MRL 122mm: BM-21
MOR 81mm: L 16; **82mm**: M-43; **120mm**: 60 Soltam
AD GUNS 14.5mm: ZPU-1/-2/-4; **37mm**: 20 M-1939
SAM SA-7
AVN no cbt act†, 3 armed hel
TRG 3†* L-39, 1 SF*-260 (non-op)
TPT HEL 3 Bell 206, 2 Bell 412, 3 Mi-17, 3* Mi-24 (only 1 Mi-17, 1 Mi-24 op)
FGA 7 MiG-21 (5 -MF, 2 -UTI) on order

Forces Abroad

DROC: ε15,000

Paramilitary ε1,800 active

BORDER DEFENCE UNIT ε600
small arms
POLICE AIR WING ε800
hel 1 *JetRanger*
MARINES ε400
8 riverine patrol craft<, plus boats
LOCAL DEFENCE UNITS ε15,000

Opposition

LORD'S RESISTANCE ARMY ε1,500
(ε200 in Uganda, remainder in Sudan)
ALLIED DEMOCRATIC FORCES ε300–500

Zambia — Z

kwacha K		1998	1999	2000	2001
GDP	K	6.3tr	7.4tr		
	US$	3.4bn	3.5bn		
per capita	US$	900	900		
Growth	%	-2.1	1.3		
Inflation	%	30.6	20.6		
Debt	US$	6.8bn	6.2bn		
Def exp	K	ε120bn	ε210bn		
	US$	64m	88m		
Def bdgt	K	99bn	186bn	ε196bn	
	US$	53m	79m	65m	
FMA (US)	US$	0.1m	0.2m	0.5m	0.6m
US$1=K		1,862	2,388	3,018	

Population			10,327,000
Age	13–17	18–22	23–32
Men	633,000	515,000	753,000
Women	622,000	508,000	778,000

Total Armed Forces

ACTIVE 21,600

Army 20,000

(incl 3,000 reserves)
3 bde HQ • 1 arty regt • 9 inf bn (3 reserve) • 1 engr bn • 1 armd regt (incl 1 armd recce bn)

EQUIPMENT†

MBT 10 T-54/-55, 20 PRC Type-59
LT TK 30 PT-76
RECCE ε60 BRDM-1/-2 (ε12 serviceable)
APC 13 BTR-60
TOWED ARTY 76mm: 35 M-1942; **105mm**: 18 Model 56 pack; **122mm**: 25 D-30; **130mm**: 18 M-46
MRL 122mm: 50 BM-21
MOR 81mm: 55; **82mm**: 24; **120mm**: 14
ATGW AT-3 *Sagger*
RCL 57mm: 12 M-18; **75mm**: M-20; **84mm**: *Carl Gustav*
AD GUNS 20mm: 50 M-55 triple; **37mm**: 40 M-1939; **57mm**: 55 S-60; **85mm**: 16 KS-12
SAM SA-7

Air Force 1,600

63† cbt ac, some armed hel. Very low serviceability.
FGA 1 sqn with 12 J-6 (MiG-19)†, 8 sqn with 12 MiG-21 MF† (8 undergoing refurbishment), 1 sqn with 8 K-8
TPT 1 sqn with 4 An-26, 4 C-47, 4 DHC-5D, 4 Y-12(II)
VIP 1 fleet with 1 HS-748, 3 Yak-40
LIAISON 7 Do-28
TRG 2*-F5T, 2* MiG-21U†, 12* *Galeb* G-2, 15* MB-326GB, 8* SF-260MZ, 8 K-8
HEL 1 sqn with 4 AB-205A, 5 AB-212, 12 Mi-8
LIAISON HEL 12 AB-47G
MISSILES
ASM AT-3 *Sagger*
SAM 1 bn; 3 bty: SA-3 *Goa*

Forces Abroad

UN AND PEACEKEEPING
DROC (MONUC): 9 obs **SIERRA LEONE** (UNAMSIL): 777 plus 11 obs

Paramilitary 1,400

POLICE MOBILE UNIT (PMU) 700
1 bn of 4 coy
POLICE PARAMILITARY UNIT (PPMU) 700
1 bn of 3 coy

Zimbabwe Zw

Zimbabwean dollar Z$		1998	1999	2000	2001
GDP	Z$	140bn	210bn		
	US$	6.6bn	6.8bn		
per capita	US$	2,300	2,300		
Growth	%	0.8	1.2		
Inflation	%	27.0	58.5		
Debt	US$	4.7bn	5.4bn		
Def exp	Z$	ε7bn	ε16bn		
	US$	327m	418m		
Def bdgt	Z$	4.2bn	6.4bn	9.0bn	
	US$	196m	168m	235m	
FMA (US)	US$	0.3m	1.0m	1.5m	1.3m
US$1=Z$		21.4	38.3	38.2	

Population			12,286,000
Age	13–17	18–22	23–32
Men	809,000	655,000	999,000
Women	799,000	651,000	997,000

Total Armed Forces

ACTIVE ε40,000

Army ε35,000

5 bde HQ • 1 Mech bde, 1 arty bde, 1 Presidential Guard gp • 1 armd sqn • 18 inf bn (incl 2 guard, 1 mech, 1 cdo, 1 para) • 1 fd arty regt • 1 AD regt • 1 engr regt
EQUIPMENT
MBT 22 PRC Type-59, 10 PRC Type-69
RECCE 80 EE-9 *Cascavel* (**90mm** gun)
APC 30 PRC Type-63 (YW-531), UR-416, 40 *Crocodile*, 260 ACMAT
TOWED ARTY 122mm: 12 PRC Type-60, 4 PRC Type-54
MRL 107mm: 18 PRC Type-63; **122mm**: 52 RM-70
MOR 81mm/82mm 502; **120mm**: 14 M-43
AD GUNS 215 incl **14.5mm**: ZPU-1/-2/-4; **23mm**: ZU-23; **37mm**: M-1939
SAM 17 SA-7

Air Force 5,000

58 cbt ac, 28 armed hel
Flying hours 100
FGA 2 sqn
1 with 11 *Hunters* (9 FGA-90, 1 -F80, 1 T-81) (in store)
1 with 8 *Hawk* Mk 60/60A (2 serviceable)
FTR 1 sqn with 12 PRC F-7 (MiG-21) (9 serviceable)
RECCE 1 sqn with 14* Reims-Cessna 337 *Lynx*
TRG/RECCE/LIAISON 1 sqn with 22 SF-260 *Genet* (9 -C, 6* -F, 5* -W, 2* TP)
TPT 1 sqn with 6 BN-2, 8 C-212-200 (1 VIP)
HEL 1 sqn with 24 SA-319, 2 Mi-35/2 Mi-35P (armed/liaison), 1 sqn with 8 AB-412, 2 AS-532UL (VIP)

Forces Abroad

DROC: up to 12,000 reported

Paramilitary 21,800

ZIMBABWE REPUBLIC POLICE FORCE 19,500 (incl Air Wg)
POLICE SUPPORT UNIT 2,300

The International Arms Trade

The international arms trade fell in 1999, with the value of deliveries estimated at $53.4bn compared with $58bn in 1998. The ranking of the regional markets remained the same as in the previous year: the Middle East continued to buy more arms than any other region, with Saudi Arabia receiving deliveries worth $6.1bn: more than any other country. However, Saudi imports in 1999 were well down on 1998 when it bought arms worth $10.8bn. The second largest regional market was East Asia and Australasia. As in 1998, Taiwan was the region's leading importer in 1999, with deliveries worth $2.6bn. Taiwan's imports were significantly down on the previous year. The overall decline in arms deliveries in 1999 does not necessarily mark a trend, but reflects the peaks and troughs of delivery programmes. For example, deliveries to Saudi Arabia from the UK under the *al Yamamah* programme passed their peak in 1997 and 1998, but Saudi imports are likely to increase significantly again as the country seeks a successor to its *Tornado* fleet, for which *Eurofighter* must be a contender. Nor does the decline in the value of arms deliveries mean that military spending overall has reduced. Arms procurement usually accounts for 20–30% of the military budgets of the larger arms-purchasing countries, while the largest portion is normally spent on operations and maintenance and personnel. Indeed, global military expenditure overall in 1999 was, at $809bn, much the same as in 1998 and available military budgets for 2000 and beyond do not indicate any decline.

Among the arms suppliers, the US increased its share of the world market to 49.1% in 1999, compared with 47.6% a year earlier. The UK was the second-largest exporter with 18.7% of the market and France was third with 12.4%. The value of French deliveries was well down in 1999 compared with the previous year when it accounted for 17.6% of deliveries but this decline can be explained by delivery cycles and will certainly be reversed in 2000. The fourth-largest exporter, Russia, increased its market share from 4.6% in 1998 to 6.6% in 1999. The country's arms exports benefited from the devaluation of the rouble in 1998 and stronger demand for its products from China, India, and some African countries

Sources

Where possible arms-trade statistics given in *The Military Balance* are obtained directly from governments, but other sources are also used. The primary source for US government figures is *World Military Expenditures and Arms Transfers 1998* released by the State Bureau of Arms Control. Its figures have been revised to include higher estimates of US direct commercial sales, including firm-to-firm export trade. Another source for 1999 data on US foreign military sales (excluding US direct commercial sales) and markets in individual countries is *Conventional Arms Transfers to Developing Nations 1992–1999* (Richard F Grimmett, Congressional Research Service, Washington DC, July 2000). Historical arms trade data are also taken from *World Military Expenditures and Arms Transfers 1997*. The UN Register of Conventional Arms is an invaluable source of information on annual equipment deliveries. Figures from the Aerospace Industries Association of America (AIAA), the Society of British Aerospace Companies (SBAC) and the European Association of Aerospace Industries (AECMA) have also been used to support the analysis.

Table 33 Value of arms deliveries and market share, 1987, 1993–1999

(constant 1999 US$m/% *in italics*)

	Total	USSR/ Russia		Warsaw Pact excl. USSR		US		UK		France		Germany		China		Israel		Others	
1987	92,494	32,444	*35.1*	5,729	*6.2*	24,987	*27.0*	7,656	*8.3*	8,291	*9.0*	2,246	*2.4*	2,670	*2.9*	1,519	*1.6*	6,952	*7.5*
1993	48,782	2,919	*6.0*	n.a	*n.a*	27,127	*55.6*	5,312	*10.9*	3,328	*6.8*	1,695	*3.5*	1,247	*2.6*	1,671	*3.4*	5,483	*11.2*
1994	44,517	3,527	*7.9*	n.a	*n.a*	23,872	*53.6*	5,160	*11.6*	3,724	*8.4*	1,563	*3.5*	803	*1.8*	1,541	*3.5*	4,327	*9.7*
1995	48,783	3,023	*6.2*	n.a	*n.a*	24,957	*51.2*	8,090	*16.6*	4,130	*8.5*	1,500	*3.1*	684	*1.4*	1,345	*2.8*	5,054	*10.4*
1996	53,121	3,728	*7.0*	n.a	*n.a*	26,042	*49.0*	10,252	*19.3*	6,108	*11.5*	713	*1.3*	634	*1.2*	1,411	*2.7*	4,233	*8.0*
1997	58,255	2,601	*4.5*	n.a	*n.a*	28,212	*48.4*	11,390	*19.6*	7,718	*13.2*	781	*1.3*	1,040	*1.8*	1,582	*2.7*	4,931	*8.5*
1998	58,006	2,688	*4.6*	n.a	*n.a*	27,584	*47.6*	9,333	*16.1*	10,200	*17.6*	868	*1.5*	521	*0.9*	1,303	*2.2*	5,509	*9.5*
1999	53,365	3,500	*6.6*	n.a	*n.a*	26,205	*49.1*	9,986	*18.7*	6,630	*12.4*	928	*1.7*	260	*0.5*	1,264	*2.4*	4,592	*8.6*

Table 34 Arms deliveries to the Middle East and North Africa, 1987, 1993–1999

(constant 1999 US$m)

	Saudi Arabia	Iraq	Iran	Egypt	Israel	Syria	UAE	Kuwait	Algeria
1987	10,309	7,596	2,388	2,528	3,230	2,809	272	281	983
1993	9,658	n.k.	1,252	2,170	1,823	307	651	1,129	152
1994	8,800	n.k.	434	1,303	1,335	153	575	899	156
1995	9,766	n.k.	543	2,062	835	184	1,031	1,411	250
1996	9,983	n.k.	434	1,736	987	98	814	1,790	272
1997	11,445	n.k.	832	1,144	868	108	868	728	488
1998	10,829	n.k.	651	1,058	1,085	120	977	543	543
1999	6,103	n.k.	481	800	1,504	120	732	314	n.k.

Table 35 Arms deliveries to East Asia, 1987, 1993–1999

(constant 1999 US$m)

	Japan	Taiwan	ROK	DPRK	Vietnam	China	Thailand	Malaysia	Singapore	Indonesia	Myanmar
1987	1,573	1,465	1,053	590	2,669	912	604	99	435	365	28
1993	2,922	1,137	1,942	5	22	654	159	307	148	102	148
1994	2,432	1,112	2,354	100	89	289	434	945	256	55	111
1995	2,496	1,303	1,856	104	217	787	1,193	814	217	184	152
1996	2,550	1,845	1,736	104	272	1,628	759	488	543	868	272
1997	2,332	7,054	1,411	104	162	434	515	326	488	434	326
1998	2,170	6,511	1,421	94	184	488	326	347	923	380	314
1999	1,866	2,604	1,847	90	174	500	410	1,200	619	767	325

Military Command, Control, Communications and Intelligence (C^3I) Systems: Drawing on Commercial Technology

The advent of powerful personal computers and Internet technology has allowed the development of integrated command, control, communications and intelligence (C^3I) systems that can link the highest level of command to the headquarters of forces in the field, enabling secure transmission of the full range of information, such as maps, pictures (including video) and other forms of data, and making better use of the limited capacity of an already crowded frequency spectrum. However, there is still much to be done. Fully digitised C^3I systems have yet to reach the forward units. Moreover, C^3I systems in multi-national operations must be become more interoperable. The technical challenges of the digitised battlefield are immense and may prove too expensive for all but the wealthiest countries. While the increasing adoption of commercial technology for command-and-control over the past ten years has led to significant advances, shared, multinational development programmes and a close partnership with private industry are essential if the battlefield-digitisation challenge is to be met.

ADVANCES IN US COMMAND-AND-CONTROL

At the strategic-to-field-force-headquarters level, the most advanced system currently deployed is the US Global Command and Control System (GCCS), which forms part of the US Command, Control, Computers, Communications and Intelligence for the Warrior (C^4IFTW) concept. One of GCCS' key features is the provision of a common operational picture of the battlefield environment, which is achieved through rich interconnection with other systems providing sensor and intelligence information. The need for systems to be interconnected has led to the concept of the common operating environment (COE), a set of standards for computer-based systems, aligned, as far as possible, with commercial standards and 'off the shelf' software. Systems built with COE core specifications and applications should be readily interoperable with other defence systems utilising the GCCS infrastructure. GCCS also provides a set of specialised applications aimed at improving the ability of the US Department of Defence (DoD) to project a force to any part of the world when needed. The physical architecture of GCCS consists of a distributed network of servers (which store data on a UNIX operating system) and workstations, which are both UNIX- and personal computer (PC)-based.

For systems to work together, it is not enough to specify the technical aspects of software applications and communications protocols. Common meanings must be given to data items used by the applications to ensure that information is interpreted in exactly the same way by both sending and receiving systems. The term 'shared data environment' has been coined to encapsulate this need for a common understanding. The shared data environment in GCCS is based on the Joint Operation Planning and Execution System (JOPES). The JOPES data model is the basis for standardising data items (data management) within the DoD.

Complementing GCCS, the Global Command Support System (GCSS) is being developed to integrate logistics support. Together, both systems give the commander an integrated set of decision aids, allowing him not only to understand the operational picture, but also to have an up-to-date view of the support available to him.

Other countries are developing similar capabilities. The UK's Joint Operational Command System (JOCS) has some commonality with GCCS, particularly in its use of software derived from that used in the US system, but it differs in other aspects. For example, JOCS is being brought into service using the incremental-procurement concept – in other words, purchasing a relatively small-scale system, learning from its operational use and then expanding it by acquiring additional system components in phases. JOCS has now completed the second phase of its procurement programme. It is fielded from the top level of defence command to joint-force headquarters overseas, and has been deployed recently in the Balkans, East Timor and Sierra Leone. The system can cope with multiple concurrent operations as well as running training exercises. Like GCCS, it is based on a mix of commercial UNIX and PC workstations.

As in the US, data standardisation is a key element of the UK programme to improve interoperability between defence systems. To achieve this, the UK has evolved a Defence Data Environment (DDE) through standardising data elements that flow across system boundaries. In addition, the Defence Command and Army Data Model (DCADM) has been developed to provide flexibility in recording operational data and to enable command systems to respond to rapidly changing circumstances. It achieves this by storing the meta-data (the data about data) as variable data items rather than fixing them, as is the practice in conventional databases. The model is a significant advance in database technology, but unless it is widely adopted by other countries, it restricts interoperability since it requires special interfaces to be developed for database-to-database information transfers. It is similar to the Army Tactical Command & Control Information System (ATCCIS) model being developed jointly by a number of nations and likely to be adopted by NATO as a database-integration standard for land operations.

NATO has also developed Crisis Response Operation in NATO Open System (CRONOS), a command system built using commercial technology. It is largely based on Microsoft NT workstations and servers and fielded NATO-wide. Most NATO countries have a number of CRONOS terminals in their joint headquarters. The system was used extensively during NATO's involvement in Bosnia and more recently in Kosovo. As well as providing basic office-automation software, CRONOS also hosts specialist applications such as the Allied Deployment and Movement System (ADAMS), used to avoid transportation bottlenecks as forces build up in the field.

COMMUNICATIONS INFRASTRUCTURE

The command systems that are being developed and fielded represent a significant advance in the use of information technology to improve the effectiveness of top level command. The difficulty facing all countries is supporting widespread deployment of commercial technology below the in-theatre command level, whether joint- or single-service. Demand for facilities such as quality-secure voice communications, video conferencing, shared operational pictures and joint planning all make demands on available communications capacity. Peacekeeping operations have also shown a demand in the field for other types of service, such as connection to the Internet to support the civilian secretariats that often accompany peacekeeping forces. For forces deployed in areas of conflict where the local telecommunications infrastructure is either severely damaged or simply absent, it is usually only possible to provide the high capacity communications circuits through ground stations linked to satellite systems, at least in the early stages of a deployment.

The US, UK, France and NATO all possess military-satellite communications capability in the form of the Defense Satellite Communications System (DSCS), *Skynet-4*, *Syracuse* and NATO IV systems respectively. All the current generation of military communications satellites have been designed to provide the maximum support for deployed forces worldwide, with steerable 'spot beam' antennae on the spacecraft that greatly improve communications capacity, along with a range of sophisticated anti-jamming features. Ownership or access to such satellite systems is essential for effective force-projection overseas. As demand increases for in-theatre information, such as intelligence imagery and detailed geographic data, it is likely that the next generation of military-satellite systems will include high capacity direct-broadcast capability, similar to that provided by commercial digital-satellite television services. Despite the almost global coverage that such satellites provide, there can be occasional difficulties that stretch to the limit military planners' abilities to provide the necessary communications to support command-and-control systems to deployed headquarters. The problem of the UK's East Timor deployment being just outside the coverage of the UK's *Skynet* satellites was ingeniously overcome by arranging for communications to be relayed through other countries in the region.

Below the level of major headquarters, it is very difficult to provide information-systems support to front-line forces. At Corps and Division level, the US communications network known as MSE/TRITAC (Mobile-subscriber Equipment/Tri-service Tactical Communications), along with the UK's *Ptarmigan* and the French Réseau Integré Transmission (RITA) networks provide some data services, but despite some mid-life improvements, these systems are relatively elderly, mainly providing voice communications, and were not designed to meet the demands of information-age warfare. To provide information-systems support to front-line units, 'digitisation of the battlefield' initiatives are underway in key NATO countries and such close allies as Australia. These initiatives are aimed at providing up-to-date local-situation awareness, friend-or-foe identification and integration of sensors and weapon systems.

Digitisation programmes are ambitious and the technical difficulties of meeting the demanding requirements of an electronically and physically hostile environment are not to be underestimated. It will be some years before programmes such as the US Battlefield Digitisation initiatives (including *Appliqué*, the generic name for components providing tactical C3I digitisation) and the UK's much-delayed *Bowman* tactical-communications project result in the next generation of digital radios and battlefield-information systems being widely available to front-line units. It may take another decade before the technologies currently being developed are in front-line service. For some digitisation programmes, user expectations, based on the facility of commercial mobile telephones and handheld computing devices such as personal organisers, are outstripping industry's ability to deliver affordable equipment providing equal functionality in a battlefield environment. It is likely that land-digitisation programmes in a number of countries will be radically reshaped in the course of 2000–01. However, for complex platforms such as ships and aircraft, advances have already been made in digitising C3I systems. The most notable is the adoption by a number of countries, including the US, the UK and Germany of the Joint Tactical Interoperable Data System (JTIDS), enabling air- and sea-platforms to share sensor data. This has already led to major reforms in air-warfare tactics. One problem with JTIDS is that it uses part of the commercial air-navigation radio-frequency band for its transmissions. This tends to restrict naval and aircrew training in regions where there are crowded airways and strict controls on radio-interference levels.

Most advances in command-and-control systems are being made in the G-8 leading industrialised countries. While the various approaches adopted have similarities, there continues to be limited interoperability between the national command systems fielded during recent NATO, UN and coalition operations. This is partly due to differences in technology, but these can be readily overcome. The main problem arises from national security rules governing interconnection between systems carrying information with national caveats and the systems of other countries.

MULTINATIONAL C3I INTEROPERABILITY

NATO has actively promoted interoperability between national systems by establishing the appropriate standards and by brokering discussions on the characteristics of the different systems. To facilitate understanding of the issues, NATO has developed a model defining four levels of interoperability between national command-and-control systems (Table 36).

Table 36 NATO's four levels of interoperability

Level	Type Description
4	**Seamless sharing of information** Integration is seamless – cooperating applications share data across national boundaries
3	**Seamless sharing of data** A common data-exchange model is used to pass data between national systems
2	**Structured data exchange** Message-based data is interpretable by both automated systems and human readers
1	**Unstructured data exchange** The exchange of free-text messages, such as word-processed documents, which are not interpretable by automated systems

At present, most interoperability between NATO countries in coalition forces is at Levels 1 and 2. As an example, most NATO countries have CRONOS terminals and NATO-secure speech systems, usually in the form of stand-alone terminals and telephones located in national headquarters, enabling word-processed documents to be sent by e-mail and printed locally for manual distribution. Most NATO countries are also able to exchange electronic messages through gateways that are connected to national systems. In coalition operations, Level 4 interoperability is the ideal, but to achieve it national systems would have to be capable of merging into a single command-and-control system, with no restrictions on information flow between countries. It is unlikely that such an ideal position will be reached. A more achievable target is Level 3 interoperability – where countries allow interconnection between command-and-control systems, but restrict access to information to that which can be shared between all or most coalition partners. Purely national information remains behind electronic barriers. Where Partnership for Peace (PfP) states and other non-NATO countries contributing forces to peacekeeping operations are concerned, interoperability between command systems is still at its most basic level, often confined to the key, or 'framework', nations lending equipment and contributing liaison officers. Such a low level of interoperability is undesirable and can lead to poor information flow and mistakes during operations.

PROMOTING INTEROPERABILITY

A number of initiatives are underway to improve interoperability between national and NATO C3I systems when forces are deployed in coalition operations. NATO operations in former Yugoslavia have highlighted the difficulty of integrating national C3I systems and the many ad-hoc interfaces and gateways that are being developed as the systems are fielded. This experience has led NATO to take both short-and long-term approaches to improving interoperability. In the short-term, the lessons learned in developing gateways have been encapsulated in a NATO 'systems interface' guide. This gives pragmatic advice on joining national command systems in coalition operations, but it is far from the most efficient solution. In the longer term, other approaches are being developed. One means of improving the integration of military messaging with command systems is the adoption of internationally agreed message-text formats (MTFs), which allow messages to be read and interpreted by computers, while they can still be read by the human eye. This approach assists new NATO members and PfP countries to achieve interoperability without hampering more advanced integration by the original NATO members. Such messages can be used to populate databases – for example, those used to generate common operational pictures. NATO countries have agreed more than 200 such formatted messages, but they still rely on English as the common C3I language. Additionally, there has been a move towards more comprehensive messaging schemes. For many years, the military messaging standard known as ACP127 has been the multinational standard. It is based on '5-hole punch-tape' telegraph codes and similar to the standard applying to commercial telegrams. It not only safeguards security, but also allows authenticity to be verified – an important matter for commanders who know that military operations may later be subject to international legal scrutiny.

The emerging new military standard, ACP123, is based on the commercial 'X400' standards with security extensions. It allows messages to contain much richer information than conventional telegrams, including the attachment of computer-generated files such as databases or pictures. However, it is not yet certain whether this standard will be widely adopted. Although it includes the formality of military messaging, it is expensive and not widely supported by software manufacturers. Within national C3I systems, 'informal' e-mail, known as SMPT (simple mail transfer protocol) is increasingly the norm. It is included with most popular operating systems and software, such as Microsoft and by the vendors of UNIX-based systems, so is inexpensive and easy to use. However it lacks the formality of military messaging and could lead to war records being lost or it being impossible to prove how a sequence of events was initiated. There are, however, improvements being made in informal e-mail messaging that reflect the commercial demand for authentication in e-commerce systems, and these could improve e-mail's acceptability for military use.

Because of the widespread use within the developing national C3I systems of commercial 'off the shelf' software, along with the few specialised military applications, NATO has recognised that achieving 'Level 3' and 'Level 4' interoperability depends on agreeing a core set of pragmatic standards for information interchange. Work is underway to develop a NATO Common Operating Environment (NCOE), based on the national COE's that have already been developed. Fortunately, the high degree of commonality that already exists between the approaches of individual countries simplifies the task of NATO's international staffs. Less easy is reaching agreement on common data definitions: for example on the units used when measurements are exchanged. A mistake in this area led to the recent failure of the National Aeronautics and Space Administration (NASA)'s Mars Lander. Based on ATCCIS, NATO is now working on a core set of agreed definitions needed if information provided by one country is to be interpreted correctly by the computer systems of another.

BEYOND NATO

Other initiatives are underway outside NATO based on practical demonstrations aimed at improving multinational interoperability. There are proposals for a 'coalition-wide area network' between Australia, Canada, New Zealand, the UK and the US. The Joint Warrior Interoperability Demonstration (JWID), originating in the US, but with increasing participation by other NATO countries and close allies, has become an annual event. Based on a multinational C³I exercise, it allows manufacturers from each participating country to demonstrate how their products function in a multinational environment. By bringing industry and military staffs together in a multinational forum, with a shared determination to solve problems, JWID has been immensely valuable in showing what can be done and in demonstrating how a coalition communications network would work in practice.

To tackle the problem of integrating C³I systems at a database level, a small number of NATO countries formed the ATCCIS forum. This gathering has also held practical demonstrations, dealing with the difficult issues of database replication (ensuring that recognised pictures are all updated simultaneously as the situation changes) and common meaning of data (data management). Working demonstrators have been produced, but what is easy to achieve in a neutral, one-off demonstration is less easy to achieve in operational systems, not least because of legacy systems with their unique interpretations of data, and the less easily solved problems of security.

Table 37 shows the initiatives that are underway in key NATO countries and among their close allies to put in place the next generation of command-and-control systems at both strategic and theatre levels.

CONCLUSIONS

The use of commercial technology in C³I systems is bringing appreciable benefits in integrating national systems, improving situational awareness and reducing decision times. It has also done much to shorten development time for strategic C³I, with new capabilities constantly added as technology develops. However, expensive development programmes are needed to 'militarise' commercial technology and to match the needs of front-line military users.

At the same time, the use of commercial technology has led national approaches to strategic C³I to converge. Radical reshaping of land-communications digitisation programmes is likely in a number of countries over the next year, which should improve interoperability in coalition operations. However, here too, there are still many problems caused by detailed variations in implementation that prevent systems being readily integrated, and by national security rules. It is particularly difficult to integrate the command systems of Russia, PfP states, and the other non-NATO countries that contribute to coalitions. Unfortunately the communications-technology gap between the more advanced NATO countries and the rest is likely to widen. A major effort is needed to tackle this problem and to improve interoperability in the next generation of systems if efficient command-and-control information infrastructures for coalition and UN peacekeeping operations are to be established more quickly than was recently achieved in Bosnia, Kosovo, East Timor and Sierra Leone. A close partnership between the military and industry is essential if the problems faced by 'digitisation' programmes are to be solved and full advantage is to be taken of rapidly developing technologies.

Table 37 **C³I initiatives and key in-service C³I systems in NATO countries and close allies**[1]

Country	Key current and planned C³I systems Comment
Australia	**GCCS, BCCS** GCCS components are being adopted at the strategic level, with a Battlefield Command and Control system (BCCS) (still in development) used in theatre. First entered service in 1999.
France	**RITA, RITA 2000, *Syracuse* SIR/SICF, FELIN** The rear-area communications infrastructure, RITA has been in service for some time. The updated RITA 2000 system is currently being deployed. *Syracuse* is based on a military communications repeater, mounted on *Telecom2*, a civil communications satellite. A replacement, wholly military-owned satellite is under consideration. SIR and SICF are the strategic-to-brigade level C³I systems; FELIN is the lowest level (front line) element of C³I. These systems are under development and currently under trial.
NATO-owned	**CRONOS, NATO SATCOM IV, NATO SATCOM V, ACE ACCIS** CRONOS and NATO SATCOM IV are mainly fielded. NATO SATCOM V is in its formative stages and may be the subject of a multinational debate on contract awards.
UK	**JOCS Phase 1 and Phase 2, *Skynet*-4 Phase 2, *Skynet*-5, *Ptarmigan, Bowman, Falcon, Cormorant*** The Joint Operational Command System (JOCS) Phase 2 is being fielded. *Skynet* 5 may be an industry-provided service under the Private Finance Initiative to replace the *Skynet-4* satellite communications system, which has been in service since the early 1980s. The rear area-to-brigade *Ptarmigan* communications system has been in service for two decades, and is to be replaced by the *Falcon* and *Cormorant* systems. *Bowman*, the 'digital' replacement for the 1970s-vintage *Clansman* radio has suffered massive development delays and is unlikely to be in service before 2007.
US	**GCCS, GCSS, DSCS, SLEP, MILSTAR, Battlefield Digitisation (*Appliqué /Land Warrior*), MSE/TRITAC, SINCGARS, FBCB2** GCCS is being fielded. The DSCS satellite system, dating from the mid-1970s, is getting a new lease of life with a service-life enhancement programme (SLEP). MILSTAR provides secure communications that are resistant to electronic attack, but has limited capacity. Similar to the UK *Bowman*, battlefield digitisation programmes such as *Appliqué* and *Land Warrior* are suffering development delays. MSE/TRITAC rear-area communication has been in service for some time and has been modified with terminal adapters to provide data services. The SINCGARS ground/air radio system has also been in service for some time. The FBCB2 system is being fielded to improve situational awareness.

CRONOS Crisis Response Operation in NATO Open System; **DSCS** Defense Satellite Communications System; **FELIN** Fantassin à Equipements et Liaisons Intégrées; **FBCB2** The Force Battle Command Brigade and Below; **GCCS** Global Command and Control System; **JOCS** Joint Operational Command System; **MILSTAR** Military Strategic and Tactical Relay; **MSE/TRITAC** Mobile-subscriber equipment/ tri-service tactical communications; **RITA** Réseau Integré Transmission; **SICF** Système d'Information pour le Commandement des Forces (Information System for the armed Forces Command); **SINCGARS** Single Channel Ground and Airborne Radio System; **SIR** Système d'Information Régimentaire

[1] The table only shows the significant C³I systems and developments. Most countries still possess a multiplicity of smaller, single service C³I systems serving specific needs.

Table 38 **International comparisons of defence expenditure and military manpower, 1985, 1998 and 1999**

(1999 constant prices)	Defence Expenditure									Numbers in Armed Forces		Estimated Reservists	Para-military
	US$m			US$ per capita			% of GDP			(000)		(000)	(000)
	1985	1998	1999	1985	1998	1999	1985	1998	1999	1985	1999	1999	1999
Canada	11,597	7,677	7,504	457	265	257	2.2	1.2	1.2	83.0	60.6	43.3	9.4
US	382,548	279,702	283,096	1,599	1,034	1,036	6.5	3.1	3.1	2,151.6	1,371.5	1,303.3	89.0
NATO Europe													
Belgium	6,100	3,737	3,445	619	370	341	3.0	1.5	1.5	91.6	41.8	152.1	n.a.
Czech Republic	n.a.	1,178	1,164	n.a.	114	113	n.a.	2.1	2.3	n.a.	58.2	240.0	5.6
Denmark	3,098	2,903	2,682	606	553	510	2.2	1.6	1.6	29.6	24.3	81.2	N.A
France	48,399	40,834	37,893	877	693	640	4.0	2.8	2.7	464.3	317.3	419.0	94.3
Germany	52,246	33,802	31,117	688	412	379	3.2	1.5	1.6	478.0	332.8	344.7	N.A
Greece	3,451	5,951	5,206	347	562	489	7.0	4.8	5.0	201.5	165.6	291.0	4.0
Hungary	3,517	673	745	330	66	74	7.2	1.4	1.6	106.0	43.4	90.3	14.0
Iceland	n.a.	n.a.	n.a.	n.a.	n.a.	n.a.	n.a.	n.a.	n.a.	n.a.	n.a.	n.a.	0.1
Italy	25,459	23,943	22,046	446	413	381	2.3	2.0	2.0	385.1	265.5	72.0	255.7
Luxembourg	95	146	138	258	352	331	0.9	0.8	0.8	0.7	0.8	n.a.	0.6
Netherlands	8,812	7,192	6,946	608	459	442	3.1	1.8	1.8	105.5	56.4	75.0	3.6
Norway	3,067	3,391	3,149	738	769	712	3.1	2.3	2.2	37.0	30.7	234.0	0.3
Poland	8,533	3,491	3,242	229	90	84	8.1	2.2	2.1	319.0	240.7	406.0	23.4
Portugal	1,816	2,382	2,280	178	241	231	3.1	2.2	2.2	73.0	49.7	210.9	40.9
Spain	11,164	7,522	7,263	289	192	185	2.4	1.3	1.3	320.0	186.5	447.9	75.8
Turkey	3,401	8,955	10,183	68	143	156	4.5	4.2	5.5	630.0	639.0	378.7	202.2
United Kingdom	47,240	38,093	36,876	835	650	628	5.2	2.7	2.6	327.1	212.4	254.3	n.a.
Subtotal NATO Europe	**226,397**	**184,192**	**174,375**	**475**	**380**	**356**	**4.0**	**2.2**	**2.3**	**3,568.4**	**2,665.0**	**3,697.1**	**720.5**
Total NATO	**620,542**	**471,572**	**469,176**	**540**	**410**	**388**	**4.0**	**2.2**	**2.3**	**5,803.0**	**4,097.1**	**5,043.7**	**818.8**
Non-NATO Europe													
Albania	280	102	140	95	28	37	5.3	3.3	3.6	40.4	54.0	155.0	13.5
Armenia	n.a.	151	159	n.a.	40	42	n.a.	8.4	8.6	n.a.	53.4	300.0	1.0
Austria	1,913	1,796	1,664	253	222	205	1.2	0.8	0.8	54.7	40.5	100.7	n.a.
Azerbaijan	n.a.	197	203	n.a.	27	28	n.a.	4.6	4.4	n.a.	69.9	575.7	15.0
Belarus	n.a.	471	466	n.a.	45	44	n.a.	4.0	5.0	n.a.	80.9	289.5	23.0
Bosnia	n.a.	405	366	n.a.	91	82	n.a.	10.2	8.4	n.a.	40.0	150.0	46.0
Bulgaria	2,425	406	392	288	49	47	14.0	3.2	3.3	148.5	80.8	303.0	34.0
Croatia	n.a.	1,089	776	n.a.	242	172	n.a.	5.7	4.1	n.a.	61.0	220.0	40.0
Cyprus	129	509	530	194	592	609	3.6	5.5	6.1	10.0	10.0	88.0	0.8
Czechoslovakia	3,472	n.a.	n.a.	223	n.a.	n.a.	8.2	n.a.	n.a.	203.3	n.a.	n.a.	n.a.
Estonia	n.a.	61	71	n.a.	42	48	n.a.	1.3	1.5	n.a.	4.8	14.0	2.8
Finland	2,226	1,929	1,695	453	375	328	2.8	1.5	1.4	36.5	31.7	540.0	3.4
FYROM	n.a.	72	67	n.a.	31	29	n.a.	2.0	2.0	n.a.	16.0	102.0	7.5
Georgia	n.a.	112	111	n.a.	21	20	n.a.	2.5	2.4	n.a.	26.3	250.0	6.5
Ireland	474	737	745	133	201	202	1.8	0.9	0.9	13.7	11.5	14.8	n.a.
Latvia	n.a.	40	58	n.a.	16	24	n.a.	0.6	1.0	n.a.	5.7	14.5	3.7

(1999 constant prices)	Defence Expenditure									Numbers in Armed Forces		Estimated Reservists	Para-military
	US$m			US$ per capita			% of GDP			(000)		(000)	(000)
	1985	1998	1999	1985	1998	1999	1985	1998	1999	1985	1999	1999	1999
Lithuania	n.a.	139	106	n.a.	38	28	n.a.	1.3	1.0	n.a.	12.1	27.7	3.9
Malta	24	30	27	66	79	72	1.4	0.8	0.8	0.8	1.9	n.a.	n.a.
Moldova	n.a.	53	6	n.a.	12	1	n.a.	4.3	0.5	n.a.	10.7	66.0	3.4
Romania	2,067	905	607	91	40	27	4.5	2.3	1.8	189.5	207.0	470.0	75.9
Slovakia	n.a.	423	329	n.a.	79	61	n.a.	2.0	1.9	n.a.	44.9	20.0	2.6
Slovenia	n.a.	323	337	n.a.	161	167	n.a.	1.6	1.8	n.a.	9.6	61.0	4.5
Sweden	4,730	5,760	5,245	566	648	588	3.3	2.4	2.3	65.7	53.1	570.0	35.6
Switzerland	2,860	3,700	3,108	443	523	439	2.1	1.4	1.3	20.0	27.7	384.9	n.a.
Ukraine	n.a.	1,415	1,437	n.a.	28	29	n.a.	3.3	2.9	n.a.	311.4	1,000.0	116.6
FRY (Serbia-Montenegro)	4,951	1,585	1,654	212	149	149	3.8	9.1	12.4	241.0	108.7	400.0	38.0
Total	**25,550**	**22,408**	**20,297**	**251**	**151**	**139**	**4.3**	**3.3**	**3.2**	**1,024.1**	**1,373.5**	**6,116.8**	**477.7**
Russia	n.a.	57,107	56,800	n.a.	390	380	n.a.	5.3	5.1	n.a.	1,004.1	2,400.0	478.0
Soviet Union	364,715	n.a.	n.a.	1,308	n.a.	n.a.	16.1	n.a.	n.a.	5,300.0	n.a.	n.a.	n.a.
Middle East and North Africa													
Algeria	1,412	3,125	3,086	64	107	104	1.7	6.5	6.6	170.0	122.0	150.0	181.2
Bahrain	224	410	441	537	669	706	3.5	7.5	7.7	2.8	11.0	n.a.	10.2
Egypt	3,827	2,888	2,988	79	47	48	7.2	3.4	3.4	445.0	450.0	254.0	230.0
Gaza and Jericho	n.a.	n.a.	n.a.	n.a.	n.a.	n.a.	n.a.	n.a.	n.a.	n.a.	n.a.	n.a.	35.0
Iran	10,523	5,879	5,711	236	95	91	18.0	6.5	6.2	610.0	545.6	350.0	240.0
Iraq	13,752	1,428	1,500	897	66	68	37.9	7.3	7.6	1,000.0	429.0	650.0	50.0
Israel	7,486	9,339	8,846	1,768	1,560	1,465	21.2	9.3	8.9	142.0	173.5	425.0	6.1
Jordan	892	559	588	255	115	117	15.9	7.7	7.7	70.3	104.0	35.0	30.0
Kuwait	2,661	3,674	3,275	1,556	1,670	1,440	9.1	14.3	11.1	12.0	15.3	23.7	5.0
Lebanon	296	586	563	111	139	132	9.0	3.6	3.4	17.4	67.9	n.a.	13.0
Libya	2,000	1,489	1,311	531	248	211	6.2	5.5	4.7	73.0	65.0	40.0	0.5
Mauritania	77	26	24	46	11	10	6.5	2.2	2.0	8.5	15.7	n.a.	5.0
Morocco	950	1,696	1,761	43	58	59	5.4	4.6	5.0	149.0	196.3	150.0	42.0
Oman	3,196	1,792	1,631	1,998	841	737	20.8	12.4	10.9	29.2	43.5	n.a.	4.4
Qatar	445	1,373	1,468	1,411	2,046	2,156	6.0	15.4	15.4	6.0	11.8	n.a.	n.a.
Saudi Arabia	26,618	21,303	21,876	2,306	1,081	1,099	19.6	16.2	15.5	62.5	162.5	20.0	15.5
Syria	5,161	986	989	491	62	60	16.4	5.8	5.6	402.5	316.0	396.0	108.0
Tunisia	618	363	348	87	39	37	5.0	1.8	1.7	35.1	35.0	n.a.	12.0
UAE	3,027	3,056	3,187	2,162	1,184	1,203	7.6	6.5	6.2	43.0	64.5	n.a.	1.0
Yemen	725	404	429	72	23	24	9.9	6.6	6.7	64.1	66.3	40.0	70.0
Total	**83,891**	**60,374**	**60,023**	**771**	**530**	**514**	**11.9**	**7.5**	**7.2**	**3,342.4**	**2,894.9**	**2,533.7**	**1,058.8**
Central and South Asia													
Afghanistan	425	255	265	24	11	11	8.7	14.5	14.9	47.0	400.0	n.a.	n.a.
Bangladesh	370	631	667	4	5	5	1.4	1.9	1.9	91.3	137.0	n.a.	55.2
Bhutan	8	19	20	18	29	31	4.9	5.4	5.3	3.0	6.0	n.k.	1.0

(1999 constant prices)	Defence Expenditure									Numbers in Armed Forces		Estimated Reservists	Para-military
	US$m			US$ per capita			% of GDP			(000)		(000)	(000)
	1985	1998	1999	1985	1998	1999	1985	1998	1999	1985	1999	1999	1999
India	9,281	13,594	14,991	12	14	15	3.0	3.2	3.4	1,260.0	1,173.0	528.4	1,090.0
Kazakstan	n.a.	508	504	n.a.	34	34	n.a.	2.2	3.5	n.a.	65.8	n.a.	34.5
Kyrgyzstan	n.a.	66	51	n.a.	14	11	n.a.	3.6	4.5	n.a.	9.2	57.0	3.0
Maldives	5	39	41	27	144	150	3.9	9.6	9.6	n.k.	n.k.	n.k.	5.0
Nepal	53	39	42	3	2	2	1.5	0.8	0.8	25.0	50.0	n.a.	40.0
Pakistan	3,076	4,078	3,523	32	29	24	6.9	6.6	5.7	482.8	587.0	513.0	247.0
Sri Lanka	338	995	807	21	53	43	3.8	6.1	5.1	21.6	115.0	4.2	103.3
Tajikistan	n.a.	102	92	n.a.	17	15	n.a.	8.3	7.6	n.a.	9.0	n.a.	1.2
Turkmenistan	n.a.	86	112	n.a.	19	22	n.a.	3.2	3.3	n.a.	19.0	n.a.	n.k.
Uzbekistan	n.a.	670	615	n.a.	29	26	n.a.	4.4	3.9	n.a.	74.0	n.a.	20.0
Total	**13,557**	**21,080**	**21,731**	**18**	**31**	**30**	**4.3**	**5.4**	**5.3**	**1,930.7**	**2,645.0**	**1,102.6**	**1,600.2**
East Asia and Australasia													
Australia	8,068	7,682	7,775	512	407	407	3.4	2.1	1.9	70.4	55.2	27.7	1.0
Brunei	304	386	402	1,356	1,217	1,240	6.0	6.7	6.7	4.1	5.0	0.7	3.8
Cambodia	n.a.	155	176	n.a.	15	17	n.a.	5.1	5.1	35.0	139.0	n.a.	220.0
China	29,414	38,191	39,889	28	31	32	7.9	5.3	5.4	3,900.0	2,820.0	1,200.0	1,000.0
Fiji	21	34	35	30	43	44	1.2	2.0	1.9	2.7	3.5	6.0	n.a.
Indonesia	3,469	967	1,502	21	5	7	2.8	0.8	1.1	278.1	299.0	400.0	200.0
Japan	31,847	38,482	40,383	264	305	319	1.0	1.0	0.9	243.0	242.6	48.6	12.0
Korea, North	6,158	2,086	2,100	302	97	98	23.0	14.3	14.3	838.0	1,055.0	4,700.0	189.0
Korea, South	9,323	10,461	12,088	227	225	257	5.1	2.4	3.0	598.0	672.0	4,500.0	4.5
Laos	81	34	22	23	7	4	7.8	2.6	2.3	53.7	29.1	n.a.	100.0
Malaysia	2,614	1,891	3,158	168	88	146	5.6	2.6	4.0	110.0	105.0	40.6	20.1
Mongolia	51	21	19	27	9	8	9.0	1.9	1.9	33.0	9.1	140.0	7.2
Myanmar	1,302	2,142	1,995	35	45	42	5.1	5.0	5.0	186.0	343.8	n.a.	85.3
New Zealand	957	898	824	294	236	215	2.9	1.5	1.6	12.4	9.5	6.3	n.a.
Papua New Guinea	53	57	59	15	12	12	1.5	1.0	1.0	3.2	4.3	n.a.	n.a.
Philippines	702	1,521	1,627	13	21	22	1.4	2.3	2.1	114.8	110.0	131.0	42.5
Singapore	1,760	4,936	4,696	688	1,275	1,174	6.7	5.6	5.6	55.0	73.0	275.0	108.0
Taiwan	9,541	14,447	14,964	492	668	687	7.0	4.8	5.2	444.0	376.0	1,657.5	26.7
Thailand	2,777	2,124	2,638	54	35	43	5.0	1.7	1.9	235.3	306.0	200.0	71.0
Vietnam	3,556	943	890	58	12	11	19.4	3.5	3.1	1,027.0	484.0	3,000.0	40.0
Total	**112,000**	**127,456**	**135,243**	**242**	**238**	**239**	**6.4**	**3.6**	**3.7**	**8,243.7**	**7,141.1**	**16,333.4**	**2,131.0**
Caribbean, Central and Latin America													
Caribbean													
Antigua and Barbuda	3	4	4	42	56	57	0.5	0.6	0.6	0.1	0.2	0.1	n.a.
Bahamas, The	14	26	26	61	90	89	0.5	0.7	0.7	0.5	0.9	n.a.	2.3
Barbados	17	13	12	77	48	44	0.9	0.5	0.5	1.0	0.6	0.4	n.a.
Cuba	2,366	765	750	235	68	67	9.6	5.3	4.8	161.5	65.0	39.0	26.5

(1999 constant prices)	Defence Expenditure									Numbers in Armed Forces		Estimated Reservists	Para-military
	US$m			US$ per capita			% of GDP			(000)		(000)	(000)
	1985	1998	1999	1985	1998	1999	1985	1998	1999	1985	1999	1999	1999
Dominican Republic	76	115	114	12	14	14	1.1	1.0	0.9	22.2	24.5	n.a.	15.0
Haiti	46	49	50	8	7	7	1.5	1.4	1.3	6.9	n.a.	n.a.	5.3
Jamaica	30	45	51	13	18	20	0.9	0.6	0.8	2.1	2.8	1.0	0.2
Trinidad and Tobago	108	44	62	91	33	46	1.4	0.7	0.9	2.1	2.7	n.a.	4.8
Central America													
Belize	6	17	17	36	73	72	1.4	2.6	2.5	0.6	1.1	0.7	n.a.
Costa Rica	43	70	69	17	19	19	0.7	0.6	0.6	n.a.	n.a.	n.a.	8.4
El Salvador	373	163	171	78	27	28	4.4	1.0	1.1	41.7	24.6	15.0	12.0
Guatemala	174	159	149	22	14	13	1.8	1.2	1.1	31.7	31.4	35.0	9.5
Honduras	107	99	95	24	15	14	2.1	1.9	1.8	16.6	8.3	60.0	6.0
Mexico	1,839	3,907	4,289	23	39	42	0.7	0.9	0.9	129.1	178.8	300.0	15.0
Nicaragua	327	30	25	100	7	5	17.4	1.1	0.9	62.9	16.0	n.a.	n.a.
Panama	133	122	128	61	44	45	2.0	1.3	1.3	12.0	n.a.	n.a.	11.8
South America													
Argentina	5,366	5,365	5,418	176	149	148	3.8	1.8	1.9	108.0	70.5	375.0	31.2
Bolivia	188	209	149	29	26	18	2.0	2.4	1.7	27.6	32.5	n.a.	37.1
Brazil	5,738	18,781	15,978	42	116	98	1.8	3.2	2.7	276.0	291.0	1,115.0	385.6
Chile	2,380	3,071	2,694	197	208	181	10.6	3.8	4.0	101.0	93.0	50.0	29.5
Colombia	628	2,574	2,164	22	63	52	1.6	3.2	2.8	66.2	144.0	60.7	87.0
Ecuador	421	543	339	45	44	27	1.8	2.6	2.3	42.5	57.1	100.0	0.3
Guyana	47	8	7	59	9	8	6.8	1.0	0.9	6.6	1.6	n.a.	1.5
Paraguay	89	135	128	24	25	23	1.3	1.5	1.4	14.4	20.2	164.5	14.8
Peru	950	1,009	888	51	40	35	4.5	1.7	1.6	128.0	115.0	188.0	78.0
Suriname	12	15	22	32	37	54	2.4	3.9	5.5	2.0	1.8	n.a.	n.a.
Uruguay	354	321	317	117	99	98	3.5	2.3	2.3	31.9	25.6	n.a.	0.9
Venezuela	1,221	1,333	1,329	71	57	56	2.1	1.4	1.5	49.0	56.0	8.0	23.0
Total	**23,055**	**38,991**	**35,447**	**63**	**52**	**49**	**3.2**	**1.8**	**1.8**	**1,344.2**	**1,265.2**	**2,512.4**	**805.7**
Sub-Saharan Africa													
Horn Of Africa													
Djibouti	47	22	22	110	31	30	7.9	5.1	5.0	3.0	8.4	n.a.	4.2
Eritrea	n.a.	297	309	n.a.	76	77	n.a.	42.1	44.4	n.a.	200.0	120.0	n.a.
Ethiopia	662	387	444	16	7	8	17.9	6.0	7.1	217.0	325.5	n.a.	n.a.
Somali Republic	68	41	40	13	7	6	6.2	4.7	4.6	62.7	50.0	n.a.	n.a.
Sudan	158	381	424	7	13	14	3.2	4.7	4.9	56.6	94.7	n.a.	15.0
Central Africa													
Burundi	52	82	69	11	12	10	3.0	7.2	6.4	5.2	40.0	n.a.	5.5
Cameroon	236	149	154	23	10	10	1.4	1.5	1.5	7.3	13.1	n.a.	9.0
Cape Verde	5	4	7	17	9	15	0.9	1.6	2.7	7.7	1.1	n.a.	0.1
Central African Republic	26	50	45	10	14	12	1.4	4.7	4.0	2.3	2.7	n.a.	2.3

(1999 constant prices)	Defence Expenditure									Numbers in Armed Forces		Estimated Reservists	Para-military
	US$m			US$ per capita			% of GDP			(000)		(000)	(000)
	1985	1998	1999	1985	1998	1999	1985	1998	1999	1985	1999	1999	1999
Chad	55	64	47	11	9	7	2.9	3.8	2.9	12.2	30.4	n.a.	4.5
Congo	83	83	73	44	28	24	1.9	3.9	3.4	8.7	10.0	n.a.	5.0
DROC	120	371	411	4	8	9	1.5	6.6	7.8	48.0	55.9	n.a.	37.0
Equatorial Guinea	4	7	10	12	14	19	2.0	1.5	1.8	2.2	1.3	n.a.	0.3
Gabon	117	135	135	117	94	92	1.8	2.2	2.1	2.4	4.7	n.a.	4.8
Rwanda	49	144	135	8	23	20	1.9	7.0	6.2	5.2	47.0	n.a.	7.0
East Africa													
Kenya	379	321	327	19	11	11	3.1	3.1	3.1	13.7	24.2	n.a.	5.0
Madagascar	80	45	43	8	3	3	2.0	0.9	0.8	21.1	21.0	n.a.	7.5
Mauritius	4	89	91	4	76	77	0.3	2.0	2.0	1.0	n.a.	n.a.	1.8
Seychelles	12	11	11	182	148	155	2.1	1.8	1.8	1.2	0.2	n.a.	0.3
Tanzania	207	146	141	9	5	4	4.4	1.8	1.7	40.4	34.0	80.0	1.4
Uganda	79	230	199	5	11	9	1.8	3.1	2.5	20.0	40.0	n.a.	1.5
West Africa													
Benin	31	33	34	8	6	6	1.1	1.4	1.4	4.5	4.8	n.a.	2.5
Burkina Faso	50	81	75	6	7	6	1.1	2.5	2.1	4.0	5.8	n.a.	4.5
Côte d'Ivoire	113	121	130	11	8	8	0.8	0.9	1.0	13.2	8.4	12.0	7.0
Gambia, The	3	15	16	4	13	13	1.5	3.6	3.5	0.5	0.8	n.a.	n.a.
Ghana	93	137	121	7	7	6	1.0	1.4	1.2	15.1	7.0	n.a.	1.0
Guinea	77	59	60	12	8	8	1.8	1.7	1.7	9.9	9.7	n.a.	9.6
Guinea Bissau	16	5	6	18	5	5	5.7	1.9	1.9	8.6	7.3	n.a.	2.0
Liberia	41	46	25	19	16	8	2.4	12.3	5.6	6.8	5.3	n.a.	n.a.
Mali	44	36	34	6	3	3	1.4	1.3	1.2	4.9	7.4	n.a.	7.8
Niger	18	26	28	3	3	3	0.5	1.6	1.7	2.2	5.3	n.a.	5.4
Nigeria	1,112	2,143	2,237	12	19	20	3.4	4.4	4.4	94.0	94.0	n.a.	30.0
Senegal	93	83	81	14	9	9	1.1	1.7	1.6	10.1	11.0	n.k.	6.0
Sierra Leone	7	26	11	2	6	2	1.0	3.3	1.5	3.1	3.0	n.a.	0.8
Togo	28	35	34	9	7	7	1.3	2.4	2.3	3.6	7.0	n.a.	0.8
Southern Africa													
Angola	959	974	1,005	109	83	83	15.1	13.6	16.5	49.5	112.5	n.a.	15.0
Botswana	55	261	259	51	161	157	1.1	5.3	5.2	4.0	9.0	n.a.	1.0
Lesotho	68	42	34	44	20	16	4.6	4.7	4.2	2.0	2.0	n.a.	n.a.
Malawi	31	26	27	4	2	2	1.0	1.5	1.8	5.3	5.0	n.a.	1.0
Mozambique	354	82	94	26	5	6	8.5	3.9	4.1	15.8	6.1	n.a.	n.a.
Namibia	n.a.	94	120	n.a.	53	65	n.a.	3.6	4.4	n.a.	9.0	n.a.	0.1
South Africa	4,256	1,900	1,755	127	49	44	2.7	1.4	1.3	106.4	70.0	88.0	8.2
Zambia	59	66	88	9	7	9	1.1	1.9	2.5	16.2	21.6	n.a.	1.4
Zimbabwe	252	334	418	30	28	35	5.6	5.0	6.1	41.0	39.0	n.a.	21.8
Total	10,206	9,682	9,830	28	26	26	3.1	4.5	4.4	958.5	1,455.1	300.0	238.1

(1999 constant prices)	Defence Expenditure US$m			Defence Expenditure US$ per capita			% of GDP			Numbers in Armed Forces (000)		Estimated Reservists (000)	Para-military (000)
	1985	1998	1999	1985	1998	1999	1985	1998	1999	1985	1999	1999	1999
Global Totals													
NATO	620,542	471,572	469,176	540 984	410 609	388 596	4.0 4.7	2.2 2.6	2.3 2.6	5,803.0	4,097.1	5,043.7	818.8
Non-NATO Europe	25,550	22,408	20,297	251 n.a.	151 120	139 108	4.3 n.a.	3.3 1.9	3.2 1.8	1,024.1	1,373.5	6,116.8	477.7
Russia	n.a.	57,107	56,800	n.a.	390	380	n.a.	5.3	5.1	n.a.	1,004.1	2,400.0	478.0
Soviet Union	364,715	n.a.	n.a.	1,308	n.a.	n.a.	16.1	n.a.	n.a.	5,300.0	n.a.	n.a.	n.a.
Middle East and North Africa	83,891	60,374	60,023	771 393	530 201	514 197	11.9 15.1	7.5 8.5	7.2 8.2	3,342.4	2,894.9	2,533.7	1,058.8
Central and South Asia	13,557	21,080	21,731	18 n.a.	31 15	30 16	4.3 n.a.	5.4 3.6	5.3 3.6	1,930.7	2,645.0	1,102.6	1,600.2
East Asia and Australasia	112,000	127,456	135,243	242 67	238 64	239 67	6.4 2.3	3.6 2.0	3.7 2.0	8,243.7	7,141.1	16,333.4	2,131.0
Caribbean, Central and Latin America	23,055	38,991	35,447	63 58	52 78	49 70	3.2 1.9	1.8 2.2	1.8 1.9	1,344.2	1,265.2	2,512.4	805.7
Sub-Saharan Africa	10,206	9,682	9,830	28 23	26 16	26 16	3.1 3.3	4.5 2.8	4.4 2.9	958.5	1,455.1	300.0	238.1
Global totals	**1253,517**	**808,671**	**808,546**	**399** **298**	**228** **138**	**221** **135**	**6.7** **5.2**	**4.2** **2.5**	**4.1** **2.4**	**27,946.6**	**21,875.9**	**36,342.6**	**7,608.3**

Note Under Defence Expenditure per capita and Defence Expenditure as a proportion of GDP, the top figure is the arithmetic mean of individual country values, and the bottom number is the arthimetic mean of the sum of regional and global totals.

Manpower and Treaty Limited Equipment: current holdings and CFE National Ceilings on the forces of the Treaty members

Current holdings are derived from data declared as of 1 January 2000 and so may differ from *The Military Balance* listings

	Manpower		Tanks[2]		ACV[2]		Artillery[2]		Attack Helicopters		Combat Aircraft[3]	
	Holding	Ceiling	Holding	Ceiling	Holding	Ceiling	Holding	Ceiling	Holding	Ceiling	Holding	Ceiling
Non-NATO												
Armenia	60,000	*60,000*	102	*220*	204	*220*	229	*285*	7	*50*	6	*100*
Azerbaijan	69,894	*70,000*	220	*220*	210	*220*	282	*285*	15	*50*	48	*100*
Belarus	83,083	*100,000*	1,724	*1,800*	2,478	*2,600*	1,465	*1,615*	60	*80*	224	*294*
Bulgaria	79,658	*104,000*	1,475	*1,475*	1,964	*2,000*	1,750	*1,750*	43	*67*	232	*235*
Georgia	26,811	*40,000*	79	*220*	113	*220*	109	*285*	3	*50*	7	*100*
Moldova	10,318	*20,000*	0	*210*	209	*210*	153	*250*	0	*50*	0	*50*
Romania	178,777	*230,000*	1,373	*1,375*	2,098	*2,100*	1,414	*1,475*	15	*120*	323	*430*
Russia [5]	584,841	*1,450,000*	5,375	*6,350*	9,956	*11,280*	6,306	*6,315*	741	*855*	2,733	*3,416*
Slovakia	44,519	*46,667*	275	*478*	622	*683*	383	*383*	19	*40*	82	*100*
Ukraine	310,000	*450,000*	3,939	*4,080*	4,860	*5,050*	3,720	*4,040*	247	*330*	911	*1,090*
NATO												
Belgium	38,785	*70,000*	140	*300*	569	*989*	242	*288*	41	*46*	135	*209*
Canada	0	*10,660*	0	*77*	0	*263*	0	*32*	0	*13*	0	*90*
Czech Republic [4]	57,735	*93,333*	792	*957*	1,211	*1,367*	740	*767*	34	*50*	110	*230*
Denmark	29,362	*39,000*	228	*335*	273	*336*	471	*446*	12	*18*	68	*82*
France	217,558	*325,000*	1,234	*1,226*	3,491	*3,700*	895	*1,192*	298	*374*	588	*800*
Germany	274,587	*345,000*	2,738	*3,444*	2,415	*3,281*	2,103	*2,255*	204	*280*	517	*765*
Greece	158,621	*158,621*	1,735	*1,735*	2,286	*2,498*	1,894	*1,920*	20	*65*	525	*650*
Hungary [4]	43,790	*100,000*	806	*835*	1,439	*1,700*	839	*840*	51	*108*	107	*180*
Italy	222,679	*315,000*	1,301	*1,267*	2,831	*3,172*	1,390	*1,818*	134	*142*	533	*618*
Netherlands	36,638	*80,000*	348	*520*	671	*864*	397	*485*	14	*50*	161	*230*
Norway	20,971	*32,000*	170	*170*	218	*275*	189	*491*	0	*24*	73	*100*
Poland [4]	205,270	*234,000*	1,674	*1,730*	1,437	*2,150*	1,554	*1,610*	107	*130*	271	*460*
Portugal	37,783	*75,000*	187	*300*	330	*430*	363	*450*	0	*26*	101	*160*
Spain	160,372	*300,000*	681	*750*	976	*1,588*	1,118	*1,276*	28	*80*	209	*310*
Turkey [5]	516,205	*530,000*	2,464	*2,795*	2,616	*3,120*	2,883	*3,523*	25	*130*	359	*750*
UK	193,688	*260,000*	584	*843*	2,330	*3,017*	424	*583*	243	*350*	520	*855*
US	100,661	*250,000*	793	*1,812*	1,572	*3,037*	345	*1,553*	136	*396*	233	*784*

Notes

[1] The adaptation of the CFE abandons the group structure (North Atlantic Group, Budapest/Tashkent Group) for a system of national and territorial ceilings. The amendment enters into force when CFE States Parties have ratified the change.

[2] Includes TLE with land-based maritime forces (Marines, Naval Infantry etc.)

[3] Does not include land-based maritime aircraft for which a separate limit has been set.

[4] Cz, Hu and Pl became NATO members on 12 March 1999.

[5] Manpower and TLE is for that in the Atlantic to the Urals (ATTU) zone only.

Unmanned Aerial Vehicles (UAVs)

Since unmanned aerial vehicles (UAVs) were last covered in *The Military Balance 1996–97*, technological developments have made them a more practical proposition. For example, the prototype US *Global Hawk* is built to stay on station conducting reconnaissance for up to 24 hours at a radius of 3,000 nautical miles. It made the first round-trip, non-stop, unescorted, unrefuelled flight by a UAV from the US to Europe on 10 May 2000.

UAVs were brought into prominence during the 1999 NATO military campaign in Kosovo. The US Army *Hunter*, Navy *Pioneer* and Air Force *Predator* conducted important reconnaissance operations, with *Predator* becoming the first US UAV to designate a target for laser-guided bombs launched from an A-10 ground-attack aircraft. German and French CL-289 UAVs (Canadian designed) and the British *Phoenix* conducted target-acquisition and battle-damage assessment missions. NATO lost 20–30 UAVs during the 78-day Kosovo air operation. They were either shot down or suffered technical failure. In assessing these losses, a major factor to take into account is that no aircrew lives were put at risk. UAVs are also an important complement to satellites in information gathering, particularly in their ability to fly below cloud cover and ability to send real-time information when needed.

As KFOR entered Kosovo on 12 June 1999, with Russian troops making their dash to Pristina airfield, staff at NATO's Combined Air Operations Centre in Italy could see, via data link from a UAV camera, Serb MiG-21s, hitherto hidden under the runway, taking off from the airfield before the Russians arrived. UAVs can now show air and ground commanders an intelligence picture in real time, allowing targets (and crucial related factors like fusing) to be changed when the aircraft are *en route*.

The value of UAV systems can be measured by their addition to many national military inventories over the past five years. Whereas reconnaissance satellites remain mostly the preserve of first-rank powers, UAVs can be an equaliser by offering immediate, independent imagery of comparable quality to a wider range of countries.

It is possible to consider a day when it may not be worth using a combat aircraft costing $50 million to deliver precision munitions when a UAV could do the whole task: real-time command and control, target verification and designation, and low-weight bomb delivery. It will be some years before that stage is reached, but the technology exists for UAVs to deliver weapons as well as identify targets. Bearing in mind that the capabilities offered by cruise missiles such as the US *Tomahawk*, it is a relatively small step to a UAV capable of identifying the target as well as delivering the weapon. An early step towards this goal is to employ UAVs in the air-defence suppression and target designation roles, flying as an integral part of a manned strike mission.

However, while UAVs keep aircrew out of harm's way, they are not a panacea. They are slow and vulnerable to ground fire, and they lack an all-weather capability because their wings can ice up in winter. Moreover, they are not cheap: the 12 British short-range *Phoenix* UAVs lost over Kosovo were valued at £3.5m. Pentagon staffs are finding that UAVs are costing four times more than they expected, with the average unit price of *Global Hawk* coming at over $15m. A mix of manned and unmanned systems will still be required for the foreseeable future. However, as commanders learn to trust UAVs, and understand all the implications of their use, these systems will become fully integrated into combined-arms operations, and their use and their capabilities will expand progressively over the next 20 years.

Table 40 **Unmanned Aerial Vehicles** page 1 of 2

Country of origin	Maker Name	Role	Payloads	Maximum Speed (km/hr)	Radius of action (km)	Endurance (hrs)	Ceiling (m)	In service
Long Range (over 500km radius of action)								
RF	Tupolev **Strizh**	recce	PC, TV, IRLS, LR/D, RS	1,100	500	n.k.	6,000	RF, Ukr
US	General Atomics **Gnat/I-Gnat**	multirole	FLIR, LLTV, IRLS, SAR, DL, NBC, ADP, LR/D, ESM, ELINT	260/220	2,700	40	7,600	Tu ▲, US ▲
US	General Atomics **Predator**	recce	RADAR, DL, SIGINT, IR, SAR	200	3,700	40	7,900	US ●
US	Northrop Grumman **Scarab**	recce	PC, IRLS, TV	850	966	n.k.	13,000	Et ●
Medium Range (100–500km radius of action)								
Ca/Ge	Bombardier **CL289**	recce, TA	PC, IRLS (RT), Video (RT)	740	190	n.k.	600	Fr, Ge
PRC	BUAA **Chang Hong**	recce	PC	800	n.k.	3	17,500	PRC
Fr	CAC **Fox AT1/AT2**	recce	FLIR (RT), IRLS (RT), RS	200	160	2/5	3,500	Fr, Indo
Fr	CAC **Fox TX**	EW	ECM, ARA, ESM	200	150	5	3,500	Fr
Fr	SAGEM **Sperwer**	recce, TA	FLIR, SAR, Video (RT)	235	150	8	5,000	Nl, Da
Il/US	AAI/IAI **Pioneer**	recce, decoy, EW, comms relay	Video, FLIR, EW, ECM, LASER designator/range finder	200	185	5	3,600	US ▲■✕
Il	TRW/IAI **Hunter**	recce, relay	TV, FLIR, LR/D, DL, ELINT, SIGINT, RP	190	150	12	4,575	Fr, Il, US, Be (on order); US ▲
Il	IAI **Harpy**	ARA	IR-seeker, ADP	n.k.	400	n.k.	n.k.	Il, Ind, ROK
Il	IAI **Searcher**	multirole	TV, FLIR, DL, RP	190	120	14	4,575	Il, Ska, ROC, Th
Il	Silver Arrow **Hermes 450S**	recce	Video, FLIR, PC, DL	200	200	20	7,600	Il ●▲
It	Meteor **Mirach 100**	recce, TA	LLTV, PC, FLIR, DL	840	250	1	9,000	It ▲
It	Meteor **Mirach 150**	recce	PC, IRLS, TV, Video, SAR, ESM, RP, LR/D, ELINT	700	250	1	9,000	It ▲
RF	Tupolev **Tu-243 Reys**	recce	COMINT, PC, TV, DL	940	180	0.22	5,000	n.k
RSA	Kentron **Seeker**	recce	TV, FLIR, Met, CWD	176	200	9	5,500	RSA ●, UAE, Ag
Tu	EES **Kirlangic/Dogan**	recce, TA, ELINT, RP	Various	175	150	8/12	6,100	Tu □
UAE	AES **Nibbio**	multirole	Various (RT)	250	200	10	6,000	UAE □
UK	Meggitt **Spectre**	recce, EW	TV, IRLS, FLIR	240	150	3–6	7,000	n.k.
US	AAI **Shadow 600**	multirole	FLIR, PC, MET, DL, TV, EW, CWD	190	200	12	4,575	R, Tu ▲ called *Falcon* 600
US	BAI **Exdrone/Dragon Drone**	multirole	TV, DL, ECM, RP, DL	185	362	2.5	3,000	US ▲✕■
US	S-TEC **Sentry**	recce, relay	Video (RT), FLIR, RP	170	370	6	4,875	US ▲

Country of origin	Maker Name	Role	Payloads	Maximum Speed (km/hr)	Radius of action (km)	Endurance (hrs)	Ceiling (m)	In service
Short Range (up to 100km radius of action)								
Ca	Bombardier **CL89**	recce	PC, IRLS, FP	740	60/70	n.k.	3,000	Ca, Fr, Ge, It, UK ▲
PRC	Xian **ASN104/105**	recce, EW	PC (RT), LLTV (RT), Video (RT)	205	60/100	2	3,200	PRC
Cr	RH-ALAN **BL-50/BLSB**	recce	TV	110	80	5	n.k.	Cr
Cz	VTUL **Sojka III**	recce	PC, TV, IRLS, RS	200	100	2	3,000	Cz ▲
Fr	Altec **Mart MKII**	recce, EW	FP, LLTV (RT), PC, Video, IRLS, ECM, ECCM	220	100	4	3,000	Fr ▲
Fr	SAGEM **Crecerelle**	recce, TA	Video (RT), IRLS (RT), FLIR, EW	240	90	5	3,500	Fr ▲
Fr	SAGEM **Ugglan**	recce, TA	FLIR (RT)	235	70	8	5,000	Swe ▲
Ind	ADE **Nishant**	recce, TA	PC, FLIR, IRLS, LR/D, ESM	200	n.k.	4	4,000	Ind ▲✕
Ir	Qods **Abadil**	multirole	PC	370	60	3	3,300	Ir
Il	IAI **Scout**	multirole	TV, PC, FLIR, LR/D, DL	175	100	6	4,600	Il ●▲, Sgp ●, RSA ●, CH ▲
Il	Silver-Arrow **Micro-Vee**	recce	Video, FLIR, PC, DL	200	50	5	4,500	Il
It	Meteor **Mirach 26**	recce	TV, FLIR, RP, ESM,LR/D, ELINT, COMINT	220	100	6	3,500	It ▲
Pak	AWC **AWC Mk 1**	recce, TA	TV, FLIR	175	15	2	3,000	Pak ▲
RF	Tupolev **Reys**	recce	COMINT, PC, TV, DL	875	95	0.22	3,000	RF, Ukr, Cz, Slvk, Syr
RF	Yakovlev **Pchela/Shmel**	recce	PC, IRLS, daylight TV, EW, FLIR	150	50	2	2,500	RF ▲✕■, Syr, DPRK
RSA	ATE **Vulture**	recce	TV, TI, Video, DL, FLIR	160	60	3	5,000	RSA ▲
Swe	TechMent **Midget**	recce	PC, TV, Video, IRLS, SAR	120	50	3	n.k.	Swe ▲
CH	Oerlikon **Ranger**	recce	TV, FLIR, LR/D, RP	220	100	6	4,500	CH ▲, SF
UK	Marconi **Phoenix**	recce	IRLS, DL	155	70	4.5	2,500	UK ▲
US	AeroVironment **Pointer**	recce (hand-launched)	TV, DL, LLTV, FLIR, CWD	80	5	1	300	US ▲■○
US	Bosch **SASS-LITE**	recce (airship)	MTI radar, PC, FLIR	72	100	12–24	6,000	US ▲
US	Lear Astronautics **SkyEye**	multirole	COMINT, ELINT, FLIR, IRLS, Video, TV	200	90	12	4,900	Et ▲, Mor ●, Th ●

Notes Modern UAV technology centres on systems rather than platforms, and the following table outlines the roles of those UAV systems that are currently in production and in operational use. The table excludes UAVs designed for purely commercial or academic use, or simply as decoys or aerial targets, and those still under development. **Payloads** Shows the type of sensors carried, but does not indicate that all the sensors can be carried at the same time.

Symbols □ Armed Forces ● Air Force ▲ Army ■ Marines ✕ Navy ○ National Guard

Abbreviations ADP Air-delivered payloads **ARA** Anti-radar attack **COMINT** Communications intelligence **CWD** Chemical weapon detection **DL** Data link **ECM** Electronic counter-measures **ECCM** Electronic counter-counter-measures **ELINT** Electronic intelligence **EO** Electro-optical **ESM** Electronic support measures **EW** Electronic warfare **FLIR** Forward-looking infra-red **FP** Flare pack **IRLS** Infra-red line scan **LLTV(RT)** Low-light television (real-time transmission) **LR/D** Laser-range finder/designator **LRT** Long-range tank (additional fuel tank) **met** Meteorological data gathering **MTI** Moving target indicator **NBD** Nuclear, biological, chemical detection **PC** Photographic cameras **RP** Relay platform (the ability of one UAV to control a second UAV) **RS** Radiation sensor **RT** Real-time **SAR** Synthetic aperture radar **SIGINT** Signal intelligence **TA** Target acquisition **TI** Thermal imaging **TV** Television

Definitions Anti-personnel mine means a mine designed to be exploded by the presence or proximity of, or contact with a person that will incapacitate, injure or kill. Typically they contain less than 1kg of explosive, however some fragmentation mines may contain considerably more. **Anti-tank mine** means a munition designed to be exploded by the presence or proximity of, or contact with a tank or other armoured vehicle, typically exerting at least 100kg operating pressure, which will disable or destroy the vehicle. This table describes only those anti-tank and anti-personnel mines that the IISS, drawing on open sources, believe to have been produced by and laid in the countries listed. Mines produced but not yet used/deployed are not included. The entries are listed by country of origin (far left column). Licensed producers are not shown. The index of country abbreviations is on pp 319–20.

Producer	Type	Diameter/ Height (mm)	Fuse	Explosive Type• Weight (kg)	Metallic Content[1]	Operating Pressure[2] (kg)	Remarks•Laid in
Anti-tank							
Arg	**FMK-3**	250/90	**FMK-1 AP**	TNT/RDX/AI•6.1	Min	150–250	•Falkland Is
Be	**PRB M3**[3]	230/130	**M30**	TNT/RDX/AI•6	Min	250	•Ang, Cha, Er, Eth, Irq, Rwa, SR, Z
C	Designation n.k.	338/117	**MUV/RO-1**	TNT•8	Det	50–100	•Ang, Nic
PRC	**Type 72**	270/100	**Type 72**	TNT/RDX•5.4	Min	300–800	•Ang, Kwt, RL, SR
Cz/Slvk	**PT Mi-Ba-II**	395/135	**RO-7-II**	TNT•6	Min	200–450	•Ang, Er, Eth, Nba, SR
Cz/Slvk	**PT Mi-Ba-III**	330/101	**RO-2/RO-7-I**	TNT•7.2	Min	200–450	•Ang, Kwt, Moz, Nba
Cz/Slvk	**PT Mi-D**	320/140	**RO-1**	TNT•6.2	Det	150–450	•Ang
Cz/Slvk	**PT Mi-K**	300/102	**RO-3/RO-5/RO-9**	TNT•5	Det	330	•Cam, Nba, Nic
Et	Designation n.k.	375/115	Similar to **P-62**	TNT•13	Det	150–300	•Ang
Et	**M/71**	315/100	Similar to **MV-5**	TNT•6.25	Det	150–300	Copy of RF TM-46•Et,SR
Ge	**Riegel Mine 43**[3]	800/120	**ZZ42**	TNT•4	Det	180–360	WW II mine•Et, LAR
Ge	**Tellermine 35**[3]	318/76	**T.Mi.Z.35** or **42**	TNT•5.5	Det	90–180	WW II mine•Et, LAR
Ge	**Tellermine 42**[3]	324/102	**T.Mi.Z.42** or **43**	TNT•5.5	Det	100–180	WW II mine•Et, LAR
Ge	**Tellermine 43**[3]	318/102	**T.Mi.Z42** or **43**	TNT•5.5	Det	100–180	WW II mine•Et, LAR
Hu	**UKA-63**	298/120	**EBG-68**	TNT•6	Det	n.k.	•Nba
Il	**No 6**	305/110	**No 61** or **No 62A**	TNT•6	Det	260	Copy of RF TM-46•Falkland Is
It	**B-2**[3]	1,067/127	**Integral**	TNT•3.64	Det	140	WW II mine•Et
It	**SB-81**[3]	230/90	Similar to **VS-1.6**	TNT/RDX/HMX•2.2	Min	150–310	•Falkland Is
It	**SH-55**	280/122	**VS-N**	Composition B•5.5	Min	180–220	•Afg
It	**TC/3.6**	270/145	**Integral**	Composition B•3.6	Min	180–310	•Afg
It	**TC/6**	270/185	**Integral**	Composition B•6	Min	180–310	•Afg, Cha

Producer	Type	Diameter/ Height (mm)	Fuse	Explosive Type• Weight (kg)	Metallic Content[1]	Operating Pressure[2] (kg)	Remarks•Laid in
It	**V-3**[3]	1,143/64	**Integral**	TNT/PETN•7.7	Det	Variable from 10	WWII mine•Et
It	**VS-1.6**	222/92	Similar to **SB-81**	Composition B•1.85	Min	180–220	•Irq, Kwt, RL
It	**VS-2.2**	230/115	**VS-N**	Composition B•2.2	Min	180–220	•Irq, Kwt
Pak	**P2 Mk2 (AT)**	270/130	**P2 Mk2 (AP)** or **P4 Mk1**	TNT•5	Min	180–300	•Afg, Er, Eth, SR
R	**MAT-76**	320/135	**P-62**	TNT•9.5	Min	200	•Ang, Kwt, Lb, Moz, Z
RF	**PGMDM/PTM-1S**	320/75	**MVDM/VGM-572**	Liquid•1–1.5	Det	n.k.	•Afg
RF	**TM-46/TMN-46**	305/108	**MV-5**, **MVM** or **MVSh-46**	TNT•5.7	Det	120–400	•Afg, Ang, Cam, Er, Eth, Irq, Kwt, RL, Moz, Nba, Rwa, SR, Z, Zw
RF	**TM-57**	316/102	**MVZ-57** or **MVSh-57**	TNT•6.34	Det	120–400	•Afg, Ang, Cam, Er, Eth, Irq, DPRK, Kwt, RL, Moz, Nba, Nic, Rwa, SR, Vn, Z, Zw
RF	**TM-62B**	315/67	**MVP-62**, **MVCh-62**, **MVZ-62**, **MVN-62/-72**, **VM-62Z**	TNT•7.5	Min	120–750	•Afg, Ang
RF	**TM-62M**	320/128	**MVZ-62**, **MVCh-62**, **MVN-62/-72**, **VM-62Z**, **MWP-62**	TNT•7.5	Det	150–550	•Afg, Ang, Cam, Er, Eth, Irq, DPRK, Kwt, Moz, Nba, Nic, Rwa, SR, Vn, Z, Zw
RF	**TM-72**	250/80	**MVN-72**, **TM-62**	TNT/RDX•2.5	Det	n.a.	•Afg
RF	**TMD-44**	320/160	**MV-5**	TNT•5–7	Det	200–500	•Afg, Ang, C, Moz, Nba, Rwa, Z
RF	**TMD-B**	320/160	**MV-5**	TNT•5–7	Det	200–500	•Afg, Ang, C, Moz, Nba, Rwa, Z
RF	**TMK-2**	307/1,130	**MVK-2**	TG-50 or TNT•6–6.5	Det	8–12	Tilt-rod•Ang, Eth, Moz, Nba
RSA	**No 8**	259/175	**Integral**	RDX/TNT•7	Min	150–220	•Ang, Nba, Z, Zw
Sp	**C-3-A/C-3-B**	290/60	**Integral**	TNT/RDX/aluminium•5	Min	275	•Ang, Falkland Is
UK	**Barmine**	1,200/82	**L89A1** or **L90A1**	RDX/TNT•8.1	Det	140	•Kwt
UK	**Mk 5**[3]	203/127	**Mk 3**	TNT•3.7	Det	160–200	WWII mine•Ang, Et, HKJ, LAR, Moz, Zw
UK	**Mk 7**	325/130	**No 5**	TNT•8.89	Det	150–275	•Afg, Ang, Et, Er, Eth, LAR, Nba, SR, Z, Zw
US	**BLU-91/B**	146/127	**Electronic influence**	RDX•0.584	Det	n.a.	Dispensed from cluster bomb unit•Kwt
US	**M15**	333/150	**M603**	Composition B•10.3	Det	160–340	•Ang, Cam, Cy, Er, Eth, Rwa, SR
US	**M19**	332/94	**M606**	Composition B•9.53	Min	160–230	•Ang, Cha, Ir, Irq, ROK, RL, Z

Producer	Type	Diameter/ Height (mm)	Fuse	Explosive Type• Weight (kg)	Metallic Content[1]	Operating Pressure[2] (kg)	Remarks•Laid in
US	**M1A1**	203/75	**Not identified**	TNT•2.75	Det	120–250	•Falkland Is
US	**M6A2**	333/83	**M601/M603**	TNT•4.45	Det	160–340	•Ang, Rwa
US	**M7A2**	64/178	**M601/M603**	Tetryl•1.62	Det	60–110	•Ang, Cam, Er, Eth, RL, SR, Z
FRY	**TM-500**	70/108	**UANU-1**	TNT•0.5	Min	10	Demolition Block•FRY
FRY	**TMA-1A**	315/100	**UANU-1**	TNT•5.4	Min	100	•FRY
FRY	**TMA-2**	260/200	**UANU-1**	TNT•6.5	Min	100	•Ang, Nba, FRY
FRY	**TMA-3**	265/110	**UTMA-3**	TNT•6.5	Min	180	•Ang, Nba, FRY
FRY	**TMA-4**	284/110	**UTMA-4**	TNT•5.5	Min	100–200	•Ang, RL, Nba,
FRY	**TMA-5**	300/110	**UANU-1**	TNT•5.5	Min	100–300	•Afg, Ang, Cha, RL, Nba, FRY
FRY	**TMD-1/-2**	320/140	**UANU-1**	TNT•5.5	Min	200	•FRY
FRY	**TMM-1**	300/90	**UTMM-1**	TNT•5.6	Det	130–420	•FRY
FRY	**TMRP-6**	290/132	**Integral**	TNT•5.1	Det	150–360	•FRY

Anti-Personnel

Producer	Type	Diameter/ Height (mm)	Fuse	Explosive Type• Weight (kg)	Metallic Content[1]	Operating Pressure[2] (kg)	Remarks•Laid in
Arg	**FMK-1**	82/42	**Integral**	TNT/RDX•152	Min	50	•Falkland Is
Be	**NR 409/PRB M409**[3]	82/28	**Integral**	TNT•80	Min	8–30	•Ang, Cha, Kwt, Irq, Nba, Rwa, SR, Z
Be	**NR-413**[3]	46/114	**NR-410**	Composition B•100	Det	2–5	•Rwa
Be	**PRB M35**[3]	65/60	**M5**	TNT•100	Min	5–14	•Ang, Er, Eth, SR
Be	**PRB M966**[3]	/244	**M605**	TNT•154	Det	4.5–9	•Moz, Nba
Bu	**PSM-1**	75/110	**RO-8, EVU** or **MVN-2M**	Hexogen•170	Det	n.k.	•Cam
C	Designation n.k.	145/50	**RO-1 type**	TNT•200	Det	1–10	•Ang
PRC	**Type 58**	60/130	**Type 58**	TNT•75	Det	1–3	POMZ-2 copy•Cam
PRC	**Type 58**	112/56	**Integral**	TNT•240	Det	5–10	PMN copy•Kwt
PRC	**Type 59**	60/107	**Type 59**	TNT•75	Det	1–3	POMZ-2M copy•Cam
PRC	**Type 66**	216/83	**Electrical** or **MUV**	Plastic•680	Det	n.a.	M18A1 *Claymore* copy•Ang, Cam
PRC	**Type 69**	61/168	**Type 69**	TNT•105	Det	7–20	•Afg, Cam, Er, Eth
PRC	**Type 72**	78/38	**Integral**	TNT•51	Min	5–10	•Ang, Cam, Irq, Kwt, Moz, SR

Producer	Type	Diameter/ Height (mm)	Fuse	Explosive Type• Weight (kg)	Metallic Content[1]	Operating Pressure[2] (kg)	Remarks•Laid in
PRC	**Type 72B**	78/38	**Integral**	TNT•51	Det	2.5	Anti-handling variant of Type 72•Cam
Cz/Slvk	**PP Mi-Ba**	102/52	**RO-7-II**	TNT•152	Min	25	•Nba, RSA
Cz/Slvk	**PP Mi-D**	135/55	**RO-1/MUV**	TNT•200	Det	1–10	•Ang, Nba
Cz/Slvk	**PP Mi-Sr**	102/152	**RO-1/RO-8/ MUV/P1/ P2**	TNT•360	Det	3–8	•Afg, Ang, Cam, CR, Et, Er, Eth, Hr, Moz, Nba, Nic, SR, Z
Et	***Claymore***	220/120	**Electrical** or **MUV**	Plastic C-4•700	Det	n.a.	•Ang
Fr	**MI AP DV 59**[3]	62/55	**AL PR ID 59**	TNT•70	None (not Det)	5	•Ang, RL, Moz
Ge	**DM-11**[3]	82/33.5	**DM-3**	TNT•122	Min	5–10	Copy of LI-11•Ang, Er, Eth, SR, Z
Ge	**DM-31**[3]	102/136	**DM-56** or **DM-65**	TNT•540	Det	8–10	•Ang
Ge	**PPM-2**[3]	134/60	**Integral**	TNT•110	Det	13	Built in former DDR•Ang, Cam, Cha, Eth, RL, Moz, Nba, Nic, SR
Hu	***Gyata*-64**	106/61	**Integral**	TNT•300	Det	5	•Ang, RL
Il	**No 4**	135/50	Similar to **MUV**	TNT•188	Det	n.k.	•Falkland Is, Il, RL
It	**AUPS**	102/36	**Integral**	Composition B•115	Min	10–20	•Moz
It	**P-40**[3]	90/200	**Not identified**	TNT•480	Det	2–10	•Kwt
It	**SB-33**[3]	85/30	**Integral**	RDX/HMX•35	Min	8	•Afg, Falkland Is
It	**TS-50**	90/45	**Integral**	T4•50	Min	12.5	•Kwt, RL, Rwa
It	**V-69**	130/205	**Integral**	Composition B•420	Det	10	•Ang, Irq, Kwt, Moz
It	**VS-50**	90/45	**Integral**	RDX•43	Min	10	•Ang, Irq, Kwt, Rwa, Zw
It	**VS-MK2**	90/32	**Integral**	RDX/wax•33	Min	12	•Ang
DPRK	**APP M-57**	205/44	Similar to **UPMAH-1**	TNT•250	Min	3	•Ang
Pak	**P4 Mk1 (AP)**	70/38	**Integral**	Tetryl•30	Min	10	•Afg, Er, Eth, SR
Por	**M/966-B**[3]	90/50	**RO-1/RO-8** or **MUV**	TNT•300	Det	1–10	•Moz, Nba, Nic, Z
R	**MAI-75**	95/61	**Integral**	TNT•120	Min	5–25	•Ang
RF	**MON-50**	226/156	**Electrical** or **MUV**	PVV-5A•700	Det	n.a.	•Afg, Ang, Cam, Er, Eth, Moz, SR, Nic, Z
RF	**MON-90**	345/202	**Electrical** or **MUV**	PVV-5A•6,200	Det	n.a.	•Afg
RF	**MON-100**	236/83	**Electrical** or **MUV**	TNT•2,000	Det	n.a.	Directional fragmentation•Afg, Ang, Cam, Er, Eth, Moz, Vn, Z
RF	**MON-200**	434/130	**Electrical** or **MUV**	TNT•12,000	Det	n.a.	Directional fragmentation •Afg, Ang, Er, Eth
RF	**OZM-3**	75/120	**Electrical** or **MUV**	TNT•75	Det	n.a.	•Afg, Ang, Cam, Er, Eth, Moz, Z

Producer	Type	Diameter/ Height (mm)	Fuse	Explosive Type• Weight (kg)	Metallic Content[1]	Operating Pressure[2] (kg)	Remarks•Laid in
RF	**OZM-4**	91/140	**Electrical** or **MUV**	TNT•170	Det	n.a.	•Afg, Ang, Cam, C, Er, Eth, Moz, Nba, Nic, Vn, Z
RF	**OZM-72**	106/172	**Electrical** or **MUV**	TNT•500	Det	n.a.	•Afg, Ang, Cam, Er, Eth, Z
RF	**OZM-160**	245/1,030	**Electrical**	TNT•4,500	Det	n.a.	Bounding fragmentation•Ang
RF	**PFM-1/PFM-1S**	120/61	**MVDM/VVGM-572**	VV VS-6D•37	Det	5	Liquid explosive•Afg
RF	**PMD-6**	190/65	**MUV**	TNT•200	Det	1–10	•Afg, Ang, Cam, Er, Eth, Hr, Irq, Moz, Nba, Nic, Rwa, SR
RF	**PMN**	112/56	**Integral**	TNT•240	Det	8–25	•Afg, Ang, Cam, Et, Er, Eth, Hr, Irq, Lao, RL, Moz, Nba, Nic, Rwa, SR, Ve
RF	**PMN-2**	120/53	**Integral**	TNT/RDX•100	Det	15	•Afg, Cam, Er, Hr, RL, Moz, Nic
RF	**POM-2S**	63/180	**Integral**	TNT•140	Det	n.a.	•Afg
RF	**POMZ-2**	60/130	**MUV**	TNT•75	Det	1–3	•Afg, Ang, Cam, C, Er, Eth, Hr, Irq, LAR, Moz, Nba, Nic, Rwa, SR, Vn, Zw
RF	**POMZ-2M**	60/107	**MUV**	TNT•75	Det	1–3	•Afg, Ang, Cam, C, Er, Eth, Hr, Irq, LAR, Moz, Nba, Nic, Rwa, SR, Vn, Zw
RSA	**Mini MS-803**[3]	220/70	**Electrical**	PE9•460	Det	n.a.	•Ang
RSA	**R2M1/R2M2**[3]	69/57	**Integral**	RDX/wax•58	Min	3–7	•Ang, Moz, Nba, Z, Zw
RSA	**Shrapnel mine No 2**[3]	220/140	**Electrical**	PE9•680	Det	n.a.	M18A1 *Claymore* copy•Ang, Z, Zw
Sp	**P-4-B**	72/43	**Integral**	TNT/PETN/wax•100	Min	10	•Falkland Is
Sp	**P-S-1**	98/189	Similar to **MUV**	TNT•450	Det	n.a.	•Ang
Swe	**FFV 013**	250/420	**Non-electric command**	n.k.•10,000	Det	n.a.	•Ang
Swe	**LI-12**	100/170	**Electrical**	Hexotol•3,000	Det	n.a.	
Th	**Model 123**	115/90	**Electrical**	RDX•250	Det	n.a.	•Cam
UK	**Mk 2**	89/200	**Integral**	Amatol•500	Det	2–3	•Ang, Et, LAR
US	**BLU-92/B**	146/127	**Electronic**	Composition B4•421	Det	n.a.	Dispensed from cluster bomb unit•Kwt
US	**M14**	56/40	**Integral**	Tetryl•29	Min	9–16	•Ang, Cha, EIS, Er, Eth, Ir, Irq, HKJ, ROK, RL, Mlw, SR, Z
US	**M16 & M16A1**	103/203	**M605**	TNT•575	Det	3.6–20	•Ang, Cam, C, Cy, Er, Eth, ROK, Mlw, SR, Z
US	**M16A2**	103/203	**M605**	TNT•601	Det	3.6–20	•C, Cy, ROK, SR
US	**M18A1**	216/82.5	**Electrical**	C-4•682	Det	n.a.	*Claymore*•Ang, Cam, Cha, CR, EIS, Gua, Irq, Mlw, Rwa

Producer	Type	Diameter/Height (mm)	Fuse	Explosive Type•Weight (kg)	Metallic Content[1]	Operating Pressure[2] (kg)	Remarks•Laid in
Vn	**MBV-78A1**	64/130	**MUV type**	TNT•75	Det	n.a.	POMZ-2 copy•Cam, Vn
Vn	**MBV-78A2**	53/80	**MUV type**	TNT•65	Det	2–5	•Cam, Vn
Vn	**MD-82B**	57/53	**Integral**	TNT•28	Det	4–5	•Cam, Vn
Vn	**MDH-10**	220/80	**MUV type** or **electrical**	TNT•2,000	Det	2–5	Directional fragmentation•Cam, Vn
Vn	**MN-79**	56/40	**Integral**	TNT•29	Min	9–16	M14 copy•Cam, Vn
Vn	**NO-MZ 2B**	57/80	**MUV type**	TNT•65	Det	2–5	•Cam, Vn
Vn	**P-40**	69/104	**MUV type**	TNT•120	Det	3–5	•Cam, Vn
FRY	***Gorazde***	58/115	**Integral**	n.k.•5	Det	10–15	•FRY
FRY	**MRUD**	231/89	**Electrical**	Plastic•900	Det	n.a.	•FRY
FRY	**PMA-1A**	140/30	**UPMAH-1**	TNT•200	Min	n.a.	•Ang, Nba, FRY
FRY	**PMA-2**	68/61	**UPMAH-2**	TNT•100	Min	7–15	•Ang, Nba, FRY
FRY	**PMA-3**	111/40	**UPMAH-3**	Tetryl•35	Min	8–20	•Nba, FRY
FRY	**PMR-1**	80/120	**UPM-1**	TNT•75	Det	3	•Nba, FRY
FRY	**PMR-2**	80/120	**UPM-1**	TNT•75	Det	3	•FRY
FRY	**PMR-2A**	66/140	**UPM-2A** or **UPM-2AS**	TNT•100	Det	3	•Nba, FRY
FRY	**PMR-3** (Old)	78/134	**UPMR-3** or **UPROM-1**	TNT•410	Det	9	•FRY
FRY	**PMR-3** (New)	77/128	**UPMR-3**	Plastic•410	Det	2–7	•FRY
FRY	**PMR-4**	130/80	**UPM-1**	TNT•200	Det	2–4	•FRY
FRY	**PMR-U**	120/75	**UPM-1**	Commercial•100	Det	2–5	•FRY
FRY	**PPMP-2**	60/140	**MUV**	Commercial•150	Det	2–4	•FRY
FRY	**PROM-1**	75/260	**UPMR-3** or **UPROM-1**	TNT•425	Det	9	•Ang, Irq, Nba, FRY
Zw	**RAP No 1** or **ZAP No 1**	36/195	**R2M1**	Pentolite•140	Min	1-5	•Nba, Zw
Zw	**RAP No 2** or **ZAP No 2**	36/195	**Chemical**	Pentolite•140	Min	1–5	•Nba, Z, Zw
Zw	**ZAPS**	200/75	**Mk-1/-2** or **electrical**	Pentolite•500	Det	2–5	•Mlw, Moz, Nba, Z, Zw

Notes

1 **Metallic Content Min (Minimum)** means that the metallic content is so low that the mine is difficult to detect using a modern, high quality metal detector. **Det (Detectable)** means that the metallic content is sufficient for the mine to be readily detected using a modern, high quality metal detector.

2 **Operating Pressure** is the amount of force in kilograms, applied to the fuse mechanism, causing immediate detonation of the main explosive charge.

3 No longer produced.

Table 42 **Designations of aircraft**

Notes

1 [Square brackets] indicate the type from which a variant was derived: 'Q-5 … [MiG-19]' indicates that the design of the Q-5 was based on that of the MiG-19.
2 (Parentheses) indicate an alternative name by which an aircraft is known, sometimes in another version: 'L-188 … *Electra* (P-3 *Orion*)' shows that in another version the Lockheed Type 188 *Electra* is known as the P-3 *Orion*.
3 Names given in 'quotation marks' are NATO reporting names, e.g., 'Su-27… *"Flanker"*'.
4 When no information is listed under 'Country of origin' or 'Maker', the primary reference given under 'Name/designation' should be looked up under 'Type'.
5 For country abbreviations, see 'Index of Countries and Territories' (pp. 319–20).

Fixed-wing

Type	Name/designation	Country of origin	Maker
A-1	AMX	**Br/It**	AMX
A-1	*Ching-Kuo*	**ROC**	AIDC
A-3	*Skywarrior*	**US**	Douglas
A-4	*Skyhawk*	**US**	MD
A-5	(Q-5)		
A-7	*Corsair* II	**US**	LTV
A-10	*Thunderbolt*	**US**	Fairchild
A-36	*Halcón* (C-101)		
A-37	*Dragonfly*	**US**	Cessna
A-50	*'Mainstay'* (Il-76)	**RF**	Beriev
A300		**UK/Fr/Ge/Sp**	Airbus Int
A310		**UK/Fr/Ge/Sp**	Airbus Int
A340		**UK/Fr/Ge/Sp**	Airbus Int
AC-47	(C-47)		
AC-130	(C-130)		
Air Beetle		**Nga**	AIEP
Airtourer		**NZ**	Victa
AJ-37	(J-37)		
Alizé	(Br 1050)	**Fr**	Breguet
Alpha Jet		**Fr/Ge**	Dassault–Breguet/Dornier
An-2	*'Colt'*	**Ukr**	Antonov
An-12	*'Cub'*	**Ukr**	Antonov
An-14	*'Clod'* (*Pchyelka*)	**Ukr**	Antonov
An-22	*'Cock'* (*Antei*)	**Ukr**	Antonov
An-24	*'Coke'*	**Ukr**	Antonov
An-26	*'Curl'*	**Ukr**	Antonov
An-28	*'Cash'*	**Ukr**	Antonov
An-30	*'Clank'*	**Ukr**	Antonov
An-32	*'Cline'*	**Ukr**	Antonov
An-72	*'Coaler-C'*	**Ukr**	Antonov
An-74	*'Coaler-B'*	**Ukr**	Antonov
An-124	*'Condor'* (*Ruslan*)	**Ukr**	Antonov
Andover	[HS-748]		
Arava		**Il**	IAI
AS-202	*Bravo*	**CH**	FFA
AT-3	*Tsu Chiang*	**ROC**	AIDC
AT-6	(T-6)		
AT-11		**US**	Beech
AT-26	EMB-326		
AT-33	(T-33)		
Atlantic	(*Atlantique*)	**Fr**	Dassault–Breguet
AU-23	*Peacemaker* [PC-6B]	**US**	Fairchild
AV-8	*Harrier* II	**US/UK**	MD/BAe
Aztec	PA-23	**US**	Piper
B-1	*Lancer*	**US**	Rockwell
B-2	*Spirit*	**US**	Northrop Grumman
B-5	H-5		
B-6	H-6		
B-52	*Stratofortress*	**US**	Boeing
B-65	*Queen Air*	**US**	Beech
BAC-167	*Strikemaster*	**UK**	BAe
BAe-125		**UK**	BAe
BAe-146		**UK**	BAe
BAe-748	(HS-748)	**UK**	BAe
Baron	(T-42)		
Basler T-67	(C-47)	**US**	Basler
Be-6	*'Madge'*	**RF**	Beriev
Be-12	*'Mail'* (*Tchaika*)	**RF**	Beriev
Beech 50	*Twin Bonanza*	**US**	Beech
Beech 95	*Travel Air*	**US**	Beech
BN-2	*Islander, Defender, Trislander*	**UK**	Britten-Norman
Boeing 707		**US**	Boeing
Boeing 727		**US**	Boeing
Boeing 737		**US**	Boeing
Boeing 747		**US**	Boeing
Boeing 757		**US**	Boeing
Boeing 767		**US**	Boeing
Bonanza		**US**	Beech
Bronco	(OV-10)		
BT-5	HJ-5		
Bulldog		**UK**	BAe
C-1		**J**	Kawasaki
C-2	*Greyhound*	**US**	Grumman
C-5	*Galaxy*	**US**	Lockheed
C-7	DHC-7		
C-9	*Nightingale* (DC-9)		
C-12	*Super King Air* (*Huron*)	**US**	Beech
C-17	*Globemaster* III	**US**	McDonnell Douglas
C-18	[Boeing 707]		
C-20	(*Gulfstream* III)		
C-21	(*Learjet*)		
C-22	(Boeing 727)		
C-23	(*Sherpa*)	**UK**	Shorts
C-26	*Expediter/Merlin*	**US**	Fairchild
C-32	[Boeing 757]	**US**	Boeing
C-37A	[Gulfstream V]	**US**	Gulfstream
C-38A	(*Astra*)	**Il**	IAI
C-42	(Neiva *Regente*)	**Br**	Embraer
C-46	*Commando*	**US**	Curtis
C-47	DC-3 (*Dakota*) (C-117 *Skytrain*)	**US**	Douglas
C-54	*Skymaster* (DC-4)	**US**	Douglas
C-91	HS-748		
C-93	HS-125		
C-95	EMB-110		
C-97	EMB-121		
C-101	*Aviojet*	**Sp**	CASA
C-115	DHC-5	**Ca**	De Havilland
C-117	(C-47)		
C-118	*Liftmaster* (DC-6)		
C-123	*Provider*	**US**	Fairchild
C-127	(Do-27)	**Sp**	CASA
C-130	*Hercules* (L-100)	**US**	Lockheed
C-131	Convair 440	**US**	Convair
C-135	[Boeing 707]		
C-137	[Boeing 707]		
C-140	(*Jetstar*)	**US**	Lockheed
C-141	*Starlifter*	**US**	Lockheed
C-160	*Transall*	**Fr/Ge**	Transall
C-212	*Aviocar*	**Sp**	CASA

Type	Name/ designation	Country of origin Maker
C-235		**Sp** CASA
Canberra		**UK** BAe
CAP-10		**Fr** Mudry
CAP-20		**Fr** Mudry
CAP-230		**Fr** Mudry
Caravelle	SE-210	**Fr** Aérospatiale
CC-115	DHC-5	
CC-117	(*Falcon 20*)	
CC-132	(DHC-7)	
CC-137	(Boeing 707)	
CC-138	(DHC-6)	
CC-144	CL-600/-601	**Ca** Canadair
CF-5a		**Ca** Canadair
CF-18	F/A-18	
Cheetah	[*Mirage* III]	**RSA** Atlas
Cherokee	PA-28	**US** Piper
Cheyenne	PA-31T [*Navajo*]	**US** Piper
Chieftain	PA-31-350 [*Navajo*]	**US** Piper
Ching-Kuo	A-1	**ROC** AIDC
Citabria		**US** Champion
Citation	(T-47)	**US** Cessna
CJ-5	[Yak-18]	**PRC** NAMC (Hongdu)
CJ-6	[Yak-18]	**PRC** NAMC (Hongdu)
CL-215		**Ca** Canadair
CL-415		**Ca** Canadair
CL-600	*Challenger*	**Ca** Canadair
CM-170	*Magister* [*Tzukit*]	**Fr** Aérospatiale
CM-175	*Zéphyr*	**Fr** Aérospatiale
CN-212		**Sp/Indo** CASA/IPTN
CN-235		**Sp/Indo** CASA/IPTN
Cochise	T-42	
Comanche	PA-24	**US** Piper
Commander	*Aero-/TurboCommander*	**US** Rockwell
Commodore	MS-893	**Fr** Aérospatiale
CP-3	P-3 *Orion*	
CP-140	*Aurora* (P-3 *Orion*)	**US** Lockheed
	Acturas	
CT-4	*Airtrainer*	**NZ** Victa
CT-114	CL-41 *Tutor*	**Ca** Canadair
CT-133	*Silver Star* [T-33]	**Ca** Canadair
CT-134	*Musketeer*	
CT-156	*Harvard* II	**US** Beech
Dagger	(*Nesher*)	
Dakota		**US** Piper
Dakota	(C-47)	
DC-3	(C-47)	**US** Douglas
DC-4	(C-54)	**US** Douglas
DC-6	(C-118)	**US** Douglas
DC-7		**US** Douglas
DC-8		**US** Douglas
DC-9		**US** MD
Deepak	(HPT-32)	
Defender	BN-2	
DHC-3	*Otter*	**Ca** DHC
DHC-4	*Caribou*	**Ca** DHC
DHC-5	*Buffalo*	**Ca** DHC
DHC-6	*Twin Otter*, CC-138	**Ca** DHC
DHC-7	*Dash-7* (*Ranger*, CC-132)	**Ca** DHC
DHC-8		**Ca** DHC
Dimona	H-36	**Ge** Hoffman
Do-27	(C-127)	**Ge** Dornier
Do-28	*Skyservant*	**Ge** Dornier
Do-128		**Ge** Dornier
Do-228		**Ge** Dornier
E-2	*Hawkeye*	**US** Grumman
E-3	*Sentry*	**US** Boeing
E-4	[Boeing 747]	**US** Boeing
E-6	*Mercury* [Boeing 707]	**US** Boeing
E-26	T-35A (*Tamiz*)	**Chl** Enear
EA-3	[A-3]	
EA-6	*Prowler* [A-6]	
EC-130	[C-130]	
EC-135	[Boeing 707]	
EF-111	*Raven* (F-111)	**US** General Dynamic
Electra	(L-188)	
EMB-110	*Bandeirante*	
EMB-111	*Maritime Bandeirante*	**Br** Embraer
EMB-120	*Brasilia*	**Br** Embraer
EMB-121	*Xingu*	**Br** Embraer
EMB-201	*Ipanema*	**Br** Embraer
EMB-312	*Tucano*	**Br** Embraer
EMB-326	*Xavante* (MB-326)	**Br** Embraer
EMB-810	[*Seneca*]	**Br** Embraer
EP-3	(P-3 *Orion*)	
Etendard/Super Etendard		**Fr** Dassault
EV-1	(OV-1)	
F-1	[T-2]	**J** Mitsubishi
F-4	*Phantom*	**US** MD
F-5	-A/-B *Freedom Fighter*	
	-E/-F *Tiger* II	**US** Northrop
F-6	J-6	
F-7	J-7	
F-8	J-8	
F-10	J-10	
F-11	J-11	
F-14	*Tomcat*	**US** Grumman
F-15	*Eagle*	**US** MD
F-16	*Fighting Falcon*	**US** GD
F-18	[F/A-18], *Hornet*	
F-21	*Kfir*	**Il** IAI
F-22	*Raptor*	**US** Lockheed
F-27	*Friendship*	**Nl** Fokker
F-28	*Fellowship*	**Nl** Fokker
F-35	*Draken*	**Swe** SAAB
F-104	*Starfighter*	**US** Lockheed
F-111	EF-111	**US** GD
F-117	*Nighthawk*	**US** Lockheed
F-172	(Cessna 172)	**Fr/US** Reims-Cessna
F-406	*Caravan*	**Fr** Reims
F/A-18	*Hornet*	**US** MD
Falcon	*Mystère-Falcon*	
FB-111	(F-111)	
FBC-1	*Feibao* [JH-7]	
FC-1	(*Sabre 2*, *Super-7*)	**PRC/RF/Pak** CAC/MAPO/Pak
FH-227	(F-27)	**US** Fairchild-Hiller
Firefly	(T-67M)	**UK** Slingsby
Flamingo	MBB-233	
FT-5	JJ-5	
FT-6	JJ-6	
FT-7	JJ-7	
FTB-337	[Cessna 337]	
G-91		**It** Aeritalia
G-115E	*Tutor*	**Ge** Grob
G-222		**It** Aeritalia
Galaxy	C-5	
Galeb		**FRY** SOKO
Genet	SF-260W	
GU-25	(*Falcon 20*)	
Guerrier	R-235	
Gulfstream		**US** Gulfstream Aviation
Gumhuria	(*Bücker* 181)	**Et** Heliopolis
H-5	[Il-28]	**PRC** HAF
H-6	[Tu-16]	**PRC** XAC
H-36	*Dimona*	

Type	Name/ designation	Country of origin Maker
Halcón	[C-101]	
Harrier	(AV-8)	**UK** BAe
Hawk		**UK** BAe
Hawker 800XP	(BAe-125)	**US** Raytheon
HC-130	(C-130)	
HF-24	*Marut*	**Ind** HAL
HFB-320	*Hansajet*	**Ge** Hamburger FB
HJ-5	(H-5)	
HJT-16	*Kiran*	**Ind** HAL
HPT-32	*Deepak*	**Ind** HAL
HS-125	(*Dominie*)	**UK** BAe
HS-748	[*Andover*]	**UK** BAe
HT-2		**Ind** HAL
HU-16	*Albatross*	**US** Grumman
HU-25	(*Falcon* 20)	
Hunter		**UK** BAe
HZ-5	(H-5)	
IA-50	*Guaraní*	**Arg** FMA
IA-58	*Pucará*	**Arg** FMA
IA-63	*Pampa*	**Arg** FMA
IAI-201/-202	*Arava*	**Il** IAI
IAI-1124	*Westwind, Seascan*	**Il** IAI
IAI-1125	*Astra*	**Il** IAI
Iak-52	(Yak-52)	**R** Aerostar
IAR-28		**R** IAR
IAR-93	*Orao*	**FRY/R** SOKO/IAR
IAR-99	*Soim*	**R** IAR
Il-14	*'Crate'*	**RF** Ilyushin
Il-18	*'Coot'*	**RF** Ilyushin
Il-20	*'Coot-A'* (Il-18)	**RF** Ilyushin
Il-22	*'Coot-B'* (Il-18)	**RF** Ilyushin
Il-28	*'Beagle'*	**RF** Ilyushin
Il-38	*'May'*	**RF** Ilyushin
Il-62	*'Classic'*	**RF** Ilyushin
Il-76	*'Candid'* (tpt), *'Mainstay'* (AEW)	**RF** Ilyushin
Il-78	*'Midas'* (tkr)	**RF** Ilyushin
Il-82	*'Candid'*	**RF** Ilyushin
Il-87	*'Maxdome'*	**RF** Ilyushin
Impala	[MB-326]	**RSA** Atlas
Islander	BN-2	
J-5	[MiG-17F]	**PRC** SAF
J-6	[MiG-19]	**PRC** SAF
J-7	[MiG-21]	**PRC** CAC/GAIC
J-8	*Finback*	**PRC** SAC
J-10	[IAI *Lavi*]	**PRC** SAC
J-11	[Su-27]	**PRC** SAC
J-32	*Lansen*	**Swe** SAAB
J-35	*Draken*	**Swe** SAAB
J-37	*Viggen*	**Swe** SAAB
JA-37	(J-37)	
Jaguar		**Fr/UK** SEPECAT
JAS-39	*Gripen*	**Swe** SAAB
Jastreb		**FRY** SOKO
Jetstream		**UK** BAe
JH-7	[FBC-1]	**PRC** XAC
JJ-5	[J-5]	**PRC** CAF
JJ-6	[J-6]	**PRC** SAF
JJ-7	[J-7]	**PRC** GAIC
JZ-6	(J-6)	
K-8		**PRC/Pak/Et** Hongdu/E
KA-3	[A-3]	
KA-6	[A-6]	
KC-10	*Extender* [DC-10]	**US** MD
KC-130	[C-130]	
KC-135	[Boeing 707]	
KE-3A	[Boeing 707]	
KF-16	(F-16)	

Type	Name/ designation	Country of origin Maker
Kfir		**Il** IAI
King Air		**US** Beech
Kiran	HJT-16	
Kraguj		**FRY** SOKO
L-4	*Cub*	
L-18	*Super Cub*	**US** Piper
L-19	O-1	
L-21	*Super Cub*	**US** Piper
L-29	*Delfin*	**Cz** Aero
L-39	*Albatros*	**Cz** Aero
L-59	*Albatros*	**Cz** Aero
L-70	*Vinka*	**SF** Valmet
L-100	C-130 (civil version)	
L-188	*Electra* (P-3 *Orion*)	**US** Lockheed
L-410	*Turbolet*	**Cz** LET
L-1011	*Tristar*	**US** Lockheed
Learjet	(C-21)	**US** Gates
LR-1	(MU-2)	**J** Mitsubishi
M-28	*Skytruck*	**Pl** MIELEC
Magister	CM-170	
Marut	HF-24	
Mashshaq	MFI-17	**Pak/Swe** PAC/SAAB
Matador	(AV-8)	
Maule	M-7/MXT-7	**US** Maule
MB-326		**It** Aermacchi
MB-339	(*Veltro*)	**It** Aermacchi
MBB-233	*Flamingo*	**Ge** MBB
MC-130	(C-130)	
Mercurius	(HS-125)	
Merlin		**US** Fairchild
Mescalero	T-41	
Metro		**US** Fairchild
MFI-17	*Supporter* (T-17)	**Swe** SAAB
MiG-15	*'Midget'* trg	**RF** MiG
MiG-17	*'Fresco'*	**RF** MiG
MiG-19	*'Farmer'*	**RF** MiG
MiG-21	*'Fishbed'*	**RF** MiG
MiG-23	*'Flogger'*	**RF** MiG
MiG-25	*'Foxbat'*	**RF** MiG
MiG-27	*'Flogger D'*	**RF** MiG
MiG-29	*'Fulcrum'*	**RF** MiG
MiG-31	*'Foxhound'*	**RF** MiG
Mirage		**Fr** Dassault
Missionmaster	N-22	
Mohawk	OV-1	
MS-760	*Paris*	**Fr** Aérospatiale
MS-893	*Commodore*	
MU-2	LR-1	**J** Mitsubishi
Musketeer	Beech 24	**US** Beech
Mystère-Falcon		**Fr** Dassault
N-22	*Floatmaster, Missionmaster*	**Aus** GAF
N-24	*Searchmaster* B/L	**Aus** GAF
N-262	*Frégate*	**Fr** Aérospatiale
N-2501	*Noratlas*	**Fr** Aérospatiale
Navajo	PA-31	**US** Piper
NC-212	C-212	**Sp/Indo** CASA/Nurtanio
NC-235	C-235	**Sp/Indo** CASA/Nurtanio
Nesher	[*Mirage* III]	**Il** IAI
NF-5	(F-5)	
Nightingale	(C-9)	
Nimrod	[*Comet*]	**UK** BAe
Nomad		**Aus** GAF
O-1	*Bird Dog*	**US** Cessna
O-2	(Cessna 337 *Skymaster*)	**US** Cessna
OA-4	(A-4)	
OA-37	*Dragonfly*	
Orao	IAR-93	

Type	Name/ designation	Country of origin Maker
Ouragan		**Fr** Dassault
OV-1	*Mohawk*	**US** Rockwell
OV-10	*Bronco*	**US** Rockwell
P-3	*Orion* [L-188 *Electra*]	**US** Lockheed
P-92		**It** Teenam
P-95	EMB-110	
P-166		**It** Piaggio
P-180	*Avanti*	**It** Piaggio
PA-18	*Super Cub*	**US** Piper
PA-23	*Aztec*	**US** Piper
PA-28	*Cherokee*	**US** Piper
PA-31	*Navajo*	**US** Piper
PA-32	*Cherokee Six*	**US** Piper
PA-34	*Seneca*	**US** Piper
PA-36	*Pawnee Brave*	**US** Piper
PA-38	*Tomahawk*	**US** Piper
PA-42	*Cheyenne III*	**US** Piper
PBY-5	*Catalina*	**US** Consolidated
PC-6	*Porter*	**CH** Pilatus
PC-6A/B	*Turbo Porter*	**CH** Pilatus
PC-7	*Turbo Trainer*	**CH** Pilatus
PC-9		**CH** Pilatus
PC-12		**CH** Pilatus
PD-808		**It** Piaggio
Pillán	T-35	
PL-1	*Chien Shou*	**ROC** AIDC
PLZ M-28	[An-28]	**Pl** PZL
Porter	PC-6	
PS-5	[SH-5]	
PZL M-28	M-28 [An-28]	**Pl** PZL
PZL-104	*Wilga*	**Pl** PZL
PZL-130	*Orlik*	**Pl** PZL
Q-5	A-5 *'Fantan'* [MiG-19]	**PRC** NAMC (Hongdu)
Queen Air	(U-8)	
R-160		**Fr** Socata
R-235	*Guerrier*	**Fr** Socata
RC-21	(C-21, *Learjet*)	
RC-47	(C-47)	
RC-95	(EMB-110)	
RC-135	[Boeing 707]	
RF-4	(F-4)	
RF-5	(F-5)	
RF-35	(F-35)	
RF-104	(F-104)	
RG-8A		**US** Schweizer
RT-26	(EMB-326)	
RT-33	(T-33)	
RU-21	(*King Air*)	
RV-1	(OV-1)	
S-2	*Tracker*	**US** Grumman
S-208		It SIAI
S-211		It SIAI
SA 2-37A		**US** Schweizer
Sabreliner	(CT-39)	**US** Rockwell
Safari	MFI-15	
Safir	SAAB-91 (SK-50)	**Swe** SAAB
SC-7	*Skyvan*	**UK** Short
SE-210	*Caravelle*	
Sea Harrier	(*Harrier*)	
Seascan	IAI-1124	
Searchmaster	N-24 B/L	
Seneca	PA-34 (EMB-810)	**US** Piper
Sentry	(O-2)	**US** Summit
SF-37	(J-37)	
SF-260	(SF-260W *Warrior*)	**It** SIAI
SH-5	PS-5	**PRC** HAMC
SH-37	(J-37)	
Sherpa	Short 330, C-23	**UK** Short
Short 330	(*Sherpa*)	**UK** Short
***Sierra* 200**	(*Musketeer*)	
SK-35	(J-35)	**Swe** SAAB
SK-37	(J-37)	
SK-60	(SAAB-105)	**Swe** SAAB
SK-61	(*Bulldog*)	
Skyvan		**UK** Short
SM-90		**RF** Technoavia
SM-1019		**It** SIAI
SP-2H	*Neptune*	**US** Lockheed
SR-71	*Blackbird*	**US** Lockheed
Su-7	*'Fitter-A'*	**RF** Sukhoi
Su-15	*'Flagon'*	**RF** Sukhoi
Su-17/-20/-22	*'Fitter-B'* - *'-K'*	**RF** Sukhoi
Su-24	*'Fencer'*	**RF** Sukhoi
Su-25	*'Frogfoot'*	**RF** Sukhoi
Su-27	*'Flanker'*	**RF** Sukhoi
Su-29		**RF** Sukhoi
Su-30	*'Flanker'*	**RF** Sukhoi
Su-33	(Su-27K) *'Flanker-D'*	**RF** Sukhoi
Su-34	(Su-27IB) *'Flanker-C2'*	**RF** Sukhoi
Su-35	(Su-27) *'Flanker'*	**RF** Sukhoi
Su-39	(Su-25T) *'Frogfoot'*	**RF** Sukhoi
Super		**Fr** Dassault
Shrike Aerocommander		**US** Rockwell
Super Galeb		**FRY** SOKO
T-1		**J** Fuji
T-1A	*Jayhawk*	**US** Beech
T-2	*Buckeye*	**US** Rockwell
T-2		**J** Mitsubishi
T-3		**J** Fuji
T-6A	*Texan* II	**US** Beech
T-17	(*Supporter*, MFI-17)	**Swe** SAAB
T-23	*Uirapurú*	**Br** Aerotec
T-25	Neiva *Universal*	**Br** Embraer
T-26	EMB-326	
T-27	*Tucano*	**Br** Embraer
T-28	*Trojan*	**US** North American
T-33	*Shooting Star*	**US** Lockheed
T-34	*Mentor*	**US** Beech
T-35	*Pillán* [PA-28]	**Chl** Enaer
T-36	(C-101)	
T-37	(A-37)	
T-38	*Talon*	**US** Northrop
T-39	(*Sabreliner*)	**US** Rockwell
T-41	*Mescalero* (Cessna 172)	**US** Cessna
T-42	*Cochise* (*Baron*)	**US** Beech
T-43	(Boeing 737)	
T-44	(*King Air*)	
T-47	(*Citation*)	
T-67M	(*Firefly*)	**UK** Slingsby
T-400	(T-1A)	**US** Beech
TB-20	*Trinidad*	**Fr** Aérospatiale
TB-21	*Trinidad*	**Fr** Socata
TB-30	*Epsilon*	**Fr** Aérospatiale
TB-200	*Tobago*	**Fr** Socata
TBM-700		**Fr** Socata
TC-45	(C-45, trg)	
TCH-1	*Chung Hsing*	**ROC** AIDC
TL-1	(KM-2)	**J** Fuji
Tornado		**UK/Ge/It** Panavia
TR-1	[U-2]	**US** Lockheed
Travel Air	*Beech 95*	
Trident		**UK** BAe
Trislander	BN-2	
Tristar	L-1011	

Type	Name/ designation	Country of origin Maker
TS-8	*Bies*	**Pl** PZL
TS-11	*Iskra*	**Pl** PZL
Tu-16	*'Badger'*	**RF** Tupolev
Tu-22	*'Blinder'*	**RF** Tupolev
Tu-22M	*'Backfire'*	**RF** Tupolev
Tu-95	*'Bear'*	**RF** Tupolev
Tu-126	*'Moss'*	**RF** Tupolev
Tu-134	*'Crusty'*	**RF** Tupolev
Tu-142	*'Bear F'*	**RF** Tupolev
Tu-154	*'Careless'*	**RF** Tupolev
Tu-160	*'Blackjack'*	**RF** Tupolev
Turbo Porter	PC-6A/B	
Twin Bonanza	*Beech 50*	
Twin Otter	DHC-6	
Tzukit	[CM-170]	**Il** IAI
U-2		**US** Lockheed
U-3	(Cessna 310)	**US** Cessna
U-4	*Gulfstream* IV	**US** Gulfstream Aviation
U-7	(L-18)	
U-8	*(Twin Bonanza/Queen Air)*	**US** Beech
U-9	(EMB-121)	
U-10	*Super Courier*	**US** Helio
U-17	(Cessna 180, 185)	**US** Cessna
U-21	*(King Air)*	
U-36	*(Learjet)*	
U-42	(C-42)	
U-93	(HS-125)	
U-125	BAe 125-800	**UK** BAe
U-206G	*Stationair*	**US** Cessna
UC-12	*(King Air)*	
UP-2J	(P-2J)	
US-1		**J** Shin Meiwa
US-2A	(S-2A, tpt)	
US-3	(S-3, tpt)	
UTVA-66		**FRY** UTVA
UTVA-75		**FRY** UTVA
UV-18	(DHC-6)	
V-400	*Fantrainer 400*	**Ge** VFW
V-600	*Fantrainer 600*	**Ge** VFW
Vampire	DH-100	
VC-4	*Gulfstream* I	
VC-10		**UK** BAe
VC-11	*Gulfstream* II	
VC-25	[Boeing 747]	**US** Boeing
VC-91	(HS-748)	
VC-93	(HS-125)	
VC-97	(EMB-120)	
VC-130	(C-130)	
VFW-614		**Ge** VFW
Vinka	L-70	
VU-9	(EMB-121)	
VU-93	(HS-125)	
WC-130	[C-130]	
WC-135	[Boeing 707]	**US** Boeing
Westwind	IAI-1124	
Winjeel	CA-25	
Xavante	EMB-326	
Xingu	EMB-121	
Y-5	[An-2]	**PRC** Hua Bei
Y-7	[An-24/-26]	**PRC** XAC
Y-8	[An-12]	**PRC** STAF
Y-12	*Turbo/Twin Panda*	**PRC** HAMC
Yak-11	*'Moose'*	**RF** Yakovlev
Yak-18	*'Max'*	**RF** Yakovlev
Yak-28	*'Firebar' ('Brewer')*	**RF** Yakovlev
Yak-38	*'Forger'*	**RF** Yakovlev
Yak-40	*'Codling'*	**RF** Yakovlev
Yak-42	*'Clobber'*	**RF** Yakovlev
Yak-55		**RF** Yakovlev
YS-11		**J** Nihon
Z-43		**Cz** Zlin
Z-226		**Cz** Zlin
Z-326		**Cz** Zlin
Z-526		**Cz** Zlin
Zéphyr	CM-175	

Tilt-Rotor Wing

Type	Name/ designation	Country of origin Maker
V-22	*Osprey*	**US** Bell/Boeing

Helicopters

Type	Name/ designation	Country of origin Maker
A-109	*Hirundo*	**It** Agusta
A-129	*Mangusta*	**It** Agusta
AB-. . .	(Bell 204/205/206/ 212/214, etc.)	**It/US** Agusta/Bell
AH-1	*Cobra/Sea Cobra*	**US** Bell
AH-2	*Rooivalk*	**RSA** Denel
AH-6	(Hughes 500/530)	**US** MD
AH-64	*Apache*	**US** Hughes
***Alouette* II**	SA-318, SE-3130	**Fr** Aérospatiale
***Alouette* III**	SA-316, SA-319	**Fr** Aérospatiale
AS-61	(SH-3)	**US/It** Sikorsky/Agusta
AS-313 – AS-365/-366	(ex-SA-313 – SA-365/-366)	
AS-332	*Super Puma*	**Fr** Aérospatiale
AS-350	*Ecureuil*	**Fr** Aérospatiale
AS-355	*Ecureuil* II	**Fr** Aérospatiale
AS-365	*Dauphin*	**Fr** Aérospatiale
AS-532	*Cougar*	**Fr** Eurocopter
AS-550/555	*Fennec*	**Fr** Aérospatiale
AS-565	*Panthar*	**Fr** Eurocopter
ASH-3	*(Sea King)*	**It/US** Agusta/Sikorsky
AUH-76	(S-76)	
Bell 47	*(Sioux)*	**US** Bell
Bell 205		**US** Bell
Bell 206		**US** Bell
Bell 212		**US** Bell
Bell 214		**US** Bell
Bell 222		**US** Bell
Bell 406		**US** Bell
Bell 412		**US** Bell
Bo-105	(NBo-105)	**Ge** MBB
CH-3	(SH-3)	
CH-34	*Choctaw*	**US** Sikorsky
CH-46	*Sea Knight*	**US** Boeing-Vertol
CH-47	*Chinook*	**US** Boeing-Vertol
CH-53	*Stallion* (*Sea Stallion*)	**US** Sikorsky
CH-54	*Tarhe*	**US** Sikorsky
CH-113	(CH-46)	
CH-124	SH-3 (*Sea King*)	
CH-139	Bell 206	
CH-146	Bell 412	**Ca** Bell
CH-147	CH-47	
CH-149	*Cormorant* (Merlin)	
Cheetah	[SA-315]	**Ind** HAL
Chetak	[SA-319]	**Ind** HAL
Commando	(SH-3)	**UK/US** Westland/Sikorsky
EC-120B	*Colibri*	**Fr/Ge** Eurocopter
EH-60	(UH-60)	
EH-101	*Merlin*	**UK/It** Westland/Agusta
F-28F		**US** Enstrom
FH-1100	(OH-5)	**US** Fairchild-Hiller

Type	Name/ designation	Country of origin	Maker
Gazela	(SA-342)	**Fr/FRY**	Aérospatiale/SOKO
Gazelle	SA-341/-342		
H-34	(S-58)		
H-76	S-76		
HA-15	Bo-105		
HB-315	*Gavião* (SA-315)	**Br/Fr**	Helibras Aérospatiale
HB-350	*Esquilo* (AS-350)	**Br/Fr**	Helibras Aérospatiale
HD-16	SA-319		
HH-3	(SH-3)		
HH-34	(CH-34)		
HH-53	(CH-53)		
HH-65	(AS-365)	**Fr**	Eurocopter
Hkp-2	*Alouette* II/SE-3130		
Hkp-3	AB-204		
Hkp-4	KV-107		
Hkp-5	Hughes 300		
Hkp-6	AB-206		
Hkp-9	Bo-105		
Hkp-10	AS-332		
HR-12	OH-58		
HSS-1	(S-58)		
HSS-2	(SH-3)		
HT-17	CH-47		
HT-21	AS-332		
HU-1	(UH-1)	**J/US**	Fuji/Bell
HU-8	UH-1B		
HU-10	UH-1H		
HU-18	AB-212		
Hughes 300		**US**	MD
Hughes 500/520	*Defender*	**US**	MD
IAR-316/-330	(SA-316/-330)	**R/Fr**	IAR/Aérospatiale
Ka-25	*'Hormone'*	**RF**	Kamov
Ka-27/-28	*'Helix-A'*	**RF**	Kamov
Ka-29	*'Helix-B'*	**RF**	Kamov
Ka-32	*'Helix-C'*	**RF**	Kamov
Ka-50	*Hokum*	**RF**	Kamov
KH-4	(Bell 47)	**J/US**	Kawasaki/ Bell
KH-300	(Hughes 269)	**J/US**	Kawasaki/MD
KH-500	(Hughes 369)	**J/US**	Kawasaki/MD
Kiowa	OH-58		
KV-107	[CH-46]	**J/US**	Kawasaki/Vertol
Lynx		**UK**	Westland
MD-500/530	*Defender*	**US**	McDonnell Douglas
Merlin	EH-101	**UK/It**	Westland/Augusta
MH-6	(AH-6)		
MH-53	(CH-53)		
Mi-2	*'Hoplite'*	**RF**	Mil
Mi-4	*'Hound'*	**RF**	Mil
Mi-6	*'Hook'*	**RF**	Mil
Mi-8	*'Hip'*	**RF**	Mil
Mi-14	*'Haze'*	**RF**	Mil
Mi-17	*'Hip-H'*	**RF**	Mil
Mi-24, -25, -35	*'Hind'*	**RF**	Mil
Mi-26	*'Halo'*	**RF**	Mil
Mi-28	*'Havoc'*	**RF**	Mil
NAS-332	AS-332	**Indo/Fr**	Nurtanio/Aérospatiale
NB-412	Bell 412	**Indo/US**	Nurtanio/Bell
NBo-105	Bo-105	**Indo/Ge**	Nurtanio/MBB
NH-300	(Hughes 300)	**It/US**	Nardi/MD
NSA-330	(SA-330)	**Indo/Fr**	Nurtanio/Aérospatiale
OH-6	*Cayuse* (Hughes 369)	**US**	MD
OH-13	(Bell 47G)		
OH-23	*Raven*	**US**	Hiller
OH-58	*Kiowa* (Bell 206)		
OH-58D	(Bell 406)		
Oryx	(SA-330)		
PAH-1	(Bo-105)		
Partizan	(*Gazela*, armed)		
RH-53	(CH-53)		
S-58	(*Wessex*)	**US**	Sikorsky
S-61	SH-3		
S-65	CH-53		
S-70	UH-60	**US**	Sikorsky
S-76		**US**	Sikorsky
S-80	CH-53		
SA-313	*Alouette* II	**Fr**	Aérospatiale
SA-315	*Lama* [*Alouette* II]	**Fr**	Aérospatiale
SA-316	*Alouette* III (SA-319)	**Fr**	Aérospatiale
SA-318	*Alouette* II (SE-3130)	**Fr**	Aérospatiale
SA-319	*Alouette* III (SA-316)	**Fr**	Aérospatiale
SA-321	*Super Frelon*	**Fr**	Aérospatiale
SA-330	*Puma*	**Fr**	Aérospatiale
SA-341/-342	*Gazelle*	**Fr**	Aérospatiale
SA-360	*Dauphin*	**Fr**	Aérospatiale
SA-365/-366	*Dauphin* II (SA-360)		
Scout	(*Wasp*)	**UK**	Westland
SE-316	(SA-316)		
SE-3130	(SA-318)		
Sea King	[SH-3]	**UK**	Westland
SH-2	*Sea Sprite*	**US**	Kaman
SH-3	(*Sea King*)	**US**	Sikorsky
SH-34	(S-58)		
SH-57	Bell 206		
SH-60	*Sea Hawk* (UH-60)		
Sokol	W3		
TH-50	Esquilo (AS-550)		
TH-55	Hughes 269		
TH-57	*Sea Ranger* (Bell 206)		
TH-67	Creek (Bell 206B-3)	**Ca**	Bell
UH-1	*Iroquois* (Bell 204/205/212)		
UH-12	(OH-23)	**US**	Hiller
UH-13	(Bell 47J)		
UH-19	(S-55)		
UH-34T	(S-58T)		
UH-46	(CH-46)		
UH-60	*Black Hawk* (SH-60)	**US**	Sikorsky
VH-4	(Bell 206)		
VH-60	(S-70)		
W-3	*Sokol*	**Pl**	PZL
Wasp	(*Scout*)	**UK**	Westland
Wessex	(S-58)	**US/UK**	Sikorsky/Westland
Z-5	[Mi-4]	**PRC**	HAF
Z-6	[Z-5]	**PRC**	CHAF
Z-8	[AS-321]	**PRC**	CHAF
Z-9	[AS-365]	**PRC**	HAMC
Z-11	[AS-352]	**PRC**	CHAF

Index of **Countries and Territories**

Index of **Country Abbreviations**

A Austria
AB Antigua and Barbuda
Afg Afghanistan
Ag Algeria
Alb Albania
Ang Angola
Arg Argentina
Arm Armenia
Aus Australia
Az Azerbaijan

Bds Barbados
Be Belgium
Bel Belarus
BF Burkina Faso
Bg Bulgaria
BiH Bosnia-Herzegovina
Bn Benin
Bng Bangladesh
Bol Bolivia
Br Brazil
Brn Bahrain
Bru Brunei
Bs Bahamas
Btwa Botswana
Bu Burundi
Bze Belize

C Cuba
Ca Canada
Cam Cambodia
CAR Central African Republic
CH Switzerland
Cha Chad
Chl Chile
CI Côte d'Ivoire
Co Colombia
Cr Croatia
CR Costa Rica
Crn Cameroon
CV Cape Verde
Cy Cyprus
Cz Czech Republic

Da Denmark
Dj Djibouti
DPRK Korea, Democratic People's Republic of (North)
DR Dominican Republic
DROC Democratic Republic of Congo

Ea Estonia
Ec Ecuador
EG Equatorial Guinea
ElS El Salvador
Er Eritrea
Et Egypt
Eth Ethiopia

Fji Fiji
Fr France
FRY Federal Republic of Yugoslavia (Serbia–Montenegro)
FYROM Former Yugoslav Republic of Macedonia

Ga Georgia
Gam Gambia, The
Gbn Gabon
Ge Germany
Gha Ghana
Gr Greece
Gua Guatemala
GuB Guinea-Bissau
Gui Guinea
Guy Guyana
GzJ Palestinian Autonomous Areas of Gaza and Jericho

HKJ Jordan
Hr Honduras
Hu Hungary

Icl Iceland
Il Israel
Ind India
Indo Indonesia
Ir Iran
Irl Ireland
Irq Iraq
It Italy

J Japan
Ja Jamaica

Kaz Kazakstan
Kgz Kyrgyzstan
Kwt Kuwait
Kya Kenya

L Lithuania
Lao Laos
LAR Libya
Lat Latvia
Lb Liberia
Ls Lesotho
Lu Luxembourg

M Malta
Mal Malaysia
Mdg Madagascar
Mex Mexico
Mgl Mongolia
Mlw Malawi
Mol Moldova
Mor Morocco
Moz Mozambique
Ms Mauritius
My Myanmar (Burma)

N Nepal
Nba Namibia
Nga Nigeria
Ngr Niger
Nic Nicaragua
Nl Netherlands
No Norway
NZ New Zealand

O Oman

Pak Pakistan
Pan Panama
Pe Peru
Pi Philippines
Pl Poland
PNG Papua New Guinea
Por Portugal
PRC China, People's Republic of
Py Paraguay

Q Qatar

R Romania
RC Congo
RF Russia
RH Haiti
RIM Mauritania
RL Lebanon
RMM Mali
ROC Taiwan
ROK Korea, Republic of (South)
RSA South Africa
Rwa Rwanda

Sau Saudi Arabia
Sdn Sudan
Sen Senegal
Sey Seychelles
SF Finland
Sgp Singapore
Ska Sri Lanka
SL Sierra Leone
Slvk Slovakia
Slvn Slovenia
Sme Suriname
Sp Spain
SR Somali Republic
Swe Sweden
Syr Syria

Tg Togo
Th Thailand
Tjk Tajikistan
Tkm Turkmenistan
Tn Tunisia
TT Trinidad and Tobago
Tu Turkey
Tz Tanzania

UAE United Arab Emirates
Uga Uganda
UK United Kingdom
Ukr Ukraine
Ury Uruguay
US United States
Uz Uzbekistan

Ve Venezuela
Vn Vietnam

Ye Yemen, Republic of

Z Zambia
Zw Zimbabwe